SMART GRID

SMART GRID

COMMUNICATION-ENABLED INTELLIGENCE FOR THE ELECTRIC POWER GRID

Stephen F. Bush
General Electric Global Research, USA

WILEY

Reprinted with corrections March 2015

This edition first published 2014

Registered office
John Wiley & Sons Ltd, The Atrium, Southern Gate, Chichester, West Sussex, PO19 8SQ, United Kingdom

For details of our global editorial offices, for customer services and for information about how to apply for permission to reuse the copyright material in this book please see our website at www.wiley.com.

Library of Congress Cataloging-in-Publication Data

Bush, Stephen F.
Smart grid – communication-enabled intelligence for the electric power grid / Dr Stephen F. Bush.
pages cm.
Includes bibliographical references and index.
ISBN 978-1-119-97580-9 (hardback)
1. Smart power grids. I. Title.
TK3105.B87 2013
621.319′1–dc23

2013036264

A catalogue record for this book is available from the British Library.

ISBN: 978-1-119-97580-9

Set in 10/12pt Times by Aptara Inc., New Delhi, India

1 2014

Contents

About the Author

Stephen F. Bush graduated from Carnegie Mellon University and worked at General Electric Information Services. From there, he obtained his PhD while working as a researcher at the Information and Telecommunications Technologies Center at the University of Kansas, participating in the design of a self-configuring, rapidly deployable, beamforming wireless radio network.

Stephen currently enjoys his role as senior scientist at the General Electric Global Research Center, where he has published numerous conference papers, journal articles, and book chapters, and taught international conference tutorials on novel communication- and network-related topics. His previous book publication, *Active Networks and Active Network Management: A Proactive Management Framework*, explained the development and operation of the intriguing and controversial active networking paradigm. Dr Bush was presented with a gold cup trophy awarded by Defense Advanced Research Projects Agency (DARPA) for his work in active-network-related research. Dr Bush has been the principal investigator for many DARPA and Lockheed Martin sponsored research projects, including: Active Networking (DARPA/ITO), Information Assurance and Survivability Engineering Tools (DARPA/ISO), Fault-Tolerant Networking (DARPA/ATO), and Connectionless Networks (DARPA/ATO), involving energy-aware sensor networks. Stephen also likes creative interaction with students while teaching Quantum Computation and Communication at the Rensselaer Polytechnic Institute and Computer Communication Networks at the University at Albany. Stephen has written "Nanoscale Communication Networks," which has helped to define this new field. He is Director of the IEEE Communications Society Standardization Program Development Board and also serves on the IEEE Smart Grid Communications Emerging Technical Subcommittee and is Chair of the IEEE P1906.1 Recommended Practice for Nanoscale and Molecular Communication Framework standards working group. Stephen also served as an IEEE Distinguished Lecturer on Smart Grid Communications.

Preface

Objective

The center of your culture is left without electric power for a few hours only, and all of a sudden crowds of American citizens start looting and creating havoc. The smooth surface film must be very thin. Your social system must be quite unstable and unhealthy.

—Alexander Solzhenitsyn at Harvard Class Day Afternoon Exercises
Thursday, June 8, 1978

It is natural for the reader who is not fully versed in both power systems and communications to be curious about many aspects of the evolving technologies. For example, how did power systems and communication develop to their present states where something like the term "smart grid" could be coined? Certainly power systems and communications are both offshoots of electrical engineering and both involve the manipulation of power. Why have the two fields diverged so radically? Thinking about these questions leads to more fundamental questions. What is the relationship between electric power and information? And more specifically, what is the fundamental relationship between power systems and communication theory? Thinking about these questions helps us address more practical questions. What are the potential impacts of communication on efficiency in electric power transmission and distribution efficiency? What types of communication are most appropriate for different portions of the power grid? It is also intriguing to consider the more distant future. What will the power grid look like in the decades to come? How could wireless power transmission revolutionize the power grid? What are the fundamental limiting factors? Is there a fundamental limit to the amount of distributed generation that is possible and, if so, how can this limit be overcome? Will communication in the power grid really enable more consumer participation, machine intelligence, and self-organization as many are predicting? What are the opportunities for your particular research to contribute to the future of the power grid? This book will provide you with the background needed to form your own conclusions to these and many other questions on this fascinating journey through the intersection of power systems and communications. It is important for information and graph theorists and network science researchers to better understand power systems to advance information theory and network analysis in order to implement fast, efficient, and realistic approaches within the power systems domain. If not, these theorists could remain in a world of simplified toy problems, not understanding how the power grid really works or simply become constrained within the strait jacket of existing theory.

The primary objective of this book is to bridge the divide between the fields of power systems engineering and computer communication. In my experience within these early stages of this round of "modernization" of the electric power grid, many power systems engineers tend to be a little overconfident in the capability of communication systems to work reliably, with sufficient capacity, and with low latency under any condition. This is not surprising given that communication networks are nearly ubiquitous and embedded within increasingly smaller devices. It is natural for noncommunication engineers to assume that communication networking is a solved problem, ready for application anywhere and everywhere. On the other hand, the power grid is also so ubiquitous and reliable that most nonpower systems engineers take electric power for granted. In fact, most of us tend to assume that an electric power socket will always be within easy reach and that our electronic devices will work perfectly once plugged into that socket. Rarely does anyone think about the complexity of the electric power grid when inserting a plug into a socket or operating their electronic devices. In a sense, both the power grid and communications have suffered from their own respective successes – the electric power grid tends to be taken for granted and communication networks are assumed to work perfectly under almost any condition and for any application. The reader will soon find that both the electric power grid and communications are each highly complex systems in their own right; the manner in which they are integrated will have far-reaching consequences.

Another objective of this book is to remove the previously mentioned dangerous assumptions: to show the complexity and operational requirements of the evolving power grid, the so-called "smart grid," to the communication networking engineer, and similarly to show the complexity and operational requirements for communications to the power systems engineer. At the time this is being written, there are few practitioners who have depth of knowledge in both power systems and computer communications and networking. Thus, another objective of this book, and probably the most important, is to provide a path towards a fundamental understanding in both power systems and communication networking. Just as power systems require a broad set of knowledge ranging from high-power device physics to protection mechanisms to power flow and stability analysis, so too communication networking requires an understanding of topics ranging from signal processing, information theory, and graph theory to cybersecurity. It is my hope that these fundamental topics of power systems and communications combine in novel ways to form far more than the sum of their parts. In other words, it would be a shame for power systems engineers to remain restricted to thinking only about their traditional discipline while directing communication engineers where to implement communications; that is, losing the chance to incorporate new ideas into their repertoire. Similarly, it would be a shame for communication engineers to blindly submit to the direction of power systems engineers and implement communications without looking at new ways to better integrate power systems and communications. My hope is that this book may serve as the impetus leading to the discovery of fundamental new relationships between the properties of electric power, energy, and information.

An overriding objective of this book is to focus on fundamentals – underlying concepts that are most resistant to change. Smart grid standards and technology are currently undergoing rapid evolution; this evolution will continue into the foreseeable future. Thus, standards and technologies as they exist today will soon change or disappear no matter how strongly their advocates may feel. Understanding more fundamental concepts that reside closer to the physics of operation will pay higher dividends for the reader. Thus, for example, understanding information entropy in the power grid per kilowatt of power delivered or the radio frequency communication power expended within the power grid per kilowatt of power delivered will be more valuable than understanding the detailed packet structure of a half-dozen supervisory control and data acquisition protocols. In particular, Section 6.3 develops the fundamental relationship among energy, communication, and computation. These are beautiful relationships that even an expert in communications or power systems should not overlook. Given rapid changes in technology, it is important to understand that technology undergoes predictable evolutionary processes, not unlike a biological organism. For example, the pressures of the market and the nature of the intellectual property processes drive technology to follow the predictable path that we see the smart grid following. It is possible, using this general knowledge of technology evolution, to predict how the power grid will evolve and thus anticipate future challenges.

Finally, many books on the topic of the smart grid, and renewable energy in particular, utilize the notion of global warming and impending environmental catastrophe to promote the importance and urgency of their topic. The reader will note that I purposely avoid discussion of this controversial topic. I believe that we can all agree upon the need for efficient power delivery, lower cost, and less reliance on imported energy. If these can be accomplished in a clean and environmentally friendly manner, then that is an added bonus.

Genesis

My prior books have always focused upon fundamental new ideas; for example, active networks or nanoscale communication networks. So readers may wonder why I have chosen to write about a topic as seemingly practical and mundane as the recent advances in power systems. As the reader will notice, I have not lost interest for thinking "outside the box." While conveying the required practical information, I have attempted to find new ways of looking at the problem wherever possible in order to add new perspectives that hopefully add deeper insight.

There is no doubt in my mind that the definition of smart grid will continue to change over time. At the time this book is being written, smart grid is synonymous with communications coupled with the power grid to accomplish novel power applications. However, smart grid will, and should, expand over time to encompass machine learning applied within the power grid and the development and incorporation of smarter power components. However, it is always important to keep in mind that without underlying communications most of the other advancements will not be possible.

There are a few fundamental trade-offs that apply to communication and networking that recur often; namely, trade-off among performance metrics such as latency, bandwidth, availability, energy consumption, transmission range, and so on. To a first order, designing the smart grid is about determining the correct trade-offs in the correct part of the power grid. For example, the so-called advanced meter reading infrastructure has very different communication requirements, and thus a different design philosophy than power protection. Understanding the reason for the different requirements is critical.

This book had its genesis in 2010 when I became involved in smart-grid-related projects and could not find a comprehensive source for communications within the electric power grid. This book also became intertwined with my IEEE Distinguished Lecture Tours in 2011. It became clear from audiences on the lecture tours that there was, and continues to be, widespread and intense interest in the "smart grid." It also became clear, as previously mentioned, that there is a fundamental lack of understanding between the fields of electric power systems and communications.

It is also evident from my experience on smart grid projects that communications is often assumed to exist when in reality it may not. Complex algorithms are developed that rely upon geographically dispersed information under the assumption that communications can be easily engineered later into the process. It is important for power systems engineers to be aware of the challenges involved in communication networking. As a simple example, establishing point-to-point wireless communication through ground clutter is a nontrivial task. Relying on a cellphone carrier introduces problems of coverage, availability, and often uncertainty regarding the bandwidth available at any specific time, in addition to excessive cost. Power line carrier suffers many problems, not the least of which is the loss of communication through a downed power line; that is, communication may be lost when it is needed most. These form only a small subset of the challenges faced by communications in the power grid; hopefully, the point that there is no simple, trivial solution will become clear.

I also noticed that, just as local regions develop their own dialects in human language, power systems and communication engineers continue to develop their own independent and unique terminology, sometimes attributing very different meaning to the same terms, causing potential confusion. For example, "security" to a power systems engineer means something that is entirely different to a communication and networking engineer. Take an "active network" or "active networking" as another example; to a power engineer it refers to a microgrid, while to a communication engineer it refers to an advanced form of programmable network. Another source of terminological confusion is "power routers": is it literally a device that routes electric power or is it just a "communication network router" that serves to control the routing of power? These and other differences in terminology are explained in detail in the text. It can be said that power systems and communications are separated by a common language. The origin of this book grew out of an attempt to understand the similarities and differences between the two disciplines.

But why is a holistic approach towards smart grid – such as that proposed in this book – expected or desired? A common example can be found in the evolution of the Internet and telecommunications, which drove exploration for the relationship between communications and information and ultimately led to information theory. The Internet and telecommunications in turn created a platform for applications that could never have been conceived at the time. A more holistic approach allows us to be more innovative – to see how components interact in a deeper manner in order to find efficiencies and develop entirely new applications. It was the drive to make the system more efficient and reduce the cost of transporting a product that drove the theory, just as the power grid is doing today in the so-called smart grid. For information theory, in Shannon's case, it was an industrial research laboratory rather than a university that created the key innovation. Again, as typically happens, it is the case today that fundamental innovation and insight are driven by, and come from, industry.

Approach and Content

The power grid is in a state of rapid evolution. Any attempt to convey a comprehensive state of the policies, standards, and even specific technologies will likely be out of date even before going to print. Thus, this book focuses upon fundamentals as much as possible; information theory and power electronics will change more slowly than policy, regulation, and standards. The reader can be confident that the material presented will always be relevant; only its implementation may change. Reliability, safety, low cost, and high efficiency have been, and will likely remain, key drivers of the technology regardless of how business models change.

In fact, the technology for what we call the "smart grid" did not suddenly appear, but has been in development for some time. Attempting to draw a precise boundary between the "legacy power grid" and the "smart grid" would not be a simple task and would perhaps not even be sensible to attempt. Communication has been an integral part of the power grid since the last century, so the idea of simply adding communication is neither novel nor does it make the grid smart. Part of the approach of this book is to explore, and perhaps sometimes debunk, why the word "smart" is in smart grid. In that regard, it has been important in writing this book to separate fact from proposed idea: what really exists and is likely to exist in the power grid from what academics often incorrectly "think" exists in the power grid.

This book covers the evolving electric power grid and its integration with communications assuming as little prerequisite knowledge as possible. We begin with a brief and intuitive introduction to the fundamentals of power systems and progressively build upon that foundation while pointing out relationships with communications and networking wherever appropriate along the way.

The book is organized into three parts with five chapters in each part. Figure 1 shows the organization of the book. Part One of the book will ground the reader in the basic operation of the electric power grid. This part covers fundamental knowledge that will be assumed in Parts Two and Three of the book. Part Two introduces communications and networking, which are critical enablers for a smart grid. The manner in which communication and networking are integrated into the power grid is an ongoing process; thus, we also consider how communication and networking are anticipated to evolve as technology develops. This part lays the foundation for Part Three, which utilizes communication within the power grid. The smart grid will ultimately become "smart" when intelligence is implemented upon the communication framework explained in Part Two. Thus, Part Three must draw heavily upon both past embedded intelligence within the power grid and current research to anticipate how and where computational intelligence will be implemented within the smart grid.

Each part is divided into chapters and each chapter has a set of questions useful for exercising the reader's understanding of the material in that chapter. The book is written so that when the chapters are read in consecutive order the material will flow well, each chapter building upon the previous chapters. However, there are other ways to read the book for readers with different backgrounds and perspectives. A power systems engineer would presumably have a strong background in the traditional power grid and would not need to read Part One of the book, so could begin reading starting with Part Two. On the other hand, a communications engineer could potentially skip Chapter 6, with the exception of Section 6.4 on power system information theory. A reader who is interested only in a summary of the technology could simply read the first chapter of each part, namely Chapters 1, 6, and 11.

One of the interesting aspects of the evolving power grid is that as it has evolved, it has become harder to neatly divide the power grid into the traditional components of generation, transmission, distribution, and consumption. These parts are becoming more interrelated. If one were interested only in generation, then Chapters 1, 8, and 15 (nanoscale power generation) would be most relevant. If one were interested only in transmission, then Chapters 3, 6, 12, and 13 might be most appropriate. If one were interested in distribution, then Chapters 4, 9, 12, and 13 would perhaps be most relevant. However, aspects of distributed generation, demand-response, and fault detection, isolation, and restoration, state estimation and stability,

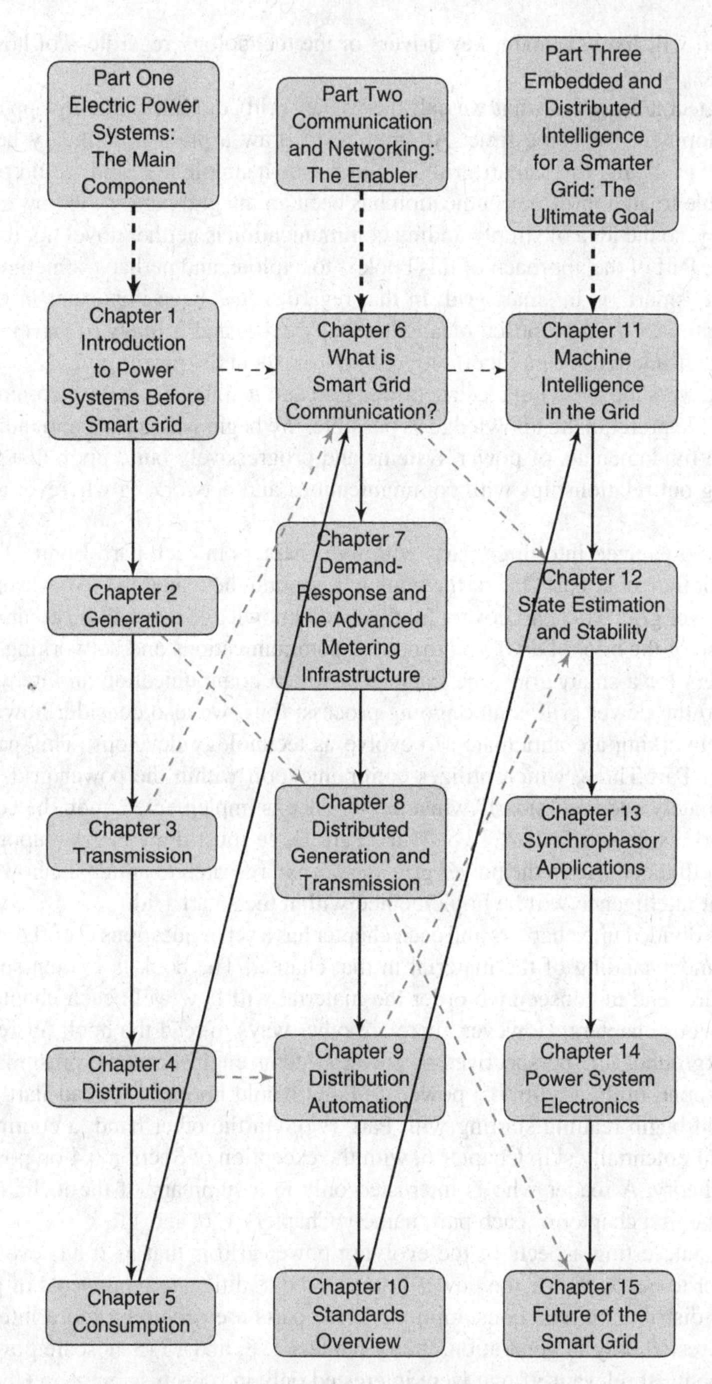

Figure 1 Selected reading arrangements are illustrated for readers interested in specific subtopics. Those interested in power generation could start from Chapter 2 and follow the dashed lines; those interested in power transmission could begin with Chapter 3 and follow the dashed lines; those interested in power distribution could start with Chapter 4 and follow the dashed lines.

synchrophasors, and so on. will all be taking place simultaneously within the power distribution network.

As will be seen throughout this work, the smart grid communication vision presented here foreshadows a high degree of integration between information theory and power systems. Specifically, fundamental relationships between information theory and Maxwell's equations could yield new insight into understanding exactly where to place communication, since entropy would be known at a low level within the power grid. Today, this placement is done in a rather ad hoc manner. We may also someday know the precise theoretical "bits per kW" needed to distribute electric power safely. Finally, we can imagine that new forms of efficiency resulting from advances in small-scale power generation could lead to widespread use of nanoscale power generation and distribution and the required nanoscale communication to support such systems. For example, a consumer's electric vehicle may be recharged by extraneous electromagnetic fields from radio transmissions. Research and references on both extracting energy from extraneous electromagnetic transmissions and nanoscale communications are provided in later chapters of this book. In the distant future, we might even imagine the quantum teleportation of energy. All of these topics are covered in the last chapter of the book.

Acknowledgements

First and foremost, I thank my wife for her kind patience and understanding. I would also like to thank the hosts and audiences on my IEEE Communications Society Distinguished Lecture Tours on smart grid in 2011. Their stimulating questions, comments, and discussions helped shape this book. I would also like to thank all the folks involved in the IEEE Smart Grid Vision 2030 Project, including Alex Gelman, Sanjay Goel, David Bakken, and Bill Ash, among many others. Stimulating discussion among smart individuals willing to explore new ideas can lead to great things.

Stephen F. Bush

Acronyms

6LoWPAN	IPv6 over low-power wireless personal area networks
ACE	area control error
ACFFI	average communication failure frequency index
ACIDI	average communication interruption duration index
ACK	acknowledgment
ACO	ant colony optimization
ACSE	association control service element
ACSR	aluminum conductor steel-reinforced cable
ADA	advanced distribution automation
ADI	advanced distribution infrastructure
ADP	adaptive dynamic programming
AGC	automatic grid control
AHP	analytical hierarchical programming
AIEE	American Institute of Electrical Engineers
AMI	advanced metering infrastructure
AMR	automated meter reading
ANSI	American National Standards Institute
AODV	ad hoc on-demand distance-vector
APDU	application protocol data unit
API	application program interface
ARQ	automatic repeat-request
ASCII	American Standard Code for Information Interchange
ASK	amplitude-shift keying
ASN.1	abstract syntax notation 1
ATM	asynchronous transfer mode
BAN	body-area network
BAS	building automation system
BCS	Bardeen–Cooper–Schrieffer
BE	best-effort
BFSK	binary frequency-shift keying
BMC	best master clock
BPL	broadband over power line
BPSK	binary phase-shift keying
BS	base station

CA	contingency analysis
CAES	compressed air energy storage
CAIDI	customer average interruption duration index
CAIFI	customer average interruption frequency index
CAN	controller-area network
CBR	constant-bit rate
CC	control center
CCITT	Comité Consultatif International Téléphonique et Télégraphique (International Telegraph and Telephone Consultative Committee)
CID	connection identifier
CIGRE	Conseil International des Grands Reseaux Electriques (International Council on Large Electric Systems)
CIM	common information model
ComSoc	IEEE Communications Society
COSEM	companion specification for energy metering
CRC	cyclic redundancy checksum
CSM	common signaling mode
CSMA	carrier-sense multiple-access
CSMA-CA	carrier-sense multiple-access with collision avoidance
CSMA-CD	carrier-sense multiple-access with collision detection
CT	current transformer
CTAIDI	customer total average interruption duration index
CVR	conservation voltage reduction
CVT	constant voltage transformer
DA	distribution automation
DAG	directed acyclic graph
DAU	data aggregation unit
DCF	distribution coordination function
DCT	discrete cosine transform
DESS	distribution energy storage system
DG	distributed generation
DHP	dual heuristic programming dielectric
DIO	DODAG information object
DLMS	device language message specification
DMI	distribution management infrastructure
DMS	distribution management system
DNP	distributed network protocol
DNP3	distributed network protocol 3
DODAG	destination-oriented directed acyclic graph
DR	demand-response
DSL	digital subscriber line
DSM	demand-side management
DSP	digital signal processor
DSSS	direct-sequence spread-spectrum
DVR	dynamic voltage restorer
EDFA	erbium-doped fiber-optic amplifier

EHV	extra-high voltage
EIA	United States Energy Information Agency
EMC	electromagnetic compatibility
EMF	electromotive force
EMS	energy management system
ENS-C	energy not served due to communication failure
EPRI	Electric Power Research Institute
EPS	electric power system
ertPS	extended-real-time-polling service
ESI	energy services interface
ETSI	European Telecommunications Standards Institute
EPSEM	extended protocol specification for electronic metering
FACTS	flexible alternating current transmission system
FAN	field-area network
FCL	fault current limiter
FCS	frame check sequence
FDIR	fault detection, isolation, and restoration
FDM	frequency-division multiplexing
FERC	Federal Energy Regulatory Commission
FET	field-effect transistor
FFD	full function device
FFT	fast Fourier transform
FHSS	frequency-hopping spread-spectrum
FN	false-negative isolated fault segment vector
FSK	frequency-shift keying
GenCo	generating company
GFCI	ground-fault circuit interrupter
GIC	geomagnetically induced current
GIS	geographic information system
GOOSE	generic object-oriented substation events
GPS	global positioning system
HAN	home-area network
HART	highway addressable remote transducer
HDP	heuristic dynamic programming
HEMP	high-altitude electromagnetic pulse
HMAC	keyed-hash message authentication code
HTS	high-temperature superconductor
HTS-ISM	high-temperature superconducting induction-synchronous machine
HVDC	high-voltage direct-current
IAE	integral absolute error
ICCP	inter-control center communications protocol
ICT	information and communications technology
IE	information element
IEC	International Electrotechnical Commission
IED	intelligent electronic device
IEEE	Institute of Electrical and Electronics Engineers

IEM	intelligent energy management
IETF	Internet Engineering Task Force
IFFT	inverse fast Fourier transform
IFM	intelligent fault management
IGBT	insulated-gate bipolar transistor
IHD	in-home display
IMF	interplanetary magnetic field
IoT	Internet of things
IP	Internet protocol
IPv6	Internet protocol version 6
IRE	Institute of Radio Engineers
ISE	integral squared error
ISO	independent system operator *or* International Standards Organization
IT	information technology
ITAE	integral time-weighted absolute error
ITIC	Information Technology Industry Council
ITU	International Telecommunication Union
IVVC	integrated volt-VAr control
L2TP	layer 2 tunneling protocol
L2TPv3	layer 2 tunneling protocol version 3
LAC	L2TP access concentrator
LAN	local-area network
LBR	LLN border router
LC	inductor–capacitor
LDP	label distribution protocol
LED	light-emitting diode
LEO	low Earth orbit
LER	label edge router
LLC	logical-link control
LLN	low-power and lossy network
LM/LE	load modeling/load estimation
LMP	location marginal pricing
LMS	load management system
LNS	L2TP network server
LPDU	link protocol data unit
LRC	longitudinal redundancy check
LR-WPAN	low rate-wireless personal-area network
LSE	load serving entity
LSR	label-switch router
LTC	load tap changing
LV	low voltage
M2M	machine-to-machine
MAC	media-access control
MAIFI	momentary average interruption event frequency index
MAN	metropolitan-area network
MDL	minimum description length

MDMS	meter data management system
MFR	MAC footer
MHR	MAC header
MIB	management information base
MMS	manufacturing message specification
MOSFET	metal-oxide-semiconductor field-effect transistor
MPDU	MAC protocol data unit
MPLS	multiprotocol label switching
MPPT	maximum powerpoint tracking
MRF	Markov random field
MR-FSK	multirate-frequency-shift keying
MRI	magnetic resonance imaging
MR-OFDM	multirate orthogonal frequency-division multiplexing
MR-OQPSK	multirate-offset quadrature phase-shift keying
MS	mobile station
MSDU	MAC service data unit
MSH-DSCH	mesh-distributed scheduling message
MSH-NENT	mesh network entry request message
MSH-NCFG	mesh network configuration
MTU	maximum transmission unit
NACK	negative acknowledgment
NAN	neighborhood-area network
NIST	National Institute of Standards and Technology
NLDN	National Lightning Detection Network
nrtPS	non-real-time polling service
OFDM	orthogonal frequency-division multiplexing
OFDMA	orthogonal frequency-division multiple access
OGW	optical ground wire
OMS	outage management system
ONR	optimal network reconfiguration
OpenADR	open automated demand response communication standards
OpenDSS	open distribution system simulator
OQPSK	offset-quadrature phase-shift keying
OSI	open systems interconnection
PAN	personal-area network
PAP	priority action plan
PCA	principal component analysis
PCB	polychlorinated biphenyl
PCF	point coordination function
PCS	power conditioning system
PDC	phasor data concentrator
PDU	protocol data unit
PER	packet error rate
PES	IEEE Power and Energy Society
PEV	plug-in electric vehicle
PHR	physical-layer header

PMP	point-to-multipoint mode
PMU	phasor measurement unit
PN	pseudorandom sequence
PPDU	physical-layer protocol data unit
PPP	point-to-point protocol
PQ	real and reactive power
PSDU	physical-layer service data unit
PSEM	protocol specification for electronic metering
PSK	phase-shift keying
PSO	particle swarm optimization
PST	phase-shifting transformer
PT	potential transformer
PTP	precision time protocol
pu	per-unit
PV	real power and voltage magnitude *or* photovoltaic
QAM	quadrature amplitude modulation
QoS	quality of service
QPSK	quadrature phase-shift keying
RAN	radio access network
RBAC	role-based access control
RDF	resource description framework
RF	radio frequency
RFC	request for comments
RFD	reduced function device
RFID	radio-frequency identification
RMS	root-mean-square
RMT	random matrix theory
ROLL	routing over low-power and lossy networks
RPC	relay protection coordination
RPL	routing protocol for low-power and lossy networks
RPM	rotations per minute
RS	relay station
RSVP-TE	resource reservation protocol for traffic engineering
RTO	regional transmission organization
RTP	real-time pricing
rtPS	real-time polling service
RTU	remote terminal unit
SA	substation automation
SAIDI	system average interruption duration index
SAIFI	system average interruption frequency index
SCA	short circuit analysis
SCADA	supervisory control and data acquisition
SCFCL	superconducting fault current limiter
SCL	substation configuration language
SDN	software-defined network
SDR	software-defined radio

SERA	Smart Energy Reference Architecture
SGIP	Smart Grid Interoperability Panel
SHR	synchronization header
SIL	surge impedance loading
SMES	superconducting magnetic energy storage
SNMP	simple network management protocol
SONET	synchronous optical networking
SPS	standard positioning service
SPSS	supervisory power system stabilizer
SRE	slack-referenced encoding
SS	subscriber station
SNTP	Simple Network Time Protocol
SUN	smart utility network
SuperPDC	super phasor data concentrator
SVC	static VAr compensator
SVD	singular value decomposition
T1	transmission system 1
TAI	international atomic time
TAM	technology acceptance model
TASE	tele-control application service element
TCC	time–current characteristic
TCP	transmission control protocol
TDD	time-division duplex
TDMA	time-domain multiple access
THD	total harmonic distortion
TLS	transport layer security
TPDU	transport protocol data unit
TSO	transmission system operator
TVA	Tennessee Valley Authority
TVE	total vector error
UCA	Utility Communications Architecture
UDP	user datagram protocol
USG	unsolicited grant service
USN	ubiquitous sensor network
UTC	universal time coordinate
VA	volt–ampere
VAr	volt–ampere reactive
VFT	variable-frequency transformer
VO	voltage optimization
VPLS	virtual private LAN service
VSI	voltage stability index
VVO	volt-VAr optimization
WACS	wide-area control system
WAM	wide-area monitoring
WAMPAC	wide-area monitoring, protection, and control
WAMS	wide-area monitoring system

WAN	wide-area network
WAPS	wide-area protection system
WASA	wide-area situational awareness
WDM	wavelength division multiplexing
WEP	wired equivalent protection
WiFi	Wireless Fidelity
WiMAX	Worldwide Interoperability for Microwave Access
WLS	weighted least squares
WPAN	wireless personal-area network
XML	extensible markup language
XPath	XML path language

Part One

Electric Power Systems: The Main Component

1

Introduction to Power Systems Before Smart Grid

This 'telephone' has too many shortcomings to be seriously considered as a means of communication. The device is inherently of no value to us.

—Western Union internal memo, 1876

Those who say it cannot be done should not interfere with those of us who are doing it.

—S. Hickman

1.1 Overview

Power systems and communications are close cousins. This may not be apparent at first, but that is how we will generally view these twin subtopics of electrical engineering. Communications and power systems are the same field with a different emphasis. Both transmit power. Communications seek to minimize power and maximize information content. Power systems seek to maximize power and minimize information content. It is particularly interesting to see what happens when these fields physically come together in technologies such as a power line carrier and wireless power transmission. In a power line carrier, communication attempts to become physically similar to power, following the same conductive path. In wireless power transmission, power seeks to become physically similar to wireless communication, propagating through space similar to wireless communication. It is at these intersections of communication and power systems that the differences between the two fields comes into sharpest contrast. The initial hyperbole regarding the fundamental shift in power systems toward what is being labeled as the "smart grid" will have died down or disappeared altogether by the time the reader has this book in hand. However, the technological change that initiated the smart grid established a platform that will enable revolutionary enhancements in intelligence for power systems. Our goal is to explore both the theoretical and technological underpinnings of this shift in power systems, focusing upon the incorporation of communications and networking technology. There are those who suggest that the integration of communications into the power

Smart Grid: Communication-Enabled Intelligence for the Electric Power Grid, First Edition. Stephen F. Bush.

grid will enable a revolution in electric power distribution, perhaps thinking of the analogy with the explosive growth of the Internet in the 1990s. The simple act of providing data interconnections (for example, via the Internet, cell phone, and other portable computing devices) has spawned new applications, ideas, and solutions that no one could have predicted. Communications in the power grid may indeed enable new and unforeseen applications in power systems. At the same time, we should temper our enthusiasm by noting that communications have already been part of the power grid for over a century, as we will see later.

As a motivation for smart grids, it has often been stated that power systems have evolved slowly while communication and networking have advanced much more rapidly: Alexander Graham Bell would not recognize the phone system of today, whereas Thomas Edison would still recognize much of today's electric power grid. However, this is, of course, not quite true. In fact, power systems has been evolving, and it is difficult to precisely define when and where the so-called smart grid began; much of the technology enabling the smart grid has existed for some time. Part One of this book covers power systems fundamentals; these are fundamentals that existed long before the smart grid and will exist long after, so they are well worth the time and effort to understand, although, as just mentioned, drawing the line between the pre- and post-smart grid is somewhat arbitrary and perhaps still ongoing, as we will see. Part Two defines what we mean by the term "smart grid" and focuses upon communications. Part Three goes on to explore what communications has enabled and could enable, including synchrophasor applications and machine intelligence.

Each new scientific discovery or advance in engineering and technology does not deplete the set of new ideas; on the contrary, it exponentially increases the number of new possibilities to be explored. This book will provide you, the student, academic or industry professional, or casual reader, with the basic building blocks of the smart grid; however, it will be you who will supply the creativity and innovation to combine these building blocks in new ways that may not yet have been considered. Please continue with the thought in mind that these are building blocks for new ideas, innovations, or even products, not as an end in themselves. One of the exciting things about the smart grid is that it is a highly dynamic and evolving system, one that you will be able to participate in, whether as a researcher, designer, developer, or consumer.

This part, Part One, consisting of Chapters 1–5, introduces the electric power grid and fundamental concepts of power systems. The goals for this part are to provide prerequisite material for understanding the power grid, to provide historical perspective on the evolution of the power grid, and to provide motivation for the concept of the smart grid.

This chapter, Chapter 1, provides a general overview of the electric power grid, including the fundamentals necessary to understand the rest of the material that will be covered. The remaining chapters in this part focus upon the topics introduced in this chapter in more detail. Because this part of the book is focused on the historical or legacy power grid, it is divided into standard electric power grid components: Chapter 2 focuses upon generation, Chapter 3 on transmission, Chapter 4 on distribution, and Chapter 5 on the consumption of electric power.

This chapter begins with an overview of the physics of electricity as it relates primarily to power systems, but also as it relates to communications as well. Then we discuss the electric power grid as it has evolved over the last century until the dawn of the smart grid; this provides us with a brief historical perspective. Then we look at the equipment in the legacy power grid; much of the equipment, or at least its functionality, will be the same or similar in the smart grid. It will be this equipment that will be monitored and interconnected via communications within the smart grid. Next, we return to basic power analysis that applies to the legacy power system, and because the fundamental physics does not change, will apply to the smart grid as well. This analysis provides insight into the operation of the power grid as well as provides our first hints at the communication and computational requirements within the smart grid. Simulation and modeling tools are introduced in the appendix; while the reader may be curious as to what tools currently exist, this information will likely become rapidly outdated and is thus not incorporated in the main text. This information may be relatively quickly outdated, but it provides a look at some of the modeling challenges for the smart grid. Next, we briefly consider blackouts in the legacy power grid. This provides us with a sobering look at what we would hope the smart grid would improve. The goals for the smart grid involve extending the capability of the power grid in many different ways, however, if the smart grid cannot reduce the likelihood of a blackout, then all of its other features are pointless. Sections 1.5 and 1.6 discuss the drivers and goals of the smart grid. Finally, we take an excursion back to fundamental theory in Section 1.8 to discuss energy and information. The goal is to intuitively motivate the reader to consider that incorporating communication and computation with the power grid may benefit from a fundamental understanding of the relationship between energy and information. The chapter ends with a summary of the important points. Finally, the exercises at the end of the chapter are available to help solidify understanding of the material.

The term "smart grid" has been used numerous times in this text already and, since it is the main topic of this book, will be used frequently throughout the remainder of the book. Before continuing further, a definition of this term is in order. Let us begin with a simple, broad, intuitive definition and refine it as we progress. The smart grid is an electric power grid that attempts to intelligently respond to all the components with which it is interconnected, including suppliers and consumers, in order to deliver electric power services efficiently, reliably, economically, and sustainably. The details of the definition and the means by which these goals are accomplished vary from one region to another throughout the world. This is in part due to the fact that different regions of the world have different infrastructures, different needs, and different expectations, as well as different regulatory systems. However, even without these differences, the power grid is a very broad system comprised of many different components and technologies. Researchers focusing on a narrow aspect, such as developing smart meters or developing new types of demand-response (DR) mechanisms, sometimes inadvertently equate their areas with the sum total of the smart grid, as illustrated in Figure 1.1. Each blind man equates the elephant with the part he can feel. The areas shown are

- advanced metering infrastructure (AMI) – systems that measure, collect, and analyze energy usage;
- distribution automation (DA) – the extension of intelligent control over electrical power grid functions to the distribution level and beyond;
- distributed generation (DG) – generating electricity from many smaller energy sources and microgrids;
- substation automation (SA) – automating electric power distribution and transmission substations;
- flexible alternating current transmission system (FACTS) – a power electronics-based system to enhance controllability and increase power transfer capability of the network; and
- DR – systems that manage customer consumption of electricity in response to supply conditions.

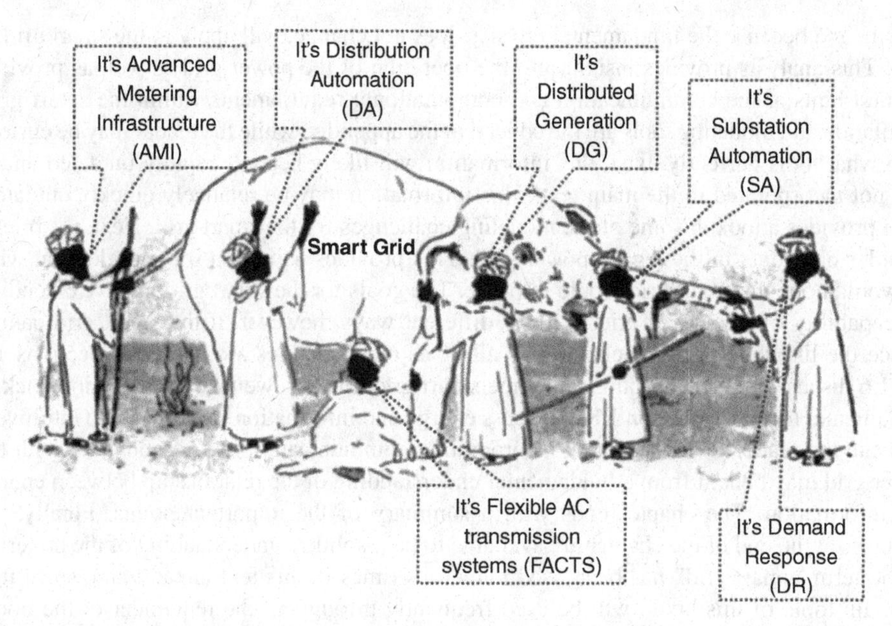

Figure 1.1 What is the smart grid? There is a risk of perceiving the smart grid as only one of many different emerging systems. It is critical for the development of smart-grid communications to understand the complete view of a smart grid. Source: Stebbins C. M. and Coolidge M. H. (1909), Golden Treasury Readers: Primer, American Book Co., New York, NY, p. 89, via Wikimedia Commons.

While these topic areas provide a feel for the smart-grid goals and we will cover these topics in detail in this book, no single subset of these areas defines the smart grid. In fact, these individual components should be viewed as only a subset of the possible components of the smart grid. Some of these components may reach maturity as planned, others may not survive, and many new ones will certainly be created as innovation continues. It is important, then, to understand the fundamentals of both power systems and communications in order to make intelligent decisions regarding how these components will progress and to identify the potential for new ones.

Smart grid is about the evolution of the power grid. In that respect, we discuss where the grid came from, its current state, and its transition into a power grid comprised of DG and microgrids. However, this book should also be of lasting value in terms of a longer term vision for the power grid; that is, how it could evolve further in the coming decades. At this point, it would be instructive to take a risk and predict how the grid could look far into the future. It is a common trend for any technology to evolve from a monolithic structure to become more dynamic, flexible, and eventually merge with its environment. The ultimate advancement is to evolve into a physical field, such as an electric or magnetic field. Figure 1.2 depicts a series of progressively more sophisticated uses of wireless power: from centralized generation and wireless transmission to hard-to-reach places today, to offshore microgrids tomorrow, and to harvesting power from literal nanogrids and stray electromagnetic radiation. Nanogrid has multiple meanings; since power systems tends to deal with power grids that extend over large regions of the earth, nanogrid can colloquially refer to a relatively small power grid such as that inside a laptop. However, here we mean power grids that are literally on the order of nanometers in volume. The concept in this vision is that any power source, including large numbers of nanoscale power generation sources, can connect to the grid and provide power that

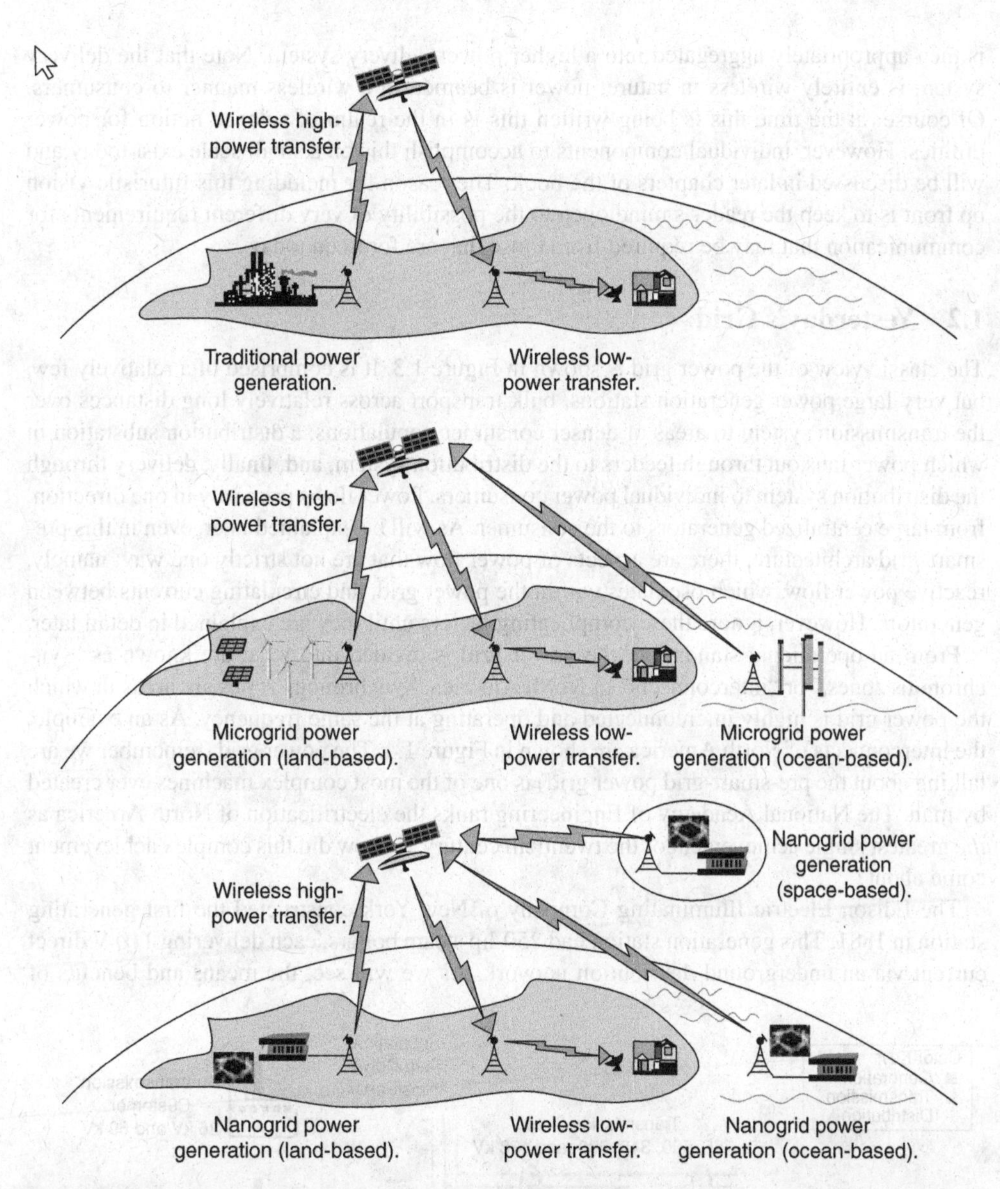

Figure 1.2 The power grid of the future may contain centralized generation with wireless power transfer, microgrids with wireless power transmission, nanogrids and electromagnetic energy harvesting among many other innovations. Smart-grid communication should anticipate how power grid technologies will be evolving so that it can remain viable for the future. Source: Bush, 2013. Reproduced with permission of IEEE.

is then appropriately aggregated into a higher power delivery system. Note that the delivery system is entirely wireless in nature; power is beamed in a wireless manner to consumers. Of course, at the time this is being written this is in the realm of science fiction for power utilities. However, individual components to accomplish this on a small scale exist today and will be discussed in later chapters of the book. The reason for including this futuristic vision up front is to keep the reader's mind open to the possibility of very different requirements for communication that may be required from those that are foreseen today.

1.2 Yesterday's Grid

The classic view of the power grid is shown in Figure 1.3. It is comprised of a relatively few, but very large power generation stations, bulk transport across relatively long distances over the transmission system to areas of denser consumer populations, a distribution substation in which power fans out through feeders to the distribution system, and, finally, delivery through the distribution system to individual power consumers. Power flows primarily in one direction, from large centralized generators to the consumer. As will be explained later, even in this pre-smart-grid architecture, there are aspects of power flow that are not strictly one way; namely, reactive power flow, which oscillates within the power grid, and circulating currents between generators. However, ignore these complicating factors until they are explained in detail later.

From an operational standpoint, the power grid is divided into what are known as "synchronous zones," or "interconnects" in North America. Synchronous zones are areas in which the power grid is highly interconnected and operating at the same frequency. As an example, the interconnects of North America are shown in Figure 1.4. The power grid (remember we are talking about the pre-smart-grid power grid) is one of the most complex machines ever created by man. The National Academy of Engineering ranks the electrification of North America as *the* greatest single achievement of the twentieth century. So how did this complex achievement come about?

The Edison Electric Illuminating Company of New York constructed the first generating station in 1881. This generation station had 250-hp steam boilers, each delivering 110 V direct current via an underground distribution network. As we will see, the means and benefits of

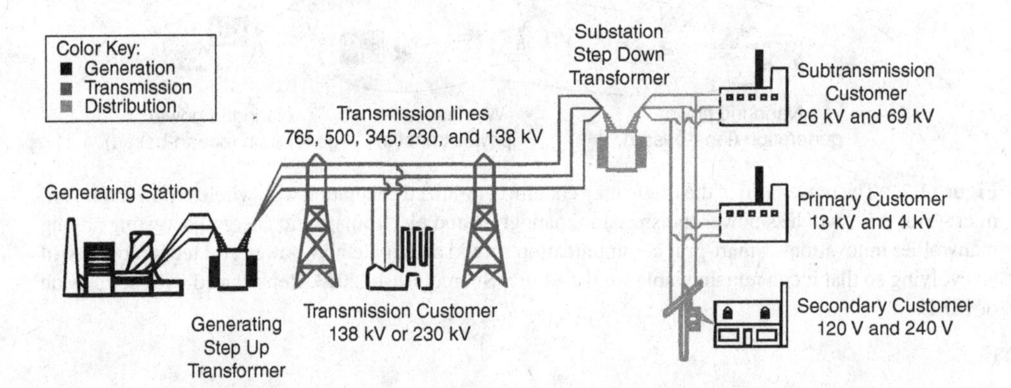

Figure 1.3 The classic view of the power grid has typically been a one-way flow of electric power from generation through transmission, distribution; to consumption. Source: U.S. Canada Power System Outage Task Force, Final Report on the August 14, 2003. Blackout in the United States and Canada: Causes and Recommendations, April 2004.

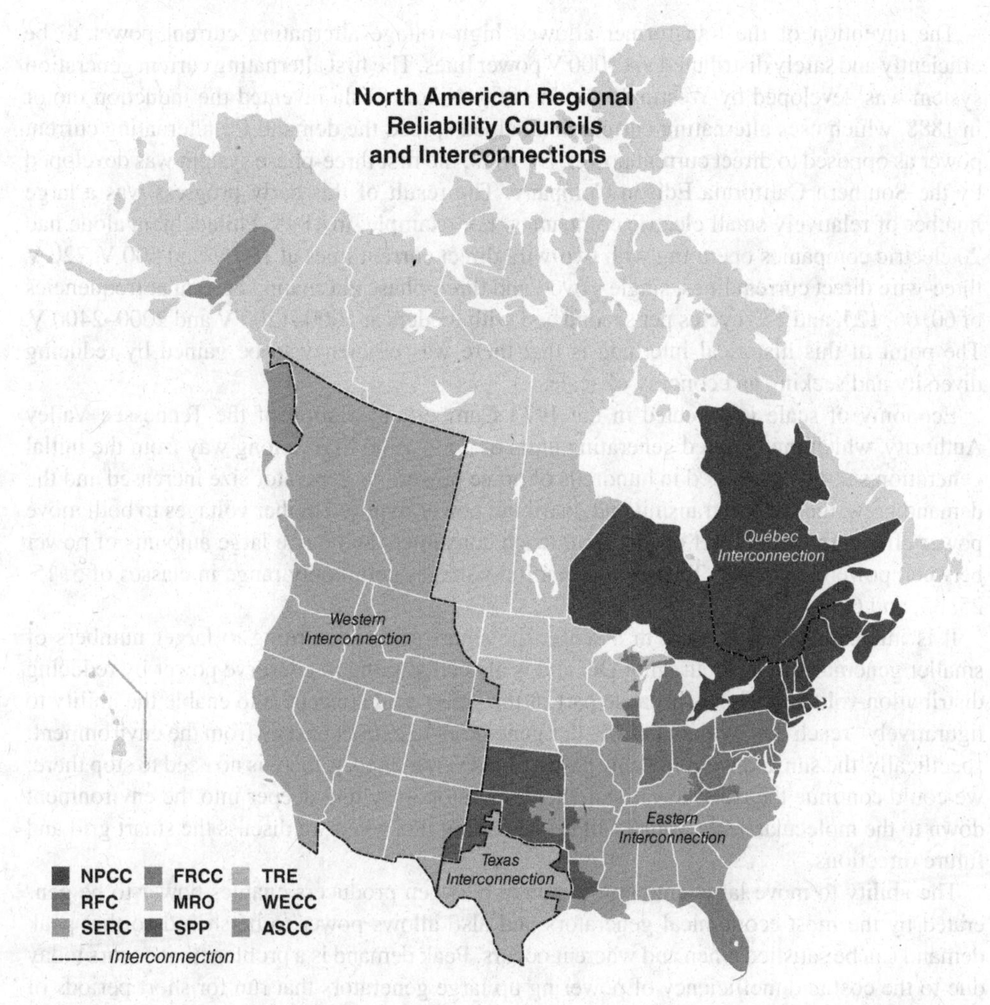

Figure 1.4 Interconnects, or synchronous zones, are areas of relatively dense power interconnectivity where the alternating current cycle is in phase. Source: By Bouchecl (Own work) [CC-BY-SA-3.0 (http://creativecommons.org/licenses/by-sa/3.0)], via Wikimedia Commons.

using alternating current had not yet been discovered. Because voltage is required to move current and the early generating stations could not transmit or distribute their power very far, many stations appeared primarily throughout urban areas to serve only local demand for power. They were, in a sense, the first primitive microgrids. The power grid grew rapidly; electric power grew by over 400 times from the early 1900s to the 1970s. This is significant if you compare it with all other forms of energy, which only grew by 50 times over the same period.

The need for communication to support the power grid was recognized almost immediately. Communication over telegraph lines for automated meter reading was utilized in the late 1800s and patents on power line carrier for meter reading were issued in Britain in 1898 and the USA in 1905 (Schwartz, 2009).

The invention of the transformer allowed high-voltage alternating current power to be efficiently and safely distributed via 1000 V power lines. The first alternating current generation system was developed by Westinghouse in 1866. Nikola Tesla invented the induction motor in 1888, which uses alternating current and helped spread the demand for alternating current power as opposed to direct current power. By 1893, the first three-phase system was developed by the Southern California Edison Company. The result of this early progress was a large number of relatively small electric companies. For example, in 1895, Philadelphia alone had 20 electric companies operating with two-wire direct current lines at 100 V and 500 V, 220 V three-wire direct current lines, single-, two-, and three-phase alternating current at frequencies of 60, 66, 125, and 133 cycles per second and with feeders at 1000–1200 V and 2000–2400 V. The point of this historical interlude is that there was efficiency to be gained by reducing diversity and seeking an economy of scale.

Economy of scale is reflected in the 1973 Cumberland Station of the Tennessee Valley Authority, which inaugurated generating units of up to 1300 MW, a long way from the initial generation stations measured in hundreds of horse power. As generator size increased and the demand grew, the need to transmit and distribute power required higher voltages to both move power efficiently over larger distances to reach consumers and move large amounts of power between power producers. Today's distribution voltages commonly range in classes of 5, 15, 25, 35, and 69 kV.

It is interesting to note that, in a sense, the smart grid is returning to larger numbers of smaller generators via encouraging DG and is also attempting to conserve power by reducing distribution voltages. This is because part of the smart-grid concept is to enable the ability to figuratively "reach out" with many smaller generators to extract energy from the environment; specifically, the sun and wind in many cases. However, in theory, there is no need to stop there; we could continue the trend toward smaller generators reaching deeper into the environment down to the molecular scale. More will be said about this when we discuss the smart grid and future directions.

The ability to move large amounts of power between producers enables power to be generated by the most economical generators and also allows power to be shared so that peak demand can be satisfied when and where it occurs. Peak demand is a problem that occurs today due to the cost and inefficiency of powering up large generators that run for short periods of time. A large amount of consideration has gone into flattening the demand; this is incorporated as a component of the smart grid. An efficient means of sharing power over long distance and between synchronous zones is the use of high-voltage direct-current (HVDC) lines. In 1954, the first HVDC power line went into operation in the Baltic. It is 60 miles long and operates at 100 kV. High voltage provides the same power with less current, and thus less power loss. The fact that power is converted to direct current before transmission and then converted back to alternating current upon reception allows the sending and receiving synchronous zones to be out of phase with each other without causing stability problems. This will be discussed in much more detail later; the point of this subsection is simply to understand the historical development of the technology.

1.2.1 Early Communication Networking in the Power Grid

As mentioned in the brief historical vignette above, communication has been vital to the power grid from the beginning. The telegraph was used for meter reading, soon followed by power line carrier at the beginning of the twentieth century. Power line communication was used for voice communication in 1918 (Schwartz, 2009). However, there are at least two key characteristics of the power grid that worked against the rapid and widespread deployment of modern communication throughout the power grid. They are economies of scale and safety. Economy of scale implies huge investment in large systems; systems that will produce large amounts of relatively low-cost power. But once constructed, they are designed to remain in operation for decades without change. Thus, they are not amenable to removing and reinstalling large amounts of equipment simply to keep pace with every incremental technological advance. These advances accumulated rapidly for communications. Also, safety and reliability in high-power electrical equipment are paramount; any perceived vulnerabilities must be mitigated or avoided. Thus, except where absolutely necessary, power grid control was designed to be accomplished locally, without the need for communication (Tomsovic *et al.*, 2005). In fact, power system engineers have, perhaps unwittingly, become experts at extracting communication signals directly from the power grid's operation and developing a self-organizing system. This statement will become more apparent as the book progresses.

We will go into great detail on the power grid technology and fundamentals soon, but for now consider a high-level discussion of the relation between communication and control in the classical power grid. By classical power grid we are referring to the power grid before the term smart grid was coined. We can consider control divided into high-level areas of protection, generation, voltage, power flow, and stability. Points to consider while reading this book are (1) how well all of these can be accomplished without communication, (2) how well they would be improved with communication, (3) what the requirements of the communication system are in terms of qualities such as cost, latency, bandwidth, and availability among others, and (4) what new features that might be added are given communication capability.

Power system protection is one of the most critical control systems because it involves personal safety. The goal of protection is to isolate faults, such as broken power lines, while keeping power flowing safely to as many consumers as possible. Protection control must be quick and accurate. The common approach is as simple as detecting excessive current flow. However, protection can be applied when needed based upon other characteristics, such as excessive frequency deviation, excessive voltage deviation, or excessive instability. Through careful design and placement, known as protection coordination, it is possible to ensure that the proper relay will open at the proper time by having each relay controlled by locally detected information. However, configuring and managing such a system is a manual process requiring manual effort to adjust every time there is a change. Differential protection relays monitor current on each end of a line and compare the current input to the line with current output from the line; any excessive difference indicates a potential fault. In this case, some form of feedback is required in order to determine whether there is a difference in input and output current (Voloh and Johnson, 2005).

Generator governor control simply refers to keeping the generator producing the required amount of power. The mechanical power applied to the shaft is adjusted to keep the rotor moving at the correct operating speed. Again, this has been done locally.

Voltage control can be accomplished by a variety of mechanisms; for example, by increasing or decreasing the current through the generator rotor coils, known as the exciter, or by changing line tap transformers, which effectively change the winding ratios, and capacitors/reactors (inductors). The exciter can be controlled locally by sensing the generator output voltage and maintaining a given constant voltage. Line tap transformer and capacitor/reactor changes are relatively slow and often predetermined by the load profile and time of day.

Power flow control involves controlling the amount of power flowing over particular power lines in the grid. The classical technique is to use a phase-shifting transformer (Verboomen *et al.*, 2005). However, similar to a line tap changing transformer, this is a relatively slow and predetermined process. A FACTS is a more advanced form of control using power electronics. However, again, it is possible to operate the system with manually configured power flow control.

Stability control is required when multiple interconnected generators lose synchronization with one another. Simply put, their electrical interconnection with each other acts to keep the generators moving together; however, the connection behaves in an elastic manner. Generators can begin swinging out of synchronization with each other to the point that they will not settle back into alignment with each other but instead swing wildly out of control. Thus, it is important to implement mechanisms to dampen such oscillations. One approach is to utilize power system stabilizers located at the generator that attempt to monitor and compensate for such oscillation by adjusting the exciter (Yang *et al.*, 2010).

As previously mentioned, communication has always been part of the power grid for relatively impromptu, specific purposes, typically to implement automated meter reading in some locations and special protection mechanisms. With the advent of the so-called smart grid, communication takes on a much larger role within the power grid. A simple example that comes initially to every layman's mind is demand response, which requires two-way communication between every consumer power meter and the utility. This application alone is a huge undertaking. However, more sophisticated uses of communication are being developed as well. The local control mechanisms mentioned previously may now be improved by expanding to wide-area control. The remainder of the book goes into detail on each of these wide-area applications after providing the necessary background required to understand them.

1.2.1.1 Why the History of Communication in the Power Grid is Important

Keep in mind that while this may seem like potentially interesting but useless ancient history, this is far from the case. First, as befits the saying "Those who cannot remember the past are condemned to repeat it," many of the so-called smart grid implementations are, in effect, a return to the way things were done in the past. This is not necessarily bad, but it would be good to avoid repeating the mistakes of the past. Second, the terminology used in today's power grid derives from terms established long ago. By understanding where terms originated, their meaning will become clearer. Finally, we know that technology follows certain patterns, and the better we know the past, the better we can project into the future. Digital communication first reached the power grid via electric power grid control centers and then substations following the introduction of digital computing in those respective components of the power grid. Both the control center and the substation are spatially relatively small areas; initial digital communication was, of course, wired. However, the inertia behind analog wired communication has been very slow to overcome; remnants of wired thinking appear even within today's terminology and standards.

1.2.1.2 Early Analog Computers

Communication exists to serve a purpose, to gather power system data for computation and then to propagate the results, either to an operator in a power grid control center or for control purposes in substations and in the field. Communication technology has been driven by computational hardware requirements and interfaces. Computation in the power grid was done in an analog manner for a significant portion of early power grid history. Before thinking to yourself how primitive this must have been, consider that analog processing techniques are superior to digital processing in terms of computational speed (Deese, 2008) and work continues on applications of analog computing to power grid analysis (Nwankpa *et al.*, 2006). Of course, industry is overwhelmingly digital in both computation and communication products; however, analog computing would likely work best with simpler, cheaper analog communication. Analog computing can be traced at least as far back as ancient Greece, from around 150 BC, when the Antikythera mechanism was estimated to have been constructed (Freeth *et al.*, 2006). The Antikythera mechanism has been a puzzle to researchers because it demonstrates astounding craftsmanship and engineering at such an early date. It is a geared mechanical computer used to compute astronomical positions. Similar technology did not reappear until the 14th century. Moving from linear mechanical to electrical components is actually straightforward since the electrical equations follow those of mechanical components such as springs and dampers.

1.2.1.3 Analog Computers and Analog Communication in the Power Grid

In 1929, General Electric and Massachusetts Institute of Technology built the first analog-computer-based alternating current transient network analyzer. In the mid 1940s, General Electric's Gabriel Kron developed an analog computer alternating current network analyzer for the power grid that he also employed to solve the Schrödinger wave equation, nuclear power problems, and the early analysis of radio communications (Kron, 1945; Tympas and Dalouka, 2008). Thus, the analog computer was utilized in the real-time operation of the power grid. Originally, the control center grew out of the control operators office (Wu *et al.*, 2005), (Dy-Liacco, 2002). The control operator was known as the "dispatcher" and that term remains in use in a variety of power systems terminology, such as "economic dispatch." Economic dispatch is the process of allocating a set of generator outputs such that the power grid is fully served at the cheapest cost.

1.2.1.4 Early Supervisory Control and Data Acquisition in Substations

By the mid-1950s, analog computers and analog communication were in use to implement automatic grid control (AGC). Load frequency control utilized the alternating current frequency as an estimate of the power balance between generator supply and demand, and this could be done locally. Details on this will be explained later in the book. In fact, early AGC was simply a flywheel to maintain constant generator speed, followed later by input from an amount proportional to the alternating current frequency deviation plus its integral. Then, penalty factors were added for transmission line power losses. Earlier forms of supervisory control and data acquisition (SCADA) existed using analog telephone lines as early as the

1930s. Thus, here we see the first use of communications for digital computers and the birth of SCADA. This replaced having humans at a substation who had to be manually called to collect information or issue control commands.

1.2.1.5 The Introduction of Digital Computers and Digital Communications

Digital computers appeared in the 1960s; remote terminal units (RTUs) were used to collect real-time measurements comprised of voltage, real and reactive power, and the status of breakers at transmission substations through dedicated transmission channels to a central computer. A blackout in the north-eastern USA in 1965 accelerated the use of digital computers to enable more real-time control of the power grid. By the late 1960s and early 1970s, smaller digital computers began to be introduced into substations. The notion of power systems security drove the need for more computational power in the 1970s. Security in power systems essentially involves answering a set of "what-if" questions regarding potential failures and estimating the ability of the system to withstand a given set of failures or contingencies. Thus, computers began by being introduced into the control center and eventually spread from there out into substations.

1.2.1.6 Intelligent Electronic Devices

Each advance in computer technology and new application within the power grid required a corresponding advance in communication technology. However, use of computer processors and communications was slow to develop for use in the power grid because extremely high reliability was required. High reliability was required both for safety and because the mindset for power systems equipment is that it should last decades before being updated. By the late 1970s and early 1980s, microprocessors became integrated into power system devices, yielding the term "intelligent electronic device." Also, by the 1980s, minicomputers were replaced by Unix and personal computers became dominant and ran on local-area networks (LANs) in the substation. Thus, we can see the tentacles of computation spreading from the control center, to the substation, and now embedded within individual devices. This spread of computational power into an increasingly distributed system opens the potential for more communication to support computational interoperability.

1.2.1.7 Impact of Deregulation on Communications

Another impact on computation and communication occurred in the second half of the 1990s when the power market became deregulated. Vertical, regulated power monopolies – that is, utilities that controlled all aspects of power generation, transmission, and distribution – turned into competitive markets where, ideally, generation, transmission, and distribution were designed to be run by different, competing, companies. Now many different competing organizations need to work closely together and share enough information to keep the system running smoothly, but not so much information that they lose competitive advantages. The goal is that the invisible hand of the free market will encourage all provider companies and the consumer to maximize their profit and create a more efficient system where the equilibrium price is closest to the true cost of power. More detail on all these topics will be provided later;

here, we are focus on the historical impact upon communication. The impact on communication is that, after deregulation:

1. The monolithic utility system became split into separate autonomous entities: independent system operators (ISOs) and regional transmission organizations (RTOs), generating companies (GenCos), transmission companies, and load serving entities (LSEs) that operate distribution systems, where each of the entities now needs to share information and related impact.
2. Business and market operation are becoming more closely integrated with communication and control of power.

1.2.1.8 Evolution into Distinct Management Systems

We can see that computation within the control center continued to become more sophisticated, involving state estimation and near-real-time solution of the power flow equation in order to more accurately combine power system stability with economic dispatch. This became known as the energy management system (EMS). Meanwhile, the substation computational sophistication was also increasing and became known as the distribution management system (DMS). Market deregulation motivated the need for business management systems in the power grid. Let us end this discussion on the historical background of power system communications by examining the current state of substation communications.

1.2.1.9 Today's SCADA System in more Detail: DNP-3 Example

This section summarizes the current state of SCADA systems using distributed network protocol 3 (DNP3) as an example. Of course, SCADA is a general term that refers generally to a wide variety of industrial control systems, such as those used in manufacturing, refining, and systems on board ships, buildings, and vehicles, as well as power systems. There are a variety of SCADA communication protocols; DNP3 is only one among many. SCADA is another example of a historical term that may be on its way to becoming obsolete. This is because SCADA and distributed control systems are evolving toward performing the same task: real-time control. Historically, before communication was fast and reliable, the term "supervisory" in supervisory control and data acquisition defined the limits of what the system could do. SCADA systems were not able to perform real-time control, but rather served as a supervisory system. SCADA systems allowed operators to see and manage what was taking place, but communication was not fast or reliable enough to enable real-time control. However, as communication performance improves, SCADA systems are becoming capable of performing real-time control; they are becoming instances of distributed real-time control systems.

Readers who are already familiar with communication will readily understand SCADA protocols by calling to mind the simple network management protocol (SNMP). SCADA protocols have some striking similarities with SNMP. This should not be a surprise, as SNMP is, in a sense, the closest de facto SCADA system for communication networks. SCADA protocols and SNMP are designed to provide quick, lightweight, communication of control and status information to a management system and a system operator. They are both concerned with an object-oriented mapping of control points and handle both query and exception-based reporting for critical control decisions.

DNP3 is a SCADA protocol that interconnects client control stations, RTUs, and other intelligent electronic devices (IEDs) (Mohagheghi *et al.*, 2009). The protocol was developed by General Electric to serve the need for a SCADA protocol while the IEC 60870-5 protocol was under development; DNP3 was released to the public through the DNP Users Group. DNP3 is a small, simple, lightweight protocol comprised of only three layers: a link layer, a transmission layer, and an application layer. It is designed to transmit relatively small messages while being quick and reliable. As we will see in communications and networking, there is always a trade-off between being fast (that is, having low latency) and reliable. Message data may be of any length, including a length of zero for commands. Message data is divided into 2048-byte packets at the application layer, creating an application protocol data unit (APDU). The APDUs are broken down into transport protocol data units (TPDUs) of 250 bytes. At the link layer, the link protocol data unit (LPDU) is created by appending cyclic redundancy checksum (CRC) sequences to the LPDU. The LPDU is transmitted by the physical layer, eight bits at a time, with an additional bit to indicate the beginning and end of a data sequence. The physical layer was originally a simple wired connection such as RS-232 for one-way communication over short distances, RS-422 for two-way communication over longer distances, or RS-485, which allowed multipoint communication. As the Internet became more widespread, the DNP3 protocol was implemented, just as described, over the internet protocol (IP) transport layer, either as a transmission control protocol (TCP) or a user datagram protocol (UDP), which enabled much longer distance, wide-area communication.

As mentioned, DNP3 is only one of a myriad of SCADA and power system-related protocols. The goal in discussing DNP3 at this point is only to provide an illustration of a representative SCADA communication protocol that is widely used. Our goal is not to attempt to exhaustively list every SCADA protocol; more protocol standards are likely to be in development as you read this, and any such list would be quickly out of date. Instead, the goal is to focus on the fundamentals as they regard communications and power systems. For example, many power system communication standards appear to be communication protocols, yet they reside upon existing communication protocols, really acting more like applications to a network engineer. The demarcation between network protocol and application standard has been blurring in the power systems field. For example, there is an inter-control center communications protocol (ICCP) that defines wide-area communication among control centers. However, ICCP resides upon the manufacturing message specification (MMS) which can, and often does, reside upon TCP/IP. MMS addresses the issue of a standardized approach to modeling physical devices as logical devices. The goal of MMS is to enable ease of implementation and code reuse by hiding the vendor-specific device behavior within a consistent, standardized, logical device. Thus, MMS creates a virtual manufacturing device with a simple, well-known interface with which many applications can easily interoperate. However, MMS typically reaches down only as far as the top of the network protocol stack (namely, the presentation and session layer) by standardizing on abstract syntax notation 1 (ASN.1) for the encoding of the machine objects and specifying issues related to opening and closing of the communication connection. Thus, ICCP is really not defining anything new below the transport layer, but rather defining how objects and services are represented. ICCP and MMS are addressing communication in a broad sense, but it is only tangential to network engineering because it is not creating or modifying existing communication network protocols. Another, more recent example of the blurring of the boundary between network and application is IEC 61850. IEC 61850 focuses primarily upon a paradigm for specifying, representing, and configuring power system information, but also resides above a set of alternative existing network protocols that actually provide communication. There are many such standards addressing power system object representation and configuration. Some of these standards are completely independent of the network protocol, while others specify in more detail how the power system objects are mapped onto existing network protocols and precisely how the network protocols are to behave. The fundamental, overarching design question, regardless of the particular protocol, involves the age-old engineering decision regarding ease of use with increased complexity versus the simplicity-and-performance curve, which will be called the functionality–complexity curve for short. DNP3 and IEC 61850 are examples on opposite ends of this curve because DNP3 is an example of a relatively simple protocol with good performance while IEC 61850 attempts to do much more self-configuration with a corresponding increase in complexity. One way IEC 61850 attempts to address the functionality–complexity curve is by providing an alternative set of network protocols, from direct encoding over an Ethernet frame intended for time-critical applications with low-latency requirements to a complete client–server architecture over MMS and TCP/IP. The functionality–complexity trade-off can be seen, for example, in the direct mapping over an Ethernet frame because, while relatively lightweight, basic functionality of the IP, such as routing and efficient handling of error over less reliable links such as wireless links, is unavailable, often requiring complex nonstandard approaches to resolve the issues.

What we learn from this is that, from a networking perspective, when you wish to understand new smart-grid communication protocols, one of the first things to consider is whether the protocol really defines a new set of lower layer network protocols or whether it is really acting as an application that utilizes existing network protocols and, if so, how it utilizes them. This book is focused on the network engineering view of communication and less on the design of power system object models (of which there are many to choose from already), although there is certainly an impact of one upon the other. Now that we have discussed the historical context of power systems and communication in some detail, Section 1.2.2 views power systems and communications from a higher level, seeking to learn through comparing and contrasting the two technologies. Following that, we will go into detail on power system fundamentals necessary to better understand the need and requirements for communications.

1.2.2 An Analogy between the Power Grid and Communication Network

In order to gain deeper insight into power systems and communications, consider a rather loose analogy between the two. The idea is to be able to use knowledge of one field to gain insight into the other. Specifically, in Table 1.1 there are three columns: (1) a general network characteristic column, (2) a power grid network column, and (3) a communication network column. We will examine the communication network as an analog to the power grid, using characteristics between the power grid network and a communication network to construct the analogy.

Let us begin with the product being transferred in each network. The content being transported through the power grid is clearly power over some time duration, which results in energy. For the communication network, the content being transported is information. Note that we do not indicate "data" here because information is more general than data. For example, executable code could be transported such that, when run, it generates the intended information. The medium over which energy is transported will be identified as electric power. Note

Table 1.1 Many analogies can be drawn between the power grid network and communication network.

Network characteristic	Power grid network	Communication network
Content	Energy	Information
Medium	Power	Power
Form	Active power	Modulation
Transmission models	Broadcast, multicast, P2P	Broadcast, multicast, P2P
Routing	Switched	Store-and-forward
QoS	Power quality	QoS
Format	Kilowatt-hours	Packets
Error correction	Capacitor banks	Channel coding
Compression	More power over same lines	Source coding
Buffering	Energy storage and DR	Memory for playback delay
Channel utilization	Reactive power	Bandwidth-delay product
Traffic shaping	DR	Leaky bucket
Network management	Power flow, state estimation	SNMP polling
…	…	…

that there is a temptation to think of the medium as a power line or some form of conductive material; however, this is not an accurate analog because power can be transferred without the need for wire. In communications, information is transferred by transmitting a signal that also involves the flow of power. The form in which the content is transferred in the power grid is active power. Active power will be explained in detail later; however, it will suffice here to understand active power as power that a consumer actually consumes. The analog in a communication network is a form of modulation; the receiver must sense a form of modulation in order to interpret a signal.

Regarding transmission models, the power grid can be thought of as broadcast (from a generator to everyone connected), multicast (from a generator to everyone connected in parallel; only those with devices turned on receive power), or point to point (from one point to another over a power line). Similarly, communication networks allow for broadcast, multicast, or point-to-point transmission. The power grid is primarily a broadcast network, with power branching out from large centralized generation sources to millions of users through the transmission and distribution network, while power lines are point-to-point links. A primary difference in contrast with communication networks is that information can be copied whereas power cannot. Thus, within a communication network, a single multicast packet can be sent toward its destination and then duplicated along its path for each branch leading to a receiver, creating more packets than originally input. Unfortunately, this is impossible for electric power – each new user increases the load on the generator.

With regard to routing through networks, electric power is circuit-switched; it generally requires switching: a circuit is modified in order to connect or disconnect power sources from destinations. Note that power flow can be adjusted using more subtle approaches, such as a FACTS, which are explained in detail later; however, we will ignore this consideration for now. Within communication networks, packets reach their destinations via a store-and-forward mechanism. In other words, packets are sent to intermediate nodes in the network that are hopefully closer to their intended destination. The packets are then sent from the intermediate nodes to other nodes along their route to the destination. Such an approach may someday be used for energy as well, utilizing wireless power transmission from one node to another until the final consumer is reached.

The notion of quality of service (QoS) exists in both power systems and communications. Ideally, the consumer should see a perfectly smooth sinusoidal waveform for both current and voltage, although this is rarely the case. In communication networks, QoS can have many different specific meanings, but they are all related to the ability of the receiver to perceive precisely what was transmitted, particularly with respect to reflecting the order and timing of the packets received. In the power grid, the format of the final product is energy, typically received and measured by the consumer in kilowatt-hours. In communications, the "consumer" receives packets; there is an effective data rate (analog of power) and a total amount of data received (analog of energy).

Both the power grid and communication networks have the notion of error correction. Error occurs in the power grid because inductive loads cause the current to lag behind the voltage, causing a change in the power factor, as we will discuss in detail later. Capacitor banks are used to correct the power factor. Analogously, in communications, bit errors occur in packets, and error correction techniques can be applied to detect and correct the erroneous bits.

Compression is involved conceptually in the power grid, and more literally in communication networks. One of the goals of the smart grid, as will be discussed later, is more efficient utilization of the existing power grid; in other words, delivering more usable power to the consumer over the same-capacity power lines. In effect, this can be somewhat vaguely viewed as "compressing" power such that more power flows over the same power lines; for example, by improving monitoring, power factor, and stability, thus allowing the system to operate closer to its physical limits. In communication, compression, also known as source coding, is often employed to send more usable information over the same-capacity communication channel. As is well known, such systems become more brittle (Bush *et al.*, 1999).

Buffering takes place in both the power grid and communication networks. It is often said that electric power supply and demand must always be perfectly balanced; while this is true on a large scale, it is not true on small scales, as we will see. There is always some "residual" power stored within inductive components of the grid, which, thankfully, allow for decisions to be made within seconds rather than instantaneously. On a larger scale, the development of energy storage within the power grid attempts to act as a large buffer, obviating the need for expensive peak generation to come online. Buffering clearly exists within communication networks; multimedia content is often buffered in order to remove jitter, allowing for a smooth, pleasant user experience.

Channel utilization is an important aspect in both the power grid and communication networks. As previously mentioned, extracting more power from the existing power grid is a significant goal for the smart grid. By utilizing power flow analysis, explained later in this chapter, power generation can be controlled in order to manage the flow of active and reactive power throughout the power grid network. Recall that the consumer only actually consumes active power; reactive power oscillates between the consumer's devices and the generator. Thus, efficiently utilizing the power line's "channel" is a key goal of the smart grid. Channel utilization in communication networks is also a common goal. Both consumers and communication providers want to make as efficient use as possible of expensive communication equipment. Traffic shaping takes place in both the power grid and communication networks. One of the well-known components that seems to have become strongly associated with the smart grid is the mechanism of DR, even though the concept was around long before the term smart grid was coined. This is the idea of enforcing the balance between supply and demand via free-market mechanisms. Ideally, consumers would set their demand–price curves and generators would set their price–generation curves and the balance occurs where these curves intersect. In this sense, the power generation rate undergoes traffic shaping. In communication networks, traffic shaping, utilizing techniques involving variations of the leaky bucket mechanism, is used to control the flow of traffic into the communication network such that the network can more efficiently handle the flow. In fact, pricing for network traffic throughput rates can be set in order to achieve an effect very similar to DR in the power grid.

Finally, a form of network management is required in both the power grid and communication networks. DNP3, ICCP, and IEC 61850 were mentioned in Section 1.2.1.9 as examples of supervisory monitoring and control protocols in the power grid. In order to properly manage and control the power grid, power flow must be either directly monitored or inferred through analysis, and is known as power flow analysis. State estimation is a technique used to infer the state of the power grid in general through many potentially erroneous measurements. Similarly, communication networks need to know the state of the communication network in order to properly manage it, including the flow of packets. Network state is often obtained via the simple network management protocol (SNMP) in communication networks.

There is a rich set of useful analogies that can be constructed. Certainly, there are many other analogies that have not been listed here, but hopefully the insight gained from the process of analogy is clear. The more obvious analogies are already being exploited by researchers applying communication to the power grid. From the analogies summarized in Table 1.1, we can see that power system and communication network engineers address similar fundamental problems and, with the proper analogy, should be able to easily understand each other's field. Much more detail will be given throughout the reminder of the book for all of the power system topics mentioned in these analogies. Section 1.3 will move on to a more thorough, but simple, introduction to power systems, beginning with fundamental physics. What is unique about this introduction and much of the rest of this book is that we will identify the relationships to communication networking along with the introduction to power systems.

1.3 Fundamentals of Electric Power

This section provides an overview of the physics required to understand power systems and communications. This can be a complex subject; however, the goal is to keep it as straightforward and elementary as possible while providing enough intuition to comfortably understand the topics covered in the remainder of the book. A reader knowledgeable in electronics could safely skip this section; however, you could miss some potentially useful intuition related to both power systems and communications. What makes this chapter and the remainder of this book unique is that an attempt will be made to discuss both power systems and communications side by side at selected points. This will hopefully stimulate the reader's thought process regarding how these topics could be integrated, perhaps in novel ways. In this section, we lay the foundation by covering charge, voltage, and ground, conductivity, Ohm's law, electric circuits, electric and magnetic fields, electromagnetic induction, circuit basics, and magnetic circuits.

1.3.1 *Charge, Voltage, and Ground*

Beginning from the bottom up, so to speak, the first topic that requires explanation related to power systems is "charge." Without charge, there can be no electric power and no electronic communication. Charge is a fundamental property of matter, yet one with which people tend to be least familiar. For lack of a simpler definition, consider charge to be the property of matter that causes particles to experience a force when placed near other charged particles. Early work in electromagnetic theory was based upon experiments related to the force experienced by a charge. This force is described quantitatively by a "field"; fields will be discussed in more detail later. To the best of our knowledge, charge is quantized; there exists a smallest unit of charge, and charge only exists in multiples of this smallest unit. Charge comes in two forms, positive and negative, as any youngster knows. Like charges repel; opposite charges attract. The smallest negative charge is an electron and the smallest positive charge is a proton, and these charges are equal. But that begs the question: What does it mean for these charges to be equal? How is charge measured? This requires the definition of another fundamental unit, the ampere. An ampere is the amount of charge in motion through a conductor past a give point per unit time. Thus, an ampere is a convenient measure for the flow of charge. The amount of charge itself is thus defined by an ampere-second, also known as a coulomb. One coulomb has the charge of 6.241×10^{18} electrons. As we will see, the interactive force among charges plays a large role in power systems, from generators to motors. When these fields are radiated, they play a large role in wireless communication; again, more detail will follow on this.

Because like charges repel and opposite charges attract, a local imbalance of charges will have a natural tendency to spread out if possible. In other words, too many charges of the same type will be under pressure to move such that the state of the system is at a lower energy level. This energy is also known as potential; it represents the potential energy of the charges, similar to the mechanical potential energy of a ball at the top of a hill. The potential energy is the work required to move the ball into position on the hill; similarly, the electrical potential energy is the work required of the electrical charge to move from or toward its minimal energy location. This work includes overcoming the difficulty of moving past obstacles in its path, which can result in the creation of heat within a wire.

This notion of pressure or potential is critical to understand power; both charge and potential are involved in defining the work done by a power system. A greater charge will exert more pressure, which exists as potential energy available to do more work. Note that a charge is unique to a particle, while potential energy, as we have described it, is related to a charge's location with respect to another location. This potential energy is known informally as electric voltage, or more formally as electric tension or electric potential difference; it is more rigorously defined as the potential energy of a charge divided by the amount of charge. We will use the less formal term, voltage. The electric potential energy is equivalent to the charge times the voltage, as shown in Equation 1.1, where U_{E} is the electric potential energy, q is the charge, and V is the voltage:

$$U_{\mathrm{E}} = qV. \tag{1.1}$$

From an intuitive perspective, voltage provides the pressure required for power to flow in power systems, while in communications a change in voltage often provides the signal representing the information to be transmitted.

Because electric potential energy is a quantity that is relative between two locations, the ideal reference location would be one that has no charge. Such a location is difficult to find in practice; the next best alternative is a location where positive and negative charges fully cancel or charges are easily dispersed. This is the case within the Earth. The reference location is thus commonly termed "ground." But keep in mind that just as there is a mechanical difference in potential energy when objects are at different heights, different electrical potential differences exist between any two locations. To recap, voltage is defined in terms of energy (joules) per charge (coulombs). Note that a joule can also be expressed as a watt-second.

1.3.2 Conductivity

Without going into detail on molecular structure, conductors are materials that have electrons that are free to travel. However, note that individual electrons typically do not travel far or fast. We can think of a charge applied to one end of a conductor as causing the conductor's electrons to realign due to their repulsion to the applied charge. This realignment is generally in the form of a slow drift, as shown in Equation 1.2, where I is the electric current, n is number of charged particles per unit volume (or charge carrier density), A is the cross-sectional area of the conductor, v is the drift velocity, and Q is the charge on each particle:

$$I = nAvQ. \tag{1.2}$$

Consider a copper wire with a cross-section of 0.5 mm^2 carrying a current of 5 A. The drift velocity of the electrons is on the order of only 1 mm/s. However, as we will discuss later, the change in position of the electrons establishes a magnetic field immediately outside the conductor. This field exists in the form of a wave that travels at a significant fraction of the speed of light. The conductor, in effect, guides an electromagnetic wave and is commonly known as a waveguide. This phenomenon is used frequently to couple communication with power lines, known as a power line carrier communication. More will be explained regarding this as we proceed.

While we tend to think primarily of metals as conductors, other materials can be conductors as well, and this has had an impact on both power systems, sometimes in negative sense, and for communications. For example, in 1837, Carl August von Steinheil of Munich, Germany, connected one leg of a telegraph at each station to metal plates buried in the ground, eliminating one wire of the circuit using a single wire for telegraphic communication. This led to speculation that it might be possible to eliminate both wires and transmit telegraph signals through the ground without wires. Attempts were also made to send electric current through bodies of water, in order to allow communication to span rivers. Experiments along these lines where done by Samuel F. B. Morse in the USA and James Bowman Lindsay in Great Britain. By 1832, Lindsay was giving classroom demonstrations of wireless telegraphy to his students, and by 1854 he was able to demonstrate transmission through a river for a distance of 2 miles, across the Firth of Tay from Dundee to Woodhaven, which is now part of Newport-on-Tay. From a power systems perspective, under the condition of strong electrical potential fields or high temperatures, the electrons of gas molecules can become free to travel. A gas in this state is known as a plasma. Air can become a conductor, and a dangerous discharge of electric current can take place through the air known as arcing.

Finally, there is a form of conduction in which there is no electrical resistance to the flow of electrons below a certain characteristic temperature. This clearly has many uses in power systems, ranging from superconducting power lines, through which there is no loss of power due to resistance or thermal effects, to superconducting magnetic energy storage (SMES), in which energy is stored in the magnetic field of a superconductor through which a direct current flows. From a communication standpoint, superconductivity can be leveraged for quantum computing, and particularly for quantum communication (Bush, 2010b). The concept of using superconducting power lines for both efficient power transfer and quantum communication for a smart grid is something to consider and is discussed again in Chapter 15.

An analogy with water is helpful. Current (measured in amperes) is the rate of flow of charge and is similar to water flow through a pipe, whereas voltage is similar to the water pressure difference between the ends of a pipe. Amperes are measured in charge per unit time, or coulombs per second. The impact of the current flow is fast, a significant percentage of the speed of light. However, as mentioned, electrons themselves do not actually travel nearly this fast; it is the impact of the electric field that is propagating through a conductor, analogous to ripples in a pond. The water does not actually flow away from a pebble; it is only the waves that are moving. This brings up the question of timing in power systems and communications. A large power grid can experience a measurable delay in the time it takes a current to propagate from one end of the grid to another. In addition, a communication signal, like all matter, is theoretically limited by the speed of light. Typically in electronics, propagation delay becomes an issue only at the extremes; that is, at the very small scale (for example, high-precision timing) and at the very large scale (namely, power grids). In most cases for power systems, propagation delay is ignored; such a circuit is called a "lumped circuit." On the other hand, for communication, overcoming, or least managing, propagation delay is a central issue. In communication systems, the bandwidth–delay product of a channel considers the communication channel as a pipe; the bandwidth times the delay of the channel is the amount of water the pipe can hold when it is full. The bandwidth–delay product is a characteristic that drives how efficiently communication protocols can take advantage of a channel. Delays – both the delay for current to flow through a line and for the communication required to control it – will remain an extremely important issue for the smart grid. From a power systems perspective, consider a 500-mile transmission line. It takes 2.7 ms to travel that distance at the speed of light (although the actual electromagnetic rate would be slower in reality). A 60 Hz alternating current has a zero crossing (that is, changes of direction) of 60 times per second, or once every 16.7 ms. Thus, we can see that propagation begins to become a significant part of the cycle time.

1.3.3 *Ohm's Law*

We have discussed current and voltage at an intuitive level, and now we need to relate them. If voltage is the pressure difference between charges and if current is the flow of charges, then voltage should cause a proportional amount of current to flow. However, as we mentioned, charges run into obstacles in their path through a typical conductor; this causes friction or resistance to the flow and maintains the pressure difference. We can think of this resistance as a constant of proportionality between voltage and current. Thus, the famous equation that most people take for granted is shown in Equation 1.3, where V is voltage, I is current, and R is resistance.

$$V = IR. \tag{1.3}$$

It should be understood that Equation 1.3 is an idealization, assuming a linear relationship between current and voltage. In reality, this is not true. For some materials, the relationship between voltage and current may change significantly with different values of voltage and current. Also, the relationship is dependent upon temperature. For example, the resistance of some metals increases as the temperature increases. Finally, the shape of the conductor plays a large role in its resistance. The term resistivity refers to the inherent property of a material to resist the flow of current, which is denoted ρ. This property can be quantified based upon the length and area of a cross-section of the conductor, as shown in Equation 1.4, where R is the resistance, ρ is the resistivity, l is the length of the conductor, and A is the area of a cross-section of the conductor:

$$R = \frac{\rho l}{A}. \tag{1.4}$$

Intuitively, it should make sense that increasing length increases resistance since the current flow has a longer path, and reducing the cross-sectional area also increases the resistance since the current is constrained within a thinner pipe. Resistance is measured in units of ohms or volts per ampere, as is apparent by rearranging Equation 1.3, while the unit of resistivity is the ohm-meter.

Conductance is the inverse of resistance, $G = 1/R$, and conductivity is the inverse of resistivity, $\sigma = 1/\rho$. Thus, conductance can be derived from resistance, as shown in Equation 1.5, where G is the conductance, σ is the conductivity, l is the length of the conductor, and A is the area of a cross-section of the conductor:

$$G = \frac{\sigma A}{l}. \tag{1.5}$$

For power systems, insulation becomes an important material property, with the goal of obtaining high resistance or low conductivity, in order to keep current contained and safely flowing. Plastics and ceramics are often used for this purpose. An old technique for estimating the amount of voltage on a power line is to count the number of ceramic insulation bells holding the power line to the tower. The number of bells grows roughly in proportion to the power on the line. The exact proportion of bells to kilovolts on the power line can depend upon the climate because, as mentioned earlier, air can become ionized and support current flow, and wetter, more humid air will more easily conduct current, thus requiring more insulation.

1.3.4 Circuits

An important point from a power systems perspective is voltage drop. As already discussed in Section 1.3.3, voltage is proportional to the current and resistance. In a power line, resistance can be significant. However, current will vary depending upon the load; that is, on the consumer and industrial appliances using the power transported by the power line. As more customers use more appliances, current flow increases, which increases the voltage drop. The result can be insufficient voltage and a brownout.

As alluded to earlier, charges collide against obstacles as they flow through a conductor, and these collisions cause friction which results in resistive heating. Heating is the work done as the charges travel toward a lower potential. This heat can be desirable for use in heating an electric heating element or it may be undesirable and potentially dangerous, as when it occurs in a power line or electrical equipment for example. A power line's goal is to transfer energy as efficiently possible; converting some of the energy into work as heat is not the desired goal for a power line. Similarly, an electronic device such as a transformer is designed to convert power from one voltage to another, not to generate heat. As we have mentioned, such heat tends to increase the resistance, increasing the voltage drop. The heat may eventually build up to the point where it may melt wiring or damage insulation, leading to power faults, which will be discussed in detail later. An entire subfield of power systems is devoted to preventing or mitigating such problems and is known as protection.

From a communication perspective, resistive heating is less of an issue; and herein lies an obvious, but fundamental, difference between power systems and communication. Communication is focused upon maximizing channel capacity while typically minimizing power consumption. Extraneous power in communications often results in noise; it reduces the signal-to-noise ratio. Communication systems strive to maximize the signal-to-noise ratio in a channel. This often implies increasing the power level of a signal, but communication power levels are many orders of magnitude lower than those in power systems. Thus, whereas power systems seek to maximize power flow as efficiently as possible, communication systems seek to maximize the reception of a signal with as little power as possible.

Resistive heating brings us to the first instance in which we have discussed useful work, in the form of the intentional generation of heat. Let us examine this further. The heat generated can be measured in terms of power, which is the energy generated per unit time. This is the rate at which energy is being converted into heat, which we assume in this case is useful work. Let us see how this relates to the electrical parameters we have discussed so far. Recall that voltage is in units of energy per charge. Thus, the units of voltage contain the energy that we require for power, which is energy per unit time. However, voltage by itself is missing the notion of time. But recall that current is flow of charge, or the rate of flow of charge. Thus, current contains the notion of time since it involves a rate, namely something changing over time. As shown in Equation 1.6, it is not too difficult to see how voltage and current are related to power:

$$\frac{\text{charge}}{\text{time}}\frac{\text{energy}}{\text{charge}} = \frac{\text{energy}}{\text{time}}. \tag{1.6}$$

Equation 1.7 shows this same relationship of power in terms of current and voltage.

$$P = IV. \tag{1.7}$$

This comes from Joule's law, which originally had another form, as shown in Equation 1.8, where I is current and R is resistance. This can be derived simply be using Ohm's law, $V = IR$, and substituting V in Equation 1.7:

$$P = I^2R. \tag{1.8}$$

Power is measured in units of watts, which we can see from Equation 1.8 is amperes^2-ohms. Note that current is squared, so it has a greater impact on power as it increases than resistance does. However, it is important to take into account other relationships when applying this equation; in other words, the equation cannot be applied blindly. As a simple example, one cannot, in general, simply increase resistance in order to increase power from the electric power grid. The utilities intentionally keep residential input voltage at a constant 120 V, regardless of the power drawn from the grid. Since Ohm's law applies and voltage is held relatively constant, then decreasing resistance causes an increase in current. The increase in current impacts power by its value squared. Thus, decreasing resistance in this case actually causes a significant increase in power. The extreme limit of this would be setting the resistance to zero, causing a short circuit, also known as a power fault.

The power line experiences a different situation: current is effectively constant while voltage changes. The current drawn by the loads on the line controls the current flow over the line; the line resistance has a negligible impact on the current relative to the consumers. Thus, from the power line's perspective, current is constant and the voltage drop along the line changes. Thus, utilities have to address this voltage drop in order to maintain a constant 120 V power supply to all consumers along the entire power line. The I^2R formula shows that resistive heating, which is lost power, is directly proportional to the resistance in the line.

Now consider the voltage level and transmission line losses in light of Equations 1.7 and 1.8. There are many choices of I and V that yield a given power P. However, since thermal losses are proportional to I^2, these can be minimized by choosing a low value of I and a correspondingly higher value for V. Thus, high-voltage transmission appears to be more efficient, and that is what is used. In the early days of power distribution, the transformer had not yet been invented; distances were limited to only a few miles due to the relatively large current and low voltage. The trade-off with modern high-voltage transmission and distribution is the greater cost and care required to keep the higher electrical potentials insulated.

1.3.5 Electric and Magnetic Fields

Fields were developed as a means to quantify how objects that are not in physical contact were able to exert forces upon one another. A field may seem like an abstract mathematical map, like a topological map of elevation, but without apparent physical reality. The attraction and repulsion of charges is a classic example of a field, although gravitation is probably a more commonly encountered field for us. We are attracted to the Earth without any apparent force pushing or pulling us to the Earth. There is a clear direction and force that objects are attracted to the Earth that can be mapped at all points in space. This mapping is a field. However, a field is more than an abstract mathematical map. During his investigations into magnetism, Michael Faraday first realized the importance of a field as a real physical object. He realized that fields have an independent physical reality because they carry energy.

Fields are extremely important in both power systems and communications. Fields create the power that flows through the electric power grid, fields pulsate along the transmission and distribution lines, and fields move the machines for the power consumer. In communications, fields are a critical component of electric oscillators, and they ultimately transport signals through space. Fields are truly the workhorse for both power systems and communications.

Recall that the electrical potential energy quantifies the potential energy of a specific charge given its attraction and repulsion to all other charges. If the other charges remain fixed, then the potential energy changes with the location of the specific charge that we are discussing. Conceptually, the charge can be moved to every location and the potential energy recorded, thus creating something analogous to a topographical map, where altitude on the map corresponds to the amount of energy at that location. In a topographical map, lines are drawn along paths of constant altitude; lines that are closer together indicate a more rapid change in altitude. The topographical map would be analogous to the electric field, but with a significant difference. The electric field maps the force on the charge at a given location, not the potential energy. The force on the charges decreases at a rate that is the square of the distance.

There is another type of field that is distinct from the electric field: the magnetic field. A charge in motion creates a magnetic field. This occurs at the subatomic level as well as the macroscopic level. Electrons are charges in motion, they have both orbital motion and spin. When these motions are aligned, a material becomes magnetic. However, unlike charge, a magnetic field has two poles, positive and negative, or north and south from its use on a compass; the magnetic field must exist as a dipole. Since moving charge creates a magnetic field, it stands to reason that a current, which is a set of charges in motion, should also create a magnetic field. This is remembered by every student as the right-hand rule: using your right hand, point your thumb in the direction of the current, then your fingers will curl in the direction of the magnetic field. One can increase the current to increase the magnetic field or, alternatively, place more wires with current near each other going in the same direction, in which case the magnetic fields will add. This can be done conveniently by looping a wire upon itself. The magnetic field lines that are concentrated near the center will add to one another, creating a strong field in the center. We can again visualize our topographical map with many lines near the center of the coil representing the strong magnetic field. The right-hand rule is shown in Figure 1.5.

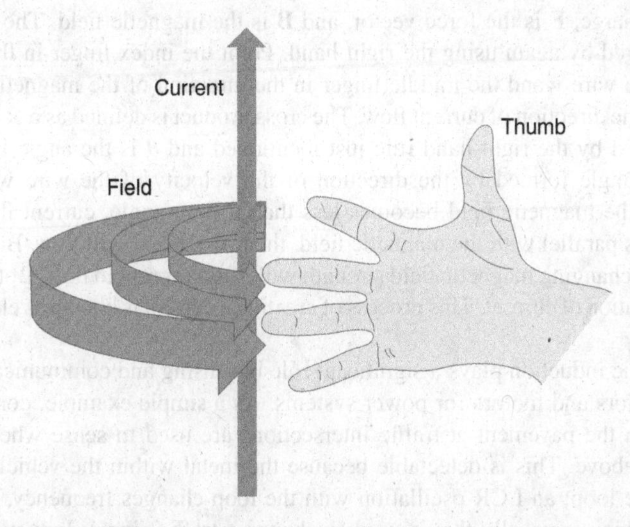

Figure 1.5 The right-hand rule, in which the straight thumb points in the direction of the current and the curled fingers in the direction of the magnetic field. Source: By Jfmelero (Own work) [GFDL (http://www.gnu.org/copyleft/fdl.html) or CC-BY-SA-3.0-2.5-2.0-1.0 (http://creativecommons.org/licenses/by-sa/3.0)], via Wikimedia Commons.

The magnetic field has a direction and is denoted by the vector $\vec{B}$ and is quantified in units of teslas (T) or gauss (G). A tesla is equal to one newton per ampere-meter. This can be seen as a particle with a charge of one coulomb passing through a magnetic field of one tesla at a speed of one meter per second experiencing a force of one newton. A tesla is simply 10 000 G. A related measure is magnetic flux, which has to do with the density of the magnetic field. Magnetic flux, denoted by ϕ, is the amount of magnetic field passing through a given surface; its unit is the weber (Wb), but it may also be commonly expressed in derived units of tesla-square meters.

1.3.6 Electromagnetic Induction

We have just seen that motion of an electric charge creates a magnetic field and, as mentioned in Section 1.3.5, the magnetic field can exert a force on a charge, causing charges to move and creating an electric current. This process of creating current is extremely important as it is the basis of electric power generation and is involved in the operation of electric motors, among other applications. However, in order to create a continuous electric current, the magnetic field must be continuously moving. Let us look at the geometry involved in this production of current. Imagine that the magnetic field lines are pointed upwards coming out of this page, like spikes. Now imagine a wire lying on this page from the top to the bottom of the page and sliding across the page from left to right. This motion of the wire across the page would cause a current to flow through the wire, forcing the negative electrons to flow out the bottom of the wire. The Lorentz equation describes this:

$$\mathbf{F} = q\mathbf{v} \times \mathbf{B}, \tag{1.9}$$

where q is the charge, $\mathbf{F}$ is the force vector, and $\mathbf{B}$ is the magnetic field. The cross product can be remembered by again using the right hand. Point the index finger in the direction of the motion of the wire v and the middle finger in the direction of the magnetic field B. The thumb points in the direction of current flow. The cross product is defined as $\mathbf{a} \times \mathbf{b} = ab \sin\theta \mathbf{n}$, where n is defined by the right-hand rule just mentioned and θ is the angle between a and b. Thus, as the angle formed by the direction of the velocity of the wire with respect to the direction of the magnetic field becomes less than a right angle, current flow decreases. If the wire moves parallel with the magnetic field, then no current will flow. Both the motion of the wire and a changing magnetic field strength will cause current to flow. Both methods are used in the generation of current. This process of creating a current is known as electromagnetic induction.

Electromagnetic induction plays a significant role in sensing and communications, as well as within generators and motors for power systems. As a simple example, conductive loops embedded within the pavement at traffic intersections are used to sense when a vehicle is located directly above. This is detectable because the metal within the vehicle changes the inductance of the loop; an LCR oscillation with the loop changes frequency, indicating the presence of a vehicle – usually this is used to change a traffic signal. Induction is used for both power and communications for implanted medical devices, such as hearing aids. As a final example, induction was used in early radio-frequency identification (RFID). An RFID reading device emits electromagnetic waves that are used to communicate with an RFID tag. The magnetic component of the electromagnetic waves of the reader can induce a current in the tag not only to transmit information, but also to supply power to the tag. Thus, the tag can be completely passive; it does not require a power supply. However, it required a conductive coil that was not always as small as users would have liked. Later, capacitive coupling was used involving carbon ink on paper. Finally, active tags were used in which tags could store a significant amount of their power and actively transmit radio-frequency signals. The goal here is not to go into detail on traffic detectors or RFID technology, but simply to highlight the relationship between power, communications, and, in this case, inductance.

We have now been introduced to enough background information to address an interesting phenomenom: the propagation of electromagnetic fields; namely, electromagnetic waves. Almost every grade-school student knows about radio waves; most of us grow up having been "taught" about radio waves and take this fact for granted, but what are electromagnetic waves? How do we know they are waves at all? Why should this energy propagate and what precisely leads us to believe that this propagation takes the form of waves? Since electromagnetic waves play such a large role in communications, it is important to take some time to address this. For now, let us continue the discussion assuming we know they are waves; the wave aspect will be discussed a bit more precisely later.

As mentioned previously, an electric charge of a field in motion creates a magnetic field. An electric field with a pulsing wavelike shape is continuously in motion; thus, its motion creates a magnetic field at right angles to the electric field. The wave is self-propagating, with each field inducing the other. The wave frequency is measured in cycles per second or hertz and the wavelength is the distance from one wave crest to the next wave crest. The frequency in cycles per second multiplied by the wavelength, or length from one cycle to another, yields the velocity of the propagating wave. The speed of a propagating wave happens to be constant: the speed of light, or 3×10^8 m/s or 186 000 miles per hour. Thus, there is a fixed relationship between wavelength and frequency, since the product is constant. A faster frequency results in a shorter wavelength. The speed of light makes physical sense since it is the speed at which the electromagnetic fields are able to induce each other at the same strength. If the wave propagation were faster, the fields would create energy, which is impossible. If they were slower, they would lose energy that could not be accounted for in the conservation of energy. Thus, from the conservation of energy, one can derive the speed of electromagnetic propagation; that is, the speed of light.

This propagation phenomenon has been utilized in many different ways for communication, from fiber-optic cable carrying light waves to wireless radio communication. Wireless communication is particularly instructive to consider. In the transmitter, electric charges move or oscillate in a manner that conveys information. This electric charge, moving through the antenna, creates an electric field. This electric field creates a corresponding magnetic field that radiates into space. When the self-propagating electromagnetic radiation reaches a wire, as discussed earlier, the motion of the magnetic field relative to the wire causes charge to flow through the wire. This charge will have the same frequency as the magnetic field that created it. Thus, the information embedded in the characteristics of the magnetic wave, such as its amplitude or frequency, will be reflected in the charge induced in the antenna. However, it is important to note that an electromagnetic field does not "know" whether a wire or other conductor happens to be a receiving antenna. It will obviously induce a corresponding current in any conductor. Thus, electromagnetic radiation from the sun occasionally impacts the Earth with enough energy to induce dangerous levels of current within power lines. This can be enough current to cause transformers to melt or breakers to trip. In other words, the power lines of the power grid become a planetary antenna.

1.3.7 Circuit Basics

Now that we have completed a very brief and intuitive look at the role of the fundamental physics, including electromagnetic wave propagation, let us return to a very brief and intuitive review of electric and magnetic circuits. Along with the rest of this section, this is a quick review and we will assume that the reader has some background in basic electronics. Thus, this section will move quickly over basic material, dwelling only on those aspects of particular importance in either power systems or communications.

We begin with resistance, or load, from a power systems perspective. Resistance is summed when the resistances are in series with one another. Conductance is the inverse of resistance. Conductance is summed when the resistances are in parallel with one another:

$$\frac{1}{R} = \frac{1}{R_1} + \frac{1}{R_2} + \frac{1}{R_3} + \cdots \tag{1.10}$$

Assuming three resistors in parallel, simple algebraic manipulation yields

$$R = \frac{R_1 R_2 R_3}{R_1 R_2 + R_1 R_3 + R_2 R_3}. \tag{1.11}$$

The solution to complex random resistor networks can be found using matrix analysis and is discussed in Bush, 2010b, pp. 102–106. This will be discussed further in Chapter 6 on network topology and complexity; the important point here is that current and voltage through a circuit are primarily a network topology problem.

In power systems, customer loads are designed to be connected in parallel with the power grid. This allows the voltage to remain constant for all customer loads. However, the loads will draw more or less current as their operation varies in order to provide the appropriate amount of power. Thus, with loads in parallel, each load can remain independent of the other loads on the same line. Interdependence among loads will only occur under abnormal conditions; that is, if such a large amount of current is drawn that it impacts the local voltage, then all other voltages on the line will be impacted.

Kirchhoff's voltage law simply states that the sum of the voltages around a closed loop must be zero. There are many intuitive ways to arrive at this law. For one example, picking any point in a circuit and measuring the voltage at points around the circuit (with reference to another common point for all measurements), it only makes sense that when we reach all the way around the circuit back to the initial point that we should measure the same voltage; there would be no reason for it to be different. Also, this is a form of conservation of energy; if the electrical potential around the loop did not sum to zero, then one could propel a charge around the loop indefinitely. This would be similar to the famous painting by Escher in which a waterway forms a loop, yet all the waterfalls along the way seem to flow down hill, an impossibility illustrated in his painting entitled *Waterfall*.

The Kirchhoff laws exist within Maxwell's fundamental equations. To see this for Kirchhoff's voltage law, start with Equation 6.40, Faraday's law of induction (this will be introduced in detail in Chapter 6) in a static field. Then consider that $\nabla \times \vec{E} = 0$. Next apply the Kelvin–Stokes theorem shown in Equation 1.12:

$$\int_{\Sigma} \nabla \times \mathbf{F} \cdot d\mathbf{\Sigma} = \oint_{\partial\Sigma} \mathbf{F} \cdot d\mathbf{r}. \tag{1.12}$$

This results in

$$\oint_{C} \mathbf{E} \cdot d\mathbf{l} = 0. \tag{1.13}$$

This equation shows that the line integral of the closed loop C around the electric field is zero. In other words, there is no potential difference around the closed loop C, where C is any arbitrary closed loop.

Kirchhoff's current law states that the sum of current entering and leaving a point in an electric circuit must sum to zero. This is simply another conservation law: charge cannot be created, destroyed, or stored in some hidden form, within the circuit. Kirchhoff's current law can be found in both Gauss's law (6.37) and Ampère's law (6.41).

Kirchhoff's voltage and current laws are extremely simple, yet they form an underlying basis for much of power systems today. Given an electric circuit, it is possible to use the Kirchhoff laws to construct a matrix of linear independent equations that can be used to solve for the currents and voltages for all components of a circuit. In fact, the Kirchhoff or admittance matrix plays a large role in studying network topology, since the network topology defines the electrical circuit. It is also worth noting here that researchers are exploring ways of using the operators that appear in Maxwell's equations, such as the divergence operator, within network graphs. The primary difference is that the operators, such as the Laplacian, that appear in Maxwell's equations are continuous operators; that is, they apply to continuous spaces and flows. The analogous operators on graphs are discrete, such as the discrete Laplacian; they apply to discrete values related to nodes within the network.

There is also a superposition principle that applies to electrical circuits. This principle states that the combined impact of multiple current or voltage sources within a circuit can be analyzed independently and the effects can be summed. Conceptually, for example, each generator in a DG system can be analyzed independently as delivering power to a set of customers; the sum of the transmission flows on a power line can then be added. Since, as we mentioned earlier, voltage is held constant, each source can be thought of as supplying current, where the currents through the transmission and distribution networks are additive. The superposition principle can be applied strategically to help simplify analysis.

1.3.8 Magnetic Circuits

We just completed a very brief review of electrical circuits by covering a few of the more important highlights. We continue the pattern that began early in this chapter of finding a magnetic analog for each electrical topic. Here, we discuss the analog to electrical circuits; that is, magnetic circuits. An electrical circuit is quite familiar to everyone; usually it is thought of as wire conductors carrying current from an electrical power source through a load and back to the power source. Current is the flow of electric charge. Recall the discussion of a magnetic field; there is no magnetic charge. The magnetic field is comprised of loops; there is no beginning and no end to magnetic field lines. The density of the magnetic field lines is described by the magnetic flux. Magnetic flux is a measure of the amount of magnetic field passing through a given surface area. Note that, technically, it is the net number of magnetic field lines passing through a given surface area; magnetic field lines that pass in the opposite direction through the surface are subtracted from the total. Also, the amount of flux ϕ depends upon the angle of the field lines through the surface; field lines perpendicular to the surface increase the flux more than lines that have a component that is parallel with the given surface.

Magnetic circuits carry magnetic flux just as electric circuits carry electric current. Since the magnetic field lines always form a loop, there is a natural circuit-like path for the magnetic flux. A magnet, as a source of magnetic field lines and thus magnetic flux, has a stronger flux closer to the magnet and a smaller flux farther away. Magnetic circuits play a significant role in generators and transformers, and generally anywhere there are strong magnetic fields and coils involved.

Reluctance $\mathcal{R}_{\mathrm{m}}$, used in the analysis of magnetic circuits, is analogous to resistance in an electrical circuit. It quantifies the difficulty magnetic flux has in flowing through a material. Similar to the way an electric field causes an electric current to follow the path of least resistance, a magnetic field causes magnetic flux to follow a path of least magnetic reluctance. The magnetic permeability is the ease with which a magnetic field flows through a substance and is denoted by μ. The reluctance is related to the physical property of a material by

$$\mathcal{R} = \frac{l}{\mu A}, \tag{1.14}$$

where l is the length of the material and A is the area of a cross-section of the material. μ is to the magnetic flux as conductivity is to electric current. A significant difference between magnetic flux and electric current is that the conductivity of metals is so much higher than other materials commonly encountered that current remains confined to the conductive element. μ_0, the permeability of a vacuum, is not zero and air has a permeability that supports magnetic flux. Thus, magnetic flux is not confined as neatly as current to conductive media. There is usually the notion of a leakage flux that must be taken into account.

While we know the **B** field, it becomes modified within a material due to magnetic fields produced by the material itself, and this modified field is known as the **H** field. As previously mentioned, the permeability μ is a property of a material's ability to permit the magnetic field to flow through it. The relationship between the **B** and the **H** fields is as follows:

$$\mathbf{B} = \mu \mathbf{H}. \tag{1.15}$$

We know that magnetic flux is the analog of current in an electric circuit, but what is the analog of voltage? There is a magnetomotive force (mmf) analogous to the electromotive force that is thought of as generating the magnetic field. As we know, the strength of the magnetic field depends upon the flow of electric current that creates it. By looping wire into a coil, the impact of the current flow adds up in creating the magnetic field. Thus, Equation 1.16

quantifies the mmf, where N is the number of turns in the coil and i is the current flow through the coil:

$$\text{mmf} = Ni. \tag{1.16}$$

As mentioned earlier, Ohm's law relates voltage to current and resistance, $V = IR$. Hopkinson's law is the corresponding analog in magnetic circuits:

$$\text{mmf} = \phi \mathcal{R}_{\text{m}}, \tag{1.17}$$

where ϕ is the magnetomotive force (mmf) across a magnetic element, ϕ is the magnetic flux through the magnetic element, and $\mathcal{R}_{\text{m}}$ is the magnetic reluctance of the element. A simple algebraic manipulation is

$$\phi = \frac{\text{mmf}}{\mathcal{R}}. \tag{1.18}$$

The reluctance is generally not linear with the magnetic flux for most materials; thus, reluctance is not as simple to use as resistance. However, the reluctance in series is additive in a similar manner to resistance.

As mentioned earlier, the magnetic flux is a measure of the magnetic field lines flowing through a given surface area. That surface can be the turn of a coil. Magnetic flux flowing through the turns of a coil is said to link the coils. Flux linkage λ is a measure of the impact of the magnetic flux through the coils:

$$\lambda = N\phi, \tag{1.19}$$

where N is the number of turns of the coil and ϕ is the magnetic flux.

Recall that the charge through the coil creates the magnetic field, and thus the flux, but also the magnetic field stores energy and can create a current flow back into the coils. In the latter process, known as induction as we have already discussed, the role of the flux linkage is shown in

$$\lambda = Li, \tag{1.20}$$

where L is the inductance measured in henries, which will be discussed in more detail in a later section. Looking at the last two equations, inductance is conceptually a measure of how much flux linkage is associated with a given amount of current.

This section has covered the foundation by explaining charge, voltage, ground, conductivity, Ohm's law, electric circuits, electric and magnetic fields, electromagnetic induction, circuit basics, and magnetic circuits. There is more to cover on the fundamentals of power systems and communications; this material will be continued in the following chapters where it more closely pertains to the chapter topic. The material covered in the above sections will be used in later chapters on generation, transmission, distribution, and consumption. For the remaining sections of this chapter, the important point is that there is a set of complex physical phenomena taking place in power systems. As the power grid becomes larger and more interconnected, complexity increases. Therefore, let us change perspective and look briefly at a study of power system failures to motivate how and why the power grid could be improved. Following that, we will discuss the driving forces and goals of the smart grid. Finally, we will complete the chapter with a foreshadowing of the notion of entropy and information as it relates to power systems.

1.4 Case Studies: Postmortem Analysis of Blackouts

We are taking a short break from the basic physics of power systems to examine how the power grid operates at a higher, more abstract level. It is important to keep both low-level operation and high-level system interconnectivity in mind when thinking about the power grid. Focusing solely on the low-level physics can cause one to miss important, potentially catastrophic, high-level characteristics. More specifically, we will look at how the power grid can fail from a system-level, network perspective. We want to look at actual faults that have occurred in the power grid in an effort to understand why they occur and how the power grid might evolve in such a manner as to lead to improvement. We want to take a critical look at smart grids and consider carefully where this could lead, whether to more reliable and improved systems or whether it could take us closer to a "tipping point." Much of the smart-grid research has focused on attempting to improve system efficiency assuming that they operate perfectly. It is also possible to learn from mistakes.

1.4.1 Blackouts Have Something to Teach Us

As the reader may begin to imagine from the introduction to the fundamentals of power systems that has been covered so far, the electric power grid is a large, interconnected, dynamic system. It has been called the most complex machine ever built by humans. It is a complex system in which the rules of complex systems should apply and provide insight into its operation. A complex system is one that lives at the boundary between order and chaos. If too ordered, the system is locked into an equilibrium state and therefore slow and unable to adapt to changing conditions. If too chaotic, it responds too quickly and breaks apart into random components and loses coherence as a single system. Thus, many complexity theories focus on systems that naturally live at the transition between order and chaos. In such systems, an attractor is a state of a system toward which the system will naturally tend to gravitate, even if perturbed.

One of the most simple and intuitive examples of a complex system has been a growing sand pile: adding a single grain of sand to a sand pile may have no effect or it may cause a catastrophic avalanche that impacts all grains of sand in the pile. A common feature of such complex systems is scale invariance, also related to self-similarity, the notion of correlation of behavior, at all scales, throughout the entire system. A complex system is able to respond globally, as a single coherent system, to local events. Somehow, information is propagated throughout the system to bring it back to its natural state, its attractor. This particular form of complex systems theory has been called self-organized criticality (Bak *et al.*, 1987). The power grid appears to show signs of this behavior, as we will see. Power grid failure is analogous to adding a grain of sand on a sand pile: there is either no effect or a large avalanche, the equivalent of a cascade event, or catastrophic failure. This is the scale-invariance aspect; there is no correlation between the perturbation and the resulting impact. Thus, even a small perturbation can lead to a large power grid cascade. This is related to brittle systems theory (Bush *et al.*, 1999), in which high-performing systems tend to fail less often, but when they do fail they tend to fail catastrophically rather than degrade gracefully. If our complex system happens to be the power grid, we would like to know if the smart grid, which seeks greater efficiency and pushes the power grid closer to its limits of operation, is unwittingly designing an attractor that takes us into this high-performing, but catastrophic, failure regime.

Scale invariance can be expressed mathematically rather simply by the power law. Given a relation $f(x) = ax^k$, changing the magnitude of the argument x by a constant amount c results in a proportionate change in magnitude of the function itself, as shown in Equation 1.21:

$$f(cx) = a(cx)^k = c^k f(x) \propto f(x). \tag{1.21}$$

In other words, scaling by a constant c simply multiplies the original power-law function by a constant value c^k. All power laws with a given scaling exponent k are equivalent within a constant factor, since each is simply a scaled version of the other. This creates a linear relationship that becomes apparent when logarithms are taken of $f(x)$ and x. Of course, one must be careful in making generalizations, because only a finite amount of real data can be plotted, the straight-line signature on a logarithmic plot indicative of a power law is a necessary, but not a sufficient, condition of a true power-law relation. So, does the power grid really exhibit the characteristics of a complex system? And if so, what does that imply for a smart grid? We can address the first question directly. The second question is one that you, the reader, who will likely be engineering the smart grid, will need to answer as we dive deeper into the power grid throughout this book.

Many networks have degree distributions that follow a power law as shown in Equation 1.22, with the exponent γ typically lying between 2 and 3:

$$P(k) \approx k^{-\gamma}. \tag{1.22}$$

A typical network analysis examines the percolation, or connectivity, of the network as nodes either fail or are maliciously attacked. The goal is to examine the robustness of the network structure; that is, the ability to keep all nodes in the network connected as nodes or edges are removed, representing faults or attacks. A typical assumption is that faults will occur to nodes at random, while attacks will occur to more highly connected nodes. The critical fraction f_c is the fraction of nodes removed from the network before the network disintegrates; that is, the network is no longer a single connected component (Cohen *et al.*, 2000). The critical fraction f_c can be determined analytically from a specific network topology. A study of the European power grid (Solé *et al.*, 2007) compared real outages with the theoretical critical fraction. For larger power law exponents $\gamma > 1.5$, the expected f_c values are very similar to those predicted by theory. These are referred to as robust networks in the discussion that follows. However, power grids having lower exponents (when $\gamma < 1.5$) strongly deviate from the predicted values. These are referred to as fragile networks in the discussion that follows. While no clear reason for this distinction was found, the suggestion was put forward regarding the growing interconnectivity in aging power infrastructures. Specifically, to increase the reliability of the power grid, more redundant connections are added, increasing the interconnectivity of the power grid network. However, while the intention of increasing reliability is good, unintended effects could be that faults can also propagate more easily through such highly interconnected

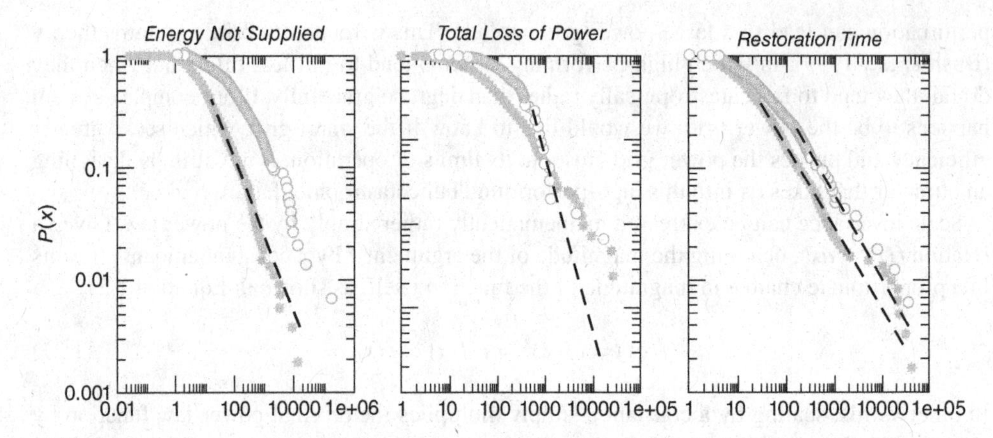

Figure 1.6 The occurrence versus severity of blackouts is expressed in the amount of energy that was not supplied, the loss of power, and the power restoration time. Energy is in units of megawatt-hours, power is in units of megawatts, and time is in units of minutes. The open circles indicate "robust" network topologies (that is, they have a relatively high degree of interconnectivity) and solid asterisks indicate more fragile topologies (that is, those with less interconnectivity). How is the evolution of smart-grid technology impacting these curves and what will be the impact of smart-grid communication? Will a more complex system, operating closer to its limits, lead to fewer but more severe catastrophes? Source: Rosas-Casals, 2010. Reproduced with permission of IEEE.

networks and that it can become a more complex process to isolate faulted segments of such highly interconnected networks. Ideally, smart-grid communication should be able to make complex wide-area fault detection, isolation, and restoration more efficient and help dampen instability problems before they propagate too far in such highly interconnected networks. However, it is always important to beware of unintended consequences.

Figure 1.6 clearly shows the scale-invariant nature of historical power faults in the European power grid. As we discussed regarding self-organized criticality, these scale-invariant features can be signs of complex, brittle systems. As smart-grid communication is added to the power grid and the power grid becomes more sophisticated they will become more complex, and we could expect the power law exponents to continue to increase. The power grid will become more efficient and operate closer to its limits; for example, its thermal limits. Power faults will become few and far between. However, in order to maintain the power law, when faults do occur, they will be larger and more catastrophic. Again, we need to watch for unintended consequences and avoid a brittle system (Bush *et al.*, 1999).

Taking a step beyond solely network structure, research has looked at how load capacity of networks interacts with network structure (Zhao *et al.*, 2004), where load is a general commodity; for example, data, power, or water flow. In other words, it is addressing the problem of how well a system can redistribute load when the network structure is perturbed. The concept here is that homogeneous loads – that is, an equal distribution of loads on all links – tend to survive better than heterogeneous loads in such systems. The notion of power flow entropy has been defined to address this issue.

In discussing communication in the smart grid, we need to be cognizant of the fact that communication is most needed and most critical when the power grid is under stress; that is, when the network is on the verge of undergoing catastrophic failure. However, when the power

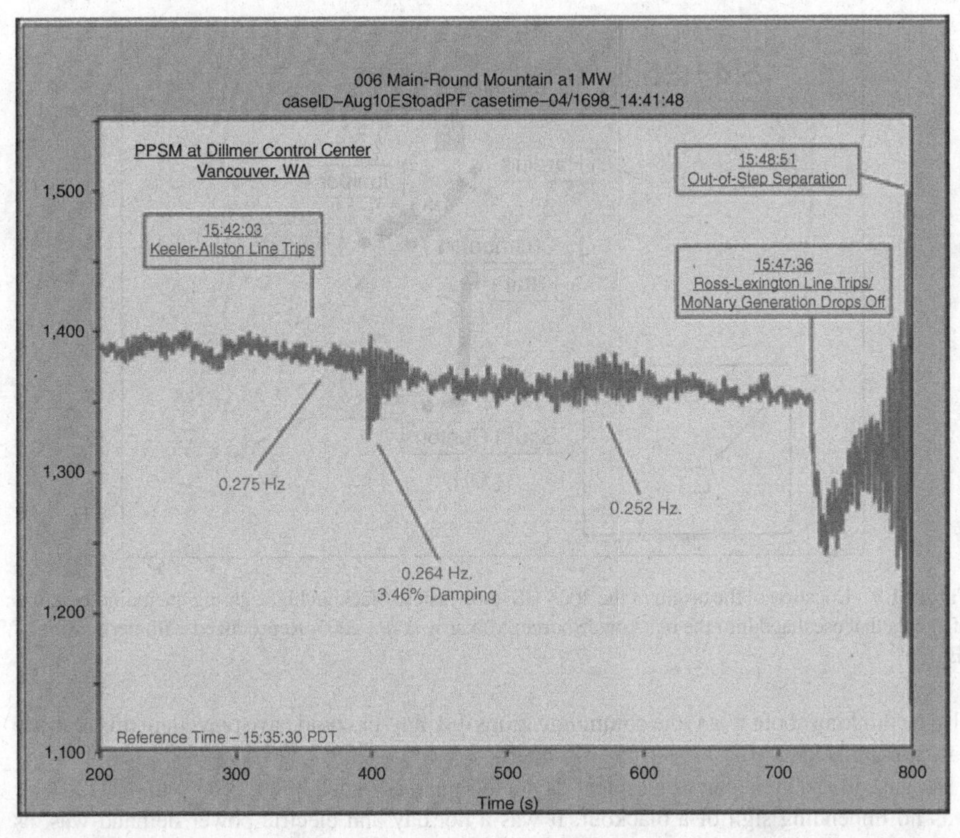

Figure 1.7 Voltage versus time is shown for a power grid that is encountering problems. Communication is most needed when the system is stressed and large amounts of difficult-to-compress data must be transferred. Source: Dagle, 2006. Reproduced with permission of IEEE.

grid is under stress, the information being measured and transmitted becomes more erratic, harder to compress, and requires more bandwidth to represent with the same fidelity. In fact, the current modus operandi, without communication, is to analyze power grid data in detail *after* a system collapse. As an example, Figure 1.7 shows voltage versus time for a power grid that is in trouble. The voltage is no longer constant as it should be, but fluctuating wildly. Since we have been previously discussing scale-free systems, scale-free data has few patterns; it is similar to white noise and nearly incompressible, potentially requiring large amounts of bandwidth. In fact, research has been done on determining the health of a system by examining the compressibility of its measured parameters (Bush, 2002).

1.4.2 A Brief Case Study

One of the largest blackouts occurred in the northeastern USA in August of 2003. We can try to understand this tragic event as a real-life example of a power system failure to stimulate thinking about what a "smart grid" might do in the future to improve the situation. We should

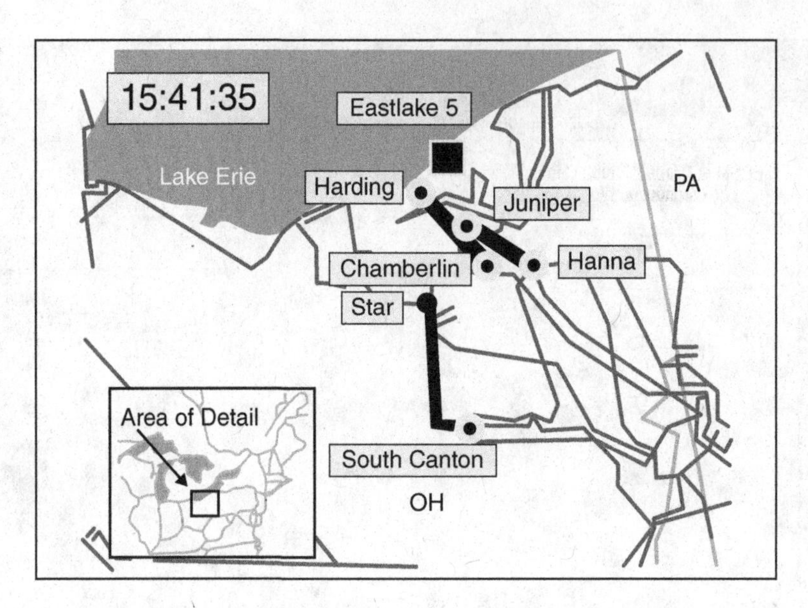

Figure 1.8 Location of the origin of the 2003 US northeastern blackout highlighting the major locations of events that escalated into the blackout. Source: Makarov *et al.*, 2005. Reproduced with permission of IEEE.

also be thinking about what role communications did play, or could have played, in this real-life cascading blackout.

August 14, 2003, began as a typical day, and remained typical until 3:05 p.m. EDT. There was no impending sign of a blackout. It was a hot day and electric power demand was, as is typical, high for use in air-conditioners. However, the demand was not excessively high. There was a heavy draw of power from both the south and west, through Ohio, to the north, including Michigan and Ontario, and the east, including New York. As we will learn later, alternating current frequency is a significant indicator of power grid health. The frequency was variable, but not excessive. Several generators were out of service that day; however, the power system operators, as they normally should, ran contingency analyses for that day and determined that there was still a safe operating margin. In fact, after the incident, investigation found that the Northeastern Interconnect was able to withstand any of at least 800 of the most likely contingencies that were modeled. In other words, the system should have had capacity to spare even with generators out of service. The epicenter of the blackout is shown in Figure 1.8.

Voltage was low in northern Ohio during the morning and early afternoon of August 14; but again, apparently not excessively so from a historical perspective. The area around northern Ohio was, in fact, operating near a voltage collapse, although this fact by itself was not found to be the direct cause of the blackout. The high power demand for air-conditioning and the lack of reactive power, to be explained in more detail later, caused the system to approach a voltage collapse. It was later determined that power system operators had not sufficiently studied the minimum voltage and reactive power supply that was required in the Cleveland–Akron area. Owing to the heavy load, northern Ohio was a net importer of power that day, importing a peak of 2853 MW, which also caused a high consumption of reactive power. Heavy use of inductive motors, such as those used in air-conditioning, causes the power factor to decrease and increases reactive power consumption. As mentioned, some generators in the area that could have helped supply reactive power were out of service due to scheduled maintenance; however, this was also not found to be a direct cause of the blackout.

Around noon on August 14 some protective relays began to trip in the area. These trips were not yet considered a direct cause of the blackout, but they had an impact. First, the Columbus–Bedford line tripped due to tree contact; tree contact was soon to play a significant role in later trips that led to the blackout. Then the Bloomington–Denois Creek 230 kV line tripped due to a downed conductor caused by a conductor sleeve failure. This was important because the loss of this line was not included in the system's state estimation procedure. State estimation, to be discussed in detail later, is used to improve accuracy when determining the state of power flows through the system given limited sampling information. Around 1:30 p.m. EDT, the Eastlake Unit 5 generator, which was capable of supplying much-needed reactive power and being pushed to its limits to meet demand, tripped, taking the system offline. This further increased the need to import more power into the area, although transmission line loads were still well within the limits of their maximum capacities.

The operators were clearly aware of the impending voltage problem and took steps to address the issue (NERC Steering Group, 2004), making the following calls.

- Sammis plant at 13:13: "Could you pump up your 138 voltage?"
- West Lorain at 13:15: "Thanks. We're starting to sag all over the system."
- Eastlake at 13:16: "We got a way bigger load than we thought we would have. So we're starting to sag all over the system."
- Three calls to other plants between 13:20 and 13:23, stating to one: "We're sagging all over the system. I need some help." Asking another: "Can you pump up your voltage?"
- "Unit 9" at 13:24: "Could you boost your 345?" Two more at 13:26 and at 13:28: "Could you give me a few more volts?"
- Bayshore at 13:41 and Perry 1 operator at 13:43: "Give me what you can, I'm hurting."
- 14:41 to Bayshore: "I need some help with the voltage … I'm sagging all over the system…" The response: "We're fresh out of vars."

The operators also requested that capacitors at the Avon substation be restored to service. We will see later in the book the impact of adding capacitance to correct the power factor.

Now we can see that the system was close to a voltage collapse, but no generators were asked to give up active power to generate reactive power. As previously mentioned, relatively large power transfers were taking place through northern Ohio from the southwest toward the northeast. One question that was considered was whether the voltage problems in northern Ohio were due to the high demand of the power transfers that were taking place. It was determined that the power transfers had minimal impact; the inability to supply reactive power given the high demand was the main cause of the voltage problem.

Analysis of the alternating current frequency in the area showed a mean drop in frequency precisely at noon; however, this was a planned event in order to implement a time correction. Frequency was in general variable, but not outside what was considered normal. There was a pattern to spikes in frequency prior to the blackout, but this can be explained by the scheduled ramping-up of generation attempting to meet demand.

Now that we have painted the picture of the situation leading up to the blackout, we now proceed to what has been determined as the direct causes of the blackout. The first cause was already mentioned: the loss of the Bloomington–Denois Creek 230 kV line and, more significantly, the fact that the line loss was not reflected in the state estimator, causing the state estimator to become inaccurate. A series of local events occurred that then led to the Sammis–Star 345 kV transmission line tripping, which began the uncontrolled cascade that spread through much of northeastern North America. So how did these events occur?

First, as mentioned, the Eastlake 5 generator tripped. There was speculation that if Eastlake 5 had not tripped there would have been a remote possibility that the cascade and corresponding blackout could have been avoided. However, the Eastlake 5 trip is not officially considered a primary cause of the blackout, because even with the loss of the Eastlake 5 generator the system was still within secure operating limits. A more direct cause of the blackout occurred at 2:14 p.m. EDT when the alarm and logging systems in the control room failed, causing operators to lose awareness of the situation as it evolved. This obviously impacted the capability of the EMS and severely degraded the ability of the operators to monitor and control the system. This only appears to have become evident around 2:30 p.m. EDT when a telephone call was received that indicated the Star–South Canton 345 kV tie line opened and reclosed. Until the phone call was received, the local operators had no knowledge of this event; this was the first clear indication to the operators that they were flying blind. The next local event leading directly to the blackout was the loss of two more key 345 kV lines in northern Ohio due to power line contact with trees. This shifted power flows from the 345 kV lines onto a network of smaller capacity 138 kV lines. Here we begin to see a cascade in action. The 138 kV lines were not designed to carry such large amounts of power and they quickly became overloaded. At the same time this happened, voltage suddenly decreased in the Akron, OH, area. Now it becomes important to understand distance relay protection, which will be explained in detail in later chapters. For now, simply put, the increased loading and decaying voltages caused 16 138 kV lines to trip sequentially between 3:39 p.m. EDT and 4:09 p.m. EDT. To put it another way, the 138 kV system in northern Ohio was in the process of a cascading failure. The heavily loaded lines were sagging into vegetation, other distribution lines, and onto other objects, causing widespread electrical faults. There was now a tremendous need for power to meet demand in the area over the only remaining transmission line, the Sammis–Star 345 kV line. The Sammis–Star transmission line finally gave way, tripping at precisely 16:05:57 on August 14, beginning an uncontrollable cascade of the Northeastern Interconnect power system. The tripping of this line was a "phase transition," or interface, between the local cascade, confined to northeastern Ohio, to the rest of the interconnect system. Loss of this heavily loaded line began a domino effect of overloaded lines that spread west to Michigan and then east, isolating New York from Pennsylvania and New Jersey.

There are many fascinating details that occurred throughout this entire ordeal that relate in various ways to communication, although communication was never mentioned as a key factor at the time. Recall that after the Bloomington–Denois Creek 230 kV line tripped, the state estimator was not updated with this information and the mismatch between the state estimator and measured values became unacceptably large. Although this problem was quickly corrected, the analyst running the state estimator forgot to reset the estimator to run automatically at 5 min intervals as it normally should. By 2:40 p.m. EDT, the failure to enable the automatic operation of the state estimator was discovered soon after the Stuart–Atlanta 345 kV line tripped. However, the estimator was restarted without the information about the last line trip. Thus, the state estimator was again giving unacceptably large errors. The state estimator was finally corrected around 4:04 p.m. EDT, too late to provide a critical warning of the impending cascade.

It is clear that by 3:46 p.m. EDT, even without the control room alerts and the incorrect state estimator, that both the ISO and the local operators knew the situation was extremely serious. At this point, the only alternative left to save the system would have been to shed load, to deliberately shut off power to local customers in order to at least keep the system running and perhaps save a portion of it. However, even if they had wanted to do so, given the alarm malfunction and the state estimation error, the local operators lacked the monitoring and control necessary to do so. In fact, even if everything had been working perfectly, the local operators had no plan in place for shedding load as quickly as would have been required. So, by this point, any hope of avoiding the impending blackout appears to have been lost.

Investigation later found that an Eastlake 5 generator configuration error caused reactive power to go to zero when the unit automatically tripped from automatic to manual control. The unit finally went completely out of service as operators tried to place it back under automatic control. Finally, a pump valve would not reseat properly, causing significant delay in trying to get the unit back online. It turns out that the net effect of the efforts to keep the generator running with reactive power set to zero actually exacerbated the loss of reactive power in the area rather than helping to alleviate it.

There is another interesting and critical detail. The key cause of the cascade, the Sammis–Star line trip, did *not* occur due to an electrical fault. Rather, high current flows above the power line's emergency rating simultaneous with the low voltage in the area caused the overload to *appear* to the protective relays as a remote fault on the system. In other words, there was no actual electrical fault, but rather the appearance of a fault that caused the protective relay to trip and drop the transmission line. The protection relay could not distinguish between a remote electrical fault and high line load condition. Again, we will discuss operation of the distance protection mechanisms in more detail in future chapters.

Eventually, as the cascade spread, the now-isolated northeast power island was deficient in generation and unstable with large power surges and swings in frequency and voltage. Many lines and generators across the area tripped, breaking the area into several smaller islands. Although much of the disturbance area was fully blacked out, some islands were able to reach equilibrium without total loss of load, including most of New England and approximately half of western New York.

So what do we make of this cascading failure? There were several tree–power line contacts that appeared to be the "grains of sand" that caused the avalanche, using our self-organized criticality metaphor. Clearly, detecting and removing potentially dangerous vegetation is a problem. There was also a computer malfunction that caused operators to lose the ability to monitor the system in near-real time. Finally, as the disaster began to escalate, no one appeared willing or able to pull the plug and purposely shut down power to customers, also known as load shedding, in order to save the system. Clearly, more reliable sensing, communication, and even visualization of the system may have enabled the operators to take action to prevent the cascade. However, even with reliable sensing and communication, network management would have been required. This is because operators were unaware that the computer alarm system was nonfunctional. They had no real-time status of their computer and communication systems in order to check the health of their monitoring and communication systems. While, as mentioned, there was a contingency analysis (CA) done for that day, it was not updated and repeated frequently enough to keep up with the loss of safe margin as the day progressed. The ISO overseeing system reliability for the region was not using real-time network topology in its state estimation. Thus, it was not estimating state properly as transmission lines were tripping.

Finally, neighboring ISO organizations lacked the ability to coordinate activities when they realized faults were occurring in the neighboring ISO's grid. Let us now consider what is driving the smart grid, but also keeping in mind how it might have worked in this tragic event.

1.5 Drivers Toward the Smart Grid

Now that we have considered power grid faults, let us try to hone in more precisely on what the "smart grid" is. This section discusses the economic and social drivers leading toward the smart grid and how the smart grid has been conceived. It is important to keep in mind that there is no absolute, technical, concise definition of the smart grid, only general policies. Trying to unearth the first use of the term "smart grid," like any claim to the oldest evidence for anything in archeology, will quickly be made obsolete by a claim of older evidence; someone will undoubtedly claim to have used the term even earlier. The term "smart grid" became current in the industry around 2003 after the Northeast Blackout of 2003. Current evidence for its first use in the academic community came later in a paper by Massoud Amin in 2004 entitled "Balancing market priorities with security issues" in the *IEEE Power and Energy Magazine* (Amin, 2004). While a few papers mentioned the term afterward, it was not until 2010 when the use of the term exploded in publication. Not surprisingly, this explosion of publications came soon after funding announced in the US federal government's Recovery Act, stimulating researchers to chase funding on the topic of the smart grid. Of course, funding is always the big driver for new research. The term "smart grid" is found in the US code, the official record of Acts of Congress, along with other terms beginning with the word "smart." Specifically, the code from 2007 stipulates:

> It is the policy of the United States to support the modernization of the Nation's electricity transmission and distribution system to maintain a reliable and secure electricity infrastructure that can meet future demand growth and to achieve each of the following, which together characterize a Smart Grid:
>
> 1. Increased use of digital information and controls technology to improve reliability, security, and efficiency of the electric grid.
> 2. Dynamic optimization of grid operations and resources, with full cybersecurity.
> 3. Deployment and integration of distributed resources and generation, including renewable resources.
> 4. Development and incorporation of demand response, demand-side resources, and energy-efficiency resources.
> 5. Deployment of "smart" technologies (real-time, automated, interactive technologies that optimize the physical operation of appliances and consumer devices) for metering, communications concerning grid operations and status, and distribution automation.
> 6. Integration of "smart" appliances and consumer devices.
> 7. Deployment and integration of advanced electricity storage and peak-shaving technologies, including plug-in electric and hybrid electric vehicles, and thermal-storage air conditioning.
> 8. Provision to consumers of timely information and control options.

9. Development of standards for communication and interoperability of appliances and equipment connected to the electric grid, including the infrastructure serving the grid.
10. Identification and lowering of unreasonable or unnecessary barriers to adoption of smart grid technologies, practices, and services.

In Europe, the smart grid was initiated in 2006 by the European Technology Platform for Smart Grids, which is supported by the European Commission.

A smart grid employs innovative products and services together with intelligent monitoring, control, communication, and self-healing technologies in order to:

1. Better facilitate the connection and operation of generators of all sizes and technologies;
2. Allow consumers to play a part in optimizing the operation of the system;
3. Provide consumers with greater information and options for choice of supply;
4. Significantly reduce the environmental impact of the whole electricity supply system;
5. Maintain or even improve the existing high levels of system reliability, quality and security of supply;
6. Maintain and improve the existing services efficiently.

There are many social drivers leading toward the smart grid. Such topics are not within the scope of this book; however, some are mentioned here more for the sake of understanding where the momentum behind the smart grid began. First, grid infrastructure around much of the world is aging; the time will be ripe for replacing power system components within the grid, and we may as well replace those components with new technology. In addition, public interest groups have been pressuring politicians for a "greener" environment by reducing carbon dioxide output and increasing the efficiency of the system. Many of the more environmentally friendly systems, such as wind and solar power, provide intermittent power generation and will require greater grid intelligence and a new grid design with features such as energy storage in order be safe, reliable, and cost effective. Almost everyone would welcome lower energy prices; this puts pressure on regulators to increase competition to reduce prices. This, in turn, requires greater communication among competing entities in order to operate together smoothly. On top of all this, the demand by society for more power is rapidly increasing. This could be exacerbated by the widespread use of electric vehicles.

1.6 Goals of the Smart Grid

The defining characteristics of a smart grid have been summarized by the US Department of Energy's Office of Electricity Delivery and Energy Reliability as:

- provide power quality for the range of needs in a digital economy;
- accommodate all generation and storage options;
- enable new products, services, and markets;
- enable active participation by consumers;

- operate resiliently against physical and cyberattack and natural disasters;
- anticipate and respond to system disturbances in a self-healing manner;
- optimize asset utilization and operating efficiency.

The following subsections briefly discuss each of the goals.

1.6.1 Provide Power Quality for the Range of Needs in a Digital Economy

The assumption is that, as the precision and coverage of power monitoring and control increase throughout the power grid, delivered power quality will come into sharper focus and control. It will become possible to provide distinct levels of power quality and at different costs. One of the challenges with this goal is that there are many definitions and aspects to power quality, as we will see in the following chapters of this book. Certainly, there is a notion of power quality, but as yet there is no clear, defining standard that everyone can agree upon and, more importantly, base a pricing model upon.

1.6.2 Accommodate All Generation and Storage Options

The smart grid should become a "plug-and-play" interface for all forms of DG and energy storage technologies. Of course, this will require a corresponding development of standards. Clearly, the goal is to ease the introduction of new forms of renewable energy onto the power grid. Large-scale energy storage will reduce peak demand for power generation, thus reducing the cost of starting up large plants for short periods of time, which is both wasteful and expensive.

Measurements of success in this area could simply be the number and coverage of interoperable DG and storage standards. The ratio of power generated by DG to centralized generation is one metric, as well as the amount of energy that can be stored. Finally, the reduction in time to install a distributed generator or energy storage device will also be a relevant metric.

1.6.3 Enable New Products, Services, and Markets

Utilities, independent power producers, ISOs, and RTOs have been the primary purchasers of power grid equipment and applications. As communication becomes a more integrated part of power products and applications, these organizations will serve as the traditional market for new smart-grid products. However, it was envisioned that, with the smart grid, it will be possible for the electric power consumer to become a direct purchaser of smart-grid products. This refers not just to large industrial consumers, but downward in scale to individual residential consumers as well. This means that the smart grid must enable enough monitoring and control by the individual residential consumer that it becomes feasible for individual residential consumers to be able to purchase useful and desirable products that are not already feasible today. This goal is tightly coupled with the next goal.

1.6.4 Enable Active Participation by Consumers

Active participation by consumers in the operation of the power grid seems to have initially been related to DR, the idea that consumers should be able to have high-resolution monitoring and control of their property's power usage and the ability to bid for precisely the amount of power they wish to purchase. This requires that every consumer will have the proper communication system to perform monitoring and issue pricing signals for their power bidding.

However, it does not take too much imagination to see that this can be extended to include consumer participation by generating and selling their own power, given the advances in microgrid technology. In this case, the consumer has been called a "prosumer," as in one who may both consume and produce electric power. The idea of zero-energy buildings and zero-energy communities becomes possible when the prosumer or community of prosumers remains self-sufficient.

Measures of active participation by consumers have typically been simply counting the number of consumers with "smart meters." However, this measure will have to become sophisticated as consumers become more savvy. For example, the number of zero-energy entities or the amount of consumer-managed load may be better measures.

1.6.5 Operate Resiliently against Physical and Cyberattack and Natural Disasters

This goal suggests that cyberattacks (malicious attacks against the power grid that are likely to exploit the grid's communication network) are included with natural disasters (natural physical events that disrupt the power grid) when considering resilience. In both cases, cyberattacks and natural disasters, the power grid should operate in a fault-tolerant manner. The notion of resiliency and its opposite, brittleness, are defined analytically in Bush *et al.* (1999). The notion is that a resilient power grid will continue to operate with a loss in performance that is proportional, hopefully a small proportion, to the severity of losses due to cyberattack or natural disaster. In other words, there is no point at which the system suddenly and catastrophically collapses. However, this goal may be competing with other goals, such as optimizing operating efficiency. This is because a system that is highly optimized, operating close to its limits, also tends to be brittle, as discussed in Bush *et al.* (1999). Thus, this goal needs to be kept in mind and given careful consideration when implementing other goals.

1.6.6 Anticipate and Respond to System Disturbances in a Self-healing Manner

A key word in this goal is "anticipate," to provide the power grid with the ability to foresee and avoid potential disturbances. It would be ideal if disturbances could be avoided in the first place. If they do occur, they should be contained quickly to have as little impact as possible. Finally, the impact of a disturbance, if it does occur, should be corrected as quickly as possible. All aspects of this goal point toward automated solutions. A "disturbance" is kept sufficiently vague to encompass any type of event that could interfere with operation of the system. This may range from instability, an abnormal oscillation in power flow, to a fault, a short circuit in the line.

1.6.7 Optimize Asset Utilization and Operating Efficiency

This is the notion of "replacing iron with bits." In other words, utilizing knowledge of power grid assets to extract the most value from them with the goal of avoiding the purchase

Table 1.2 Illustration of some of the technologies' capabilities to fulfill the smart grid definition. The technologies are GISs, AMIs, outage management systems (OMSs), DMSs, DA, and SCADAs.

Definition	GIS	AMI	OMS	DMS	DA	SCADA
Power quality (Section 1.6.1)	✓			✓	✓	✓
DG + DER (Section 1.6.2)	✓			✓		✓
New services (Section 1.6.3)	✓	✓	✓	✓	✓	✓
Participation (Section 1.6.4)	✓	✓	✓	✓		
Cyber + disaster (Section 1.6.5)	✓	✓	✓	✓	✓	✓
Self-healing (Section 1.6.6)	✓			✓	✓	✓
Optimize assets (Section 1.6.7)	✓		✓	✓	✓	✓

and deployment of heavy equipment. This can include something as simple as an accurate geographic information system (GIS) that keeps track of where assets are located. It could also include analysis (for example, optimization techniques) and modeling to utilize the available assets in the most efficient manner. This could also include prognostics and health monitoring in order to determine precisely when and how to extract the most value from an asset before replacing it. Clearly, improving efficiency of the power grid is a theme that is apparent throughout this book.

So how are all of the goals being implemented? Table 1.2 maps the goals of the smart grid to a select set of high-level technologies that are being developed to help meet those goals.

1.7 A Few Words on Standards

For the practical-minded reader eager to see how the smart grid phenomenon can be implemented today, international standards are incorporating and codifying smart grid concepts. Chapter 10 reviews the state of smart grid standards in more detail. Some of the standards-related bodies developing smart grid standards include:

- National Institute of Standards and Technology (NIST) Smart Grid Interoperability Panel (SGIP)
  www.nist.gov/smartgrid/
- Institute of Electrical and Electronics Engineers (IEEE) Standards Association
  http://grouper.ieee.org/groups/scc21/2030/2030_index.html
- International Telecommunication Union (ITU) ITU-T Focus Group on Smart Grid (FG Smart)
  www.itu.int/en/ITU-T/focusgroups/smart/Pages/default.aspx
- Microsoft Power and Utilities Smart Energy Reference Architecture (SERA)
  www.microsoft.com/enterprise/industry/power-utilities/solutions/smart-energy-reference-architecture.aspx
- International Electrotechnical Commission (IEC)
  www.iec.ch/smartgrid/roadmap/
- Conseil International des Grands Reseaux Electriques (International Council on Large Electric Systems) (CIGRE)
  www.cigre.org/

- Electric Power Research Institute (EPRI)
  http://smartgrid.epri.com/
- United States Department of Energy
  www.oe.energy.gov/smartgrid.htm

1.8 From Energy and Information to Smart Grid and Communications

There is a deep relationship between the electric power grid and communication; much of it has yet to be discovered. Hopefully, this book will stimulate the reader to think beyond simply adding communication links to today's power grid or even the power grid as it has been projected to operate when it becomes "smart." Instead, new ways of thinking about the power grid and communication might be inspired by the relationship between energy and information. Power is the flow of energy just as communication is the flow of information, and there has been a long and deep history between and energy and information.

Researchers concerned with the energy requirements and battery life of large sensor networks have explored this topic, as illustrated in Figure 1.9. From their point of view, a unit of information requires energy to create, transmit, route, and be received. The goal has been to understand the energy requirements for each of these steps and minimize the overall energy consumption. Now consider a unit of energy instead of a unit of information. The unit of energy requires information to generate and manage, route, protect, and be purchased by the consumer, as illustrated in Figure 1.10. In other words, if we were to scale up the power normally used to create a communication signal, it could instead become energy supplied to consumers. This reinforces the notion mentioned at the beginning of this chapter and throughout the book that the power grid and communication are related closely enough to draw inspiration from one another.

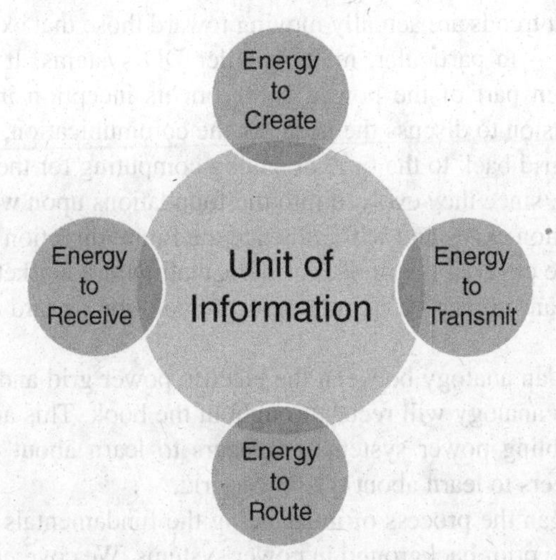

Figure 1.9 The notion of minimizing the amount of energy used to communicate is illustrated. Here the goal has been to minimize the amount of energy needed to communicate. Source: Bush, 2010.

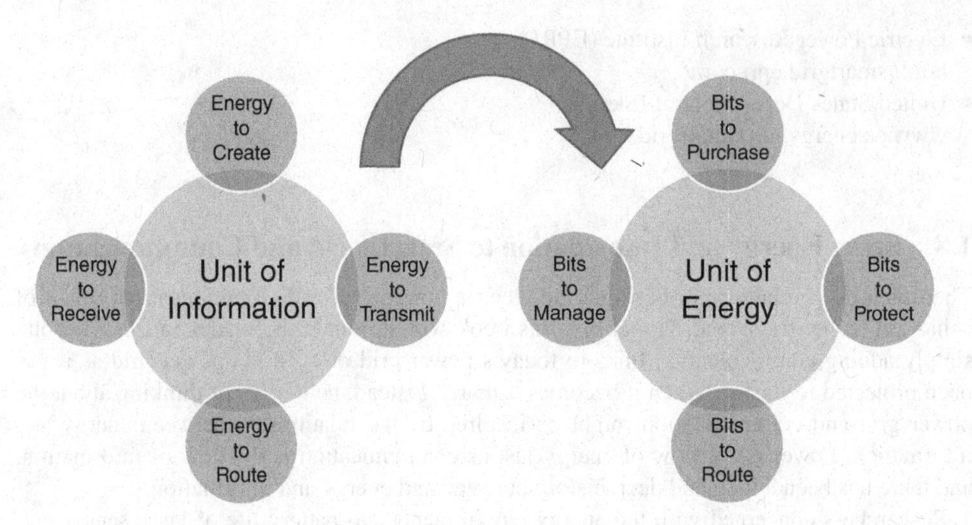

Figure 1.10 The notion of communicating information about energy is illustrated. Here the goal is to minimize the number of bits required to communicate information about energy. More on this topic is covered in Chapter 6. Source: Bush, 2010.

1.9 Summary

This chapter discussed the background and context in which the smart grid is emerging. It is important to realize the depth and breadth of the evolving power grid; many technical components of the grid are emerging independently while being claimed as part of the smart grid. Later in this chapter we saw how legislation and public policy defined the goals for the smart grid.

Historically, we saw how the power grid has evolved into its current state. It was noted how many of the current trends are actually moving toward those that existed when the power grid was in its infancy – in particular, many smaller DG systems. It was also noted how communication has been part of the power grid from its inception in the late 1800s. We also took a small digression to discuss the need for the communication, namely early control systems for the power grid back to the time of analog computing for the power grid. We also covered early SCADAs, since they evolved into the foundations upon which much of today's power grid communication exists and will influence the future direction of communication in the power grid. Next we covered the impact of deregulation and market forces on the power grid; this had a significant impact on its more recent evolution toward our current notion of the smart grid.

Then we constructed an analogy between the electric power grid and communication networks. Elements of this analogy will recur throughout the book. This analogy has been very helpful in quickly enabling power systems engineers to learn about communications and communications engineers to learn about the power grid.

This chapter also began the process of introducing the fundamentals of the physics of the power grid, assuming no prior background in power systems. We covered the topics of charge and voltage to electromagnetic circuits and fields. Knowledge of this material is crucial to understanding the rest of the book. It applies equally well to communications.

Following the introduction to power grid physics, which will be picked up in Chapter 2, we transitioned to a discussion of a blackout in the power grid. This brought us head-on with real problems that have been manifested in the power grid. While many may feel that the smart grid is about making a great leap and redefining electric power transport, this section was a sobering reminder that we had better make sure the changes are addressing real and potentially life-threatening problems. In particular, we noted that introducing complexity into the smart grid can have unintended, adverse consequences on a trend toward less frequent but larger and more devastating blackouts.

The next two sections discussed the drivers and goals of the smart grid. We noted that the goals for the smart grid involved extending the capability of the power grid by introducing new features and services; however, if the smart grid cannot reduce the likelihood of a blackout, then all of its other features and services are pointless. In the next section we considered both where the term "smart grid" originated and how it has been shaped.

Next we considered in more detail public policy concerning the goals of a smart grid. By considering these goals in detail, we can gain an intuitive, nontechnical sense of what the smart grid is supposed to be.

The power grid is too large an entity and there are too many entities involved for it to be constructed without standards to guide how the pieces fit together. Both the power electronics and communication components will require standards and guidelines in order to ensure that components are interoperable and that the architecture takes shape in the desired manner. Later in the book there is an entire chapter devoted to smart grid standardization efforts. However, in order to understand and appreciate the standards, the fundamentals covered in this book are required.

Next, we took an excursion back to fundamental theory to discuss energy and information. The goal was to intuitively motivate the reader to consider that incorporating communication and computation within the power grid may be done effectively only from a fundamental understanding of the relationship between energy and information. Specifically, we considered the derivation of the channel capacity from communication theory and then applied it to an analogy with the compression of an ideal gas. Gases and fluids turn turbines that produce electricity; this analogy inspires us to consider more deeply the relationship between information and communication and power efficiency.

Exercises at the end of this and every chapter are available to help test and solidify understanding of the material covered. Some higher numbered questions require information from later chapters of the book and are designed to motivate the reader to think about issues that will be dealt with later.

It should be clear from this chapter that communication and power systems have had a long history together. Power systems are complex dynamic systems that can be understood from an information theory and communication theory vantage point for the smart grid. Now that we have laid the context for the smart grid, let us look at each component of the power system in a little more detail in the succeeding chapters following the classic description of the power grid; namely, starting from generation to transmission, fanning out to distribution and finally reaching the consumer.

Chapter 2 will focus upon pre-smart-grid power generation. In particular, we will build upon the fundamentals of electric power physics established in this chapter to include aspects related to generation, particularly centralized power generation. This includes basics of alternating current, the notion of complex and reactive power, and finally a review of how generators operate. This basic knowledge is required in order to hope to understand the smart grid; these are fundamental physical aspects that will remain valid with DG. Communication and control play a central role in generation regardless of whether it is centralized or distributed. Thus, in Chapter 2 we go into detail on the SCADA systems used to control generation. Finally, Chapter 2 ends with a discussion of pre-smart-grid energy storage systems. These systems and their characteristics will also remain valid in the smart grid. An understanding of electric power generation and storage will build the foundation for later communication and control concepts for the smart grid.

1.10 Exercises

Exercise 1.1 Resistance

1. Consider one power cord that is twice the diameter of another power cord. If the cords are of the same length and material, how do their resistances compare?
2. Consider the current in a 20 foot, 16-gauge cord is 5 A. What is the voltage difference between the two ends of each conductor?

Exercise 1.2 Heat Dissipation

1. Consider an oven that draws 8 A at a voltage of 120 V. How much power does it dissipate in the form of heat?

Exercise 1.3 Power Drawn by a Load

1. Consider two incandescent light bulbs, with resistances of 320 Ω and 500 Ω. How much power does each draw when connected to a 120 V outlet?

Exercise 1.4 Series Voltage

1. A string of Christmas lights in a 120 V outlet has 60 identical bulbs connected in series. What is the voltage across each bulb?

Exercise 1.5 Current Flow

1. Consider several appliances operating from the same power cord: one at 40 Ω, one at 80 Ω, and one at 20 Ω. The cord has a resistance of 0.1 Ω on each wire. What is the current through each device when all are in use?

Exercise 1.6 RMS Voltage

1. If 120 V is the standard for RMS voltage, what is its magnitude?

Exercise 1.7 Impedance

1. An electrical device contains a resistance, an inductance, and a capacitance, all connected in series. Their values are $R = 1\ \Omega$, $L = 0.01$ H, and $C = 0.001$ F, respectively. At an a.c. frequency of 60 cycles, what is the impedance of the device?

Exercise 1.8 Reactance

1. Consider a transmission line that has a resistance $R = 1\ \Omega$ that is small compared with its reactance $X = 10\ \Omega$. What are the approximate conductance and susceptance?

Exercise 1.9 Ohms Law

1. Consider an incandescent bulb rated at 60 W. This means that the filament dissipates energy at the rate of 60 W when presented with a given voltage, which we assume to be the normal household voltage, 120 V. The current is 0.5 A. Now compute the power used by the bulb in terms of its resistance.

Exercise 1.10 Power Factor

1. Consider an appliance that draws 750 W of real power at a voltage of 120 V a.c. and a power factor of 0.75 lagging. How much current does it draw?

Exercise 1.11 Reactive Power

1. For the preceding example, how much reactive power does the appliance draw?
2. What is the impedance of the appliance?

Exercise 1.12 Power Factor and Line Loss

1. For the preceding example, how much of a reduction in line losses could be achieved by improving the power factor to 0.9, assuming that real power remains unchanged?
2. How much capacity is freed up on the substation transformer that supplies this line?

Exercise 1.13 Voltage Conservation

1. Compare the power consumption of a 100 W light bulb at 114 V versus 126 V.

Exercise 1.14 Resistance and Heat Dissipation

1. A 100 W incandescent light bulb in a 120 V circuit is dimmed to half its power output using an old-fashioned rheostat, or variable resistor, as a dimmer. What is the value of the resistance in the dimmer at this setting, and how much power is dissipated by the rheostat itself?

Exercise 1.15 Phase-to-Phase Voltage

1. A service to a business comprises three conductors – Phase A, Phase B, and neutral – where each phase-to-ground voltage is 120 V. The 120 V loads around the store, such as light and regular outlets, may be alternatively connected to Phase A or B. Heavier loads, would be connected phase-to-phase between A and B. What is the voltage that the heavier loads see?

Exercise 1.16 Wye-Delta Transformer

1. A wye-delta transformer with a 10:1 ratio steps voltage down from a 230 kV transmission line to a distribution circuit. What is the voltage on the secondary side?

Exercise 1.17 Clock Frequency

1. Suppose the 60 Hz frequency remains low, at 59.9 Hz for 1 day. How much time is lost in clocks that depend directly upon the alternating current cycle to keep time?

Exercise 1.18 Vehicle Battery Storage

1. A car battery has a storage capacity of 80 Ah (amp-hours) at 12 V. Assuming losses are negligible, how many such batteries would be required to supply a residential load of 5 kW for 24 h?

Exercise 1.19 Per Unit

Consider a three-phase power transmission system that deals with power at 500 MW and uses a voltage of 138 kV for transmission. Arbitrarily select $S_{\text{base}} = 500$ MVA and use the voltage 138 kV as the base voltage V_{base}. We then have:

$$I_{\text{base}} = \frac{S_{\text{base}}}{V_{\text{base}} \times \sqrt{3}} = 2.09 \text{ kA}, \qquad Z_{\text{base}} = \frac{V_{\text{base}}}{I_{\text{base}} \times \sqrt{3}} = \frac{V^2_{\text{base}}}{S_{\text{base}}} = 38.1\ \Omega,$$

$$Y_{\text{base}} = \frac{1}{Z_{\text{base}}} = 26.3 \text{ mS}.$$

1. If the actual voltage at one of the buses is measured to be 136 kV, then what is its per-unit (pu) value?

Exercise 1.20 Symmetric Components

The symmetrical components of a set of unbalanced, three-phase voltages are $V_0 = 0.6\angle 90°$, $V_1 = 1.0\angle 30°$, and $V_2 = 0.8\angle -30°$.

1. Obtain the original unbalanced phasors.

Exercise 1.21 Transformer

1. How does a line tap transformer work?
2. There are many sources of energy loss in a transformer. Describe at least three.

Exercise 1.22 Types of Power

1. Explain how real, reactive, and complex power are drawn on a complex plane. How is the phase of voltage relative to current represented on such a diagram?

Exercise 1.23 FACTS

Recall that a FACTS can manipulate the impedance of a line X in order to control or optimize power flow.

1. Show mathematically why reactive power provided to a FACTS-controlled transmission line must be increased as the impedance is decreased.

Exercise 1.24 Fault Analysis

1. Explain how symmetric components can be used to analyze a fault.
2. If the above process is automated, what communication is required?

Exercise 1.25 Energy and Information

1. How is Shannon channel capacity similar to compression of an ideal gas?

Exercise 1.26 Substation

1. Explain the following hierarchical levels in a power grid and the types of communication each would require: control center level, substation level, bay level, and process level.
2. What are the general communication requirements for each level?

Exercise 1.27 Energy Efficiency

1. What is the difference between maximizing power transfer and maximizing efficiency? Why are these not the same?
2. What is the parallel, or similarity, with communication systems in utilizing maximum power transfer and maximizing efficiency?

2

Generation

Electricity is really just organized lightning.

—George Carlin

2.1 Introduction to Generation

This chapter is an introduction to pre-"smart-grid" power generation. As always, it is important to keep in mind that the term "smart grid" is more of an ill-defined academic term and does not reflect the true, continuous evolution of the power grid. The grid has been continuously evolving to become "smart." Also, remaining consistent with the goal of this book, the topic of communication within the power grid remains the focal point of this chapter. However, we will review selected power systems topics in depth, as that will be necessary to understand the rest of the material in this book. The high-level goal of this chapter is to convey the state of the power grid before large-scale distributed and renewable generation became a popular topic and, at the same time, to introduce fundamentals of electric power and power generation that will be required in later chapters.

The topic of power generation is introduced and the basics of centralized generation and alternating current are covered. This includes issues such as complex and reactive power, as well as power quality. From a communication perspective, it is important to have a fundamental understanding of communication and control in centralized generation systems. Thus, this chapter covers fieldbus protocols used in generating plants because these supervisory and data acquisition protocols have played an important role in communication for power systems and will continue to do so. Much can be learned from the experience of developing and using legacy supervisory and data acquisition protocols; microgrids and smart-grid protocols will have to communicate with these protocols for a long time to come. Since one of the main goals of power systems has been improving the efficiency of power generation and delivery, this chapter introduces power generation efficiency and the environment. The technologies related to energy storage have been usurped under the umbrella of the smart grid; however, many forms of energy storage systems existed long before the smart grid came along. Energy storage is discussed in this chapter along with power generation because the line between generation and storage can sometimes be blurred. Energy storage involves the conversion of

Smart Grid: Communication-Enabled Intelligence for the Electric Power Grid, First Edition. Stephen F. Bush.

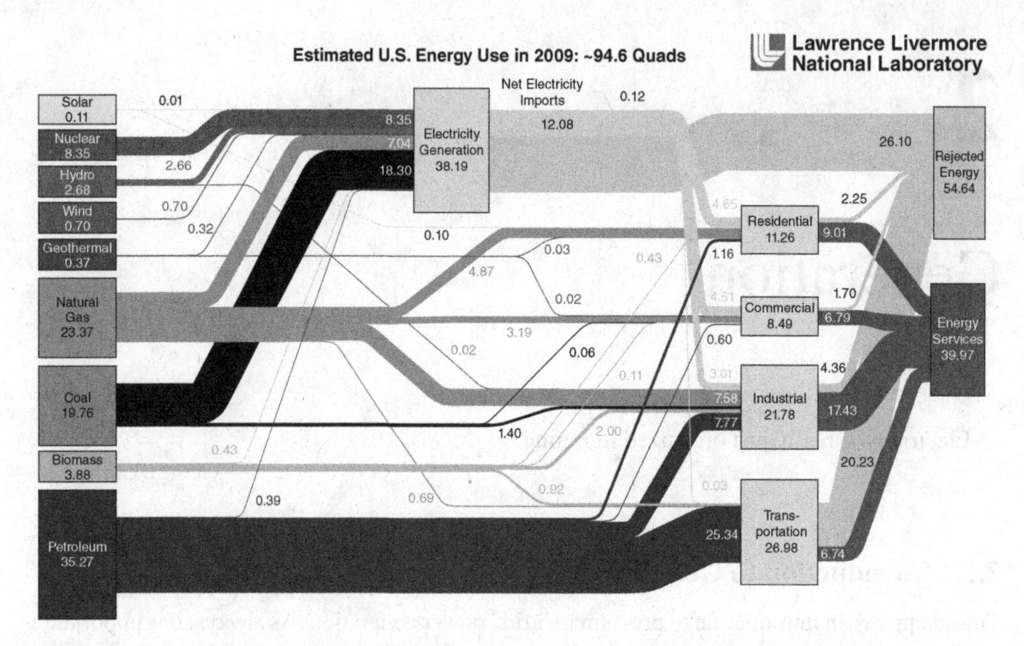

Figure 2.1 A snapshot of the sources and uses of energy is illustrated for the year 2008. This places electric power in context with other forms of energy. It also shows the rejected energy, which is the energy that does not perform useful work. This may be due to the second law of thermodynamics or the use of inefficient equipment or processes. The fundamental question addressed by this book is how smart-grid communication fundamentally relates to electric power efficiency. (http://www.nap.edu/reports/energy/supply.html). Source: LLNL 2008; data based on DOE/EIA-0384(2006), June 2007. Credit is given to the Lawrence Livermore National Laboratory and the Department of Energy, under whose auspices the work was performed.

energy from electrical to another form, such as chemical, potential, or kinetic energy, which can later be transformed back into electrical form with minimal loss. One could argue that any electric power generation (for example, solar, wind, steam, and hydroelectric power plants) is simply converting naturally stored forms of energy into electric power and thus performing half of the power storage process, namely converting from a stored form into electric power. The chapter ends with a brief summary and exercises for the reader.

Figure 2.1 places the power grid within the context of overall energy production and consumption in the USA. Measurements are in quads, which are a quadrillion (10^{15}) British Thermal Units or 1.055×10^{18} J (joules). This is a conveniently large unit for discussing amounts of world and national energy. To get a sense of the amount of an energy in a quad, it is equivalent to the energy in 8 007 000 000 US gallons of gasoline, 293 083 000 000 kWh of electricity, 52 000 000 tons of TNT, or five times the energy released in the Tsar Bomba nuclear test, which was the most powerful nuclear weapon ever detonated. Looking at Figure 2.1, we can make several high-level observations. On the left of the figure there are two main sources of energy: (1) coal and natural gas and (2) oil. Coal and natural gas supplies the energy for buildings and industry, while oil supplies the energy for transportation. There is some overlap in industrial usage. From the right side of thc figure it is clear that more than half of the energy is rejected energy; that is, wasted energy. Transportation appears to be the single largest sector consuming energy from oil in this particular snapshot of energy consumption. Electric power produces the largest amount of rejected energy of any energy source, although it has an efficiency of approximately 32% compared with approximately 25% for oil. Unless electric power production can be made significantly more efficient or transition to new sources, transitioning vehicles from oil to electric power would still yield significant wasted energy. From this energy flow snapshot, renewable energy use for electricity is negligible. Some take-aways from this figure are that there is plenty of room for improvement in electric power efficiency, from building more efficient generators to reducing the power loss when transporting power. There is also room to encourage greater use of renewable energy sources for electric power generation. Another item of note is that residential and commercial building energy use, while important, is relatively small in comparison with other sectors using energy. Finally, as the electric vehicle market attempts to grow, we may see the flow of oil in the above figure being replaced with electric power in the transportation sector. The bottom line is that much is riding on our ability to improve electric power efficiency and utilize renewable power sources in order to improve overall energy efficiency.

2.2 Centralized Generation

Large, centralized generation has been the dominant form of electric power generation for many decades, and the reasons were discussed in the first chapter. The high-level characteristics of centralized generation have included primarily one-way power flow, from large, centralized generators through high-power transmission lines to lower power distribution systems, and then to individual consumers. Also, generation capacity must follow demand; any deviation from a balanced supply and demand results in potentially catastrophic failures. This means that sudden abrupt changes in the system, such as generator rotational inertia causing frequency swings, if not anticipated and mitigated quickly, can lead to a dangerous, unstable system. These characteristics of electric power generation must be better understood in order to comprehend the role communication can play with the smart grid. Thus, this chapter focuses upon introducing power system fundamentals related to power generation assuming no prior power system or communication knowledge.

2.2.1 Alternating Current

The goal of this section is to introduce the fundamentals of alternating current and electric power. In Chapter 3 we will discuss power generation by applying the fundamentals from this chapter. Power generation results, in the vast majority of cases, in alternating current, for reasons that will be discussed later. Thus, a large part of the smart grid will be involved with the most effective forms of alternating current power transmission and distribution. The goal of this section is to explain fundamentals unique to alternating current that will be required to understand the later chapters on power transmission and distribution, both in the legacy power grid and in the smart grid.

Alternating current is current that switches direction, or alternates, 60 times per second in North America and 50 times per second in many other parts of the world. Both current and voltage switch direction, which creates a wave pattern. As we now know from earlier discussion on waves and Maxwell's equations, this changing current will cause electric and magnetic fields to form, as well as other interesting phenomena that will be discussed later in this section. One of the primary benefits of this automatic creation of fields is that transformers are easily constructed to change voltage levels. A direct-current system will not create the changing current flow required to create a magnetic field.

In the early years of electric power, direct current was used; transformers had not yet been invented. This meant that voltages had to be kept relatively low for safety reasons. The larger current required to achieve the power levels necessary created huge losses on the power lines. Thus, electric power could not be distributed very far until transformers allowed voltages to be stepped up and down safely. On this historical topic, it took some time to converge on a standard frequency defining the rate at which alternating current switched. It is simpler for generators to create lower frequency current, since fewer magnetic poles are required within the generator. It also reduces power line loss to keep the frequency low, as we will see later. However, on the consumer side, higher frequency is generally better. For example, there is less noticeable flicker in light bulbs at higher frequencies. As we know, the USA eventually standardized on a frequency of 60 Hz and Europe on 50 Hz.

Alternating current, in its ideal form, is simple enough to be described by basic trigonometric functions such as the sine or cosine. Alternating current is described by its amplitude, frequency, and phase. This is no different than the communication perspective of typical signaling approaches; information can be encoded by changing the amplitude, frequency, or phase. The difference from power systems is that while communication engineers seek to modify oscillation as a means to efficiently encode information, power system engineers desire a nice clean sinusoidal waveform in order to maintain a high level of power quality. More detail will be discussed about this later; now, we are simply introducing the basics of alternating current for power systems.

The amplitude is the height of the sinusoidal wave. The frequency is the number of oscillations per unit time. The inverse of the frequency is the wavelength, period, or cycle, or the distance from one location within the wave to the same location within the next wave, usually considered from the crest of one wave to the crest of the next. The phase can be thought of as describing how the wave starts, whether the wave begins at the crest, the trough, or anywhere between. The phase is important because two waves with the same amplitude and frequency can start at different points within a wave's cycle. All else being equal, the waves will be what is known as "out of phase"; that is, they will be slightly shifted copies of each other.

While we might normally think of a wave as a function of position versus time – see the wave equation in Section 6.6.2 – it is often more convenient to mark the horizontal axis in terms of angle rather than time. We can take the sine or cosine of an angle, but not time. Understanding the basic cyclical properties of alternating current is critical to understanding more advanced topics such as phasors and synchrophasors, which will be discussed in much more detail later. Thus, we take time to explain angular analysis of alternating current carefully here.

The angle is most often represented in radians. Consider that a circle has a circumference of $2\pi r$, where r is the radius of the circle. With radians, the angle of rotation around a circle is 2π. Any fraction of 2π represents the corresponding fraction of a complete rotation around the circle. The radius r is irrelevant, since we are not concerned with the size of the circle. The angular frequency ω is the number of rotations per second, where rotation is measured in radians.

The angular frequency for 60 Hz alternating current used in the USA is

$$\omega = 60\,\text{cycles/s} \cdot 2\pi\,\text{rad/cycle} = 377\,\text{rad/s}. \tag{2.1}$$

It is found by taking 60 cycles per second, where there are 2π radians per cycle, giving us the result in radians per second.

Keep in mind that we are talking about alternating current; however, voltage alternates along with the current. The current and voltage may not always be in phase with one another on the same power line; however, that is something that will be discussed later as we develop the background required to deal with it.

The current at any point in time can be represented as a sine function:

$$I(t) = I_{\text{max}} \sin(\omega t + \phi_I). \tag{2.2}$$

This is a sine function that has some of its properties adjusted: the amplitude – that is, the height of the wave crest – above zero is I_{max} and below zero is $-I_{\text{max}}$; the frequency is ω, as discussed in Equation 2.1; and the phase, or starting offset, is ϕ_I. This allows us to represent any form of oscillating or alternating current. The phase ϕ_I can range from zero, which indicates no offset, to just under 2π. If it were offset by 2π, the wave would be offset (that is, slid over) exactly on top of the next wave and would look indistinguishable from having no offset.

The voltage can be described in a similar manner:

$$V(t) = V_{\text{max}} \sin(\omega t + \phi_V). \tag{2.3}$$

As mentioned earlier, the current and voltage phases may differ, even when measured at the same location. Thus, ϕ_V does not necessarily equal ϕ_I; this has very important implications that will be discussed later.

It is important to keep in mind that these equations describe the ideal case of perfectly smooth waveforms. The actual waveform at any point in the power grid may differ significantly due to many different potential complicating factors, particularly transient, or short-term, conditions such as switching, or harmonics. Harmonic distortion occurs when different fundamentals, or multiples of the base frequency, are combined with the original signal, thus distorting it. For example, as discussed much earlier, loads are connected to the power grid in parallel with the main grid current. Thus, voltage tends to maintain its shape while more or less current may be drawn by a load. Some loads may draw current in a nonlinear manner, thus distorting the current relative to the voltage.

While direct current is easy to measure, it only has a constant value, some equivalently simple way to measure alternating current is required that results in a single scalar value. Alternating current is continuously oscillating, so there is no constant value that can be measured. However, it is possible to represent its value as the RMS, or RMS value. The procedure is to first square the function that represents the sinusoidal value. This removes the negative component of the value. Otherwise, just taking the average of any sinusoidal function would result in a value of zero, since it is equally above and below zero. After squaring the function, we can find the nonzero average value. Then, to compensate for squaring the value, the square root is taken of the average. RMS values are used ubiquitously throughout the alternating current world of power systems. In fact, the 120 V power from the outlet that you may be using is an RMS value.

Recalling our earlier discussion of Maxwell's equations, an existing electric field will resist a change in voltage, while an existing magnetic field will resist a change in current. An alternating current and voltage is continuously changing. The faster it changes (that is, the more rapidly it oscillates), the greater the resistance to change. Although entirely different from the microscopic physical obstructions of real electrical resistance, this resistance appears in a similar manner to alternating currents and voltages. Thus, it is treated similarly to electrical resistance and is known as reactance. Impedance takes into account both true resistance and reactance.

There are two kinds of reactance: inductive reactance, due to magnetic fields, and capacitive reactance, due to electric fields. The inductive reactance is

$$X_L = \omega L. \tag{2.4}$$

It is apparent from this equation that the reactance will increase as either the frequency or the inductance increases.

Since we are talking about both the inductance and the rate of change of current, it is possible to apply Ohm's law $V = IR$ to this situation. Equation 2.5 uses the inductance L and the rate of change of current to compute the voltage drop across an inductive element in a circuit:

$$V = L\frac{dI}{dt}. \tag{2.5}$$

Capacitive reactance, impacted by an electric field, is best illustrated by thinking of two charged metal plates separated by a gap, but with an electric field between the plates. A direct current simply builds up a charge on the plates, but cannot cross the gap between the plates. Recalling our earlier discussion that current flow is really caused by electromagnetic fields "pushing" charges a short distance, the initial "push" created by the electric field across the plates when the charge first forms is enough to create a current. Alternating current continuously switches direction across the plate gap, creating a continuous series of "pushes," which appears as current flow through the circuit. This is shown by the following equation:

$$X_C = -\frac{1}{\omega C}, \tag{2.6}$$

where the reactance is reduced as frequency increases. A larger capacitance C allows the plates to hold more charge and increases current flow; thus, a larger value of C reduces the reactance. Finally, the capacitive reactance is negative in order to indicate that it is the opposite of inductive reactance. Capacitive and inductive reactances tend to cancel one another. In fact, in a circuit in which the inductive and capacitive reactances cancel each other perfectly, there is no reactive resistance and a phenomenon known as resonance can occur. Resonance can occur in a circuit that has both inductance and capacitance. Current in the inductor creates a magnetic field in which energy is stored, then an electric field is created across the plate of the capacitor, which stores energy; the flow of energy switches between the electric and magnetic fields of the inductor and capacitor, creating an oscillating current and voltage. Because there is little resistance when the inductive and capacitive reactances cancel each other perfectly, the oscillation can continue long after power has been initially applied and then removed. In fact, there are pathological conditions in the power grid in which resonance can create extremely large voltages or currents, known as ferroresonance (Dugan and McDermott, 2002; Rye, 2007).

In a similar manner to inductance in Equation 2.5, a version of Ohm's law can be applied as shown by

$$V = C\frac{\mathrm{d}V}{\mathrm{d}t}. \tag{2.7}$$

The impedance Z is the combined impact of resistance and reactance. Because it combines both resistance and reactance, it is the most general form of "resistance" for any component even though the component may have negligible resistance or negligible reactance. The impedance is not simply the sum of the resistance and the reactance; it is typically represented in a more clever manner that distinguishes the impact of the reactance from resistance. The impedance is represented as a complex number; the resistance is the real part of the complex number and the reactance is the imaginary part of the complex number:

$$\mathbf{Z} = R + \mathrm{j}X \tag{2.8}$$

This is a clever way to represent impedance because it not only distinguishes the resistance from the reactance, but also, if displayed on the complex plane – in which the x-axis is the real number line and the y-axis is the imaginary number line – the angle formed by the impedance given the resistive and reactance components represents the phase shift between the current and voltage. This can be seen in the Euler equation:

$$\mathrm{e}^{\mathrm{j}\theta} = \cos\theta + \mathrm{j}\sin\theta. \tag{2.9}$$

There is an inverse form of impedance, known as admittance Y. It is composed, conceptually, of the inverse components of the impedance, namely the conductance G and the susceptance B:

$$\mathbf{Y} = G + \mathrm{j}B. \tag{2.10}$$

However, we have to be careful when dealing with complex numbers. Consider the complex number simply as a vector of two components, a real and an imaginary component. In that case, we are dealing with vector algebra, where the inverse is

$$\mathbf{Y} = \frac{1}{\mathbf{Z}}. \tag{2.11}$$

The magnitudes should be the inverse of one another, but the angle should remain the same, as shown in Equation 2.12, where Equation 2.13 shows the values of G and B.

$$\begin{aligned}
Y &= \frac{1}{Z} \\
&= \frac{1}{R + \mathrm{j}X} \\
&= \frac{R - \mathrm{j}X}{(R + \mathrm{j}X)(R - \mathrm{j}X)} \\
&= \frac{R - \mathrm{j}X}{R^2 + \mathrm{j}RX - \mathrm{j}RX + X^2} \\
&= \frac{R - \mathrm{j}X}{R^2 + X^2} \\
&= \frac{R - \mathrm{j}X}{Z^2} \\
&= \frac{R}{Z^2} - j\frac{X}{Z^2} \\
&= G - \mathrm{j}B.
\end{aligned} \tag{2.12}$$

Notice that G is not simply the inverse of R, and B is not simply the inverse of X, but rather the same corresponding value normalized by Z^2. Thus, G and B remain proportional to R and X; an impedance with a high resistance will have an admittance with a high conductance, and similarly for the reactance and susceptance.

$$G = \frac{R}{Z^2} \qquad B = -\frac{X}{Z^2}. \tag{2.13}$$

Now we can move on to what many people see as the heart of power systems analysis, fundamentals directly related to computing and analyzing power. Let us start with the most fundamental aspect, the physical definition of power. Power is defined as energy per unit time. Thus, power is a rate, the rate at which energy is produced, consumed, or transmitted. Also, because power is energy divided by time, given power and the amount of time, one can compute the total energy; energy is power multiplied by time. As we will see later, if the unit of power is the watt, then energy is measured in watt-hours. Recall way back in Equation 1.8, power was discussed as a function of I^2R:

$$P = IV = I(IR) = I^2R. \tag{2.14}$$

As a means of review, let us recall why the above equation should yield power. Back in Equation 1.6, we saw why the units lead to the dimensions of power. However, from our discussion of Maxwell's equations, electric power flows in conjunction with electric and magnetic fields that exist together. From an electromagnetic field standpoint, the integral of the cross product of the electric and magnetic field vectors can also define power, as shown in Equation 2.15, where the cross product yields a vector known as the Poynting vector:

$$P = \int_S (\mathbf{E} \times \mathbf{B}) \cdot \mathbf{DA} \tag{2.15}$$

The Poynting vector is in the same direction as the propagating **E** and **B** fields of the electromagnetic wave. However, Equation 2.15 takes the power passing through a surface area surrounding the Poynting vector, so that the result is a scalar value. We will not go into the field aspect any deeper than this; the remainder of the section will deal with simpler, non-field, aspects of power analysis.

2.2.2 *Complex and Reactive Power*

With alternating current, rather than direct current, the simple $P = IV$ definition of power becomes more complex (literally) because of the previously discussed reactances. However, let us ignore those complications for now and gradually progress toward the development of a complex form of power; that is, one that expresses power using complex numbers. To start with, the simple $P = IV$ definition of power that holds for direct current also holds for power produced by alternating current when measured at specific points in time. Recall that alternating current and voltage obviously have no constant value, they are continuously changing over time. Thus, to correctly evaluate power with alternating current, it must be a function of time:

$$P(t) = I(t)V(t). \tag{2.16}$$

Now recall that the RMS value represents a form of an average value of an oscillating value. Also recall that if a load is purely resistive (that is, it has no reactance or reactive component), then we can use the RMS values of the voltage and current to represent power as shown by

$$P_{\text{ave}} = I_{\text{rms}} V_{\text{rms}}. \tag{2.17}$$

Again, it should be remembered that this only holds if there is no reactance in the circuit.

Now consider the case of reactance in the circuit. This is where the complication begins to arise. Recall that there is inductive and capacitive reactance and these impede either the current of voltage respectively. This impediment causes the current or voltage to shift with respect to each other. Think about this carefully; if current and voltage are always in phase with one another, as for a purely resistive load, then the product of the current and voltage is always positive, because they are always the same sign, either both positive or both negative. What happens when they begin to shift with respect to one another? Clearly, their product is no longer always positive. As they shift with respect to each other – that is, they become out of phase with one another – a larger portion of the cycle becomes negative power. In other words, there is a larger amount of time, or greater angle, over which the current may be in a position where the voltage is negative and vice versa.

We now discover that we can have both positive and negative power, or, more precisely, positive and negative power (flow). The positive power is flowing from the generator to a load and the negative power is flowing back from the load to the generator. Phasor notation will be discussed shortly; however, even though it may not yet be perfectly clear until phasors are understood, the current and voltage can each be thought of as vectors, starting at zero and of length equal to their magnitude, rotating around the complex plane. The dot product is the projection of one vector onto the other, where $\mathbf{I} \cdot \mathbf{V} = |\mathbf{I}||\mathbf{V}| \cos(\theta)$ is the definition of the dot product. If in perfect alignment (that is, parallel with one another), $\theta = 0$ and the dot product is the product of the vector magnitudes:

$$P_{\text{ave}} = I_{\text{rms}} V_{\text{rms}} \cos(\phi). \tag{2.18}$$

The $\cos(\phi)$ is known as the power factor.

Recall that RMS values are computed by first squaring the periodic signal, taking the mean, and then taking the square root. If the amplitude, or maximum value, of the original periodic signal is A, then the RMS value is $A/\sqrt{2}$. Thus, power is also

$$P_{\text{ave}} = \frac{1}{2} I_{\text{max}} V_{\text{max}} \cos(\phi). \tag{2.19}$$

Let us think about the meaning of ϕ a bit more. If $\phi = 0$, then the $\cos\phi$ is purely resistive. There is no reactance. The power is simply the product of the current and voltage. All power is positive and flows in the expected direction, from the power source to the load. On the other hand, if $\phi = 90°$, the load is completely reactive. Then the current and voltage are shifted completely out of phase. The factor $\cos(\phi) = 0$; thus, there is no power flow. This means that power flow is oscillating back and forth; it is equally positive and negative, but no power is actually dissipated.

Positive power flow is known as real or active power. It is power that is actually transmitted and consumed. However, based upon our mathematical descriptions, there are several other forms of power. For example, one can simply ignore the phase shift and compute the power using

$$S = I_{\text{rms}} V_{\text{rms}}. \tag{2.20}$$

In other words, the power is computed as though the voltage and current are in phase even if they are not. This form of power is known as apparent power; and to further distinguish it from other forms of power, its units are in volt–amperes (VA). Since apparent power is independent of the $\cos(\phi)$ power factor, it is often used in power equipment ratings, where the amount of current is often the parameter that varies; the voltage is generally held constant and the phase difference is less important.

Finally, note that power can be defined using all complex values as shown in Equation 2.21, where power is represented as $\mathbf{S}$ instead of P and all values are now complex numbers:

$$\mathbf{S} = \mathbf{I}^* \mathbf{V}. \tag{2.21}$$

There is yet another form of power. We have discussed real power (which measures the extent to which current and voltage are aligned yielding positive power flow) and apparent power (which is power assuming perfect current and voltage alignment). Now there is reactive power, which is the negative power seen earlier. This is power that oscillates back and forth between electric and magnetic fields in the power system and is shown by

$$Q = I_{\mathrm{rms}} V_{\mathrm{rms}} \sin(\phi). \tag{2.22}$$

To keep this form of power distinguishable from others, its units are specified in VAr. For the reader unfamiliar with power systems, "VAr" is not a typographical error. The lowercase "r" may look odd, but that is intentional; it is an acronym for volt-ampere reactive, the amount of reactive power in an alternating current electric power system. Both the real and reactive power are components of the apparent power. It should be easy to see that real power, being a function of $\cos(\phi)$, is a projection onto the x-axis of the apparent power, while the reactive power, being a function of $\sin(\phi)$, is a projection onto the y-axis of apparent power. When $\phi > 0$, reactive power is positive and the current is lagging behind the voltage. Recall that this occurs when inductive reactance dominates. If $\phi < 0$, then the current is leading the voltage, and this is due to capacitive reactance dominating.

Terminology is often used in which inductive loads "consume" reactive power and capacitive loads "supply" reactive power. These terms cannot be taken literally; reactive power is not literally consumed or supplied here. Inductive loads cause the lag in current relative to voltage previously discussed; in that sense, reactive power appears to be "needed" and "used" by inductive loads. Similarly, capacitive loads can counteract the inductive load reactance current lag by shifting the voltage and creating a leading current. Thus, the capacitive load appears to be compensating for the reactive power "used" by the inductive loads and thus "supplying" the required reactive power. In reality, utilities often simply increase real power to compensate for reactive power requirements. Here, we can begin to see a possible need for communication to control not only real power, but also reactive power.

In reality, there must always be enough real and reactive power present to meet load requirements. More about this will be discussed later; however, it is important to note that if power requirements are not met, the laws of physics will require that conservation of energy be maintained. Thus, if either real or reactive power requirements are not met, the system will begin to compensate in undesirable ways. If real power is not met, the frequency of oscillation will decrease from the standard 50 or 60 Hz. If reactive power requirements are not met, the voltage will begin to drop in order to compensate. In other words, standard values of the system that should be kept constant will begin to change in order to maintain the energy balance between supply and demand. Also, the fact that reactive power and voltage are so intimately related appears in integrated volt-VAr control (IVVC) discussed in Section 3.4.1.

As mentioned, the majority of loads in the power grid are inductive, thus causing a lagging power factor. For any given interconnected power network, it is possible to compute the total load power factor. This can be done by simply computing the load impedances based upon whether the loads are in series or parallel. The most important parts of impedance are the reactive components; thus, the aggregate impedance involves working with complex impedance values. Recall that the same aggregate angle ϕ in the aggregate complex impedance will be the power factor.

As a practical example, consider the ballast in older fluorescent lights, which plays a role in causing a low power factor. The ballast limits current flow and steps up the voltage in the light. The ballast can be implemented as an inductor coil or using a semiconductor; either way, the result is a power factor of 0.5 or 0.6, creating significant lagging current. An ideal incandescent bulb is purely resistive and has a power factor of 1.0. Thus, while a fluorescent light may use less total power, it can use more reactive power than an old-fashioned incandescent bulb. Of course, the consumer is only charged for real power usage, but in the end, we all have to pay, one way or another, for utilities to install the required reactive power compensation.

2.2.3 *Power Quality*

Power quality takes into account the voltage magnitude, frequency of the current and voltage, and the shapes of their corresponding waveforms. The voltage as seen by the consumer should be within the range specified by the standard. The frequency should hold steady at 60 Hz in North America and 50 Hz in Europe. The current and voltage should form nice, ideal, sinusoidal waves. From the standpoint of the waveform, to remain ideal, it should be free from harmonic distortion. However, the power grid has not been able to meet all of these conditions all the time, particularly the ideal waveform. It is often the case that meeting the minimum power quality requirements for use by consumer electronics is good enough. The case for providing a spectrum of power quality, or keeping it consistently high, is an ongoing debate.

From a voltage standpoint, as load increases due to higher demand, more current is drawn through the power line and the voltage drop along the line increases as per Ohm's law. Thus, voltage will vary with demand. The utility can use various techniques to adjust the voltage; however, maintaining a perfectly stable target value for all consumers is not feasible. For the USA, the standard voltage is allowed to range ±5% in magnitude; that is, from 114 to 126 V for a 120 V line. Brownout occurs when demand becomes high enough to cause the voltage to dip to the point where lights become dim. Obviously, a voltage that is too high will damage equipment.

Consider that the utilities' revenues are based upon the kilowatt-hour meter readings from consumers. One way to keep the meters running faster is to maintain the voltage on the high end of the allowed tolerance, which essentially pushes more power through the system. On the other hand, it is possible to save energy, increasing the efficiency of delivery, by reducing the voltage as much as is safely possible. This is known as conservation voltage reduction (CVR) and has been proposed for use in energy conservation since at least the 1970s. The ability of utilities to monitor and maintain control of the the voltage levels has been limited, and the tendency, as previously mentioned, has been to err on the side of higher rather than lower voltages, which also avoids the possibility of brownouts and customer complaints. One of the hopes and goals of the smart grid and communication is to be able to provide an advanced level of monitoring and control that would allow the voltage to be safely reduced with confidence. Again, we see an example of communication being used in the power grid to attempt to operate the system as close to the limit – a lower limit in terms of voltage magnitude in this example – as possible.

Voltage sags and swells can occur at any time, where one example is on the distribution network due to a fault. Until a fault is detected and isolated, there is an abnormally large current and power being drawn through the fault. This is one example of the cause of a voltage sag, as there will be a large voltage drop across the faulted power line until the fault is detected and disconnected from the circuit. Reclosers, introduced in more detail in Section 4.1, have a time–current characteristic curve such that lower fault currents take longer to detect than higher ones. Thus, lower voltage sags will tend to last longer than larger ones. Voltage sags are the most common power quality problem in the USA, with an estimated loss of $5 billion dollars per year (Von Meier, 2006). While the utility should be able to record and publish results on power quality issues, they rarely do so for both image and potential liability reasons. Power quality is also something that is not entirely in their control, as an act of nature or simply a backhoe digging in the wrong spot can be responsible for power quality problems.

Communication and networking have dealt with stability control similar to voltage stability in the power grid; for example, to implement congestion control. In communication, congestion control is concerned with controlling the rate at which packets are transmitted in order to avoid excessive delays, queues filling up, and packets being dropped, and ultimately the collapse of the communication link.

In the case of sudden load loss in the power grid, the system will have excess power. Traditional techniques to address this include resistive breaking and generator dropping. The resistive breaking process involves switching a bank of resistors onto the three-phase circuit, which adds a large load onto the system. The appearance of this large load should be enough to slow the generator's rotor back to an appropriate speed. The resistors in the breaking mechanism are generally cheaper, shorter life resistors; if the overspeed condition persists for more than several cycles – for example, the time it would take for a breaker to clear – then the breaks should be disconnected and generation capacity reduced. Generator dropping is simply switching a generator offline. This is a more dramatic action; resistive breaking can provide a smoother response when it can be utilized. In some cases, inserting series capacitance can help alleviate stability problems by reducing the reactance and improving the magnitude and quality of the waveform. Finally, to relieve an overloaded and unstable system, more power can be imported over high-voltage direct-current lines. In the case of instability due to undercapacity, the options appear to be more limited; namely, load shedding of some form, as well as the obvious reaction of increasing mechanical power to the rotor if at all possible.

2.2.4 *Generation*

A generator produces power by utilizing electromagnetic induction, described in detail earlier. An electric charge within a changing magnetic field experiences a force, where the force is in a direction that is perpendicular to the motion of the magnetic field and the direction of the magnetic field lines, as explained in Equation 1.9. This force acting on many charges in a conductor is an electromotive force that causes a voltage drop along the conductor and thus causes current to flow through the conductor. The key element in the creation of this current is the changing magnetic field: the change of the magnetic field relative to charges in the conductor. This change can be mechanical motion or simply a change in the intensity of the magnetic field. Large-scale power generation is focused upon creating the mechanical motion necessary to induce current.

The same mechanism used to create current in a generator is used in reverse to create mechanical motion in an electric motor. Instead of the force from the rotation of a shaft causing motion of a magnetic field used to induce current, current induces a force against a magnetic field to turn a shaft. In fact, the same device can operate either as a motor or a generator.

Consider now a generator constructed as a magnet rotating within a coil of wire. Recall from earlier discussions of magnetic flux that it is the product of the magnetic field and the area that it crosses. The current induced in the wire surrounding the magnet is directly proportional to the magnetic flux flowing through the surface area formed by the wire loop. One does not need to keep track of the magnetic flux versus the wire loop at all points around the wire loop; only the total flux through the wire loop need be known.

As the magnet rotates within the wire loop, the magnetic flux continuously varies in relation to the plane of the wire loop. When the field lines are parallel with the wire loop surface, there is no flux through the wire loop. As the magnet continues to rotate, the magnetic field lines become perpendicular to the wire loop and the magnetic flux through the loop is at a maximum. As the magnet continues to rotate, magnetic field lines again become parallel with the wire loop and there is no magnetic flux through the wire loop. This continuous rotation of the magnet creates a sinusoidal change in the magnetic flux through the loop.

The actual current induced in the wire loop is proportional to the rate of change of the magnetic flux through the wire loop. The derivative of the sine is the cosine; that is, the voltage is offset by 90° from the magnetic flux. This can also be seen simply by thinking about the rate of change of flux. The magnetic flux is changing most rapidly as it crosses zero. This is the point when the voltage is highest; that is, the top of the voltage wave. The magnetic flux is changing least rapidly at its crest, which corresponds to a zero crossing of the voltage wave.

The current induced in the wire loop will create its own magnetic field. However, this magnetic field will oppose the rotating magnetic field of the magnet. Obviously, the magnetic fields must be in opposition, otherwise the fields would reinforce one another, causing the magnetic rotation to increase and an infinite supply of power could be created without additional energy, violating the conservation of energy. The fact that the magnetic field of the induced current is in opposition to the field of the rotating magnetic field is extremely important. The force required by the rotating magnet to push against the induced current's magnetic field and maintain the proper frequency of rotation is precisely the act required to convert mechanical energy to electrical energy; in other words, this is the essence of power generation. As loads connected to the generator draw more power, they are in reality drawing more current, because, as discussed earlier, voltage is generally held constant. As the loads draw more current, the induced-current magnetic fields become stronger, requiring more force from the rotating magnet to maintain the same frequency of rotation. The force between the rotating magnetic field and the induced current's magnetic field is also what helps keep multiple generators balanced; this will be discussed in more detail later. Let it suffice for now that communication is not explicitly necessary to keep multiple generators spinning in phase with each other; the interacting magnetic fields act as a natural force to maintain balance. The technical name for the magnetic field induced in the wire loop of our generator is the armature reaction.

A real generator is not likely to have a permanent bar magnet, but rather an electromagnet in the form of a wire coil with a direct current flowing through it. An advantage of such an electromagnet is that its field strength can be controlled by the amount of current allowed to flow through the coil. This is known as excitation current, and the source of this current is known as an exciter. Also, a real generator can increase its voltage by increasing the number of turns in the wire coil surrounding the rotating magnetic field. The coils are additive in terms of increasing the output voltage.

Also, in a real, large-scale production generator, three-phase power is generated. The alternating current from a three-phase generator is comprised of three separate power lines, each with an alternating current waveform that is 120° shifted in phase from each other. This is accomplished by having three sets of armature windings in the generator that are physically located at 120° locations around the rotating magnetic field. Each armature winding becomes its own individual output for the generator. The three-phase generator (and motor) distributes the force caused by the armature reaction magnetic fields uniformly around the rotating magnetic field, thus creating a smoother application of force on the mechanical components of the system. The armature reaction of the three-phase generator thus appears as a magnetic field that rotates precisely with the rotor's magnetic field created by the electromagnet. Because both magnetic fields rotate in a fixed position relative to one another, this type of generator is called a synchronous generator.

Consider some of the design aspects that govern the characteristics of electric power at the generation stage. If a generator is to produce 60 Hz alternating current and it uses an internal electromagnet with two poles, then the electromagnet must turn at a rate of 3600 rotations per minute to create alternating current at 60 cycles per second. To rotate at this rate, the velocity on the outside edge of a turbine blade may exceed the speed of sound. Thus, construction to withstand such stress is a tremendous engineering challenge. In order to allow for a slower rate of rotation, more poles are added to the electromagnet; this creates more cycles of alternating current per rotation. The frequency is shown in Equation 2.23, where r is the rotational frequency, p is the number of poles of the electromagnet:

$$f\,\text{cycles/s} = \frac{r\,(\text{rotations/min})\; p\,(\text{poles/electromagnet})}{60\,(\text{s/min})\; 2\,(\text{poles/electromagnet})}. \tag{2.23}$$

The typical voltage at the generator terminals is 10 kV. The voltage value is a trade-off between two competing factors: a higher voltage requires more insulation to protect from arcing and also requires more turns of wire in the internal windings. Ultimately, there are space limitations on the number of windings and the amount of insulation. On the other hand, a lower voltage causes a larger current for the same amount of power. As we discussed, a larger current results in a higher I^2R power loss and more heat is generated due to the greater resistance. Thus, just as in transmission, there is less power loss with higher voltage. There are many other design constraints and objectives, some of which include the need for cooling, optimizing the geometry to best utilize magnetic fields, and reducing eddy currents.

Recall that the electromagnet within the generator that supplies the internal rotating magnetic field needs its own source of power in order to create the magnetic field. It can get this in a few different ways. One way is to obtain this smaller, direct current from another generator known as an exciter. There is then the question of how the exciter starts up, since it is a generator that requires its own internal electromagnet. It turns out that the exciter can be self-excited: it can use its own current to create its own magnetic field. Even from a complete stop, the exciter is able to start using the small residual magnetic field that remains within the metallic structures after the generator has been turned off. The application of mechanical power to turn the rotor and the residual magnetic field create a current that causes a self-amplifying effect. Some generators without an exciter require alternating current from the grid, which is then rectified to direct current, instead of an exciter. However, this means that such a generator would not have a black-start capability; in other words, it could not start unless the power grid were already running in order to supply the necessary exciter current.

Recall that a synchronous generator is synchronous because the rotor's magnetic field travels precisely with the stator's magnetic field; the fields rotate together in synchrony. This means that the generator's output alternating current waveform corresponds directly to the physical orientation of the rotor. If there is a synchronous generator, then there must be a corresponding asynchronous generator, and there is. The asynchronous generator is sometimes called an induction generator, corresponding to its use as an induction motor. In the asynchronous generator, the rotor is passive, in the sense that it has no external current source; there is no exciter. When multiphase current flows through the windings of the stator of an asynchronous motor, the rotor's passive windings experience a moving magnetic field. This induces a current in the rotor's windings that creates an opposing magnetic field that causes the rotor to turn. However, there is a critical difference between the synchronous and asynchronous generator: in order for current to be induced in the rotor's coils, the rotating magnetic field in the stator must be faster than the rotation of the rotor's coil and its corresponding magnetic field. Keep in mind that in both the synchronous and asynchronous generators the stator is always physically stationary; the combination of the multiphase (usually three-phase) alternating current creates a moving magnetic field. Because the stator field must necessarily rotate faster than the rotor field, and thus the rotor and stator fields are no longer synchronized, this type of generator earns the name "asynchronous."

The ratio of the velocity of the rotor magnetic field to the speed of the stator magnetic field is called the slip. The alternating current frequency will be slightly lower from an induction generator than from a corresponding synchronous generator due to the slip. Also, the alternating current waveform will have no precise correlation with the location of the rotor relative to the stator. The trade-off between synchronous and asynchronous generators is primarily one of complexity versus efficiency and ease of use. The synchronous generator is more complex, requiring an exciter, but its frequency is easier to regulate. The asynchronous generator is less complicated to construct, but it requires a connection to the power grid in order to start. The speed of the asynchronous generator can vary, for example, with the speed of the wind in a wind turbine. Thus, it does not require precise mechanical control to control the output frequency. Overall, the asynchronous generator can be cheaper and more flexible for DG. However, because of the inductive process that it relies upon, it has the same problem as inductive loads: it "consumes" reactive power.

Let us assume a synchronous generator from now on, unless specified otherwise. Communication, in some manner, is required to control a generator. The primary controls for a generator are the frequency of rotation of the rotor, related to the real power generated, and the voltage at the generator terminals, which is related to the reactive power produced by the generator. The relation between real power and the frequency of rotation of the rotor is quite simple and relates to the instantaneous balance between supply and demand of electric power. Consider what happens when more load is added to the power system. The voltage remains constant but more current is drawn to provide power to the new load. This additional current flow ultimately comes from the generator's armature windings, which increases the internal magnetic field that the rotor is pushing against. Unless provided with more mechanical power, the rotor would reduce its rate of rotation. This would be seen as a change in frequency of the current from the nominal 60 Hz requirement. Thus, simply monitoring the frequency of current in the transmission network can provide information about the generator's ability to supply the required real power.

Voltage and reactive power are controlled by the current of the exciter. Increasing the direct current of the exciter increases the magnetic field of the rotor and creates a larger electromotive force across the generator terminals, causing current to flow. The electromotive force is the voltage of the generator. However, the actual current flow is controlled by the loads: more load draws more current. An inductive load – that is, a load with a lagging current – impacts the electromagnetic fields within the generator. Because it is a synchronous generator, the rotor and stator fields move in unison. However, the inductive load will cause a change in the relative angle between the rotor and stator fields. This change in angle, due to the inductive load, will weaken the impact of the rotor's field. This will reduce the electromotive force and reduce the generator voltage. Thus, the need for reactive power cause by the inductive load has caused a reduction in generator voltage. The way to counteract this effect is to increase the exciter current to bring the voltage and real power back to normal. The important point to note is that generator voltage and the reactive power that it produces cannot be controlled independently: voltage and reactive power change together as the exciter current is adjusted. Thus, volt-VAr control is often seen in the literature. To drive home the point, if the generator voltage is to remain constant, as it typically is, then the power factor and reactive power are determined by the type of load; that is, purely resistive, inductive, or capacitive.

The electric power grid has many generators simultaneously supplying power to meet load requirements. The requirements for the grid in the USA, for example, are 120 V alternating current at 60 Hz frequency. Simultaneously operating generators must work together to maintain this constant frequency. However, there is no master clock communicating rotor frequency to all generators; rather, internal magnetic fields naturally act to maintain a constant frequency among simultaneously operating generators. In a sense, there is a shared magnetic field among all the interconnected generators: the stator magnetic field in each generator is produced from the same current flowing between all the interconnected generators. If one of the interconnected generators reduces the mechanical power applied to its rotor, the other generators will do work to attempt to maintain the correct frequency. Assuming the other generators have the power required to maintain the standard frequency, the generator with the reduced rotor power will not need to apply as much force to move its rotor against the rotating stator magnetic field and may be able to maintain its rate of rotation with reduced power. However, the other generators are taking over the required work for the generator with the reduced rotor power. Similarly, a generator that attempts to increase its rotation by applying more power to its rotor will be pushing against its stator field which is supplying current to all the loads. The increased force of the faster generator will be relieving all the other generators of having to apply an equivalent amount of power. In other words, for a single generator in a multigenerator system to change its frequency from the constant frequency of the other generators would generally require a significant amount of effort. Barring any large abnormal disturbance, there is a natural tendency for the simultaneously operating generators to reach an equilibrium frequency.

Given the natural equilibrium in rotor speed for all interconnected synchronous generators, each generator's rotor should be in the same position at the same time. Any offset from this equilibrium position is known as the power angle. Thus, if the power angle is larger for a given generator, then it is pushing harder and producing a larger share of power. Of course, if the power angle deviates too much, the generators are no longer in synchrony with each other and they could lose the ability to reach an equilibrium frequency. A very rough mechanical analogy to multiple generators would be a set of wheels, one for each generator. The wheels are lined up, like the front wheels of a car but with as many wheels as there are generators. The wheels are interconnected by a flexible axle. The axle represents the shared magnetic fields and its flexibility represents the magnetic field's ability to apply force to the rotor; that is, to tug against the rotor's field. The axle is just rigid enough that if one smoothly turns one wheel then the others will turn as well, eventually turning at the speed of the first wheel. However, consider a sudden change, such as suddenly stopping or speeding up one of the wheels. The force of the sudden change will be transferred through the flexible axle, causing the other wheels to rock back and forth uncontrollably as the flexible axle stretches one way and then the other. While this is not a perfect analogy, it provides an intuitive feel for how a sudden change in one generator could cause the others to lose synchrony or even become chaotic. The ability of generators to keep from losing equilibrium with one another is a form of stability.

When one generator's frequency changes from that of another, this implies that their voltage waveforms shift relative to one another. This voltage differential between generators causes a current to flow between the generators, which is known as circulating current because it does not travel to a load but rather from generator to generator. Because this current is traveling to and through the windings of the armature of the destination generator, it is a current with a nearly completely inductive load. Thus, the current is a full 90° out of phase with its generating voltage. This circulating current is what increases the magnetic field at the stator of the faster generator, causing its rotor to experience more force and require more mechanical power to operate.

In order to maintain stability, a generator when brought online to support an operational grid is first powered up offline until its frequency precisely matches that of the grid where it will connect locally. Only when the generator frequency precisely matches that of the grid will it connect with the grid. This is known as synchronization. Once the generator is online, after precisely matching the grid frequency, it has zero load; that is, it is not yet producing any power. However, at this point, the rotor power is then gradually increased further until the generator is actually contributing its required share of power to the grid.

The law of conservation of energy holds throughout the electric power grid; at all times, electric power supplied must equal the electric power consumed by the loads. If anything changed to attempt to violate this law, bad things would happen in order to compensate and enforce this law. First, the frequency would change in order to enforce the balance between supply and demand. If more power was supplied than consumed, the frequency would increase to compensate; if too little power was generated, the frequency would decrease. Thus, wide-area communication is useful for such frequency monitoring. If the frequency could not change enough to enforce the balance between supply and demand, the voltage would begin to change. Obviously, if the power supplied were too little to meet demand and the frequency could not slow down further, then the voltage would begin to drop.

The reactive power must also remain balanced at all times. Recall that reactive power is an oscillating flow of power from one field to another. Thus, there is no net flow in one direction; however, there must be a balance in the oscillating reactive power flow. The amount of reactive power, as mentioned earlier, is controlled by the loads; loads may be purely resistive, inductive, or capacitive, with the large majority being inductive, or lagging. Also, as mentioned earlier and will be explained in detail later, voltage and reactive power (that is, VAr) are coupled and cannot be changed separately. Thus, reactive power is controlled by adjusting the generator voltage.

If two generators are in phase with one another and the voltage of one generator is increased to produce more reactive power, it will again create a circulating current among generators due to the voltage difference. Recall that the voltage can be increased by increasing the current in the exciter. The circulating current will be 90° out of phase with the voltages, which are in phase with one another. Again, the circulating current travels from the generator with the higher voltage to the one with the lower voltage and will experience an inductive load through the destination generator's windings. Thus, the higher voltage generator is "supplying" reactive power to the lower voltage generator. The reactive power transmitted by the higher voltage generator is experienced by the lower voltage generator as a leading current, and thus capacitive power. This changes the field angles internally in order to support the rotation of the rotor, and the lower voltage generator's voltage will rise. The lagging current from the higher voltage generator has the opposite effect: it changes the angle of the field internally, weakening the force of rotation. Thus, all else remaining constant, the voltage of the two generators tends to balance with one another. In a larger, more complex system of generators, raising or lowering the generator voltages is used to control how reactive power is allocated among generators. Note that the circulating current between generators is a real phenomenon and adds to the stress and congestion on transmission lines. It experiences I^2R power loss and adds to the thermal heating on power and transmission lines. Thus, circulating currents do not come for free.

However, it should be stressed that total reactive power is determined by the types of loads. Ultimately, enough reactive power to match that required by the loads must be created by the generating system as a whole. Reactive power production can be allocated among generators based upon business and market requirements, as well as by physical requirements. Locally, the generator operator can only measure the real and reactive power, or power factor, of the generator. The system-level real and reactive power, or power factor, will generally be different from each individual generator's real and reactive power output. Thus, communication is required to provide a complete system-level view of a complex interconnection of generators.

A generator undergoes mechanical stress as it operates; its operating limits cannot be exceeded without some form of damage, some of which is reversible and some irreversible. A typical form of stress encountered is overheating, which can be addressed by increasing the cooling mechanism used and/or shifting load to other generators. If a generator continues to run beyond its rated capacity, there can be abrasion and damage of insulation and finally many forms of irreversible damage, such as bending shafts and extreme vibration causing structural damage. Thus, sensor communication for generator monitoring is vital. A trend throughout power systems, and for generators as well, is that as monitoring and communication improve, equipment is being pushed closer to its operating limits. In other words, confidence in monitoring and communication is enabling operators to accept more risk by pushing equipment closer to its breaking point.

2.3 Management and Control: Introducing Supervisory Control and Data Acquisition Systems

Management and control of a centralized power plant is certainly not, by itself, a smart grid technology, since centralized generation has been the backbone of power generation for

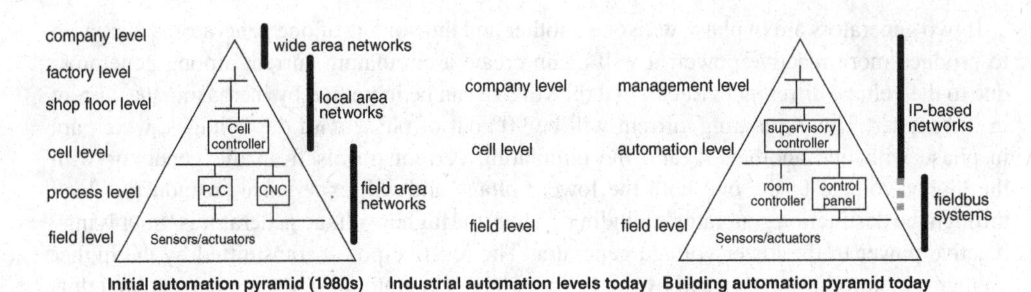

Figure 2.2 SCADA networks derive from the concept of a hierarchical network framework designed to cope with the anticipated complexity of communication in an integrated environment that covers a wide area while also extending deep into the real-time operation of a system. This originated in 1980s factory process automation. The power grid and its associated communication network are simply an example of process automation applied to the generation, management, and transport of power. Source: Sauter, 2010. Reproduced with permission of IEEE.

many decades. However, it is worth reviewing management and control of these centralized systems because it has been the trend to extend the communications used there to the smart grid because such communication is considered proven, safe, reliable, and well tested over many years of use.

Industrial communication encompasses a broad array of hardware and software required for robust, reliable, and often real-time operation and control of dangerous and expensive equipment. Typically, an industrial communication architecture is organized in a hierarchical manner with the human–machine interface at the top, allowing a human operator to view and control the system from a high-level perspective, as illustrated in Figure 2.2. Underneath the human–machine interface (that is, within the middle layers) are programmed-logic controllers. The programmed-logic controllers are interconnected via non-time-critical communications. Finally, at the bottom layer resides the fieldbus that interconnects the programmed-logic controllers with the physical devices that directly perform the work, such as sensors, actuators, electric motors, switches, valves, and many other possible devices.

An industrial network formed by this hardware and software is often called a fieldbus. The origin and meaning of the term "fieldbus," even after at least 20 years of development, is unclear. The English word "fieldbus" appears to have originated circa 1985 during IEC standardization efforts. However, it was a translation of the word "Feldbus" from German, where the German word was used at least as far back as 1980. Interestingly, much of the early work on this topic exists only in German. The term "field" is used in the process industry to refer to the "process field," meaning the location in a plant or factory where distributed sensors and actuators are in direct contact with the process being controlled. The early topology of such systems was a star with analog, wired links from each device in the field arriving at a central control center. This, of course, led to a significant amount of confusing and expensive wiring, as well as systems that were monolithic in nature; components, particularly software-based components, were not modularized and could not be replaced with a different vendor's component.

It was the desire to replace the aforementioned wired star topology with a simpler, more effective single bus system that led to the notion of the fieldbus. Because fieldbus systems were developed in parallel in many diverse process control industries, this has led to a large number of diverse standards maintained by many diverse organizations. They are also distinct from the technology developed within information technology (IT), in which Ethernet and wireless technologies are dominant. It was recognized that it would be ideal for the IT and plant automation communication technologies to become better integrated. However, they developed with different requirements in mind; plant automation required communication amenable to real-time control; that is, low latency with small packet size. The IT realm was focused on high bandwidth for large data transmission without the strict need for low latency or tight timing control. Also, technology increased relatively quickly in the IT world, while plant automation needed to maintain compatibility with a large amount of expensive equipment over a longer period of time.

As shown in Figure 2.3, there are large numbers of diverse fieldbus technologies and they have a significant influence on the smart grid. It would be a mistake to look at the smart grid as only the integration of the IT view of communication technology without considering the role of plant automation communications. For example, power plants and DG will continue to utilize fieldbus protocols long into the future. Substation automation is largely an outgrowth of fieldbus technology, particularly IEC 61850. However, there would be little value in examining each fieldbus protocol in detail; our goal is simply to understand their basic characteristics and trends in relation to the smart grid. One trend that is clear from Figure 2.3 is that fieldbus protocols are moving from proprietary standards toward international standards. Another trend, one that is less apparent in the figure, is that fieldbus technologies are merging computer network and telecommunication technology with fieldbus technologies.

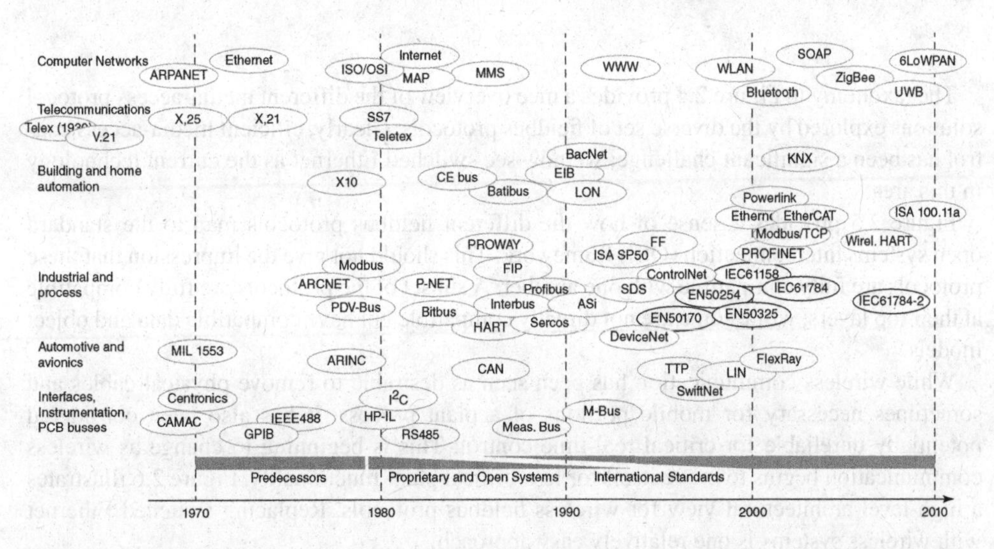

Figure 2.3 The fieldbus time line really begins with demand for a reduction of cabling weight in avionics and space technology that led to the development of the Military Standard 1553 bus, which can be regarded as the first "real" fieldbus in 1970. It demonstrated characteristic properties of fieldbus systems, including (1) serial transmission of control and data information over the same line, (2) master-slave communication architecture, (3) the possibility to cover longer distances, and (4) integrated controllers. Eventually, only "open" systems would survive and gain substantial market share. The time line ends with international standardization for broad market acceptance. Source: Sauter, 2010. Reproduced with permission of IEEE.

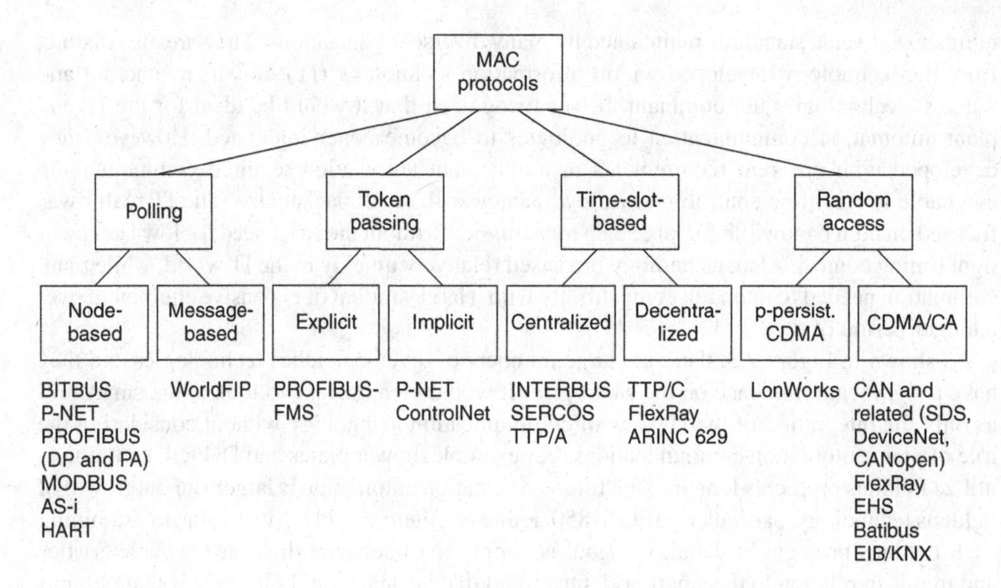

Figure 2.4 The mid 1980s saw significant growth in SCADA systems enabled by power line carrier coupled with more intelligent sensors and actuators. These new systems were tailored to different applications. The large variety of medium-access mechanisms are shown for fieldbus systems that still have a significant market share. Source: Sauter, 2010. Reproduced with permission of IEEE.

The taxonomy in Figure 2.4 provides a nice overview of the different media-access protocol solutions explored by the diverse set of fieldbus protocols. Clearly, efficient media-access control has been a significant challenge; we now see switched Ethernet as the current technology in this area.

Figure 2.5 provides a sense of how the different fieldbus protocols map to the standard open systems interconnection (OSI) framework. This should not give the impression that these protocols are fully compatible with one another. A subset of the protocols are fully compatible at their top layers, while some are not directly compatible but have compatible data and object models.

While wireless communication has been seen as desirable to remove physical cables and sometimes necessary for mobile portions of a plant process, it has also been considered potentially unreliable for critical real-time control. This is beginning to change as wireless communication begins to prove itself for less-critical plant functionality. Figure 2.6 illustrates a high-level architectural view for wireless fieldbus protocols. Replacing switched Ethernet with wireless systems is one relatively easy approach.

One of the earliest forms of digital industrial communications is the RS series; namely, RS-232, RS-422, and RS-485. These standards do not define software communication protocols, but rather the electrical characteristics required to make a connection. RS-232 is the simplest and is limited to point-to-point communication. However, owing to advances in cabling and interface drivers, RS-232 can perform better in terms of speed and distance than its original standard rating. RS-422 uses a differential line encoding, thus allowing it to be more resilient to noise and interference and capable of transmitting reliably over longer distances. This is

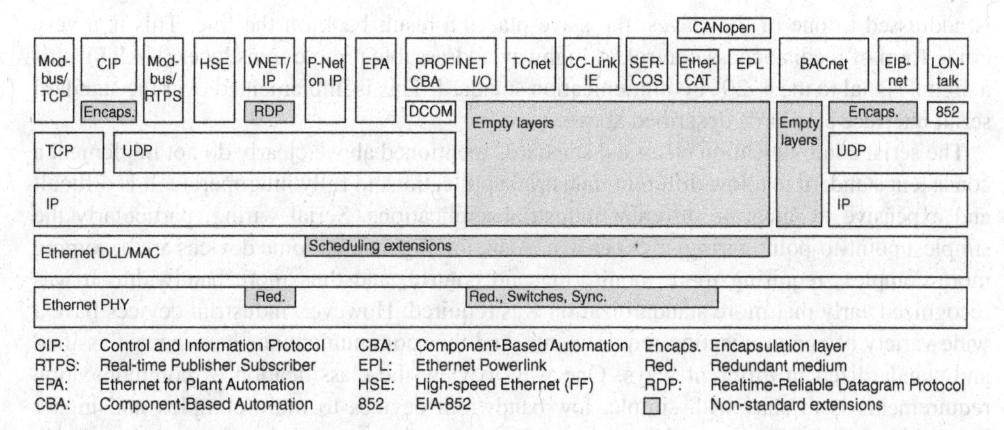

Figure 2.5 Fieldbus architectures derived from research that, at least initially, carefully attempted to avoid concepts that violated the Ethernet standard. However, this hope was futile. Interoperability between different industrial Ethernet solutions remains a challenge and typically cannot be accomplished without customized solutions. This figure illustrates the variety of different medium-access control techniques in use. Source: Sauter, 2010. Reproduced with permission of IEEE.

very important in a power plant, in which the potential for electromagnetic interference is a significant concern. Also, RS-422 allows up to 10 receivers to tap into one line, known as multidrop. RS-485 further increases performance and allows up to 32 transceivers on a line implementing multipoint communication. Because these basic communication standards do not specify the higher layer protocol stack, addressing and protocol data unit (PDU) structures are defined by the application. However, a common architecture is a master–slave relationship, in which a master device transmits to all other devices (slaves) on the line and, if the message

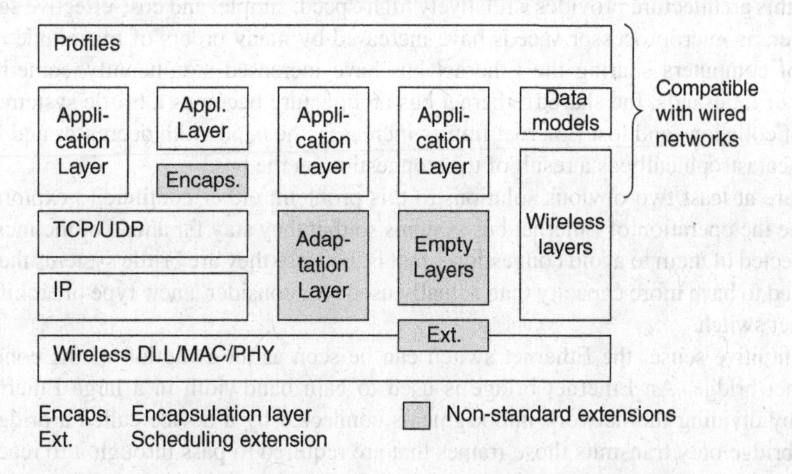

Figure 2.6 Wireless fieldbus architectures attempt to maintain compatibility with wired networks by using only the lower layers of the wireless technologies. This allows higher protocol layers from wired, field-level networks to be compatible over wireless systems. Source: Sauter, 2010. Reproduced with permission of IEEE.

is addressed to one of the slaves, the slave places a result back on the line. This is a very crude form of media-access control. A highway addressable remote transducer (HART) adds a digital signal to the 4–20 A communication standard. This is implemented over the standard serial interface standards described above.

The serial communication electrical standards mentioned above clearly do not implement a consistent standard to allow different industrial applications to fully interoperate. It is difficult and expensive to integrate different industrial applications. Serial wiring, particularly the simplest point-to-point wiring, is expensive. Also, intelligent electronic devices are becoming more complex, requiring more monitoring and control, and thus more bandwidth. It was recognized early that more standardization was required. However, industrial devices have a wide variety of communication requirements, and thus communications have been classified and standardized in different ways. One way to view the classification of communication requirements is to start with simple, low-bandwidth devices to more complex and higher bandwidth devices. For example, a sensor bus has small messages sizes but a fast response time. A device bus has a moderate message size and moderate response time, and a fieldbus has large message sizes and a longer response time. Moving from the simple electrical serial interface standards to industrial network protocols requires moving from simple American Standard Code for Information Interchange (ASCII) command/response sets to more complete and interoperable protocols.

Ethernet was originally designed using a bus architecture; multiple computers shared the same Ethernet media, known as a bus. Because microprocessors were slow enough relative to the capacity of such a bus architecture, this form of Ethernet provided more than sufficient capacity and was widely used; it was almost ubiquitous in the IT sector. Because multiple computers share the same bus and communication is not scheduled or synchronized, it is possible for more than one computer to transmit at the same time, resulting in a collision between their Ethernet frames. This results in a garbled, or corrupt, frame for everyone and a loss of usable bandwidth. However, as long as the traffic load is well below the maximum capacity, this architecture provides a relatively high-speed, simple, and cost-effective solution.

However, as microprocessor speeds have increased by many orders of magnitude and the number of computers sharing the Ethernet bus have increased significantly, sometimes to hundreds or thousands, the shared Ethernet bus architecture becomes a brittle system; as the number of collisions and lost Ethernet frames increases, the bandwidth decreases and latency increases catastrophically as a result of this congestion on the bus.

There are at least two obvious solutions to this problem: either continue to explore ways to increase the operation of Ethernet bus systems so that they stay far ahead of the increasing loads expected of them to avoid congestion – that is, because they are brittle systems they must be designed to have more capacity than actually used – or consider a new type of architecture, an Ethernet switch.

In an intuitive sense, the Ethernet switch can be seen as an evolution of the concept of an Ethernet bridge. An Ethernet bridge is used to gain bandwidth in a large Ethernet bus network by dividing the network into segments connected by a device called a bridge. The Ethernet bridge only transmits those frames that are required to pass through it to reach their destination segment. In other words, they block frames that do not need to pass and thus prevent unnecessary collisions from occurring. However, each bridge adds a small amount of delay and cost to the network. Placing a traditional Ethernet bridge on every segment would make an efficient network, but would be very expensive and could result in significant delay if not needed. Determining the optimal location of Ethernet bridges is an interesting optimization problem; the goal is to cluster microprocessors that communicate with one another often on the same Ethernet segment while adding bridges between such clusters, in which communication occurs less often. One can then minimize latency and/or cost while maximizing available bandwidth.

A switched Ethernet network does not literally utilize bridges as explained above; however, it does allow each computer to appear as though it were on its own dedicated segment. A very simple intuitive analogy, using vehicles on local roads, can illustrate the concept. At night or during non-rush-hour it is relatively easy to travel from one location to another. During the rush hour congestion increases rapidly and can reduce traffic to a crawl. In other words, the load has increased but the shared Ethernet bus network remains the same. If we were to make the analogy of creating a switched Ethernet network, each driver would ideally have a direct road leading from the driver's current location to their destination; there would be no competing traffic and the driver could travel as fast as desired.

There are a few reality checks to consider in the previous idealized scenario. The main one is that any large road system would become nonplanar; concrete bridges or overpasses would be required to avoid physical collision with other vehicles. Similarly, in building an electronic switch, the architecture of the switch fabric becomes crucial. We have effectively moved the problem of collision on the Ethernet bus to designing an efficient switch fabric. A cost-effective switch will have queues, either internally or on its input or output ports; thus, there remains a stochastic queuing delay that depends upon the characteristics of the offered traffic load. If the switch's queues become full, frames are typically dropped. Thus, Ethernet switches are not a perfect Ethernet solution, and research into traffic shaping, scheduling, and mechanisms for implementing frame priority still takes place in order to minimize internal switch blockage.

There are three general types of Ethernet switches: store-and-forward, cut-through, and hybrid. They operate conceptually as previously described with minor variations. The store-and-forward switch receives each frame in its entirety through its input port and then checks the frame for errors via its CRC. If no error is found, the frame is then forwarded along to its proper destination port. This means that each switch only passes through uncorrupted frames. However, checking the integrity of the frame takes time, adding to latency. Thus, store-and-forward is most suitable for networks with relatively high error rates. A cut-through switch begins forwarding immediately upon determining its destination address, and thus its corresponding output port. The result is fast forwarding of frames, but at the cost of wasting bandwidth if the frame becomes corrupt. A hybrid switch attempts to gain the best of both store-and-forward and cut-through switching. It does this by simply changing dynamically from cut-through to store-and-forward depending upon the error rate. The concept is for the switch to sample the frame error rate: if it is relatively low, cut-through switching is used; if it exceeds a given bound, then store-and-forward is used.

2.3.1 *Efficiency and the Environment*

The same improvements that make the power grid efficient and reduce cost also help the environment. There are many different processes for generating electricity, including coal, nuclear, gas, oil, and hydroelectric. Each power generation facility will use a SCADA system for communications, as just discussed regarding fieldbus networks. The technologies that convert mechanical, chemical, or nuclear energy to electric power are becoming more efficient (Beer, 2007). They are converting more energy to electric power while reducing the amount of wasted energy, usually expelled as heat, and potential pollutants. However, current smart grid efforts focus more upon increasing power transmission efficiency, load control via DR and demand-side management (DSM), and incorporating many, smaller renewable DG units into the power grid. From an environmental standpoint, the largest volume of pollution comes from coal-burning power plants. The pollutants include carbon dioxide, nitrogen oxides, sulfur oxides, and mercury, along with solid particulates such as fly ash.

A study by the United States Energy Information Agency (EIA) has made some approximations extrapolating demand to the year 2035 (Conti, 2010). This study assumes a nominal growth in electricity demand and the retirement of 45 GW of existing capacity. It expects that 250 GW of new generating capacity will be needed by 2035. The breakdown for this new capacity is expected to be natural gas-fired plants at 46%, 37% for renewables, 12% for new coal-fired plants, and 3% for nuclear plants. The point, of course, is not to assume these values are precise or accurate, but rather to provide a sense of expected trends. A study by EPRI estimates a savings in power plant production of 12% by 2030 if the smart grid as envisioned today is fully implemented (Siddiqui, 2008). These savings are not only cost savings, but an environmental benefit as well. The specific reduction of pollutants is expected to come from the following aspects of the smart grid. DR and DSM will enable consumers to manage their loads in order to conserve power. Electric vehicles will take their power from the grid, eliminating gasoline engine pollutants, but adding to electric power pollutants, which will be discussed shortly. Renewable and distributed energy sources are anticipated to generate power with limited impact on the environment.

Pollution from gasoline and diesel fuel used in transportation includes carbon dioxide, unburned hydrocarbons, carbon monoxide, and nitric oxides. While lead has been removed from the gasoline in cars and light trucks, it is still a component of fuel in the aviation industry. Chemical reactions caused by these pollutants produces ozone. Note that ozone is a respiratory hazard and pollutant when it resides near ground level; however, the ozone layer, which is a portion of the stratosphere with a relatively high concentration of ozone, prevents damaging ultraviolet light from reaching the Earth's surface, and is thus beneficial in that location. Thus, by moving from gas and diesel to electric vehicles, the pollutants specific to gas and diesel would be greatly reduced, but the electric generation pollutants may be correspondingly increased depending upon the power generation sources utilized. It should also be noted that the power grid also has other sources of potential pollution beyond generation. For example, polychlorinated biphenyls (PCBs) and other oils were used in transformers. Many of these transformers are still in operation. Similarly, switches, relays, and reclosers may also use various fluids to mitigate the impact of high-voltage arcing during operation.

Another environmental benefit to consider for the smart grid is that, to the extent that self-healing becomes reality, the number of utility vehicles dispatched to read meters and correct problems may be greatly reduced, thus reducing their pollution to the environment. Similarly, if faults can be avoided or power restoration can occur more rapidly after a fault, this will reduce the pollution from diesel generators that start up in islanded operation to supply temporary power. Another environmental relationship is that between the outdoor temperature and transmission thermal ratings. Transmission capacity is limited by the thermal rating of the power line, which changes with temperature. It is thus possible to increase the load through a transmission line when the temperature goes down, rather than assuming the lowest capacity, worst-case rating.

2.4 Energy Storage

This section provides an introduction to energy storage techniques. Advanced smart-grid-related energy storage techniques are discussed in Chapter 8. Energy storage systems existed long before the term smart grid was coined; however, they tended to be unique, specialized systems with no standard means of connection to the power grid or operation and control (Mohd *et al.*, 2008). In fact, storage systems have been treated as black boxes by the utilities, which depend upon the storage system vendor to provide proprietary installation, communication, and control mechanisms. One of the goals of the smart grid is to standardize the operation of energy storage systems so that they become a more natural and integral part of grid design and operation.

Energy storage systems have been used in the past as an ancillary power source in conjunction with intermittent power generation to smoothen power output. For example, an electric generator could be used to periodically recharge a battery to be used as a relatively small, local energy source in a microgrid. A similar concept applies to smoothen power output from intermittent renewable energy sources. Thus, from a communications perspective, energy storage is analogous to a buffer that can compensate for the stochastic behavior of the power system. Another potential application could be the ability to use lower power transmission lines; a storage device at the end of a power line could be recharged continuously, allowing loads to occasionally draw more power than the power line is rated to carry.

There are many different types of energy storage systems with different characteristics that can be important to consider for their particular application within the power grid. Examples include power and energy density; that is, the amount of power or the amount of energy stored per unit volume. There can be a significant difference between the amount of power per unit volume and the amount of energy per unit volume. The lifetime of the storage mechanism is another important consideration; for example, whether the storage system will break down or leak over time. The recharge time is another particularly important characteristic; a storage system that takes too long to charge may not be convenient or useful for many applications. Another related characteristic is the dynamic response: How quickly can the storage system respond to inject power into the grid when its needed? A storage system that requires a relatively long ramp-up time to release energy may not be suitable for many applications. For example, a storage system intended to improve power quality will need to inject power at precise points in time. The maintenance cost can be another issue: a storage system that meets or exceeds all the technical requirements but is too expensive to maintain may not be economically feasible. A related characteristic is, of course, the cost per kilowatt of storage; this should include the cost to charge, store, and discharge. Other considerations are potential environmental effects, a nonpolluting, renewable energy source paired with a potentially harmful or toxic storage device may defeat the purpose of creating an environmentally friendly generation system. Finally, another characteristic is known as the round-trip efficiency; that is, the percentage of energy lost during the charging, storage, and release processes. Storage system technologies vary widely over all of these characteristics and its important to match the storage technology with its proper application requirements.

A form of energy storage to which everyone can immediately relate is the electric battery. The purchase of a battery for its electric power to drive a flashlight, laptop, or cell phone is an event that most people have experienced. We have also experienced first hand how long they last and the time they take to recharge. The principle of operation of a battery involves a chemical reaction that results in the buildup of electrons at the anode; that is, the negative electrode of the battery. This results in an electrical potential difference between the anode and the cathode, which is the positive electrode of the battery. Electrons have a natural tendency to avoid such an imbalance in their distribution; the only way to do this would be for the electrons to flow from the anode to the cathode of the battery. However, the electrons cannot flow through the electrolyte, which is the material residing between the anode and cathode within the battery. The only way to resolve this imbalance is for the electrons to flow through an external circuit from the anode to the cathode and, in the process, drive a load. The chemical process creating free electrons is limited; this is the limit on the amount of storage the battery can provide. In many types of batteries the chemical process can be reversed; applying power to the battery reverses the chemical process, allowing the battery to recharge. There are many different types of batteries, with different anodes, cathodes, and electrolyte materials. In some batteries, the anode electrolyte and cathode electrolyte are different; each electrode with its own electrolyte in such a battery is known as a half-cell. The half-cells are placed in series connection to form a whole cell.

The strength of the chemical reaction to provide free electrons results in an electromotive force. This is measured across the electrodes of the battery and is known as the terminal voltage. If this measurement is taken when the battery is neither charging nor discharging, it is known as the open-circuit voltage. However, note that a battery has internal resistance; that is, a resistance between its anode and cathode. This causes a battery that is discharging to have a lower terminal voltage than its open-circuit voltage. Conversely, a battery that is charging will have a higher terminal voltage than the battery's open-circuit voltage. Ideally, a battery would have no internal resistance and a constant terminal voltage that would drop to zero when the battery is fully discharged. Of course, batteries are nonideal and their internal resistance increases as they discharge, which decreases the open-circuit voltage as they discharge as well.

Peukert's law, derived in the late 1800s, is still used today as an approximation of the capacity of a battery given that its terminal voltage decreases during discharge:

$$I^k t = \text{constant}. \tag{2.24}$$

In this equation, I is the discharge current, t is the actual time to discharge the battery in hours, and k is the Peukert coefficient. k typically has a value between 1 and 2; it is specific to a given battery's make and model. A larger load will draw more current and deplete the battery at a faster rate; thus, the effective capacity is lower than the same battery driving a smaller load. Stated another way, the battery capacity has fewer ampere-hours when the amperes drawn are higher because a higher amperage drags down the voltage more than a lower amperage load, so the endpoint is reached sooner.

If $k = 1$, then the capacity of the battery would not depend upon the current discharge rate. Lead–acid batteries have been the most widely used battery storage system for utility storage. They have a Peukert coefficient $k > 1$. Battery manufacturers will typically indicate the capacity of their batteries over a specified discharge time n in hours. With a little algebra, Peukert's law (Equation 2.24) can be manipulated to obtain the Peukert equation:

$$C_{n_1} = C_n \left(\frac{I_n}{I_{n_1}} \right)^{k-1}, \tag{2.25}$$

where n_1 is a new time period (in hours) over which the current is drawn and I_{n_1} is a new discharge rate (in amperes).

The point is that efficient use of the capacity of a system that is as simple as a battery is more complex than might be at first suspected due to nonideal behavior. Often, battery storage may serve as a power source for the communication supporting the smart grid as well as for buffering energy for the power grid. Such interdependent complex networks of networks need to be analyzed carefully.

The pumped-hydro (that is, pumped water) energy system is the oldest and appears to be the most widely used system, perhaps because it can cost-effectively store large amounts of energy, of up to 1000 MW. Operation is relatively simple. Water can be contained at two different levels. During storage, water is pumped to a higher level. During energy release, water falls to the lower level while turning a turbine.

Compressed air energy storage (CAES) stores energy by compressing air into a large contained volume such as an underground reservoir. When power is needed, the compressed air is released through a turbine to create electric power. Compressed air storage systems capable of providing power (non-electric) over a city-wide area have been around since the late 1800s. Compressed air storage systems can be massive; one of the earliest compressed air storage systems designed to store electric power was completed in 1978 in Bremen, Germany. This facility, known as the Huntdorf Plant, provides a storage capacity for 290 MW for 3 h utilizing space within a salt mine. Since then, larger capacity CAES plants have been developed.

A flywheel storage system accelerates and then maintains a rotor at a high rate of speed. The amount of energy stored is proportional to the square of the angular momentum. Energy is released by simply reversing the process; the energy of the angular momentum is used to power the motor, turning it into a generator, thus creating power. Flywheels have been typically used for power systems in the 150 kW–1 MW range and to enhance power quality and improve reliability.

SMES systems have the ability to sustain high currents within a superconducting material. These storage systems allow an exchange of power with the grid with extreme sensitivity; a single cycle can be injected when needed, which is ideal for power quality improvement as well as for bulk storage.

Super-capacitors have the ability to directly receive and hold a charge, like a battery, but are able to charge and discharge much more quickly because there is no intermediate chemical reaction. The charge is held within an electric field between two electrodes, just like a capacitor; however, the energy density is much higher than a regular capacitor.

Fuel cells are another storage mechanism that is popularly discussed for electric power storage. However, the term "fuel cell" covers a broad spectrum of storage techniques. The common characteristic is that fuel, in a variety of possible forms, is converted to electricity through a chemical reaction with oxygen or an oxidizing agent. Hydrogen is often assumed as the fuel; however, natural gas and alcohols are also possible fuel sources. Like a battery, the fuel cell has a positive side, the cathode, a negative side, the anode, and electrolyte in between. The electrolyte allows charge to move between the fuel cell's anode and cathode. An electrolyte is typically a liquid, gel, or gas that can be used to create ions by means of electrolysis. The fuel cell's electrolyte determines the fuel that is required. The anode catalyst breaks down the fuel into ions and free electrons; this can consist of platinum powder, for example. The cathode catalyst turns the ions into spent by-products such as water, if hydrogen is the fuel, or carbon dioxide. An example cathode catalyst is nickel. Any load placed across the anode and cathode of the fuel cell can be driven by the resulting current. A typical fuel cell produces about 0.7 V; however, many fuel cells can be stacked together to increase current and voltage. What we have just described is the simplest concept of operation for a fuel cell; more advanced types of fuel cells exist, such as proton-exchange membrane fuel cells and solid oxide fuel cells, among others. The hydrogen–oxygen proton-exchange membrane fuel cell uses a proton-conducting polymer as the electrolyte. This membrane is an insulator; it does not allow electrons to pass through. The fuel, hydrogen, diffuses to the anode catalyst where it separates into protons and electrons. The protons react with oxidants and then travel through the membrane to the cathode. Because the membrane is an insulator, electrons must flow through the external circuit to reach the cathode. Once at the cathode, the protons and electrons meet again to form water. The solid oxide fuel cell is quite different; it uses yttria-stabilized zirconia as the electrolyte and must operate at a high temperature, 800–1000 °C.

Hydrogen energy storage is yet another possibility; however, it is still in the early stages of development. Hydrogen can be stored in gas, liquid, or solid forms. Electrolysis is typically used to obtain the hydrogen to be stored and the resulting hydrogen is released by a fuel cell.

In the recent past, all of these energy storage approaches, with the exception of pumped-hydro, have been considered too expensive to be used on a wide scale. However, another way to look at it is that the cost of not using these techniques has been cheaper; that is, simply providing reserve power when necessary. However, this assumes that the cost of generation remains relatively cheaper than storage and that sufficient transmission and distribution capacity is available to handle the additional generation, transmission, and distribution. When the cost of wide-scale energy storage decreases and the cost of generation and transmission increases, a price threshold will exist that will tip the scales in favor of energy storage and its potential benefits.

The benefits of energy storage are numerous. Perhaps the largest and most obvious use of energy storage is known as load leveling or peak shaving. Spinning up new generation units can be very costly; minimizing the number of times they have to be spun up and shut down can significantly reduce cost. Thus, the goal is make the demand as constant, or level, as possible. This can be accomplished by storing power during times of low demand and releasing the stored power during high demand. This is also known as a form of "peak shaving."

Another benefit of utilizing power storage is known as grid voltage support. As the term implies, the goal is to maintain voltage levels as constantly as possible, or at least within their specified tolerance. Recall that loads use real power but can also consume or create reactive power. Both forms of power impact the voltage level, and energy storage for this application should be capable of supporting both forms of power when needed.

Recall that when demand exceeds generation capacity, the alternating current frequency decreases. Grid frequency support can be provided by energy storage systems so that they can mitigate any sudden, transient reductions in frequency. Grid angular stability can be maintained via stored energy by injecting power at the correct time to dampen oscillations due to instability in the grid; for example, due to generators that become out of synchronization due to the sudden loss of a transmission line as the result of an electrical fault. Energy storage systems can also act as spinning reserve; that is, a fully operational source of power that has not yet been committed to produce power.

Clearly, if enough stored energy is available, power reliability can be improved by being able to use the stored energy to provide power during power outages. The term "ride-through" refers to the ability of a power system device, such as a generator, to remain connected to the grid through an abnormal occurrence or disturbance. As one example, stored energy could provide voltage support during a voltage sag, allowing a generator to ride-through. Otherwise, many devices may be required to disconnect temporarily from the grid, particularly during low voltage or faults.

Power quality can be improved through the use of stored energy as well. There are many sources of "noise" that impact the shape of the voltage and current waveforms, such as power factor, transient instability, flicker, voltage sags or swells, and harmonics. Stored energy can be injected to correct these problems. Finally, unbalanced load compensation can be accomplished by stored energy systems. Three-phase systems are designed assuming all three phases are equal in magnitude and equally offset in phase. If the phases become temporarily unbalanced, stored energy can be used to bring the phases back into balance.

2.5 Summary

Generation before the smart grid era has been dominated by large, centralized power plants to gain economy of scale. The bulk of the power grid has been designed around this architecture. For example, large generators capable of synchronizing among themselves has led to a focus on understanding how large power systems synchronize, and the rest of the power grid can synchronize to the large power generation systems. The fundamentals of alternating current power were covered, which carries with it the need to understand and manage reactive power. Large, centralized generation has driven the evolution of its corresponding communication systems. They tend to be rather old-fashioned but simple and reliable, hierarchical, centralized, SCADA systems. Finally, this section concluded with a review of pre-smart-grid era energy storage systems. An understanding of pre-smart-grid power generation is useful in order to gain insight into where the grid has come from, what some of the archaic terminology means, and where generation is likely to be headed in the future.

Chapter 3 covers the transmission system before the smart grid. The purpose and benefits of the power transmission system are reviewed as well as communication typically used with the transmission system. It is important to understand where progression of the transmission system before considering how it will evolve in the future. Power grid system components that play a role in the transmission system and throughout the power grid are introduced, including the control center, transformers, capacitors, relays, substations, and inverters; the better one understands the needs and requirements of these devices for communication, the better the communication network can be designed and implemented. Power system concepts such as frequency analysis, the phasor, and the per-unit system are introduced. Power systems control has typically been designed to sense and control locally; explicit communication is avoided to maximize reliability. Networked control is introduced and evaluated as a potential communication alternative for control. A description of the challenges in the transmission system, such synchronization, reducing power loss, and mitigating the impact of geomagnetic storms is given. The promise of wireless power transmission is introduced, to be picked up in more detail in later chapters. Overall, Chapter 3 provides information prerequisite to understanding smart grid.

2.6 Exercises

Exercise 2.1 Types of Generators

1. What is the difference between a synchronous and asynchronous generator?
2. Which type of generator would be best for a wind turbine and why?

Exercise 2.2 Synchronization

1. How do multiple generators maintain synchronization with one another?
2. Does the answer to the above question involve a form of communication between generators?
3. What happens if the generators become significantly out-of-synchronization with one another?

Exercise 2.3 Generation Control

1. How is generator power controlled? What communications are involved?

Exercise 2.4 Frequency

1. Describe what happens when generated power falls below demand?
2. Describe what happens when generated power exceeds demand?

Exercise 2.5 Generator Synchronization

1. What is circulating current?
2. Why is circulating current undesirable?
3. How can circulating current be minimized?

Exercise 2.6 Types of Power

1. Explain the relationship among real, apparent, and reactive power.

Exercise 2.7 Power Generation Networks

1. What type of communication network is used in power plants?
2. How do power plant communication networks differ from the Internet?

Exercise 2.8 Fieldbus Networks

1. What changes need to be made to classical Ethernet in order to enable it to operate as a fieldbus network?
2. What is the difference between the classical Ethernet bus architecture and the switched Ethernet architecture?
3. Was the development of switched Ethernet a solution that solved the problem of allowing Ethernet to implement a fieldbus? If so, why? If not, why not?

Exercise 2.9 Peukert's Law

Assume a battery has a Peukert constant of 1.2 and is rated at 100 Ah when discharged at a rate that will fully discharge the battery in 20 h.

1. If the battery is discharged at a rate of 5 A, how long would it take to fully discharge?
2. If the battery is discharged at a rate of 10 A, how long would it take to fully discharge?

Exercise 2.10 Energy Storage

1. Which energy storage mechanisms would be suitable for mitigating sudden, transient power quality problems for applications that are extremely sensitive to power quality problems?

Exercise 2.11 Distributed Generation

DG suggests construction of more, smaller decentralized electric power generators.

1. What is the role of the transmission system in a highly DG environment?

Exercise 2.12 Microgrid

1. How does a microgrid know how to synchronize its alternating current with the main power grid?

3

Transmission

> Like a flash of lightning and in an instant the truth was revealed. I drew with a stick on the sand the diagrams of my motor. A thousand secrets of nature which I might have stumbled upon accidentally I would have given for that one which I had wrestled from her against all odds and at the peril of my existence.
>
> —Nikola Tesla

3.1 Introduction

This chapter covers the process of transporting electric power from the generator, discussed in Chapter 2, toward its path that ultimately leads to the consumer. This is the first step in terms of power transport within the power grid; many fundamental components and concepts used throughout the book are first introduced in this chapter. Remember that this section of the book is an introduction to the power grid as it existed prior to the notion of a smart grid. The transmission system takes large amounts of bulk power from the generator, typically a large, centralized generator in a power plant located in a less-populated area, and transports the power, typically over long distances, to a distribution system located near the consumer in more densely populated areas.

As we continue, keep in mind the relationships between the power grid and communication. One way to classify communication networks is to divide them into field-area networks (FANs), wide-area networks (WANs), metropolitan-area networks (MANs), and home-area networks (HANs), as shown in Table 3.1. One way to think about how communication networks are classified is to consider the manner in which power and energy are utilized in each type of network. The FAN, discussed in Chapter 2, is a network that applies power and energy toward enabling fast, deterministic communication and control over a large factory setting, such as a power plant. A WAN is a communication network whose power is applied toward transporting information over relatively long distances. Thus, a WAN applies well to the long-distance electric power transport characteristic of the transmission system. In a MAN, the area of coverage is roughly that of a city; thus, it is most amenable to an electric distribution system. In a MAN, the network's power and energy are applied both to conveying information over a distance and to handling contention among multiple nodes. Finally, the consumer is often

Smart Grid: Communication-Enabled Intelligence for the Electric Power Grid, First Edition. Stephen F. Bush.

Table 3.1 Network architectures and their corresponding suitability within the power grid.

Network type	Power grid system	Characteristic
FAN	Plant/microgrid	Maintaining determinism for control
WAN	Transmission	Long distance
MAN	Distribution	Moderate distance and density of nodes
HAN	Consumption	Short distance within small area

viewed as operating in an HAN environment; that is, controlling energy within the home. Here, one anticipates a relatively small area with many network devices.

While the above correspondence between grid component and network type may sound like a nice, clean, and well-defined architecture, it is less than ideal. This is because precise definitions of FAN, WAN, MAN, and HAN are lacking and there are many possible technologies within each network type. In addition, there are some network technologies that span multiple network types. However, even though it is oversimplistic, Table 3.1 can be used as a rough initial guideline as we proceed. Since this chapter covers transmission, it should be intuitive that the type of communication required for a transmission system that carries large amounts of power over long distances is a WAN.

Next, basic, tangible power grid system components that play a role in the transmission system and throughout the power grid are introduced, including control centers, transformers, capacitors, relays, substations, and inverters. In the past, each of these components have typically been designed to sense and be controlled locally; their connection via a networked control system is typically considered an aspect of the smart grid. Keep in mind that the physical composition of these components is undergoing change; high-power solid-state device physics are being developed to replace many of these relatively simple mechanical components. However, it is valuable to understand these devices as they exist in the power grid today because much of their basic function and terminology will be carried into the future. After introducing basic power grid components, the chapter moves on to explain analytical techniques involving these components. These are pre-smart-grid analytical techniques. It is important to understand these analyses, which are required for operation of the power grid and how it will evolve, because communication is required to provide much of the data input to these algorithms, and the results from these algorithms may need to be communicated to remote locations. The better one understands the need and requirements for communication, the better the communication network can be designed and implemented. This also provides a foundation for understanding the evolution of the power grid and the supporting communications that will evolve with it. Concepts are presented assuming no prior power system background and include simple frequency analysis, the pu system, and phasor notation. Then discussion proceeds to symmetrical component analysis, power flow analysis, and fault analysis. Finally, state estimation and the concepts behind flexible alternating current transmission systems are introduced. Next, we look at challenges involved in the transmission system. The basic challenge, as noted earlier, revolves around the most efficient transport of large amounts of power over long distances. As explained later, high-voltage, low-current, alternating current transmission is often optimal over such long distances. However, high-voltage direct-current transmission also has benefits that will be discussed in this chapter. Reactive power, explained later in this chapter, is another aspect of power transport that must be taken into account, and its management plays a role in the need for communication and control. Long transmission lines are essentially a large antenna capable of generating current induced by geomagnetic storms; these unexpected currents pose another challenge to power transmission. Yet another challenge is that interconnected components of the power grid must remain synchronized: the current in each component must alternate at precisely the same frequency and in phase with one another. Maintaining synchronization over large areas is another challenge. This book is focused on the fundamental physics of the power grid and communications, not on financial or economic aspects. However, a brief discussion of the power transmission market is included, because the market has interactions with power grid communication. The last topic in this chapter is wireless power transmission. While this may appear to be a futuristic concept, the idea has a long and fascinating history dating back to at least Nikola Tesla in the 1890s. This topic is briefly introduced here and covered in detail in Section 8.4. The chapter ends, like all chapters, with a brief summary of the main points of the material covered in this chapter on the classical power transmission system. It is called "classical" because it refers to the power grid architecture before the advent of the smart grid. As the smart grid develops, the notion of many, smaller renewable energy sources may begin to supersede the current implementation of relatively few, large centralized generation plants. This means that, in the future, the need for large, long-distance power transport may be greatly reduced. Exercises at the end of the chapter are worth reviewing not only to review your understanding of the chapter, but also to help stimulate creativity and new ideas using the ideas presented.

Transmission demonstrates the power of networking, or more specifically the networking of power. In the 1880s, during the early days of electric power, generators were connected directly to specific loads, forming many independent, isolated power systems, not too different from the microgrid concept evolving today. After the early 1900s, the trend began of interconnecting these independent systems. There are many benefits of doing this, as described earlier; for example, the ability to share generators and the ability to gain diversity in loads. Today, there are only three large independent synchronous systems remaining: the Western Interconnect, the Eastern Interconnect, and the Texas Interconnect. Western Europe has its own interconnect system. Economies of exchange – as opposed to economy of scale – leverage the ability for power to be bought and sold, which requires a highly interconnected system. The main benefits of highly interconnected systems are economy of scale, improvement in the load factor, and the ability to pool generation reserves.

An alternative term for interconnection is the "wide-area synchronous grid" or "synchronous zone." These are regions in which power transmission is interconnected and operating synchronously; that is, at the same frequency due to their direct interconnectivity and circulating currents and magnetic pull on one another. Note that this is quite different from interconnects or synchronous zones that are connected via high-voltage direct-current links, because such connections are not synchronized; the alternating current is converted to direct current, transported, and then converted back to alternating current at whatever frequency is required. More will be discussed about high-voltage direct-current interconnections in Section 3.4.4. There are also other methods of transferring power between asynchronously operating interconnects or synchronous zones; for example, the variable-frequency transformer (VFT), which will be discussed in more detail next.

The VFT is a relatively recent invention, first commercially operated in 2004. It is somewhat of a brute-force approach to interconnecting asynchronous interconnect systems. The VFT is essentially a generator, with a three-phase rotor winding connected to one of the interconnects and the three-phase stator winding connected to the other interconnect. If the interconnects happen to be synchronous with each other (that is, in-phase with one another), then the rotor would remain stationary. In this case, the generator becomes a large transformer with the rotor as the primary winding and the stator as the secondary winding. If the interconnects are out of phase with one another, then the rotor will move to compensate for the difference in phase between the interconnects while the system continues to act as a transformer.

Returning to the notion of interconnects, the largest interconnect or synchronous zone in terms of generating capacity appears to be either the North American Interconnect or the synchronous grid of continental Europe; both interconnects appear to generate and manage approximately the same power capacity. The largest interconnect in terms of land coverage appears to be the interconnect serving the countries of the former Soviet Union. China plans to interconnect all its existing grids by the year 2020, which would be, by far, the largest single interconnect when the project is completed.

Recall that economy of scale simply means that it is cheaper and more efficient to build one large generator rather than multiple smaller ones. The costs for construction of a large generator are generally fixed; they depend not upon the final megawatt output of the generator, but rather the fixed costs of laying a foundation and utilizing heavy equipment. Operational costs are similarly relatively fixed for any large generator. Thus, if the costs are relatively fixed regardless of size, then it makes sense to build the largest, highest output, generator possible. There was a trend toward construction of larger and larger generators until around the late 1960s. It seems that the upper limit of economy of scale was reached around that time, and there appears to be some evidence that the limit was not anticipated to arrive so early. It was expected that building yet larger generators would continue to push down the cost of energy; unfortunately, it did not. However, the large generation systems that exist require a significant power transmission and distribution network in order to reach the widely dispersed customer base that they service.

As discussed earlier, the load factor is the ratio of the average to maximum power consumption for any or all loads. Utilities must build their systems to anticipate the highest possible demand; that is, maximum instantaneous power consumption. However, they are only paid for the actual power consumed. A low load factor, meaning a relatively high maximum and low average consumption rate, means that a large, expensive generator is required to be available to meet the maximum demand, which may only last for a short period of time. If the load factor is low, then much of the time large expensive generation capacity is idle and not "generating" money for the utility. A large, highly connected power transmission and distribution network helps to ensure the load factor will be higher than it would be otherwise. This is due to the statistics of having a large customer base with a mix of many different usage patterns; that is, maximizing the entropy of usage so that the probability of usage is evenly distributed.

Finally, reliability can be increased as the power network becomes more highly interconnected. If a generator or power line goes down, there will be alternate generators and paths to take over for the failed component. At the power distribution level, an electrical fault on any segment of the power network can be isolated from the power grid and customers can be reconnected to the same feeder or to a different feeder through new paths in the power network. This is a very familiar scenario in communication networking. The ability to route packets around links that are down or unavailable is central to the notion of ad hoc and mesh networking. However, it is not only a reliability benefit, but a business benefit, to have a highly interconnected power network. It becomes easier for power generators to sell their electricity not only to consumers, but also to other utilities. It also offers more alternatives for customers seeking to purchase their power from cheaper or perhaps "greener" power producers.

There are, however, drawbacks to larger power networks; namely, there is more I^2R power loss due to the more numerous, longer power lines of a larger network. As previously mentioned, increasing the voltage reduces the current and the corresponding power loss, thus driving up the incentive to increase voltage for long-distance power transmission. Long transmission lines introduce stability issues. Long lines between generators introduce circulating current that must travel long distances and introduces delay in the potential electromagnetic synchronization among generators. Longer lines are also more susceptible to geomagnetic storms, discussed in Section 15.2.

3.2 Basic Power Grid Components

This section introduces components of the traditional power grid. There are several reasons for doing this. First, the non-power system engineer, in particular the communications engineer, will likely require an introduction to these components. Second, keep in mind that this chapter is about the traditional power grid, but with a focus as to how the traditional power grid will transition into the so-called smart grid. Thus, these components are discussed with an eye toward how they relate to communication and how they are likely to transition within the smart grid. More detail on high-power solid-state electronics for the smart grid is given in Section 14; this section serves as an introduction to that topic so that the reader will have the fundamental information necessary for the remainder of the book.

As we introduce a selected set of components of the power grid, keep in mind that these are only a select set of components; explaining all components would require an entire book of its own. Also, the goal is to keep in mind how these components relate to communication and how they will evolve within the smart grid. Another aspect to keep in mind is to what degree the component could undergo fundamental change within the smart grid. In other words, does the smart grid version of the component basically remain the same, but with the addition of some sensing and control added to it, or is there the prospect for a fundamentally different type of device in the near future?

3.2.1 Control Centers

We begin with the control center, which, as the name implies, is the highest level monitoring and control location for large sections of the power grid (Zhang *et al.*, 2010). This is where the "big picture" of the health of the power grid comes together and also where strategic decisions are made. As in any industrial control process, the design of power grid operations has to balance the requirements of control with the latency of communication. The remaining chapters of this book will dive into the details of this trade-off, but it is important to keep this general concept in mind. Control decisions that have to made quickly and can be made locally, without communication, have always been preferable in general. Droop control, explained in detail in Section 5.5.1, is one example. The idea is that key power control parameters can be measured locally and control taken locally to react to changing conditions. When this is done throughout the power grid, a self-organizing distributed system is created. Thus, the idea that the control center actually controls every detail of the power grid is not true; rather, the control center presents the big picture and makes higher level strategic decisions because it has a wider

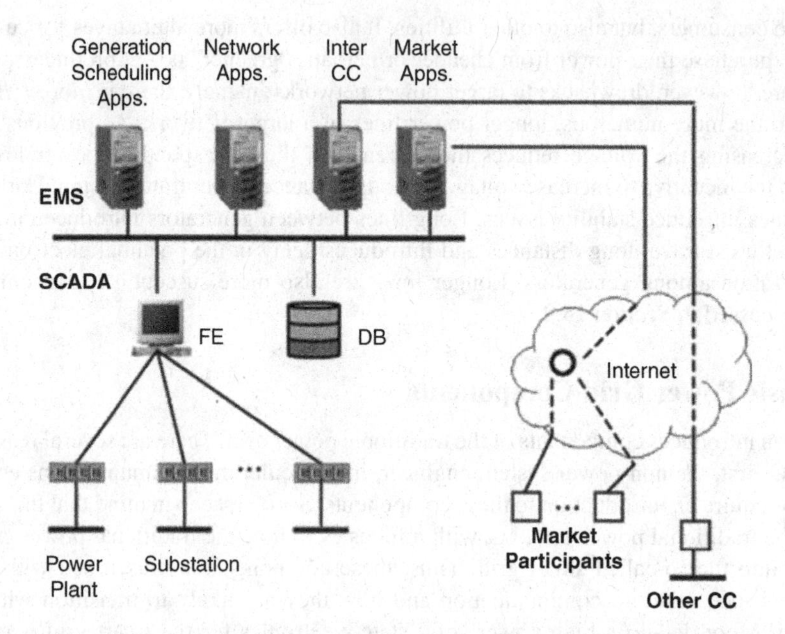

Figure 3.1 Typical pre-smart-grid control center requires communication among components of the EMS, SCADA process control systems described in Section-2.3, energy market participants, and other CCs. Note that the thick horizontal bars represent LANs. All other lines represent point-to-point links. Source: Wu *et al.*, 2005. Reproduced with permission of IEEE.

view of the grid in space and time and because it keeps a history of key information from across the power grid. As wide-area communication becomes more ubiquitous, the control center may be able to provide higher resolution and more direct control. The real-time use of synchrophasors at the control center is one example; synchrophasors are described in detail in Section 13.2.

Figure 3.1 shows a sketch of the communications that can be found in a control center. The EMS is a critical component used by operators of electric utility grids to monitor, control, and optimize the performance of the generation and transmission system. This is done utilizing the communication network and involves scheduling generators to meet market demand. The control center also requires communication connections to generators and substations. There are also potential connections to other power grid control centers.

3.2.2 *Transformers*

Transformers are one of the most basic and ubiquitous devices within the power grid other than power lines themselves. As discussed in prior historical sections, transformers were a critical invention that enabled long-distance transportation of alternating current; the ability to reduce current and increase voltage while maintaining the same amount of power significantly reduces I^2R line losses. A single power distribution network can have hundreds of distribution transformers. The transformer can serve not only to step up and step down voltages over large magnitudes for power transport, but they can also serve as voltage regulators, adjusting the voltage over smaller magnitudes or maintain required constant voltage levels. In fact, in conjunction with the evolution of the smart grid, the transformer, like almost all other power system devices, is becoming smarter and gaining more features and capabilities, although at this time these features are still in the early stages of development.

Transformers are the work horse of the power grid. They transform electric power into forms optimized for transmission and use wherever required. However, they are not always the static, double-coiled induction transformer that most people first imagine. They have incorporated some amount of dynamic behavior and, with solid-state transformers covered in Section 14, are becoming extremely dynamic and capable of many functions for the power grid. A line tap transformer is illustrated in Figure 3.2. The line tap is a mechanical switch that can change the contacts, or taps, within the coil so that the relative length of a coil can be dynamically adjusted. This allows the transformer to regulate the degree to which voltage is increased or decreased in order control the voltage. Many such line tap transformers are still in the field and

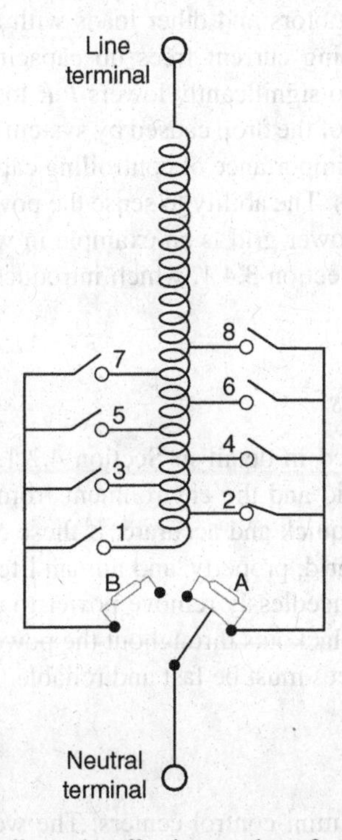

Figure 3.2 A line tap transformer allows the effective number of windings to be varied by mechanically contacting the coil at different locations along its length. Changing the tap while current is flowing requires careful design because of arcing. The components in the figure labeled A and B are large resistors that are contacted by the rotary switch near the bottom, which is driven by a motor. Taps 1 through 8 are changed by opening or closing taps on the opposite side of the rotary switch position before the rotary switch changes sides to close the connection for the new tap switch position. The goal of this multistep switching process is to avoid arcing. However, switching frequently may still shorten the life of the transformer; there is concern that the smart grid may attempt to switch more frequently as it tries to maintain accurate voltage regulation. Source: BillC at the English language Wikipedia [GFDL (www.gnu.org/copyleft/fdl.html) or CC-BY-SA-3.0 (http://creativecommons.org/licenses/by-sa/3.0/)], via Wikimedia Commons.

have a long lifetime remaining before an upgrade is likely to occur to solid-state transformers. An interesting concern is that as the power grid becomes "smarter" it will react more quickly to adjust voltage and cause significant wear on the mechanical tap switches of these transformers.

The simplest view of a transformer is as a pair of windings or coils such that the number of individual windings in coil 1 is N_1 and in coil 2 is N_2. An alternating current flowing through the second coil at voltage V_2 will induce a voltage in coil 1 V_1 that is proportional to the turns ratio according to

$$V_1 = \frac{N_1}{N_2} V_2. \tag{3.1}$$

3.2.3 *Capacitor Banks*

A capacitor bank is a large row of capacitors used to improve power quality in the grid. By canceling the reactive power to motors and other loads with a low power factor, capacitors decrease the line current. Reducing current frees up capacity; the same circuit can serve more loads. Reducing current also significantly lowers I^2R losses. Capacitors also provide a voltage boost, which cancels part of the drop caused by system loads. Switched capacitors can regulate voltage on a circuit. The importance of controlling capacitance will become apparent throughout the remaining chapters. The ability to sense the power factor and insert the correct amount of capacitance into the power grid is an example in which communication can play a significant role. For example, Section 3.4.1, which introduces integrated volt-VAr control, covers this aspect in detail.

3.2.4 *Relays and Reclosers*

Relays and reclosers are discussed in detail in Section 4.2.1. These are protection devices designed to protect the power grid and the environment from the impact of excessive fault current. Their operation must be quick and accurate. If these devices do not operate quickly, significant damage to the power grid, property, and human life can occur. On the other hand, if they operate too quickly, they needlessly remove power to customers or create instability, potentially leading to cascading blackouts throughout the power grid. Thus, any communication in support of protection devices must be fast and reliable.

3.2.5 *Substations*

Substations might be considered mini-control centers. The word substation comes from the days before the distribution system became part of the power grid. The first substations were connected to only one power station where generators were housed and were considered subsidiaries of the local power station. Currently, they house equipment, typically switches, protection and control equipment and one or more transformers at the intersection of different transmission areas or between transmission and distribution systems. A transmission substation connects two or more transmission lines. The simplest case is where all transmission lines have the same voltage. In such cases, the substation contains high-voltage switches that allow lines to be connected or isolated for fault clearance or maintenance.

A distribution substation transfers power from the transmission system to the distribution system. It is uneconomical to directly connect electric power consumers to the main

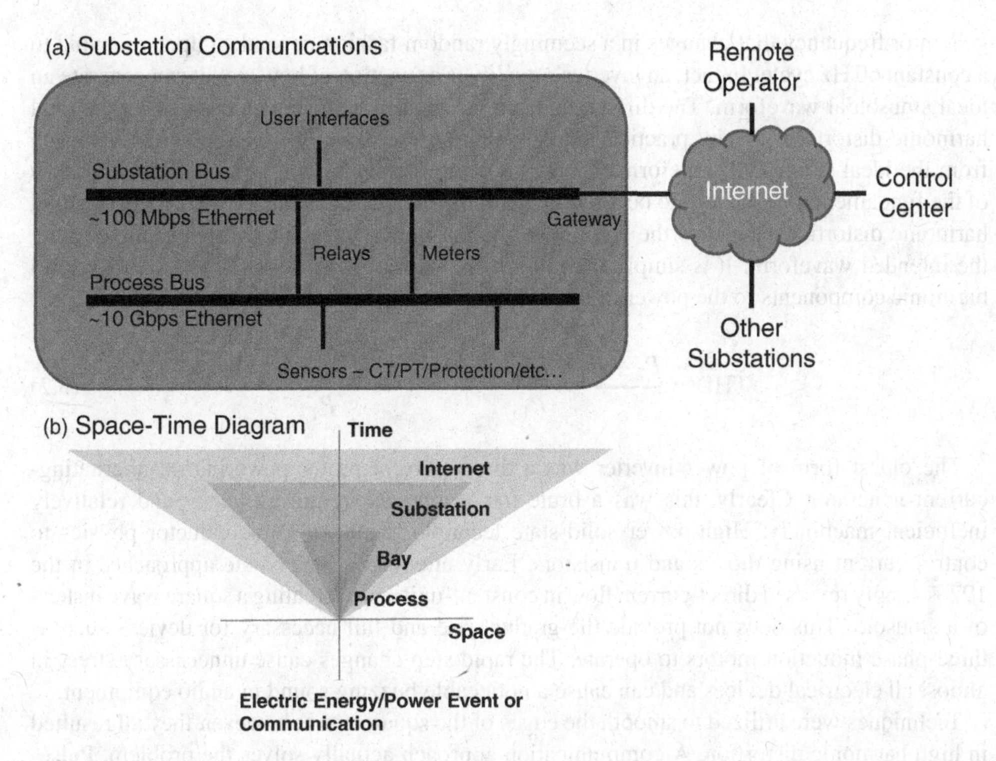

Figure 3.3 This is a more detailed view of the LANs inside a substation. Note that the process bus runs at a faster rate than the substation bus in (a) allowing fast response for time-critical operations, while the slower substation bus handles oversight and more general interface and control activity. In (b), a space-time diagram illustrates the maximum distance communication reaches versus time. The communication space-time profile must match or exceed that of the power event being monitored or controlled. The process bus is represented by the smallest inverted triangle while the substation bus and Internet are the slowest, but reach longer distances, as represented by the largest inverted triangles.

transmission network unless they use large amounts of power, so the distribution station reduces voltage to a value suitable for local distribution. In DG projects such as a wind farm, a collector substation may be required. A collector substation somewhat resembles a distribution substation although power flow is in the opposite direction, from many wind turbines to the transmission grid. An example of a substation is shown in Figure 3.3. Notice from Figure 3.3 that there is a significant amount of communication that takes place within the substation. In fact, the substation is one of the first places where advances in digital networking started.

3.2.6 Inverters

Inverters convert direct-current power to alternating-current power. This means that they must transform the steady, unidirectional, direct current flow into a bidirectional, sinusoidal waveform. Also, since many direct-current sources are relatively low power and low voltage, inverters need to increase the voltage to the standard 120 V. Inverters are also used in situations where power quality is poor in order to improve quality. An example is to convert wind generator frequency that changes in a seemingly random fashion dependent upon the wind to a constant 60 Hz cycle. In fact, an inverter's quality is a function of how well it can generate an ideal sinusoidal waveform. The difference from the ideal waveform is measured by the total harmonic distortion. In any practical implementation there will always be some distortion from the ideal sinusoidal waveform. There is a natural tendency for harmonics, or multiples of the fundamental frequency, to occur along with the intended generated waveform. The total harmonic distortion measures the amount of the harmonic waves inadvertently mixed with the intended waveform. It is simply the ratio of the sum of the powers (RMS values) of all harmonic components to the power of the fundamental frequency, as shown by

$$\mathrm{THD} = \frac{P_2 + P_3 + P_4 + \cdots + P_\infty}{P_1} = \frac{\sum_{n=2}^{\infty} P_n}{P_1}. \tag{3.2}$$

The oldest form of power inverter was a direct-current motor powering an alternating-current generator. Clearly, this was a brute-force approach requiring heavy and relatively inefficient machinery. High-power solid-state technology allows semiconductor physics to control current using diodes and transistors. Early attempts at solid-state approaches in the 1970s simply reversed direct-current flow in constant-unit steps, creating a square wave instead of a sinusoid. This does not provide the gradual rise and fall necessary for devices such as three-phase induction motors to operate. The rapid step changes cause unnecessary stress in almost all electrical devices and can cause a noticeable buzzing sound in audio equipment.

Techniques were utilized to smooth the edges of the square wave; however, they all resulted in high harmonic distortion. A communication approach actually solves the problem. Pulse-code modulation is used to send a sinusoidal waveform over a digital communication link. An ideal sinusoidal waveform can be sampled at least at the Nyquist frequency; that is, at least half the frequency of the ideal waveform to be transmitted. The digital samples can be transmitted and an approximate ideal waveform reconstructed by playing and holding the sampled values for a short duration; the reconstructed waveform is actually comprised of many, very tiny, step values. Pulse-width modulation accomplishes a similar result in power systems. A high-power semiconducting device can sample the original waveform and generate a series of short pulses at each voltage value, creating a series of tiny steps that approximate the original ideal waveform. Pulse-width modulation is a simpler approach. Here, there are only two voltage output values: the positive maximum voltage value and the negative maximum voltage value. The difference that is used to approximate the waveform is the length of time the voltage is in the "on" position. As the original sinusoidal voltage value rises, the maximum voltage step values are held in the on-state longer. As sinusoidal waveform values decrease, the maximum voltage values are turned on for shorter durations. While the resulting voltage waveform looks very little like a sinusoid, the power that it carries, when averaged over time, does approximate a sinusoid. The narrow pulses carry little power, while the larger pulses carry more power.

3.3 Classical Power Grid Analytical Techniques

This section introduces common power grid analytical techniques used prior to the smart grid. The purpose is to understand what types of computation and intelligence might be needed within the smart grid and what requirements may be needed for communication.

3.3.1 Frequency Analysis

The sudden addition or removal of large amounts of load or generation in a power system leads to changes in frequency. For example, a generator trip causes a decline in frequency, whereas load shedding results in an increase in frequency. The change in frequency is proportional to the size of the tripped generator or the amount of load shed. These changes propagate in both space and time throughout the grid (see part (b) of Figure 3.3 for a space-time diagram). Devices known as fault detection, isolation, and restorations (FDIRs) have been used within the power grid to measure and record such changes in frequency. By correlating the results with a map of the power grid, interesting results regarding the spread of events throughout the power grid have been observed. Because FDIR results are essentially a frequency reading with a global positioning system (GPS) timestamp, it is basically a synchrophasor. An interesting example of a frequency event is shown in Figure 3.4, where lower frequencies are represented by darker shading and higher frequencies by lighter shading.

Figure 3.5 shows a specific frequency reading in which a significant event impacting the frequency has occurred. Frequency is represented along the *y*-axis and time along the *x*-axis. Clearly, a sudden event occurred in which the power supplied was not able to meet demand; however, the system was able to gradually restore power to normal.

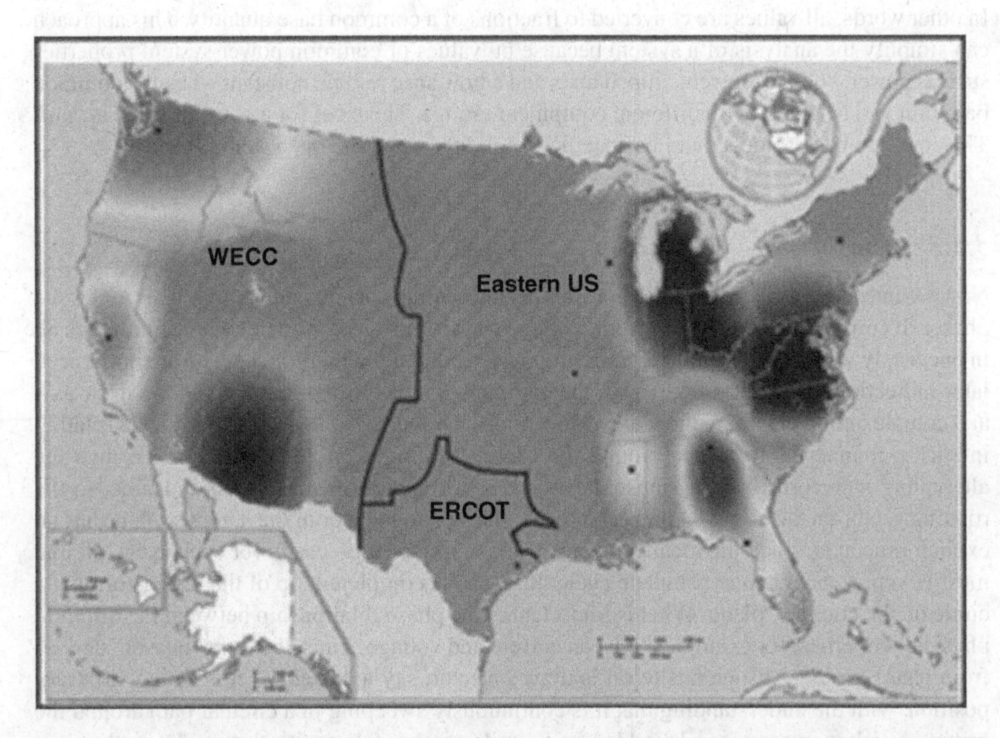

Figure 3.4 A frequency map illustrates the relative alternating current frequencies over a wide area. A darker shading indicates a lower frequency, and thus indicates regions where load may be exceeding supply. Communication plays a critical role in collecting data over a wide area in order to keep the map updated. Source: Zhong *et al.*, 2005. Reproduced with permission of IEEE.

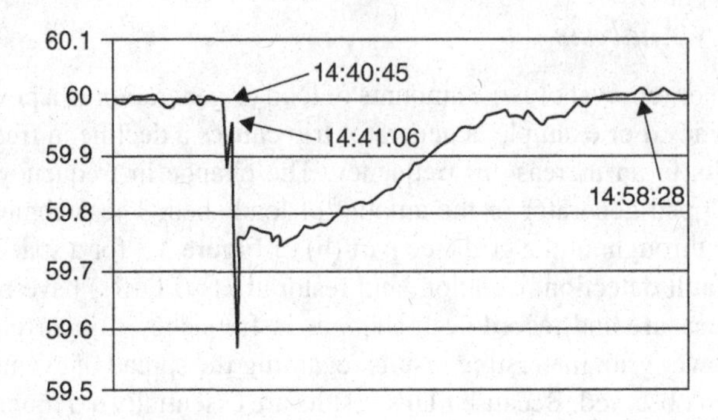

Figure 3.5 The details of a particular event are captured in a frequency-versus-time plot. The first two times indicated are events where relay trips occurred. The last indicated time is restoration of the system. Source: Zhong *et al.*, 2005. Reproduced with permission of IEEE.

3.3.2 Per-Unit System

A significant difference from typical communication analyses is that the pu system is widely used in power systems. It is simply the expression of power system values as normalized values. In other words, all values are converted to fractions of a common base quantity. This approach can simplify the analysis of a system because pu values of common power system properties such as power, voltage, current, impedance, and admittance remain constant when the common base unit varies widely with different equipment ratings. The label for a per-unit value is "pu." The common base value should be specified explicitly when the pu system is used.

3.3.3 Phasors

Now we introduce a useful tool for representing alternating current and voltage. This tool is the phasor. It combines alternating current, voltage waveforms, vectors, and complex numbers all in one handy visual representation. Synchrophasors build upon this concept and are introduced later in Section 13.2. Let us start with the complex plane with real numbers along the x-axis and complex numbers along the y-axis. Now think of a vector located at the origin and rotating in such a manner as to sweep around the origin forming a circle. The goal is to map the alternating waveform from a sinusoidal wave, as a function of magnitude and time, to the rotating vector on the complex plane. Note that the rotating vector in the complex plane has no explicit time axis; since the idealized waveform is replicated precisely for each cycle, all that must be represented is one complete cycle; that is, one complete loop of the vector forming a circle on the complex plane. What is important is the phase relationship between the different physical properties; for example, between current and voltage. Thus, the term "phasor" derives from phase vector. All one has to do is draw a vector, say a current vector, in any arbitrary position, with the understanding that it is continuously sweeping in a circular path around the origin. A voltage vector can be added in a similar manner, but with the condition that it is positioned at a specific angle relative to the current. This angle is the power factor ϕ. If this angle were zero, $\cos(\phi)$ would be one, and the power factor would be one; the angles would be located directly on top of one another.

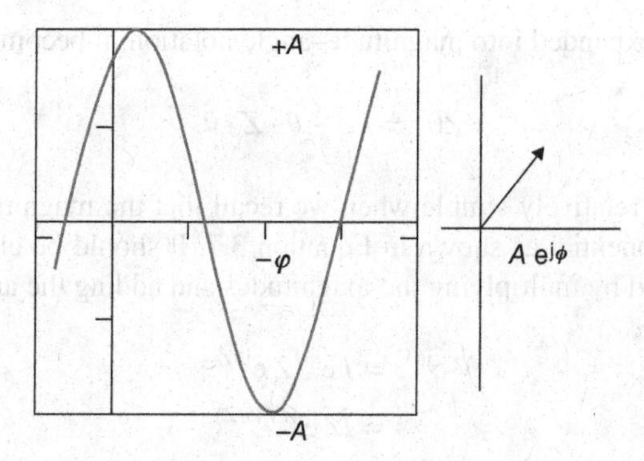

Figure 3.6 A sinusoidal waveform on the left, with amplitude A and angle ϕ, is shown with its corresponding phasor representation on the right. Source: Martin, 2010. Reproduced with permission of IEEE.

The Euler equation

$$e^{j\phi} = \cos(\phi) + j\sin(\phi) \tag{3.3}$$

indicates how the phasor is mapped to the complex exponential. The two terms of the Euler equation can be thought of as projecting the value onto the real number axis of the complex plane with the $\cos(\phi)$ term and the imaginary axis with the $j\sin(\phi)$ term. As ϕ increases, the real and imaginary components become alternately larger and smaller so as to map out a path around the circle formed by a vector represented by $e^{j\phi}$.

Figure 3.6 illustrates a sinusoidal wave and its associated phasor, where the amplitude is A and the angle is ϕ. We still have not explained how a wave, with real axes values, gets mapped to a complex plane, in which one axis is real and the other imaginary. This is an artifice; in other words, the imaginary axis is a convenience used along with the Euler equation (3.3) to compactly represent a cycle. The imaginary axis is not attempting to represent reactance, for example, which may be a cause for some confusion.

Note that because the phasor representation does not include an explicit notion of time, all phasors plotted on the same graph are assumed to be operating at the same frequency ω; the only difference between phasors may be their magnitude or their relative phase with one another. Phasor notation thus has only two components: amplitude and phase. The amplitude can be expressed either as the maximum value or the RMS value. The phase angle is represented by $\angle\phi$. Thus, an example of complete phasor notation would be $\vec{V} = V_{\text{rms}}\angle 0°$.

A feature that makes phasor notation even more useful is that impedance can be represented consistently with power, current, and voltage phasors as well. Recall that impedance **Z** is comprised of a resistive (real) and a reactive (imaginary) component. When represented on the complex plane, impedance has a magnitude and angle. Because this angle is the power factor, it is a consistent representation with other phasors. Ohm's law in phasor notation is

$$\mathbf{V} = \mathbf{IZ}. \tag{3.4}$$

When this equation is expanded into magnitude–angle notation, it becomes

$$V\angle 0^\circ = I\angle -\theta \cdot Z\angle\theta. \tag{3.5}$$

Phasor arithmetic is relatively simple when we recall that the magnitude and angle come from the complex exponential as shown in Equation 3.7. It should be clear that multiplying phasors is accomplished by multiplying the magnitudes and adding the angles (exponents):

$$\begin{aligned} V\,\mathrm{e}^{\mathrm{j}\theta V} &= I\,\mathrm{e}^{\mathrm{j}\theta I} Z\,\mathrm{e}^{\mathrm{j}\theta Z} \\ &= IZ\,\mathrm{e}^{\mathrm{j}(\theta I+\theta Z)}. \end{aligned} \tag{3.6}$$

Similarly, dividing phasors requires dividing magnitudes and subtracting the angles (exponents).

Now we can use phasors to derive power as shown in Equation 3.7. **I** and **V** are the current and voltage phasors, which yield a complex power phasor **S**. The asterisk indicates taking the complex conjugate of the complex current. As the reader may recall, the complex conjugate results in simply changing the sign of the imaginary component of a complex number. This is done, because, by convention, the current is generally lagging the voltage due to inductive loads in the system and is assigned a negative phase with respect to the voltage. Taking the conjugate changes the current's phase to a positive value, as shown in Equation 3.10, where the equivalent magnitude–phase angle multiplication is worked out in detail.

$$\mathbf{S} = \mathbf{I}^*\mathbf{V} \tag{3.7}$$

$$S = I_{\mathrm{rms}}\angle -\theta^* \cdot V_{\mathrm{rms}}\angle 0^\circ \tag{3.8}$$

$$= I_{\mathrm{rms}}\angle\theta \cdot V_{\mathrm{rms}}\angle 0^\circ \tag{3.9}$$

$$= I_{\mathrm{rms}} V_{\mathrm{rms}}\angle\theta. \tag{3.10}$$

3.3.4 Power Flow Analysis

The goal of power flow analysis is to determine the amount and characteristics of power transmitted through each path within a network of power lines. Each node within the power network is commonly called a bus, which comes from the word busbar, a wide strip of conductive material, typically metal, from which additional power lines can be connected in order for power to branch out to other locations. Thus, a bus, being a common conductive point within the power network, has a single common current and voltage. A bus, or node in our network, can have multiple connections with power entering and leaving the node. We can analyze power in terms of current and voltage using

$$S_i = V_i I_i^*, \tag{3.11}$$

where the index i keeps track of which bus is being described. The characteristics of the power that we will analyze include the ability to distinguish between real P and reactive power Q. We can analyze such a network, conceptually, as we would any other electrical circuit, by using Ohm's law and Kirchhoff's laws. Recall from explanations much earlier that the general form of Ohm's law, taking reactance into account, which we must do since we need to distinguish between real and reactive power, is $V = IZ$. Solving for I yields $I = V/Z$. Since we are planning to eventually use matrices to describe the network, it is easier to avoid division by using admittance instead of the impedance. The admittance is defined as $Y = 1/Z$ and we have $I = VY$. Note that the absence of a connection between nodes, which would have been indicated by infinite impedance, is now indicated by zero admittance, which is also more convenient for matrix analysis.

Getting back to the reactance used to distinguish real from reactive power, recall that admittance is of the form $Y = G + jB$, where G is the conductance (the real component related to resistance) and B the susceptance (the imaginary component related to reactance). We can use a matrix **Y** to describe the complete network of admittances, where the rows and columns of the matrix indicate the connections, or in a less esthetically pleasing way use subscripts i and k to explicitly indicate connections from i to k, $y_{ik} = g_{ik} + jb_{ik}$. The node connections are then

$$I_{ik} = V_{ik} y_{ik}. \tag{3.12}$$

Now all we need to do is replace the value for I in Equation 3.12 in the equation for power and add up all the connecting power flows, as shown in Equation 3.13. Remember that we can do this because Equation 3.12 is current and the sum of the current at a point must be zero – the net flow into a point must equal the net flow out of that same point:

$$S_i = V_i I_i^* = V_i \left(\sum_k 1_n y_{ik} V_{ik} \right)^* . \tag{3.13}$$

We can proceed to simply replace the admittance with its expanded form:

$$S_i = V_i \left[\sum_k 1_n (g_{ik} - jb_{ik}) V_{ik} \right]^* . \tag{3.14}$$

Next, we continue expanding the equation by writing the voltages in complex exponential form:

$$S_i = \sum_k 1_n |V_i| |V_{ik}| \, e^{j(\theta_i - \theta_k)} (g_{ik} - jb_{ik}). \tag{3.15}$$

The important point to note in the above equation is that the asterisk, representing the complex conjugate, has been implemented by the phase difference $(\theta_i - \theta_k)$. While the phase difference between current and voltage determines the reactive power, we have represented the current as the product of voltage and admittance.

The next step is to further expand the equation by replacing the complex exponential with its Euler equation form:

$$S_i = \sum_k 1_n |V_i||V_{ik}|[\cos(\theta_i - \theta_k) + \mathrm{j}\sin(\theta_i - \theta_k)](g_{ik} - \mathrm{j}b_{ik}). \tag{3.16}$$

The final step is to perform the multiplication of terms in Equation 3.16 and gather the real and imaginary terms together. The real terms are P and the imaginary terms are Q:

$$P_i = \sum_k 1_n |V_i||V_{ik}|[g_{ik}\cos(\theta_i - \theta_k) + b_{ik}\sin(\theta_i - \theta_k)], \tag{3.17}$$

$$Q_i = \sum_k 1_n |V_i||V_{ik}|[g_{ik}\sin(\theta_i - \theta_k) - b_{ik}\cos(\theta_i - \theta_k)]. \tag{3.18}$$

Now we need to take a step back and think again about why we have derived these equations and how they are used. We started with a network of power flows. The assumption is that some of the variables in this equation are measurable in the actual system and that we then use these equations to derive the remaining variables in order to describe the real and reactive flows throughout the entire network. Recall that each node is a bus in the actual power system. The bus may be directly connected only to loads, in which case it is known as a load bus. A load bus has no directly connected power generator. A bus that has a directly connected generator, or potentially many generators in a DG system, is called a generator bus. One of the buses that has a generator is arbitrarily selected and called the "slack bus."

It is assumed that for all load buses, the real and reactive powers are known; thus, these buses are also called real and reactive power (PQ) buses or PQ nodes, since P and Q are assumed to be known. For generators, it is assumed that the amount of real power being generated is known as well as the magnitude of the voltage; thus, these are known as real power and voltage magnitude (PV) buses, or PV nodes. The generator bus that we chose as the slack bus is special; here, we assume that the voltage magnitude and voltage phase are known. Thus, we can see that each load bus, or PQ node, requires a solution for the voltage magnitude and angle, which are assumed to be unknown. For each generator, or PV node, the voltage angle is known and must be found. For the one arbitrary slack bus, everything is assumed to be known and nothing needs to be found. We can use the equations previously derived, as shown in Equations 3.19 and 3.20:

$$0 = -P_i + \sum_{k=1}^{N} |V_i||V_k|(g_{ik}\cos\theta_{ik} + b_{ik}\sin\theta_{ik}), \tag{3.19}$$

$$0 = -Q_i + \sum_{k=1}^{N} |V_i||V_k|(g_{ik}\sin\theta_{ik} - b_{ik}\cos\theta_{ik}). \tag{3.20}$$

These are the real and reactive power balance equations for each PQ or load bus and the real power balance equation for each generator, or PV, bus. Only the real power balance equation is included for a generator bus because the reactive power is not assumed to be known at the generator. Note that including the reactive power balance equation would create an additional unknown variable. Since all the variables are assumed to be known for the slack bus, there are no equations written for the slack bus either.

It is important to note that there is no closed-form solution for these equations. Instead, an iterative technique is used in which a guess is made for the solution and then the results are examined to see how closely the results match the known values. A refinement in the guess is made based upon these results and the process continues until the results converge to a value that is suitably close to the known values.

3.3.4.1 Power Flow Analysis Solution Methods

To start the solution, assume that the unknown values are the voltage angles and magnitudes at every bus except the slack bus. Assume they are all PQ buses. In a *flat start*, assume that all the voltage angles are zero and the magnitudes are all at their nominal values, or 1 pu. These estimated values are entered into the power flow equations with the knowledge that the results will initially be incorrect. That is, the results will not match the observed P and Q values at the slack bus. The solution procedure is to reduce the error by using information from these results to choose a better set of estimates for the next iteration. Standard techniques for doing this are the Newton–Raphson method, the Gauss method, and the Gauss–Seidel method.

No matter which technique is used, the general idea is to determine the amount and direction of the error using the estimated values. This is similar to a sensitivity analysis, because the goal is to determine precisely how sensitive the results are to changes in parameters so that a good estimate can be made for new parameters that will lead to a result that is more consistent with the observed parameters; that is, the slack bus values in our case. The Jacobian matrix for real power P versus voltage angle θ neatly contains the combinations of sensitivities of real power at the slack bus to the voltage angle at each bus:

$$\frac{\partial P}{\partial \theta} = \left\{ \begin{array}{ccc} \frac{\partial P_1}{\partial \theta_1} & \frac{\partial P_1}{\partial \theta_2} & \frac{\partial P_1}{\partial \theta_3} \\ \frac{\partial P_2}{\partial \theta_1} & \frac{\partial P_2}{\partial \theta_2} & \frac{\partial P_2}{\partial \theta_3} \\ \frac{\partial P_3}{\partial \theta_1} & \frac{\partial P_3}{\partial \theta_2} & \frac{\partial P_3}{\partial \theta_3} \end{array} \right. . \tag{3.21}$$

Equation 3.21 is a simplified example showing three buses. There would need to be three more Jacobian matrices; namely, $\frac{\partial P}{\partial V}, \frac{\partial Q}{\partial \theta}, \frac{\partial Q}{\partial V}$. Each of these four Jacobian matrices is combined into one large Jacobian matrix $\mathbf{J}$. $\mathbf{J}$ can essentially be used as a large, complicated derivative. We can let $\mathbf{f}(\mathbf{x})$ be a vector difference between the $P(\theta, V)$, $Q(\theta, V)$ as computed and the actual values. Thus, the goal is to achieve a value as close to zero as possible for $\mathbf{f}(\mathbf{x})$. $\mathbf{f}(\mathbf{x})$ is known as the mismatch equation.

Before continuing, it can help to review some basics using a simple scalar parameter x and a simple generic function f. First, Equation 3.22 should be obvious; adjusting the x by some value is equivalent to moving along the slope of the function $f'(x)$ by that value. If we happen to want a parameter that drives the function to zero, our mismatch equation, then this is shown in Equation 3.23. Finally, we can rearrange Equation 3.23 into Equation 3.24. These basic steps can be followed, conceptually, for a more complex matrix analysis.

$$f(x + \Delta x) = f(x) + f'(x)\Delta x, \tag{3.22}$$

$$0 = f(x) + f'(x)\Delta x, \tag{3.23}$$

$$\Delta x = \frac{-f(x)}{f'(x)}. \tag{3.24}$$

The basic idea is quite simple: to use the function and its derivative to determine the amount of adjustment required to the parameters. If we simplify things by considering all the Vs and θs as a single vector $\mathbf{x}$, with a superscript to indicate the iteration, then $\mathbf{x}^0$ is our initial guess. Equation 3.25 shows conceptually what we would like to do, which is simply to apply Equation 3.24 to our matrices:

$$\Delta\mathbf{x} = \frac{-\mathbf{f}(\mathbf{x})}{\mathbf{J}} = -\mathbf{f}(\mathbf{x})\mathbf{J}^{-1}. \tag{3.25}$$

Having obtained the change in $\mathbf{x}$, this value is simply added to the previous value of $\mathbf{x}$ and the process is repeated until $\mathbf{f}(\mathbf{x})$, the amount of mismatch, approaches close enough to zero to be acceptable. Thus, the values of θ and V have been determined for each PQ bus. For PV buses, the values of θ and Q are determined. Recall that PQ buses are load buses and PV buses are generator buses, where V specifies the voltage at the generator. Thus, from the power flow analysis, the power at the generator and loads is determined and the difference is the power lost. Simply applying Ohm's law allows one to find the power flow on each power line. From the values determined for θ and V and in the power flow analysis, the real and reactive power flows can be determined for each line. The system state has now been fully specified for a given amount of power generation and load demand.

3.3.4.2 Decoupled Power Flow Analysis

Working from two assumptions, we can make some simplifications to the prior analysis. The first assumption is that transmission lines are primarily reactive, rather than resistive. Thus, their reactive component of impedance dominates over their simple resistive component. The second assumption is that the difference in voltage angle between power lines and between buses tends to be small. The question is what do real and reactive power each depend upon most, voltage angle or magnitude? The answer can be found by looking at the derivatives of P and Q with respect to each of θ and V. Recall that these derivatives are in the Jacobian when we performed the power flow analysis. We are interested in the derivatives with respect to the relevant parameter from *adjacent* lines or buses, not on the same line or bus, because we are interested in power flow from one bus to another. Thus, using a simplified set of buses 1, 2, and 3, we can examine the derivatives.

Equations 3.26 and 3.27 show the derivatives of real power with respect to voltage magnitude and reactive power with respect to voltage angle. We can now begin to reason about the sensitivities of real and reactive power to each of the parameters.

$$\frac{\partial P_2}{\partial V_3} = |V_2|[g_{23}\cos(\theta_2 - \theta_3) + b_{23}\sin(\theta_2\theta_3)], \tag{3.26}$$

$$\frac{\partial Q_2}{\partial \theta_3} = |V_2||V_3|[g_{23}\cos(\theta_2 - \theta_3) + b_{23}\sin(\theta_2\theta_3)]. \tag{3.27}$$

Recall that g is conductance and b is susceptance. Going back to our assumptions, if transmission lines have negligible resistance compared with reactance, then the conductance will be negligible compared with the susceptance. This means that only the term with the sine in each of the above equations is non-negligible. However, going back to our other assumption, that the voltage angles are small, the sine term will also be negligible in each equation. Thus, we can conclude that there is little dependence of real power on voltage magnitude or of reactive power on the voltage angle.

Now consider Equations 3.28 and 3.29. In these equations, the negligible conductance and sine values are in the same term and the non-negligible susceptance and cosine values are in their own term, since the cosine of our assumed small angles will result in a value close to one. Thus, real power is most sensitive to the voltage angle and reactive power is most sensitive to voltage magnitude.

$$\frac{\partial P_2}{\partial \theta_3} = |V_2||V_3|[g_{23}\sin(\theta_2 - \theta_3) + b_{23}\cos(\theta_2\theta_3)], \tag{3.28}$$

$$\frac{\partial Q_2}{\partial V_3} = |V_2|[g_{23}\sin(\theta_2 - \theta_3) + b_{23}\cos(\theta_2\theta_3)]. \tag{3.29}$$

Based on our initial assumptions and the sensitivities just determined, it is possible to make some simplifying approximations in the equations by removing the negligible terms and approximating the cosine of the small angles as one. The approximations are shown in Equations 3.30 and 3.31:

$$\frac{\partial P}{\partial \theta_3} \approx -|V_2||V_3|b_{23}, \tag{3.30}$$

$$\frac{\partial Q}{\partial V_3} \approx -|V_2|b_{23}. \tag{3.31}$$

These approximations mathematically decouple the parameter pair P and θ from the pair Q and V. Only derivatives comprised of these pairs would have values in the Jacobian when doing the power flow analysis; all other derivative combinations in the Jacobian would be zero. Because this is an approximation, the iterative power flow process may not proceed as efficiently as it would otherwise. However, with enough iterations, it will reach a similar result.

3.3.5 Fault Analysis

Fault analysis involves understanding the different types of faults that can occur, designing systems to efficiently protect against the most likely or severe faults, and determining where faults have occurred and restoring the system or repairing the fault. There are many different types of faults, the most general being either the well-known short circuit – that is, an unintentional return of power to the source before reaching the load(s) – or an open-circuit fault, in which power flow is stopped before reaching the intended load(s). A common example in both cases is a broken transmission line touching the ground or another power line in a short circuit or a power line that is simply cut without making contact with an extraneous object in the open-circuit case. Of course, there are many more opportunities for power faults to occur other than in power lines.

A less-commonly known fault condition is called a high-impedance fault. In this type of fault, a power line may be down and touching the ground or an extraneous object, but the impedance is so high that the fault current flow is not excessive. Most fault detection schemes depend upon detecting an abnormally high current flow to detect the fault. If a high current flow does not occur due to the high impedance of the extraneous object or the type of soil in the area, then the fault may be difficult to detect quickly.

Given that most power is transmitted in a three-phase system, it is possible for power lines carrying different phases to accidentally come into contact, creating a short circuit between different phases. Thus, short-circuit faults can be classified into "phase faults" as distinct from "ground faults." Phase faults can be further classified into "symmetrical" or "asymmetrical" faults. If the fault impacts all three phases equally, it is known as a symmetrical fault. If it impacts phases differently, it is asymmetrical. Three-phase current is designed such that all current magnitudes are equal; however, an asymmetrical fault causes that condition to be violated. Thus, the simplification of treating all phases as one power line, such as in the one-line diagram, cannot be utilized; each phase must be analyzed independently. However, a technique known as symmetrical components can be used to simplify the analysis of such complex faults by decomposing an unbalanced system into a superposition of balanced currents or voltages. Section 13.2.1 discusses symmetrical components in detail. Asymmetric faults include:

- line-to-line faults, in which power lines of different phases come into contact with each other;
- line-to-ground faults, in which a short circuit is formed between only one phase and ground;
- double line-to-ground fault, in which only two power lines come into contact with ground or each other.

Another aspect of faults is their time duration. Many faults are transient. A transient fault is one that either disappears on its own or if the power is briefly removed and restored. Typical examples are animals in contact with a power line, lightening strikes, or wind gusts and tree contact. In high-voltage systems, an electric arc can occur between a conductor and ground. Such arcs are dangerous but difficult to detect because they are a form of high-impedance fault. Since these types of transient faults are common, the process of clearing them has become automated by the use of reclosers. These are protection devices that automatically sense a fault and then briefly switch power off and then restore power. On the other hand, a persistent fault is one that does not eventually disappear on its own and cannot be corrected by toggling power off and on.

In fault analysis, many techniques can be used to help simplify the faulted system. As mentioned, symmetrical components can be used to obtain a balanced system from one that is not balanced. Generators are assumed to be in phase, and motors can be viewed as power sources since they can actually participate in supplying power during a fault. The fault location can be viewed as being supplied with a negative voltage source when all other sources are removed from the system for the analysis. Also, the time evolution of the fault should be considered. This includes a beginning sub-transient phase, in which the largest currents initially flow through the fault. Then a transient phase occurs before the fault reaches a final state and finally a steady-state fault condition is reached.

Some of the classical techniques for detecting the location of a fault include attempting to utilize the change in impedance that occurs in a conductor at the fault location. For example, a time-domain reflectometer can transmit a pulse into a conductor and evaluate the return signal to estimate the location of the fault. A more aggressive approach is to inject a brief, high-voltage pulse into a conductor and listen for where a load pop occurs along a damaged cable.

3.3.6 *State Estimation*

State estimation is a technique used to estimate the internal state of a system from measurements taken of the system. Clearly, any understanding and control of the power grid requires knowledge of its actual state; namely, active and reactive power, phase differences, voltages, currents, and many other properties of the system. The need for state estimation arises because the actual state of the entire power grid cannot be directly measured. That is beginning to change with the smart grid. The deployment of more sensors and communication capability is allowing for more ubiquitous collection of a greater variety of information throughout the power grid. However, it remains the case that for cost and efficiency reasons it can be beneficial to utilize fewer, known measurements to derive the state of the power grid in regions that are not fully instrumented with sensors. If the system is observable, then its state can be fully derived from such partial state measurements.

State estimation is used throughout the power grid, including the transmission system. Traditionally, transmission state estimation has been focused upon estimating voltage magnitudes and angles. Information about the interconnection of transmission lines is generally assumed to be known, and some voltage readings are taken and transmitted to a control center. The control center can utilize knowledge of the readings, and a priori knowledge of the transmission topology can be used to estimate the value of voltages in all areas of the transmission system. From this information, the health of the transmission system can be derived, such as safety margins; that is, how close transmission line capacities are to their limits. This, of course, also includes deriving information such as the health of equipment and the need for operator action to avert problems. The state estimation process not only enables the state of a system to become observable with great accuracy and high probability, it allows for errors in the observed information to be detected and corrected. In fact, this is a technique used in communications for error correction and determining the state of the physical layer in a communication channel.

It is important to keep in mind that the power grid is a dynamic system; its actual state is changing as it is being estimated. However, if state estimation is run quickly enough and often enough, a reasonably accurate model of the state can be constructed and kept up to date. Classical uses for state estimation are power load flow analysis and CA, where CA uses the current state of the power grid along with potential pre-computed problems that could arise in order to determine how safe – that is, the safety margin – the power grid is at the time of the state estimation. CA allows power grid operators to understand quickly what they need to do to keep the power grid running safely at all times. State estimation is covered in more detail in Section 12.3.

3.3.7 Flexible Alternating Current Transmission System

A FACTS utilizes advances in high-power solid-state electronics to change some of the underlying fundamental physical properties of the power system in order to control power flow. This is not simply discrete switching of power through transformers or capacitors, but rather instantaneously changing the underlying impedance using solid-state electronics. Applications include reactive power compensation, phase shifting, and power flow control.

Consider a long transmission line with a generator on each end of the line. Recall that stability may be compromised if the generators attempt to share too much power over the line. This will cause the power angle to increase to a point where the system becomes unstable. Also recall that the power angle is also simply the voltage angle or, in other words, the voltage phase difference on each end of the transmission line. A FACTS device can shift the voltage phase angle toward a smaller difference, allowing more power to flow without loss of stability. Extremely fast high-power semiconductor switching allows for many operations to occur within the space of a single cycle. The use of a FACTS provides an interesting fundamental change to the way power systems have operated in the past. This could involve new applications and ways of integrating communication into power systems that will hopefully be stimulated in the reader's mind as we proceed. Next, let us review the challenges involved in the power grid transmission system and later consider some of the ways that communication can help.

3.4 Transmission Challenges

The are several challenges involved in power transmission. Because transmission lines are usually very long and carrying significant amounts of power, even what might appear to be small problems can be greatly magnified. One of the most obvious challenges is keeping losses on the transmission line to a minimum. There are simple resistive I^2R losses. This is the reason the voltage is usually stepped up, to reduce the current I and thus reduce resistive losses. However, this can bring about other problems, to be discussed soon.

Another type of loss is due to the "skin effect," which is caused by alternating current flow. When direct current flows through a wire, the flow is uniform, using the entire cross-section of the conductor. However, alternating current creates fields that are continuously expanding and collapsing. The electrons closer to the center of the conductor experience the impact of fields from surrounding electrons. This creates self-induction. Self-induction results in the electrons near the center of the wire experiencing a greater changing flux density; their velocity through the conductor is reduced compared with the electrons flowing along the outside of the conductor. This results in much of the current actually flowing along the outer wall, or skin, of the conductor, rather than the center. In fact, at very high frequencies, the center of the conductor could be removed without impacting current flow. However, this means that most of the current is flowing through a smaller cross-section of conductor, causing the resistance to increase. Thus, increasing frequency will result in more electrons having to flow through a smaller outer layer, resulting in increasing resistance and greater resistive losses.

There can also be induction and radiation losses. If the field surrounding the conductor happens to pass through an extraneous conducting object, a current will be induced in that object, resulting in power flowing through the extraneous object instead of the transmission line. This could be a malicious attempt to steal power with a coil or an accidental induction into a stray object. Although rare at low frequencies used in the power grid, it is possible for some of the electromagnetic field surrounding the transmission line to radiate into space rather than collapse back onto the transmission line. This electromagnetic radiation is lost power that is dissipated into space, further increasing transmission losses. Corona power loss is caused by the high voltages used by transmission lines. The ionization of air molecules around the transmission line provides a conductive path for the power. This type of loss is highly dependent upon weather conditions, such as humidity and temperature.

Another challenge with transmission lines is to control the growing congestion of power flowing through these lines. The use of FACTS technology (Section 3.3.7) is a potential solution. Another challenge with transmission lines has to do with their great length. Long conductors are more susceptible to geomagnetic storms. In a nutshell, the transmission line acts as a long antenna into which power from space weather is induced. Such apparent random injection of current can cause severe problems and has been known to cause blackouts. However, power from such storms may be captured for useful purposes. The potential for harnessing power from geomagnetic storms is discussed in more detail in Section 15.2.

The final challenge comes from the evolution of the power grid itself. If distributed and renewable generation becomes successful enough to replace or minimize the use of centralized power generation, then the question of whether a transmission system is even needed may be raised. In other words, a new type of power grid architecture could evolve, powered entirely by microgrids. In this case, the need for long-distance power transmission lines may no longer be required. Next, we consider a specific power system application that highlights the importance of communications.

3.4.1 Integrated Volt-VAr Control

Communication is utilized in support of control systems that reduce power loss and improve power quality within the power grid. A notional illustration is shown in Figure 3.7 in which the thickness of the black edge represents power loss in the grid and the thickness of the gray edge represents the communication channel capacity utilized. This figure illustrates the relation between power grid efficiency and information flow through communication channels. A fundamental understanding of this relationship is lacking. Instead, communication links tend to be added in a haphazard, ad hoc manner. Only a small subset of power applications is shown in the illustration, namely stability control, IVVC, FDIR, and AMI. Each of these applications implements independent control mechanisms to optimize grid operation.

The IVVC (Borozan *et al.*, 2001) application serves as an example allowing the reader to gain insight into how communication has been applied in a specific power system scenario. As

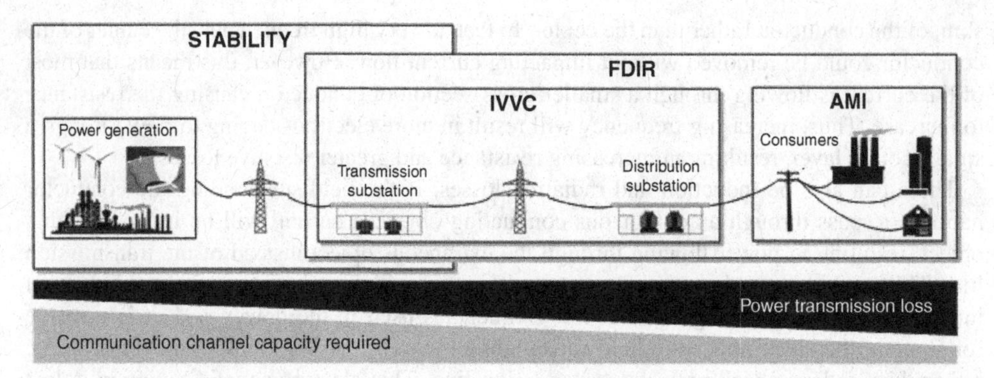

Figure 3.7 A simplified, conceptual illustration of the trade-off between communication bandwidth (gray) and power loss (black) during power transport. Power is generated on the left side of the figure and transported to the right side of the figure. The black wedge indicates the accumulated power loss through the grid. More communication bandwidth enables better control and a reduction in power loss. There is a fundamental trade-off among communication (this could be measured in communication energy or power), channel capacity requirements, and reduction in power loss due to efficient operation enabled by communication. Source: Bush, 2013. Reproduced with permission of IEEE.

the term implies, IVVC is the joint control of both voltage and VAr, where VAr is an acronym for volt-ampere reactive, which is also known as reactive power. Reactive power exists when a power system contains a reactive component, such as an inductance or capacitance, that causes the current to lag or lead the voltage. When considered as a phasor, power can be represented in the complex plane with both a real and imaginary component. Reactive power is the imaginary component of complex power. In physical terms, reactive power remains in the power grid and flows or pulses back and forth between reactive components within the grid. While reactive power does not result in energy that can do useful work for the consumer, it is required in order to support the flow of real power. In fact, reactive power is strongly correlated with voltage levels throughout the power grid. Reactive power flow and voltage levels must be carefully controlled to allow a power system to operate within acceptable limits. There are devices within the power grid to regulate voltage and there are separate devices to control the amount of reactive power. There are many forms of reactive power control, including shunt capacitors, shunt reactors, and static VAr compensators. Shunt capacitors, in the form of capacitor banks that can be switched in and out of the circuit, are common. A tap changing transformer (load tap changer) is a common form of voltage regulator in which the transformer winding ratio can be changed by a simple mechanical switch. The result is a change in voltage required to maintain the proper operating voltage. Thus, there are many voltage-regulation devices and reactive power support devices that are spatially separated throughout the power grid; they must operate harmoniously together in order to keep both the voltage and reactive power in an optimal operating range. This is a complex task given that voltage will tend to drop as demand for power increases. The amount of voltage drop will also increase for loads located farther from the power source. The dynamics of reactive power are also complex in a real operating environment because the size of the reactive component depends upon the mixture of types of loads that consumers are operating. Motors are highly inductive, and many solid-state electronic devices are capacitive. The result is a complex, dynamically changing voltage and reactive power profile that requires constant monitoring and control of the jointly interacting voltage and reactive power compensation devices. An additional consideration is that reducing voltage (within safe limits) reduces the amount of power that devices consume. CVR is a technique in which voltage is purposely reduced for the power consumed by loads; however, this brings the voltage level closer to its lower limit, reducing the safety margin for handling fluctuations in voltage. Similarly, reducing reactive power, when possible and within safe limits, also frees resources for real power, which is the product for which utilities are actually paid. The important point from a communication standpoint is that the voltage and VAr compensation devices must be located at specific points within the power grid that are spatially separate. Monitoring and control information must be continuously and reliably exchanged among the devices in order to keep the power grid operating safely; that is, avoiding a voltage collapse or causing an electrical fault in extreme cases.

Because there are a variety of distinct metrics that IVVC can optimize, IVVC can be viewed as a multiobjective optimization problem. For example, IVVC can attempt to minimize power loss through the transmission and distribution system, maximize the power factor, which is the ratio of real to apparent power (the magnitude of the complex power mentioned earlier), or maintain the voltage profile, which is the voltage level as a function of time and distance from the power source. When applied in the transmission system, stability is another objective function to be maximized, where stability is the difference in voltage angle across a transmission line that, if too large, could cause generators to lose synchronization. As is typical in multiobjective optimization, there is no single point at which all of these objectives are simultaneously maximized. Instead, there is a Pareto-optimal front, which is a set of solutions such that it is impossible to improve one objective without degrading the others. Thus, there is a significant amount of computation required to perform IVVC, including solving the optimal power flow equation, which is highly dependent upon the network structure of the power grid. IVVC requires the communication of large amounts of up-to-date state information, as well as fast, reliable communication for control. Synchrophasors are becoming widely deployed throughout the power grid in order to supply the required information. Keep in mind that this is only one relatively simple example of a power system application that requires communication networking. Hopefully, this relatively simple example of one particular application provides insight, without requiring a power systems background, as to the potential interaction among the power system, communication channel capacity and information theory, and network science. In this example, a poor communication channel could either (1) slow the rate at which IVVC operation receives data or issues commands, thus reducing its reaction time and causing incorrect operation, or (2) drop packets, causing incorrect state to be inferred, also leading to incorrect operation. Next, let us consider the impact of a truly fascinating topic on the transmission system: the impact of space weather and communication.

3.4.2 Geomagnetically Induced Currents

There is a natural phenomenon that has caused havoc with the power grid, communication, and other electrical and conductive systems. This phenomenon involves large, naturally occurring, magnetic fields moving through space. These fields are large enough and sometimes extend close enough to conductors on the Earth to induce unexpected current flows. This phenomenon is known as geomagnetically induced current (GIC). While most people perceive this phenomenon as something abnormal and perhaps bizarre, as well as a potentially dangerous nuisance, and try to mitigate it, its power has been harnessed in the past to do useful work. More on the research involving harnessing GIC will be covered in Section 15.2. For now, we review the basics of the phenomenon. We begin with the well-known concept that the Earth itself is, in essence, a large generator. The Earth's magnetic field is generated via the combination of a molten iron core, thermal motion of the core, and the Earth's rotation. The result is the formation of a planetary magnetic dipole with field lines running roughly from the north pole to south pole.

GIC originates from geomagnetic storms, which are, in turn, caused by space weather. The term "space weather" did not come into use until the 1990s. It deals with changes in plasma, magnetic fields, radiation, and matter in the region of space between the sun and the Earth's atmosphere, much of which is driven by the Sun's chromosphere and corona. The solar wind is comprised of electrically charged particles emitted by the Sun that perturb the Earth's geomagnetic field lines, flattening them in the direction of the Earth's surface facing the Sun and stretching them out behind the Earth; that is, toward the night-side of Earth. However, the actual events that take place are much more complex; we will only hint at some of the complexity here (Lui, 2000).

An illustration of a charged particle trapped by the Earth's magnetic field is shown in Figure 3.8. The particle travels along the magnetic field bouncing from pole to pole. However, an electrically charged particle subject to a magnetic field will follow Maxwell's equations; this means that the particle will have a velocity not only along the magnetic field line, but also a circular, or spiral, component, similar to an electron flowing through a magnetron in a microwave generator. The total energy of the particle is conserved, so that the circular motion will slow down when the particle's velocity along the field increases and the circular motion will increase when the particle's motion along the magnetic field decreases. Because the charged particles follow the Earth's magnetic field lines, which look like rings extending from the poles, the motion of the particles is sometimes referred to as "ring" current. The net effect of many charged particles and their motion is to temporarily depress the Earth's magnetic field. There is a standard index used to record space weather and geomagnetic storms known as the disturbance storm time index, or simply Dst, which can provide some insight

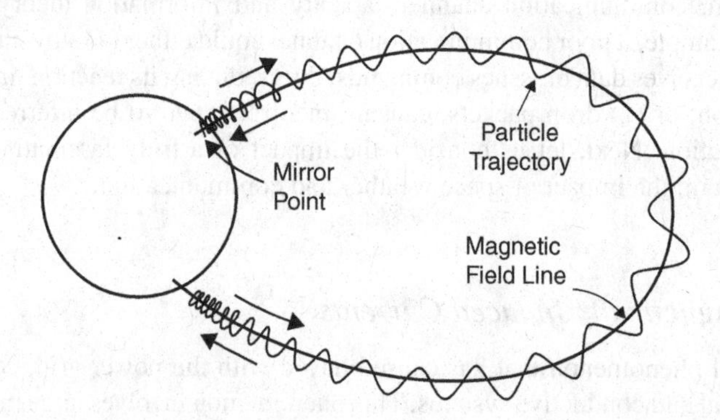

Figure 3.8 A charged particle is shown trapped in the Earth's magnetic field. The particle trajectory forms a spiral shape due to the interaction of the particle's charge and the Earth's magnetic field. Source: Lui, 2000. Reproduced with permission of IEEE.

into what happens during a geomagnetic storm. First, we need to understand the components of the Earth's magnetic field.

The magnetic field can be characterized by many different components; however, the three most relevant components for Dst are: (1) the horizontal intensity H, (2) the vertical intensity Z, and (3) the total intensity F. The horizontal intensity is tangent to a point on the Earth and can be thought of as the component of the field that would most impact a compass reading. The vertical component is the field component pointing toward the center of the Earth. The total intensity is just that: the sum of the H and Z component vectors.

Knowing the horizontal intensity, we can move on to the definition of the disturbance storm time index Dst. First, the Dst is meant to be a standard measure of geomagnetic activity and is used to quantify the intensity of geomagnetic storms, allowing for a standard comparison of storm severity. The Dst is comprised of the average value of the horizontal component of the Earth's magnetic field measured at hourly intervals from four geomagnetic observatories located near the equator. The unit of measurement is the nanotesla. Note that there are other currents contributing to the H field, such as the magnetopause current; these currents, as well as the quiescent time ring current, are subtracted from the Dst index value.

The strength of the Earth's magnetic field near the surface is inversely proportional to the ring current mentioned earlier; this is known as the Dessler–Parker–Sckopke relation. Generally, a geomagnetic storm will begin with a sudden increase in the H component of the geomagnetic field, known as "storm sudden commencement." This is then followed by a period in which the field remains strong, known as the "initial phase," which is next followed by a duration of field strength that is significantly below normal and known as the "main phase." Finally, the field returns to its normal quiescent value, known as the "recovery phase." To provide an idea of the time durations for each phase of the storm, the storm sudden commencement, initial phase, and main phase are typically on the order of hours or days, whereas the recovery phase can last from days to weeks. Like any familiar weather phenomenon, the precise duration and changes in field intensity of geomagnetic storms are very difficult to predict.

As changes in the magnetic field cut through conductors such as power lines, metallic phone wires, and metal pipes, a current is induced in the conductor, as illustrated in Figure 3.9. It is interesting to note that the effect is more significant for long conductors, such as power lines in Canada and the USA for example, while the impact may be slightly less for the shorter lines typically used in Europe. However, damage from GIC has occurred in all locations. The induced current has a strong direct-current component that can cause core saturation in transformers, as well as cause most equipment to experience heating that could exceed safe thermal limits. Also, protection devices could be induced to trip, generating cascading effects throughout the power grid. There are many documented outages caused by GIC. Communication plays a significant role in helping utilities mitigate the impact of GIC. Space weather satellites transmit information regarding events that are precursors of geomagnetic storms, thus allowing utilities to take preventive action.

3.4.3 Quality Issues

Given that we have just discussed a number of challenges involved in power transmission that impact not only transmission efficiency, but also the quality of the power that is transmitted, we now consider power quality. Quality in the transmission system is a broad topic and can include all the general power quality properties ranging from availability of power, variation in voltage magnitude, transient currents and voltages, harmonic content in the power waveform,

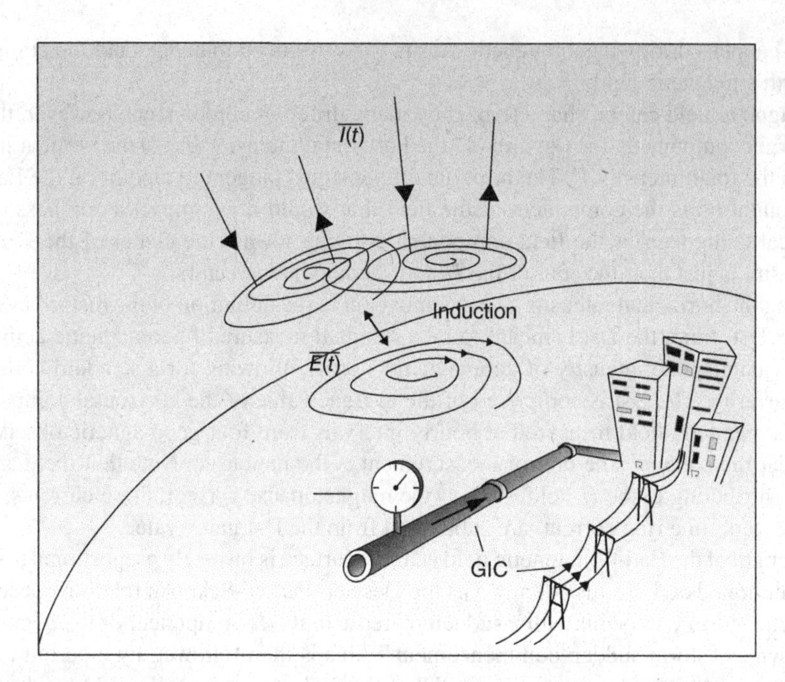

Figure 3.9 GICs occur in long conductive structures such as pipes and power lines as they pass through magnetic fields that originate in space. Source: By Axpulkki at en.wikipedia (Original text: Antti Pulkkinen) [Public domain], via Wikimedia Commons.

and variations in frequency to name a few. These are all unwelcome and undesirable aspects that reduce power quality.

The transmission system is experiencing the advances of power electronics. These include FACTS devices, covered in Section 8.3.1, and HVDC transmission interconnections, discussed in Section 3.4.4 and in progressively more detail in later chapters. FACTSs allows for transmission power flow to be controlled because it can enhance controllability and increase the power transport capability of the transmission system. These technologies both utilize power electronics with switching components. The switching components rapidly turn power on and off through different parts of a circuit or shape the power output. The problem is that they are somewhat like fingers plucking a string causing the string to vibrate not only at a desired fundamental frequency, but also with additional frequency components known as harmonics, which are integer multiples of the fundamental frequency. These harmonics travel through the power grid and cause distortion of the power signal unless filtered. Because large amounts of power pass through the transmission system to reach consumers, the impact upon power quality at the transmission system level could potentially ripple down to large numbers of consumers. Another related component of power quality is the challenge of maintaining adequate synchronization among many different power sources over long-distance transmission lines, which will be discussed next.

3.4.4 Large-scale Synchronization

Because the transmission system can transport power between large-scale interconnection systems, careful consideration has to be given to the implication of joining such large, independent systems. One concern is that each interconnection is supplied by a relatively tightly coupled system of generators that have settled upon their operating frequency. There is a tremendous amount of momentum in the mechanical motion of the generator rotors. If a transmission line runs between interconnection zones that are both operating at the same frequency, but even slightly out of phase, there can be a significant impact on the operation of the transmission line.

Several methods have been developed to allow the transmission of power given such a potential difference in phase. One approach involves the VFT, discussed in Section 4.1.1. The VFT can be thought of as an electric motor in which the armature is connected to the destination of the power transmission line and the rotor to the incoming power from the transmission line. With no applied torque on the motor, the rotor will turn based upon the power difference in phase between the source and destination of the transmission line. If torque is applied to the rotor, the direction of power flow can be changed. A benefit of the VFT is the fact that, compared with HVDC transmission, no power-electronic inverters are required; thus, there is less noise introduced into the system.

Another approach to transmission between different large-scale interconnects is HVDC transmission. The basic idea to handling phase difference in HDVC is quite simple: convert the alternating current power to direct current, transport the direct-current power at high voltage in order to reduce the I^2R loss, then convert the direct current back to alternating current at the destination and ensure that it is at the same phase as the destination. Figure 3.10 illustrates an HVDC connection. One of the problems with HVDC transmission is that the inverters required at both ends of the transmission line introduce distortion, potentially reducing power quality, as discussed earlier.

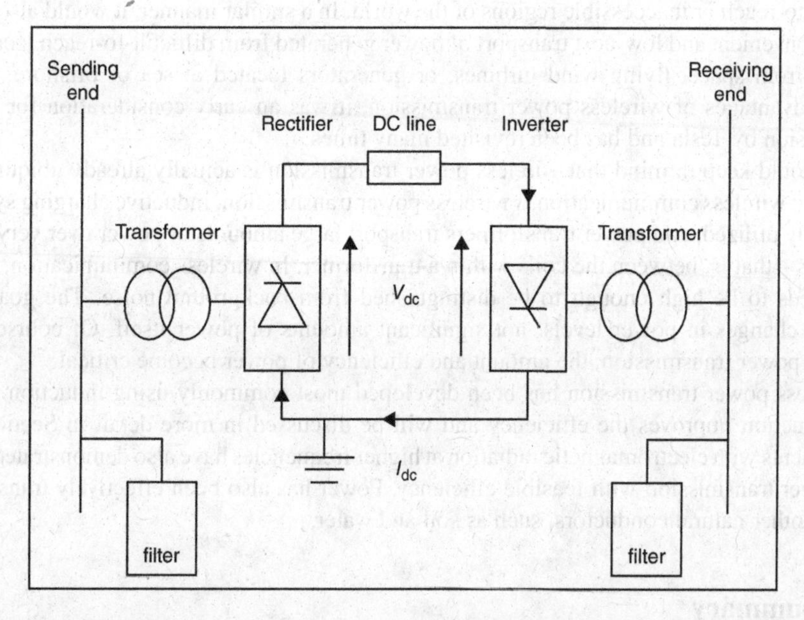

Figure 3.10 A HVDC interconnect connects two alternating current systems. The concept is simple: alternating current at the sending end is rectified to direct current and transmitted to the receiving end. At the receiving end, the direct current passes through an inverter to convert it back to alternating current. The frequencies at each end may differ in phase. The rectifier and inverter can introduce harmonics that may need to be filtered. Source: Wang, 2010. Reproduced with permission of IEEE.

3.4.5 The Power Transmission Market

It is important to keep in mind that, just as power generation and consumption have been turned into commercial markets via deregulation and DR, the transmission system is also a market activity. A transmission system operator (TSO) handles the transmission of electrical power to distribution operators. The TSO is generally independent of both the generation and distribution companies. TSOs form their own market and can charge a fee that is in proportion to the amount of power they carry. This book focuses on power and communication technologies and not the markets, with the exception of any requirements the markets might place upon the technology.

In the next section we discuss another emerging technology that involves rethinking the fundamentals of how power is transported. It goes back to the idea of Tesla and involves the ability to remove or minimize the ubiquitous power line.

3.5 Wireless Power Transmission

Wireless power transmission is the transmission of power over a distance without the requirement for a conducting cable. Energy is the accumulation of power over time, so wireless *energy* transmission is also a term used for this topic. The ability to transmit significant amounts of power wirelessly would be a tremendous advance for the power grid, reducing the cost for power lines, increasing flexibility of the power grid, and allowing for many, convenient, mobile power system applications, such as charging electric vehicles while they are in operation as one particular example. It would also allow for electric power to be easily transported to difficult-to-reach or inaccessible regions of the world. In a similar manner, it would also allow for the convenient and low-cost transport of power generated from difficult-to-reach locations, such as from space, flying wind turbines, or generators located at sea or offshore. Given all the advantages of wireless power transmission, it was an early consideration for power transmission by Tesla and has been revisited many times.

We should keep in mind that wireless power transmission is actually already ubiquitously deployed: wireless communication *is* wireless power transmission, inductive charging systems are widely utilized, and power transformers transport large amounts of power over very small distances – that is, between the coils within a transformer. In wireless communication, power only needs to be high enough to be distinguished from background noise. The goal is to transmit changes in power levels, not significant amounts of power itself. Of course, with wireless power transmission, the amount and efficiency of power become critical.

Wireless power transmission has been developed most commonly using induction. Resonant induction improves the efficiency and will be discussed in more detail in Section 8.4. Experiments with electromagnetic radiation at higher frequencies have also demonstrated wireless power transmission with feasible efficiency. Power has also been effectively transmitted through other natural conductors, such as soil and water.

3.6 Summary

This chapter has covered the fundamentals of the pre-smart-grid era transmission system with an eye toward building a foundation for understanding smart grid concepts developed in later chapters. We have reviewed basic power grid components found in the transmission system and developed some of the fundamental techniques for power flow analysis, as well as introduced phasors, fault analysis, and state estimation. Each of these concepts will also be progressively developed in later chapters.

We also covered some of the many challenges involved in the power transmission system and highlighted where communication and communication expertise could play a role. IVVC was highlighted as an example application involving communication. However, we also covered GICs, quality issues, and large-scale synchronization. Finally, we introduced the notion of wireless power transmission, which would obviously benefit from a corresponding development of wireless communication for the power grid.

The next chapter builds upon these ideas and discusses the distribution system in the pre-smart-grid era. The distribution system must partition large amounts of power into relatively small amounts for many different consumers. Techniques to increase efficiency protect the public from power faults and self-heal require challenging and unique approaches. While the general rule of local sensing and control applies to the distribution system, communication through the distribution system using techniques such as distribution line carrier for applications such as automatic meter reading, protection, and many other distribution system applications introduced here has a long history. As more communication and control are incorporated into the power grid, the closer the grid can come to its operating limits. However, this can make the system more brittle; brittle system analysis is introduced as a means of studying this phenomena and its impact on the communication system. Thus, Chapter 4 provides information that lays the foundation for understanding many of the smart grid applications taking place in the distribution system that will be covered in Part Two of the book.

3.7 Exercises

Exercise 3.1 Reactive Power

1. What is reactive power?
2. Why is reactive power necessary in the power grid? What happens if there is insufficient reactive power?
3. How does communication play a role in controlling reactive power?
4. Can you think of an analogy between reactive power and communication overhead?

Exercise 3.2 Transmission Power Flow

1. In simple terms, how is power flow controlled over a power line?
2. What roles can communication play in the transmission system?
3. What characteristics of the transmission system are unique with respect to communication?

Exercise 3.3 Integrated Volt-VAr Control

1. What is IVVC?
2. How is voltage control related to reactive power control?

Exercise 3.4 Geomagnetically Induced Current

1. What causes GIC?
2. How is the disturbance storm time index defined?
3. What is the typical pattern of geomagnetic field intensity changes during a geomagnetic storm?

Exercise 3.5 Conservation Voltage Reduction

1. What is conservation voltage reduction?
2. What are the pros and cons of conservation voltage reduction?
3. How will conservation voltage reduction increase reliance upon communications?

Exercise 3.6 Power Quality

1. What is power quality?
2. Should it be designed to match the power quality required by the load; that is, be no better or worse than the quality required? What are the pros and cons of engineering such a match?
3. What is the relationship of information entropy to power quality? Is there a theoretical optimum amount of communication required to maintain a given level of power quality? Sketch the relationship between information entropy and power quality.

Exercise 3.7 HVDC Interconnect

1. What are the advantages of HVDC transmission interconnections?
2. What are the disadvantages?
3. What are the roles and requirements of communication in managing such interconnections?

Exercise 3.8 Goubau Line

In 1950, Georg Goubau discovered that adding a dielectric layer surrounding a wire will slow the surface transmission wave, reducing the amount of energy lost through radiation. The waves are at UHF and microwave frequencies, which differs from power line carrier operating at much lower frequencies and up to hundreds of watts.

1. To what does Goubau line communication refer?
2. What are the advantages and disadvantages of Goubau line communication?
3. How does Goubau line communication differ from power line communication?
4. What makes using the power line as a waveguide in the transmission system easier than using power lines in the distribution system?

Exercise 3.9 Transmission Line Sensing

1. What are parameters that should be monitored by sensors in the transmission system?
2. What will determine the requirements for communication of these values?

Exercise 3.10 Wireless Power Transmission

1. What is wireless power transmission?
2. What are the advantages and disadvantages of wireless power transmission?
3. What would you consider to be the main challenges of wireless power transmission?

4

Distribution

And God said, “Let there be light” and there was light, but the Electricity Board said He would have to wait until Thursday to be connected.

—Spike Milligan

4.1 Introduction

This chapter is about the next step in the classical power grid toward reaching the consumer after the transmission system. The transmission system delivers large amounts of power at high voltage and low current over long distances to population centers where power is required. The next step in the pre-smart-grid system is to distribute the power to consumers, either residential or industrial. From a high level this involves two key steps: transforming the power to lower voltage and distributing or fanning the power out to individual consumers.

Referring back to Table 3.1, the communication involved could approximate a MAN. The transmission system ends at a substation, which distributes power through feeders to populated areas. This chapter begins by discussing transformers in the distribution system, feeders, a quick discussion of power line, and distribution system topologies. Because the distribution system reaches out to the consumer, this part of the grid most exposes the power system to the public. This raises safety concerns and requires that power protection receives a high priority. The next part of the chapter focuses upon power protection mechanisms. While many of these mechanisms can be configured to operate based upon local information, some protection techniques have relied upon communication as a fundamental part of their operation. The issue with communication for protection mechanisms is that speed and reliability are extremely important in handling a dangerous electrical fault when it occurs. Communication can sometimes be viewed as adding another level of complication and vulnerability if not implemented perfectly. Thus, it is important to understand precisely when and how communication adds benefit to protection mechanisms. Each of the protection mechanisms previously discussed fits into a larger system that handles FDIR. The goal is to rapidly identify when and where an electrical fault occurs, isolate the fault by removing power to the fault, and then restore power safely to as many consumers as possible. Clearly, communication plays a role in

Smart Grid: Communication-Enabled Intelligence for the Electric Power Grid, First Edition. Stephen F. Bush.

automating this process. While fully automated FDIR may be considered a property of smart grid self-healing, FDIR has been handled manually in the past. Another technique that is on the evolutionary boundary between the classical power grid and smart grid is conservation voltage reduction, which is simply the reduction of voltage to as low a value as safely possible while still delivering power as needed to consumers. By reducing the voltage, power usage is reduced. However, the downside is that the power grid is operating close to its lower voltage limit and any unanticipated dips in voltage could cause problems. This is one example of a broader issue with the smart grid; namely, more efficiency may be squeezed out of the power grid, but it also means that the power grid has less room to degrade gracefully if problems occur. In other words, the system becomes brittle (Bush *et al.*, 1999). At this point, the power line carrier for the distribution system is introduced. This is sometimes called distribution line carrier. The distribution system contains many transformers, relays, and noise with which the distribution line carrier must contend. Given the capability of communication throughout the distribution system, DA can become a reality. While DA is likely to be claimed as an evolving smart-grid activity, it is introduced in this chapter. IVVC, discussed in the previous chapter, can be implemented in the distribution system and is another example of distribution automation. Finally, a summary reviews the main points in this chapter. Exercises follow to help the reader consolidate the material and explore new ideas related to communication and the distribution system.

As we have seen, the power transmission and distribution system is comprised of a hierarchical structure of power flows at different voltages. The longest distance lines are the transmission lines transporting power on the order of hundreds of kilovolts. Once the power reaches an area that is relatively near to the consumer, the voltage is stepped down before entering the distribution system. While the job of the transmission system is to carry the power over relatively long distances, the job of the distribution system is to allow customers to tap into that power; that is, to spread or distribute the power to individual consumers. The distribution system can be divided into the primary distribution system at tens of kilovolts and the secondary distribution system, which is the final step to the consumer, typically at 120 V. Thus, it is the transformer that demarcates the boundaries of transmission, primary, and secondary distribution systems. The boundary between transmission and distribution usually exists within an enclosed structure called a substation. Power leaves the substation at the primary voltage of low tens of kilovolts through power lines called feeders. Note that feeders carry three-phase current, which has been assumed to flow all the way from the source generator to this point in the system. From these main feeders, lateral feeders carry one or two phases of the current to within a block of the target consumers. From the lateral feeders, distribution transformers further step down the current to the secondary distribution voltage of 120 V and supply power to roughly a block of homes. An industrial site, or sub-transmission customer, can receive three-phase power directly from the distribution primary system. Figure 4.1 shows a simplified overview of the classical, "pre-smart-grid" power grid. The figure illustrates generation, a step-up transformer, transmission to a substation, and then distribution to an industrial sub-transmission consumer, a primary customer, and a secondary customer.

The power transmission system is designed to be a bidirectional network structure. Power can flow in any direction through the transmission system depending upon load and available generator capacity. The distribution system, on the other hand, tends to be radial in nature; that is, current is expected to enter from the transmission lines in the substations and to branch outwards toward the consumers (Figure 4.3). Distribution protection mechanisms are

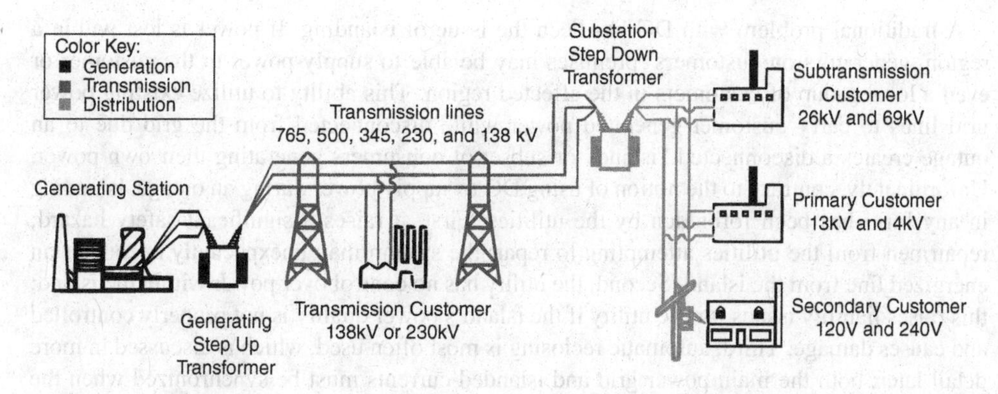

Figure 4.1 The next step in the transport of electric power is the distribution system, illustrated on the right here. Source: By MBizon [CC-BY-3.0 (http://creativecommons.org/licenses/by/3.0)], via Wikimedia Commons.

configured assuming there is a one-way downstream current flow. More will be discussed about protection later; however, one can assume that the load drawn from the feeder will be monotonically decreasing as one approaches the end of the distribution line, and thus the amount of current drawn will also be decreasing. One can also set the protection mechanisms to trip in a desired manner that minimizes the impact to customers assuming load decreases near the end of the radial feeder lines.

Another type of distribution network structure is a loop distribution system, in which customers are connected to radial branches as in the radial network structure, but in the loop system there is a normally open "tie" switch between the radial branches from two different feeders. In Figure 4.4 the concept is to provide redundant paths in the event of an electrical fault. The distribution system is arguably the most exposed portion of the entire power system since, by design, it reaches out to every location that requires power. This exposure also means that the distribution network is susceptible to accidents causing short circuits. These could range from a car hitting a pole, to a squirrel fried on or between power lines, to lightning hitting a power line. The result is that the power distribution system is particularly concerned with power protection of human life and property. Thus, the distribution system is designed to be particularly sensitive to electrical faults and break the circuit at any sign of an overcurrent condition. However, breaking or opening the circuit stops the power flow to all downstream customers. It is desirable for the utilities, who have to pay a penalty based upon the number and duration of customers without power, to provide alternate paths for power flow to reach customers. The loop distribution system does this by allowing the normally open tie switch to close in the event of a power fault. This enables power to flow from a radial branch with power to one without power due to a fault. Distribution systems can be even more highly interconnected, with more redundant paths available to bypass faults that may occur. These are known as network or mesh distribution systems. While loop and mesh distribution systems are more reliable than simple radial branch systems, they are also more expensive and complicated to configure as they require protection mechanisms that allow current to flow bidirectionally as well as switching hardware that allows the paths to open and close as needed.

A traditional problem with DG has been the issue of islanding. If power is lost within a region, generators on customers' premises may be able to supply power to the customer or even a local group of customers in the affected region. This ability to utilize existing power grid lines to carry customer-generated power while disconnected from the grid due to an outage creates a disconnected "island" or subset of consumers generating their own power. Unfortunately, contrary to the notion of using DG to supply power during an outage, islanding in any form has been forbidden by the utilities. First, it raises a significant safety hazard; repairmen from the utilities attempting to repair the system may unexpectedly encounter an energized line from the island. Second, the utility has no control over power within the island; this raises liability issues for the utility if the island's power quality is not properly controlled and causes damage. Third, automatic reclosing is most often used, which is discussed in more detail later; both the main power grid and islanded currents must be synchronized when the connection is reestablished, and this cannot always be guaranteed. Therefore, islanding has been prevented by requiring DG to automatically disconnect from the grid in the event main power is lost.

The issue of loop flow arises when there are multiple paths that current can travel. Unlike packets transported through a communication network whose direction through the system can be controlled via routing, current must follow Kirchhoff's laws, in which the impedance determines power flow. Section 3.3.4 on power flow analysis discussed this in more detail. Also, as discussed at the end of Section 2.2.4, circulating currents can exist among generators. The results are often not obvious; increasing generator output can actually reduce the congestion on some lines. This becomes important when power lines approach their capacity limits or become congested.

The European distribution system differs from North America owing to general population differences. In North America, where the population tends to be more diffuse, or spread out, the primary portion of the distribution system tends to be larger, maintaining a higher voltage to reach longer distances. The distribution secondary then reaches out to a few customers or a block per distribution transformer. In Europe, where population is denser, the primary distribution system is smaller in scope and a single secondary transformer connects to many more customers. Also, the European secondary distribution lines more often run underground rather than overhead.

Generally speaking, transmission and distribution substations house equipment needed to interface different architectural regions of the power grid. For example, substations may house the equipment to interface different transmission systems or to interface a transmission and primary distribution system. Larger substations may be staffed and have a control room, while smaller substations may be unstaffed. The main components of a substation are the transformers; typically, the substation, as an interface between different architectural regions of the grid, has the main goal of stepping up or stepping down the voltage between different transmission systems or between the transmission and distribution systems. The substation may also contain capacitors to counteract inductive loads. Since power is flowing as three-phase current, there are typically three sets of each device, referred to as "banks." Substations will also typically have switches to modify the current's path through the system if necessary. The components in the substation are protected by overcurrent relays. These devices detect the duration and amount of current and take action to open a connection to protect a circuit if the current exceeds a given value for a given length of time.

One of the most direct forms of controlling the power grid involves switching power flows through high-power connections. This can take place manually, automatically, or through SCADA systems. Protection systems operate automatically for the most part. For example, a recloser is a device that automatically opens a circuit when an overcurrent is detected for a given duration of time. The recloser will automatically and repeatedly open, wait for the fault to clear, and then attempt to close. It will automatically repeat this process several times. If the fault does not clear after all attempts to reclose, the circuit is left permanently open and a signal may be automatically sent to a tie switch to attempt to restore power to as many customers as possible. However, this process, particularly the restoration part, may take place manually. Crews may be assigned to an outage area to manually find the fault and restore power. Power restoration over a large area needs to be done with care so as not to destabilize the system with the introduction of a sudden, large load. Instead, loads are gradually reconnected to the main power system. During the restoration process, islanding may be allowed as utility crews gradually synchronize and reconnect islanded sections.

Switching of power can take place to reduce congested lines or overheating transformers. Switching can also be utilized in the more drastic case, which is a last resort, of load shedding, where customers are deliberately switched off from the power grid in order to keep the rest of the grid running. Finally, switching can be used in the distribution system to perform load balancing. The idea behind load balancing in the distribution system is to minimize resistive power line losses by equalizing current flow through all feeders and transformers. Recall the I^2R power loss; the sum of these losses will be minimized when the currents are equal.

Power system diagrams will often show only one power line in a schematic when three-phase power is assumed. This is known as a one-line diagram. Three-phase power is beneficial for driving three-phase motors. With three stator windings, each 120° apart, it is possible for the rotor to experience a smooth, continuous force as it rotates. With fewer than three phases, this is not possible. Instead, there would be a discontinuous, pulsating torque on the rotor. Another advantage of three-phase power applies to transmission. Three-phase power allows for power to be transmitted more efficiently and with less conductive material required than any other number of phases. For direct-current power, it is sometimes possible to use a ground return; that is, only require one line to supply power because ground is utilized to complete the circuit. This technique is sometimes used in rural areas; however, it is not always feasible or reliable, depending on the impedance through the ground.

Each phase of three-phase power originates from each of the stator windings in the generator. Each phase may be considered a separate circuit, each having a single line carrying current to a load and another line returning to the generator to complete the circuit. Thus, for three phases, there would be a total of six lines. However, six lines are not required. As we will see, three of the return lines may be combined into one or may even be potentially eliminated, leaving only three or four lines required. The reason that a single, common return line from the load can be used is that sinusoidal currents have precisely the same magnitude and are precisely 120° shifted in phase. Using the superposition principle from Kirchhoff's laws, the sum of the currents is zero, and similarly for the voltages. If this holds true, then the common return line could be eliminated completely. If this does not hold true – in other words, if a phase differs in magnitude slightly or its phase shifts slightly – then there will be a net current requiring a return line. If the generator and loads are grounded, then, for small imbalances in the phases, any residual current can flow through the earth.

An argument can be made that one advantage of choosing three phases, rather than a smaller number of phases, is that there are more phases available to spread out error that may occur in magnitude or phase. Another argument in favor of three phases is that more than three phases would require more complex and expensive equipment. Dividing power onto four or more lines would reduce the diameter of the conductors required; however, it would require four or more transformer windings, four or more capacitors in capacitor banks, and so on. Going back to the idea of economy of scale, it is cheaper to have fewer and larger devices rather than many small ones.

Now we can see the importance of keeping loads balanced across all three phases; the ideal case is to maintain equal magnitude and phase at all times. This requires keeping loads balanced. For industrial sites with large, three-phase motors, this is relatively easy since each phase is used equally. However, for residential consumers, only one phase is used; there is no guarantee that each phase will be used equally by different sets of customers. Many residential devices are either on or off, causing discrete changes in load. However, as the loads become aggregated as one moves away from residences and closer to the substation, differences in load tend to cancel one another and discrete changes have an increasingly smaller effect. In the USA, the trend is to use three wires, one for each phase, and a neutral return wire, resulting in a four-wire system. In Europe, only three wires tend to be used; there is no neutral return wire. If there is any imbalance in the phases, current must flow through the ground.

4.1.1 Transformers in the Distribution System

Given three wires, one for each phase, there are two distinct ways in which loads can be connected. One way is to connect a load between each phase and ground. This is known as a wye connection, since the connection has the shape of the letter 'Y.' The ground connection is at the common point in the middle of the 'Y.' An alternative way to connect the loads does not involve a ground connection; instead, the loads are connected between each pair of phase wires. This is known as a delta connection because it looks like the Greek letter 'Δ.'

For the wye connection, each load's voltage waveform will be 120° apart, since each load is connected to one, and only one, phase. In the delta connection, each load connects to two different phases; thus, the load–phase relationship is less clear. The voltage at the load will be the difference between the voltage at the phases to which it is connected. The resulting difference turns out to be a sinusoidal waveform that is shifted 120° from the voltage at the other loads and has a larger amplitude by a factor of $\sqrt{3}$. Transformer connections are almost always either delta or wye configurations. An interesting note about the delta configuration is that, because it is not grounded, its voltages are all relative and floating. Thus, an accidental connection to ground will simply set the delta configuration to ground level rather than draw an overcurrent and cause a fault, as would happen in a non-delta connection. Thus, the delta connection has an added degree of reliability. However, the ground fault needs to be found and corrected, otherwise an additional fault will cause an overcurrent and subsequent damage. Thus, the additional reliability of a delta connection comes at the cost of "hiding" a potentially dangerous fault condition.

In order to simplify analysis, one can use a one-line diagram to represent a three-phase system. This assumes the ideal condition that the phases are balanced and that no fault exists. To compute the power transmitted, one can simply multiply the voltage by the current in phase and then multiply by three for the full three-phase system. One complicating factor is the difference between delta and wye connections and the $\sqrt{3}$ factor. Equation 4.1 shows the power delivered to loads in the wye configuration. V_{rms} is the phase-to-phase voltage and I_{rms} is the current through the conductor:

$$S = 3I_{\text{rms}}\frac{V_{\text{rms}}}{\sqrt{3}}. \tag{4.1}$$

Equation 4.2 shows the power delivered to loads in a delta configuration:

$$S = 3\frac{I_{\text{rms}}}{\sqrt{3}}V_{\text{rms}}. \tag{4.2}$$

As shown in Equation 4.3, the power delivered to a single load in either the delta or wye configuration is the same ($3/\sqrt{3} = \sqrt{3}$):

$$S = \sqrt{3}I_{\text{rms}}V_{\text{rms}}. \tag{4.3}$$

On the other hand, direct-current transmission has advantages, particularly over long distances. First, it avoids the stability limit, discussed in more detail shortly, relevant to Equation 4.8. Second, there is no inductance when using direct current. Direct-current power lines are recognizable by having two rather than three lines overhead. While direct-current transmission was a problem in the early power grid, suffering heavy losses, the difference now is that modern direct-current transmission uses very high voltage, which greatly reduces I^2R losses. Direct-current transmission is ideal for bulk power transmission over long distance and, because its output is converted into alternating current, it can easily connect two asynchronous systems.

The key to the transmission and distribution system is the transformer. The transformer is comprised of two coils: the primary, or input coil, and the secondary, or output coil. At a fundamental level, everything reverts back to Maxwell's equations. The alternating current supplied to the primary coil of the transformer establishes a magnetic field within the core of the coil. The core may be empty (that is, comprised of air) or, more likely, a material that is magnetically susceptible. The strength of the magnetic field is related to both the magnitude of the input current and the number of turns in the primary coil winding. Recall from the earlier fundamental review that magnetic circuits are analogous to electric circuits. As shown in Equation 4.4, the magnetomotive force is the product of the current I and the number of turns in the winding n:

$$\text{mmf} = nI. \tag{4.4}$$

The magnetomotive force generates the magnetic flux Φ within the winding core. As shown by

$$\Phi = \frac{\text{mmf}}{R}, \tag{4.5}$$

the flux is related to both the magnetomotive force and the core material's reluctance $\mathcal{R}$. Recall that the reluctance is somewhat analogous to resistance in an electric circuit and is the inverse of the core material's magnetic permeability.

A design goal of the transformer is to minimize leakage of flux, which is accomplished by utilizing a core material with a low magnetic reluctance. The 60-cycle per second change in current through the primary causes a corresponding change in flux that is also linked to the secondary coil. This induces an electromotive force within the secondary coil winding proportional to the number of secondary windings. The current that flows through the secondary winding will be determined by the impedance of the load connected to the secondary coil. Note that this process is quite similar to the operation of a generator, with the significant and obvious exception that there is no rotor and there are no mechanically operating parts. In fact, the VFT is essentially a generator used as a transformer to connect two independent asynchronous zones, because the rotor, acting as the transformer primary, can rotate in order to compensate for the difference in frequency between the asynchronous zones.

The simple relationship between the turns ratio, the input and output voltages, and the input and output currents is

$$\frac{V_1}{V_2} = \frac{n_1}{n_2} = \frac{I_2}{I_1}. \tag{4.6}$$

It should be clear that the turns ratio is trading off voltage for current such that power through the transformer remains constant. Of course, efficiency is never perfect and all the usual forms of energy loss apply, including inductive impedance, resistive losses, and corresponding heat buildup.

A legacy transformer has some degree of ability to support change in configuration; namely, variable "taps," or connections into different locations of the secondary winding, are used to change the effective number of windings and thus the turns ratio shown in Equation 4.6. This is used, in particular, in the power distribution network in order to increase or decrease the voltage as necessary in order to adapt to a changing load. Thus, they are also known as "load tap changing" LTC transformers.

The resistive heating within the transformer is referred to as "copper losses." However, there is another form of energy loss within the transformer known as "iron loss." This is due to the constant and continuously rapid change in the magnetic field. The continuous change causes microscopic iron core particles to rapidly reorient themselves, thus producing heat. The average efficiency of a transformer is approximately 90%. For small transformers, this heat loss is readily observed by touch, with the heat simply dissipating into the air. However, a larger transformer of many megawatts will generate tremendous amounts of heat. This may be seen in the design of the transformer with many vanes to help radiate the heat, as well as active oil-cooled systems. The capacity limit of a transformer is determined by its ability to handle heat buildup. When the capacity approaches the thermal limits of the transformer, careful temperature monitoring is required. Here is where sensor networks provide a useful function for the power grid.

The discussion above regarding the operation of the transformer assumed a simple, single-phase connection. The three-phase discussion regarding loads connected in a delta or wye also applies to three-phase current through transformers. In this case, the loads are the transformer coils. Thus, it is possible to have a delta–delta transformer connection in which three primary coils are connected in a delta configuration and the secondary coils are also connected in a delta configuration. In this case, current and voltage are determined only by the coil winding turns ratio from Equation 4.6. Similarly, in a wye–wye transformer connection the primary and secondary windings are each connected in wye configurations, and again the voltage and current are determined only by the turns ratio. However, two other configurations are possible: a wye–delta and delta–wye configuration. In these cases, it is not only the turns ratio that determines the voltage and current output, but also the factor of $\sqrt{3}$. Specifically, recall that the delta, or phase-to-phase, connection has a $\sqrt{3}$ increase in voltage compared with a wye, or phase-to-ground, connection. Thus, for a delta–wye transformer, each coil of the wye configuration will see a $\sqrt{3}$ increase in voltage above and beyond the voltage determined by the coil winding turns ratio. Similarly, the opposite is true for a wye–delta transformer configuration; the voltage will be reduced by a factor of $\sqrt{3}$. Also, it is important to note that, as mentioned earlier, the phase-to-phase delta connection induces a 30° phase shift in the voltage. It is important to transform each phase equally; transforming only one phase would cause problems if the three phases pass through another delta transformer or connection since the ideal 120° phase difference would no longer apply. Instead, it could cause an imbalance in the phase-to-phase difference that would induce a wasteful circulating current to exist. Next, we consider the feeder, which distributes power from the substation.

4.1.2 Feeders

The job of a feeder in the distribution system is to safely extend the power flow from the substation toward customers. Beginning from the substation, the feeder typically starts with a transformer and circuit breaker. The transformer steps the voltage down and the circuit breaker provides protection against any severe fault that may occur along the feeder. However, the feeder is also partitioned into sections where each section of the feeder is protected with an automatic reclosing device and sectionalizer, to be discussed shortly. The automatic recloser detects fault current along its section of the feeder and can automatically open and close the connection to attempt to clear a transient power fault. The automatic recloser typically works in conjunction with a device called a sectionalizer. The sectionalizer can be located at the opposite end of each section of the feeder from the automatic recloser. The sectionalizer senses the voltage. If a fault is persistent, then the recloser will eventually end its attempts to close and remain open, causing a zero voltage. When the sectionalizer senses this, it also opens in order to ensure that the section of the feeder protected by the recloser is isolated. The sectionalizer is not capable of interrupting fault current, and is thus less expensive to construct than a recloser. It is possible that once the sectionalizer has opened the circuit to isolate a section of the feeder, the recloser can close again to supply power to other parts of the feeder. There may also be additional equipment on a feeder, including additional fuses, capacitors, and voltage regulators. Next, we consider distribution power lines in more detail.

4.1.3 Power Lines

The conducting material for overhead lines is typically made of aluminum, which has the advantage of being lightweight. Steel strands are incorporated in order to increase strength. Underground cables are generally made of copper, which is actually a better conductor than

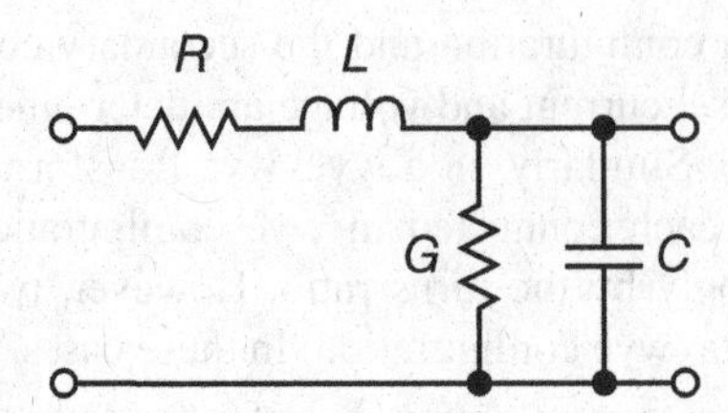

Figure 4.2 The model for a unit length of a transmission line is shown. The resistance *R*, inductance *L*, conductance *G*, and capacitance *C* are per-unit of transmission line length. The transmission line model approaches reality as the size of the above unit length of transmission line is decreased and the number of units is increased to form the total length of the line. Source: By User: Omegatron [CC-BY-SA-3.0-2.5-2.0-1.0 (http://creativecommons.org/licenses/by-sa/3.0) or GFDL (http://www.gnu.org/copyleft/fdl.html)], via Wikimedia Commons.

aluminum, but also more expensive. As previously discussed in connection with Equation 1.4, a larger diameter conductor is better in order to reduce power loss. However, this needs to be balanced by weight and cost.

When analyzing a power line to first order, the dominating factor is the inductive reactance of the line rather than the resistance. In fact, power lines in which resistance is ignored and only inductance is considered are called *lossless lines*. Recall that inductive reactance is caused by conductor loop linkage. Power lines are generally straight-line connections, not loops. However, a straight line can be thought of as being part of an infinitely long loop with a small magnetic flux that adds up over the length of the line. In fact, there are two contributions to the magnetic flux: the power line flux of a single line as just described, and also mutual inductance shared among the lines of the three phases. There is also capacitance; capacitance exists between power lines, where each line can be thought of as the plate of a capacitor, as well as between the power line and the surface of the Earth. Thus, a power line model is comprised of resistance, capacitance, and inductance. In the power line model, Figure 4.2, resistance and inductance are in series and capacitance in parallel over many small segments. Adding enough small model segments to compose an actual line length involves adding the resistances, inductance, and capacitance.

The capacitance and inductance in the power line impact its efficiency in transmitting power. Specifically, the ratio of the series impedance (*R* and *L* in Figure 4.2) to the shunt, or parallel resistance–capacitance (*G* and *C* in Figure 4.2), of the power line model determines a value known as the *characteristic impedance* of the line. The characteristic impedance can also be thought of as the ratio of the amplitudes of a voltage and current wave pair propagating along the line. When the inductance and capacitance of the line cancel, the line has only real impedance, or resistance. In this case, power transmitted on the line is not dissipated in the line itself. A transmission line of finite length with a resistance at the end of the line that is equal to the characteristic impedance will appear to the source as an infinitely long transmission line. If the resistance in the power line model is negligible, then the characteristic impedance ratio reduces to the square root of the ratio of the series inductance to the parallel capacitances; this is known as the *surge impedance* of the line. This is important in communications because when the resistance at the end of the line is equal to the surge impedance a signal suffers minimal loss over the line. The surge impedance loading (SIL) is the square of the transmission voltage divided by the surge impedance, which results in real power. It is the amount of real power transmitted when the power line's inductive and capacitive values cancel. If power is transmitted below the surge impedance loading, then the line appears as a capacitive load, one that injects reactive power into the system. If the power transmitted on the line is above the surge impedance loading, then the line appears as an inductive load and consumes reactive power.

The characteristic impedance of a line is defined by

$$Z_0 = \sqrt{\frac{R + \mathrm{j}\omega L}{G + \mathrm{j}\omega C}}, \tag{4.7}$$

where R is the resistance per unit length of the line, L is the inductance per unit length, G is the conductance of the dielectric per unit length, C is the capacitance per unit length, and ω is the angular frequency of the current.

From a wave perspective, the voltage and current phasors on the line are related by the characteristic impedance as $\frac{V^+}{I^+} = Z_0 = -\frac{V^-}{I^-}$, where the superscript plus and minus signs represent forward-traveling waves and backward-traveling waves, respectively. For a lossless line, R and G are both zero, so the equation for characteristic impedance reduces to $Z_0 = \sqrt{L/C}$. The imaginary term j has also been canceled, making Z_0 a real expression that is thus purely resistive with magnitude $\sqrt{L/C}$. The characteristic impedance of a transmission line is expressed in terms of the SIL, or *natural loading*, which is the power loading at which reactive power is neither produced nor absorbed, $\mathrm{SIL} = V_{\mathrm{LL}}^2/Z_0$, in which V_{LL} is the line-to-line voltage in volts. Loaded below its SIL, a line supplies lagging, reactive power to the system, tending to raise system voltages. When loaded above the SIL, the line absorbs reactive power, tending to reduce the voltage.

The relationship of surge impedance loading as a function of power loading on the line can be seen another way. As loads draw more power, they are really drawing more current; voltage is held constant. Inductance increases with increasing current; thus, higher loading draws more current and increases inductance and reactive power consumption. Capacitance, on the other hand, is dependent upon the line voltage; at lower loading levels, the current is lower and the capacitance dominates, leading to injection of reactive power.

The Ferranti effect, first observed in underground cables in Sebastian Ziani de Ferranti's 10 000 V distribution system in 1887, describes the voltage gain toward the remote end of a lightly loaded (or open-ended) transmission line. Underground cables normally have a low characteristic impedance, resulting in a SIL that is typically above the thermal limit of the cable. Therefore, the cable is almost always a source of lagging reactive power. The effect is exhibited as a rise in voltage occurring at the receiving end of a long transmission line relative to the voltage at the sending end. This effect occurs when the line is energized but there is a very light or disconnected load. It is due to the voltage drop across the line inductance when the drop occurs in phase with the sending-end voltage. This effect is more pronounced for longer lines and higher voltages. The voltage rise is proportional to the square of the length of the line. The Ferranti effect tends to be greater in underground cables, even at shorter lengths, due to the higher capacitance in underground cables.

As previously mentioned for transformers, power lines are limited in capacity by the amount of heat they can withstand. As power lines become loaded with more current, their energy loss in the form of heat increases with the square of the current. As power lines heat up, they can begin to sag. If the current continues to increase and overheating continues, power lines can actually melt. Thus, even though power lines are rated by the amount of current they can handle, this is only an approximation, because weather conditions play a significant role as well. Thus, the ability to sense and communicate temperature and predict weather conditions can play a significant role in safely operating closer to the thermal limits of power lines.

Another limit on the capacity of power lines is related to stability and the length of a power line. Recall the discussion of power generation and the mechanism by which generators tend to naturally remain in synchrony with one another. A generator that wants to take more of the load will tend to push harder with its rotor; its rotor position will be slightly ahead of the other generators' rotors. The angle by which the generator's rotor differs from the others is known as the *power angle*. It is reflected in the difference between the generator's voltage peak and the voltage peak of the system as a whole. Generally, the power angle is measured between any two points in the system by measuring the difference between the peaks.

The amount of real power transmitted on an ideal lossless line is shown by

$$P = \frac{V_1 V_2}{X} \sin \delta_{12}, \tag{4.8}$$

where δ_{12} is the difference in power angle between the sending and receiving ends of the line, V_1 and V_2 are the magnitudes of the voltages at either ends of the line, and X is the reactance of the line.

One way to transmit a large amount of power over the line is to increase the power angle δ_{12}. Recall the discussion on generator synchrony in Section 2.2.4. When the power becomes large, there is a greater chance of the system becoming unstable and losing synchrony. If X is relatively small, that is, for a short line, a small power angle δ_{12} can still yield a large power transfer. This transfer can potentially exceed the thermal limit of the line. Thus, a question revolves around which will be exceeded first: the thermal limit or the stability limit. This is due to the fact that a long transmission line could have a large reactance X allowing for a large power angle δ_{12}, a power angle that is small enough to be under the thermal limit but large enough to cause the system to become unstable. The thermal limit is expressed in either current or apparent power, since both contribute directly to heating, while the stability limit is expressed in units of real power, because Equation 4.8 defines real power.

As mentioned several times prior to this section, voltage is typically held constant while current changes depending upon the power required by loads. However, this statement needs to be refined; typically, the goal is to hold the voltage constant. In practice, voltage levels can vary up to a few percent from their target values as the load and generation balance changes. The voltage at any location is determined by the voltage drop caused by resistive losses and the amount of reactive power consumed or generated within the local area. A good example of the resistive voltage drop effect can be seen in radial distribution networks. In a radial distribution system a feeder provides power at one end of the line at a given voltage while the voltage level drops along the line as one progresses further from the substation. Quite simply, the power line resistance increases with length, causing a larger voltage drop as the length increases. Since by Ohm's law $V = IR$, R is increasing with length. However, as loads increase their demand for power, I increases as well. Increasing both terms also increases the voltage drop. Guidelines in the USA require that the voltage remains within $\pm5\%$ of the target value. Thus, for a long enough line or a line with enough load, the voltage level may drop below the acceptable guideline value. In this case, either the voltage at the feeder end can be raised to compensate for the drop along the linc or a transformer may need to be added somewhere along the line to restore the voltage to its target value. Of course, loads fluctuate throughout the day, so voltage adjustments may need to change accordingly. This is typically handled by an LTC transformer. If the voltage drop is due to too much inductance, or lagging current, then adding capacitance to balance the inductance is another possible solution. If the load tap changer transformer is placed away from the feeder along the distribution power line, it is called a *voltage regulator*.

As mentioned, reactive power may be added to increase voltage if there are highly inductive loads causing the voltage drop. The way to add reactive power is by adding capacitors, synchronous condensers, or static VAr compensators (SVCs). Capacitors were discussed earlier under fundamental electronics. A synchronous condenser is fundamentally a synchronous generator with zero real power output. The rotor spins freely while its excitation current can be increased to add reactive power to the system. The advantage of the synchronous condenser over a simple capacitor bank is that it can be continuously adjusted, the energy stored in the rotor can help to stabilize the system during short transient conditions, and its operation is independent of the line voltage. The last advantage is in reference to the fact that the power through a simple capacitor increases or decreases with voltage level; the synchronous condenser can increase current as voltage decreases. The disadvantage of a synchronous condenser is the inefficiency due to mechanical operation. An SVC is a solid-state device for providing fast-acting reactive power and is part of the FACTS device family. The term "static" refers to the fact that the SVC has no moving parts. If the power system's load is capacitive, or has a leading current, the SVC will use reactors, meaning components similar to inductive coils, typically in the form of thyristor-controlled reactors, to consume VArs from the system and lower the voltage. Under inductive, or lagging-current conditions, simple capacitor banks are automatically switched into the circuit, providing a higher voltage.

For all of the techniques above, adding reactive power (that is, compensating for inductive loads) is equivalent to bringing the power factor from a value less than one back toward one. The techniques discussed above – namely, use of capacitors, synchronous condensers, and SVCs – are distributed approaches to adding reactive power to the system. Recall that reactive power can be created at the generator, but that reactive power is also a continuous oscillation of stored energy in the form of electric and magnetic fields. Thus, there is an associated circulating current involved with reactive power. If the reactive power can be supplied locally, then the "bandwidth" or power line capacity required to carry the circulating current can be reduced, which results in increased efficiency.

4.1.4 Distribution Topologies

Just as in communication networks, the topology of the power flow in an electrical distribution system plays a significant role in its cost and reliability. The simplest and least-costly topology utilizes a radial feeder system, shown in Figure 4.3. In this topology, power radiates outward from the substation in a tree structure. The substation is the root of the tree, and branches located farther from the substation are at progressively lower voltages. The problem with this topology is that isolating a fault will also block power to all downstream customers.

The opposite of the radial system is a mesh distribution system. This is essentially a radial system with additional power lines running between the branches of the tree. Switches allow the additional lines to be used if necessary. This provides more reliability at the cost of constructing additional power lines. The concept is obviously similar to a mesh communication

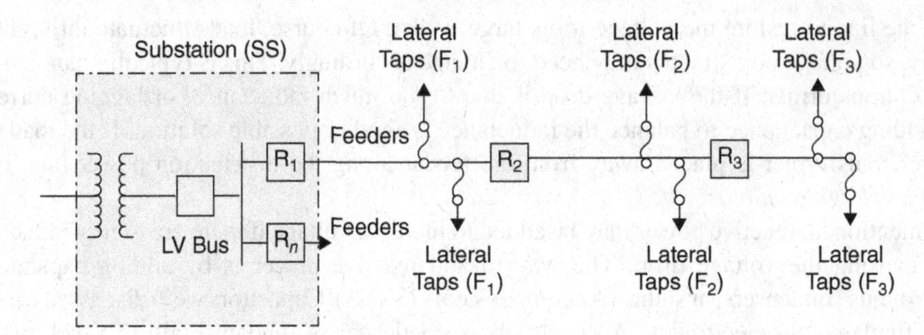

Figure 4.3 A radial distribution topology is shown emanating from a substation. The low voltage (LV) bus is fed by a transformer and fuse and provides power to two feeders each protected by a relay. Lateral lines extend from the feeder with relays included to protect each segment of the feeder.

network, which increases reliability by allowing multiple paths through the network. With a mesh topology, fault isolation can occur while allowing additional paths to keep customers supplied with power.

Finally, a loop system provides a compromise between the cost of redundancy and the need for reliability. A loop topology is illustrated in Figure 4.4. In the loop approach, two feeders are

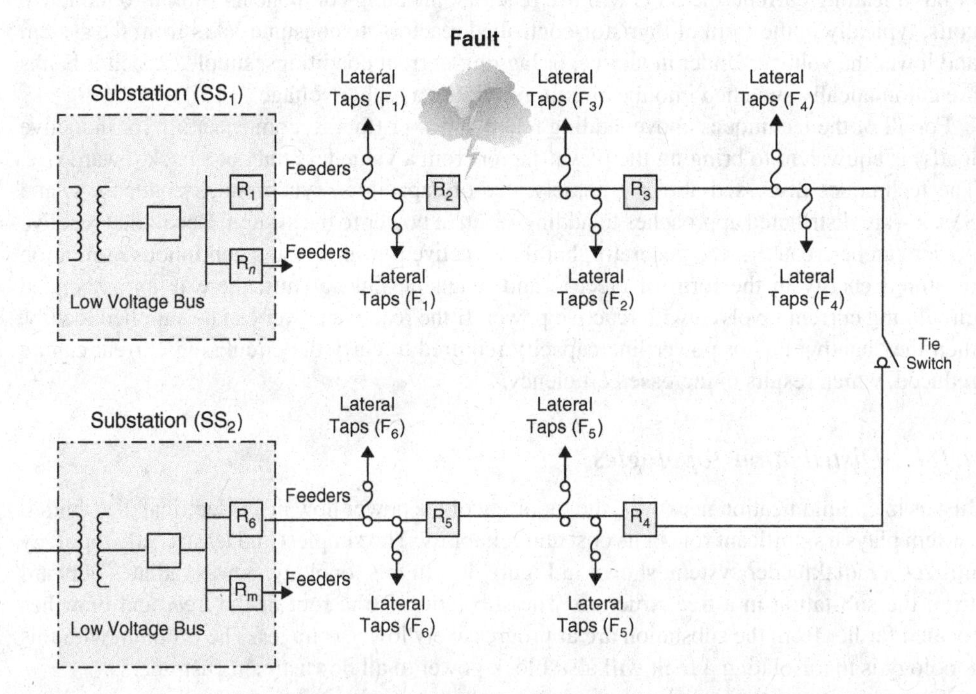

Figure 4.4 A loop distribution topology is shown emanating from two substations. Each half of the loop is a radial topology. However, there is a tie switch, normally open, that can close if necessary to allow a substation's feeder to supply power to the other substation's feeder. This process is known as restoration and occurs after a fault – illustrated by the lightning strike here – has been isolated.

connected by a tie switch. The tie switch is normally open, keeping the feeders isolated from one another. However, if a fault occurs along one feeder, the tie switch can close, allowing power to flow in the opposite direction through the feeder with the fault in order to keep as many customers as possible supplied with power after the fault is isolated.

4.1.5 Designing Distribution Systems

Protection and topology are key aspects in the design of distribution systems. Protection in power systems refers to mitigating the impact of an electrical fault. A fault is an inadvertent flow of current serious enough to cause damage to life or property, including power grid components themselves. A fault is colloquially known as a short circuit. Two general types of faults are phase-to-ground and phase-to-phase faults. A phase-to-ground fault is an accidental current flow between an energized conductor and ground, which can occur through intermediate objects such as trees. A low-impedance connection with ground will essentially act like a load, drawing a tremendous amount of power in the form of a large current flow. A phase-to-phase fault occurs when two or more phases inadvertently come into close enough proximity to draw current between them. As with any load, fault current is a function of the impedance of the fault and the ability of the system to maintain the target voltage during the fault.

The goal of protection mechanisms is always to first accurately detect a fault, then attempt to either clear or disconnect the faulted segment from the power network while minimizing the number of customers impacted. A fault can always be cleared by turning off power to the entire power grid or distribution system, but that would be unnecessarily annoying to a large number of customers. Detecting a fault is not unlike cybersecurity, particularly intrusion detection, where one looks for abnormal patterns; distinguishing "self" from "non-self." For power systems, detecting faults can be based on patterns such as abnormally high current over an extended period of time, abnormally low voltage, abnormal changes in reactive power or frequency, or a combination of many other parameters.

One of the simplest protection devices is the fuse. However, the fuse provides a fixed level of protection; it is not configurable. Also, the fuse requires manual intervention to replace. The next higher level of sophistication is the circuit breaker. This is a simple relay that opens when current exceeds a given value for a given length of time. Some circuit breakers have variable sensitivity settings. The amount of current and time required to open a relay is defined in a time–current characteristic (TCC) curve that plots the amount of current versus the current duration designed to cause the relay to open.

It possible that a fault may occur in which the impedance is relatively high, high enough to appear similar to a legitimate load. This is known as a high-impedance fault. In this case, in order for the TCC curve to detect the fault, it would overlap with normal operating current levels. A more direct technique to detect a fault in such a case is to measure the current going into the system in question and the current coming out of the system. If there is a difference, then, by Kirchhoff's current law, some current must be exiting through a fault. This technique is known as a differential relay; this is a relay set to open when there is a significant difference between the current entering one side of the power line and the current exiting on the other, assuming there is nothing else along the line drawing current. However, this technique requires communication of current information over the distance of the line. In other words, communication is clearly required for this approach.

Another form of protection is a ground-fault circuit interrupter (GFCI), which works somewhat like a sensitive, fast differential relay. The GFCI resides in the power receptacle where it can differentially monitor current between the hot and neutral wires. Normally, all current should flow from the hot wire through the load to the neutral wire. However, if current finds an easier path to ground, perhaps in a dangerous situation through a human, there will be a difference between the hot and neutral current. The GFCI is designed to be sensitive and to operate quickly and reliably as a life-critical protection system to open the circuit in such a situation.

The term "switchgear" is used for any power system equipment capable of isolating power line segments. This includes fuses, circuit breakers, and reclosers. This equipment must be able to handle abnormally high power; that is, the abnormally high current caused by a fault. Switchgear must be able to safely interrupt large currents while controlling and mitigating dangerous arcs that occur as contacts separate from one another. High voltage enables large currents to jump across space between the contacts. A typical approach to control arcing is to place the switch within a tank of nonconducting fluid, such as transformer oil or sulfur hexafluoride (SF_6). The chemical design of SF_6 is such that the molecules of the nonconducting material have no free electrons to conduct current; they are symmetric, so that they are nonpolarized. A vacuum works to suppress arcs as well. Another interesting variation is an airblast or puffer interrupter, in which a compressed, nonconducting fluid is injected quickly at the location of a potential arc in order to quench the arc. In an alternating current circuit breaker, as the contacts move apart an arc will form during the positive or negative portion of the voltage cycle. However, as the cycle approaches the zero crossing, current flow will disappear because the voltage will approach zero. During this brief interval, the switch has the potential to open without an arc. If the contacts cannot open wide enough in this period of time, there is a potential for the cycle to approach its maximum in the opposite direction and causing a new arc. This is known as a restrike. The contact-opening rate must exceed the sinusoidal cycle time from zero to maximum voltage; mechanical switches have to be accelerated to high speeds within milliseconds. Reactance in the circuit must be taken into account as that will change the timing between current and voltage. Pressurized quenching gases must be emitted at supersonic speeds. Thus, it should be clear that switchgear devices are much more complex and expensive than ordinary switches.

A recloser is a special type of switchgear device that automatically opens and attempts to close again after a delay in the hope that the fault will clear. This device was constructed because many faults are transient; that is, they last only for a short time and then disappear. An animal may accidentally touch two phases, be electrocuted, and fall from the power line, leaving a normal circuit again. A power line may be blown against a tree, causing a momentary fault. Thus, the recloser generally operates using a TCC curve. When an overcurrent condition exceeds a given time duration, the recloser will open for a preset time and then close again. If the fault has cleared, then everything returns to normal. If the fault is still present, the recloser will immediately open and repeat the sequence. This entire open–close operation may be repeated a specified number of times, typically three. This results in the lights blinking that people may witness during a storm, often immediately before a blackout. If the fault does not clear after the recloser has exhausted its attempts to reclose, it will remain permanently open. This is known as a lockout. At this point, it typically requires a repair person from the utility to arrive and manually clear the fault. In many portions of the grid, the utility will not know of the outage until a customer calls to complain – this has been the rather primitive, but effective, method of sensing and communication. Setting the TCC curve, duration of time between reclosings, and the total number of reclosing operations is somewhat of an art. For example, reclosing times on a transmission line tend to be shorter than on distribution lines. Transmission lines involve more power, more customers, and tend to be placed higher in the air. The hope is that any transient faults along a transmission line will, therefore, be of short duration given the fact that they are less exposed to interference from the ground than the distribution system. The hope is also that the transmission transients will be short enough to go unnoticed by customers. The TCC curve for reclosers in the distribution network are set to allow for a longer time between reclosings. This is to anticipate the fact that the distribution system is closer to the ground and more exposed to animals and trees and will likely require more time for faults to clear.

Protection coordination is the topic of utilizing and configuring the protection devices just discussed in order to quickly isolate faults while reducing the impact of the fault on the rest of the system. This includes breaking the circuit as close to the fault as possible to minimize the impact to the rest of the system, as well as utilizing protection devices that act as backups for other protection devices, in case the device closest to the fault fails to operate. When all protection mechanisms have been configured to provide the desired level of protection for the entire system, protection is known as *coordinated*. Each device protects an area of the system known as a protection zone. A protection zone is the section of the power system that will be isolated by a given protection device. It is possible for protection zones to be nested within one another. In a radial system – that is, one in which a single power line extends from a feeder to all customers – the coordination settings are relatively simple to illustrate. Such a power line's protection setting will be designed to increase in sensitivity with the distance from the feeder. There are two reasons for this. First, the goal is to break the circuit, when a fault occurs, as far from the feeder as possible. This allows as many customers as possible to continue to remain connected to receive power through the line. Second, since more current is flowing closer to the feeder, there is a larger aggregate load as seen by the line closer to the feeder. Thus, in the ideal case, *only* the fuse, breaker, or recloser upstream and closest to the fault will operate to isolate the fault. Communication among the protection equipment, tie switches, and the substation can effectively isolate the fault and potentially restore power quickly and efficiently.

Topologies more sophisticated than a simple radial or loop distribution system can quickly become complex. If the current is bidirectional, protection configuration needs to handle faults in both directions. If current from sporadic DG comes and goes, it makes distinguishing normal from abnormal current difficult.

Even when protection involves a relatively simply radial system, there are interesting subtleties to consider. First, a fault current is not simply an instantaneous increase in the magnitude of alternating current. Instead, depending on the amount of inductance in the line, the fault current will initially start with a *larger* magnitude than the final steady-state fault current. This is due to the fact that the inductance in the line will initially impede the sudden change in current flow. In effect, there will be a direct-current component of the alternating current that decays rapidly and exponentially over time. Thus, when there is significant inductance relative to resistance, the fault current will continue to have an alternating waveform of the same frequency, but shifted initially upward and then gradually fall back to a zero direct-current bias. The inductance-to-resistance ratio X/R for the line determines the time constant for the decaying direct-current bias of the alternating fault current. Second, an issue that can impact protection systems is known as back-electromotive force (EMF). As mentioned earlier, a motor and a generator are basically the same device. When a large motor loses power, its rotor can continue to spin due to inertia. In effect, the motor becomes a generator for a short time, emitting current back into the system. Thus, protection systems should anticipate less-obvious factors such as a direct-current component to the initial fault current and back-EMF from motors.

Another issue for protection mechanisms that will become more significant for the smart grid is DG. Distributed generators connected throughout the distribution system will inject current into the distribution system at nontraditional locations, such locations normally occupied by consumers. This can raise the level of fault current as well destroy the protection coordination if configured assuming there were no DG. This will be discussed in more detail in the Part Two.

4.2 Protection Techniques

This section continues the introduction of power protection with emphasis on techniques used in the distribution system. Because the power distribution system is highly exposed to the environment and the public, power faults are more likely to occur and to do more damage in this part of the power grid. There are a myriad of power protection techniques. They involve the coordinated effort of many different protection schemes to detect a short circuit and stop or limit the fault current while minimizing the impact on power flow to consumers. Protection is about blocking or limiting the right current at the right time. In this sense, power system protection has some similarity to cybersecurity: it involves detecting abnormalities and blocking suspicious flows of power just as network cybersecurity systems attempt to detect suspicious information flows and block potentially dangerous activity. In this section, different general types of power protection devices and methods are introduced, including fuses and relays, distance relays, pilot protection, to special protection schemes.

4.2.1 Fuses, Breakers, Relays, and Reclosers

Fuses, breakers, relays, and reclosers, as well as variations of these devices, are the principle mechanisms of protection for the power grid. Fault current limiters are another, advanced technique discussed in Section 14.5. Fuses, breakers, relays, and reclosers are based upon detecting a fault current and reacting to it by breaking the circuit when the amount of current passes an amperage–time threshold. The only difference among these devices is their ability to automatically recover from the fault. Fuses are the least automated and require manual replacement to recover. Circuit breakers do not require replacement, but generally require manual intervention to recover. Relays, particularly digital relays, can be programmed to be more sophisticated in how they detect and automatically recover from a fault. Finally, reclosers are typically the most sophisticated of these devices, with the ability to automatically perform multiple opening and reclosing operations in an attempt to resolve a transient fault. Let us consider each of these devices in more detail.

The fuse is a sacrificial device that offers low resistance under normal operation but quickly melts when current exceeds a given threshold and thus permanently interrupts current flow. The fuse, like all of the devices mentioned in this section, operates using a time–current relationship that ensures it will only melt when the fault current magnitude and duration are large enough to cause damage, but will not melt for high currents of short enough duration that they would not cause damage. If the fuse were to melt too soon at too low a current level, it would needlessly cause manual replacement of the fuse and cause customers to lose power in the process.

The type of fuse is defined by several parameters. The rated current I_N is the maximum current that the fuse can continuously conduct without melting. The speed at which a fuse melts depends upon the amount of current flowing through it and can be controlled by the material of which the fuse is made. The TCC for a fuse, as well as all the devices in this section, is not a fixed value; instead, time decreases as current increases. A fuse is relatively primitive in terms of implementing a TCC curve compared with a relay or recloser. Fuses may be fast-blow, slow-blow, or time-delay. The time and current required for a fuse to blow depends upon the type of load being protected and the relationship of the fuse to other protection devices in the circuit. A sensitive device requires the fuse to blow quickly with less current, while a circuit with other protective devices, such as relays or reclosers, may take longer and require more current to blow in order to serve as a "last resort," allowing other automated protection devices to operate first.

Another property of a fuse is its I^2t value, where I is current and t is time. If we assume voltage is constant, then this is proportional to the amount of energy required by the fuse to melt and thus break the connection. There are two general I^2t values: the melting value and the clearing value. The melting value is the energy required to *begin* melting the fuse, and the clearing value is the total energy the fuse lets through before melting enough to sever the connection and thus clear the fault. This type of energy rating is useful because it provides an indication of the amount of energy that a fault could deliver, and thus the amount of thermal and magnetic forces that would be experienced by protected loads before isolation from the circuit by the fuse. Another property of the fuse is the breaking capacity. This is the absolute maximum current that the fuse can safely interrupt. This is important because some fuses may simply not be capable of interrupting an extremely large current. Similarly, the voltage rating is the maximum voltage that can be handled by the fuse. A fuse may simply not be capable of blocking an extremely high voltage. Another property is the voltage drop. The resistance of a fuse may change with current and temperature. In particular, a high current can cause resistive heating, thus increasing the resistance of the fuse. This will cause a corresponding increase in voltage drop across the fuse. If large enough, the voltage drop may need to be considered in the design of the system. The time–current behavior of the fuse can change with temperature. It can effectively become more sensitive at higher temperatures and less sensitive at cooler temperatures. These characteristics are important not just for fuses, but apply generally to all the protective devices discussed in this section.

A circuit breaker operates in a manner similar to a fuse, but has the benefit of not requiring replacement. Instead, the breaker can be reset back to normal operation. The relatively small circuit breaker that the average homeowner is familiar with is generally contained within a single enclosure. Large power system breakers can utilize current transformers to sense the current and a separate, battery-powered solenoid to break the connection. Also in large breakers, a motor can be used to reset the springs. Significant arcing can occur when the circuit breaker operates, wearing down the contacts, which may have to be replaced periodically. A large circuit breaker must be able to manage the arc that forms during operation by keeping it contained, cooling it, and extinguishing it by using a variety of media inside the breaker, including various gases, a vacuum, or oil as examples. Many tricks have been developed to decrease arcing. For example, the arc can be deflected, cooled, techniques can be used to divide the arc into multiple, smaller arcs, or the current waveform can be monitored and the contact only opened when the current is at, or near, a magnitude of zero during its cycle.

As discussed previously, three-phase power can experience asymmetric faults. A device known as a residual current device or ground fault interrupter can determine when the current in one of the phases and the neutral return wire are not balanced, indicating that there is a short circuit and current is leaking to ground; this can be a potentially serious situation and shock hazard. The device will then trip, isolating the circuit.

Protective relays are the main operating component of many power protection devices and continue to grow in sophistication. They measure characteristics of the power flowing through them and determine when to open to protect the system from fault current. Protection relays are able to measure and respond to excessive current, excessive voltage, reverse power flow, and anomalous frequency conditions. Distance relays, discussed later, can be programmed to respond to fault conditions that are located precisely within a certain range of distances from the relay. All of this intelligence within the relay allows it to better recognize fault conditions and only respond to true faults. It also allows the relays to work effectively together, known as protection coordination, to be discussed later. Even as relays have advanced and incorporated digital processing capability, they still utilize power system terminology used decades ago, and it is important to understand the terminology for communication purposes.

The most commonly used relay was the induction disk overcurrent relay. The terminology from this relay is in use today. Section 9.2.1 covers the induction disk overcurrent relay in detail. The basic idea is that torque is applied to a rotating metal disk by a magnetic field. The field creates eddy currents in the disk and the interaction between the magnetic field and eddy currents on the disk creates the torque that turns the disk. This is similar to the operation of a homopolar motor explained in detail in Section 14.5. The interaction between the strength of the electromagnetic field and mechanical action of the mass and damping of the disk can be used to create a "programmed" response to open and close a contact. In some sense, this is a crude analog computer.

Digital, or "numeric," protective relays are now widely used. In these relays, the electromechanical "programming" is replaced by a microprocessor that can be programmed to sense and respond as required. Of course, the current, voltage, and any other measurement must be sampled and converted to digital representation. One advantage is that there are fewer moving parts in a digital relay and so its lifetime may be longer. Digital relays also provide a direct interface to the communication network for the sampled values.

The fact that digital relays allow for more sophisticated operation can be seen in the distance relay and differential relay explained later in this section. Here, we start by explaining the simple overcurrent relay. This type of relay operates based upon the inverse TCC curve. Section 9.2.1 discusses time–current curves in more detail. A current transformer is used to measure the current; given the amount of current and the duration of time that the current exceeds a given value, the relay can be programmed to open.

A recloser is similar to a relay, with the additional feature of being able to automatically close. The idea here is to balance the convenience of not having to manually reset a relay after a fault has occurred with the need to ensure that the fault can be, or has, really cleared. In other words, they are designed to allow a transient fault to clear itself.

As we have seen, reclosers are used in feeders within the distribution system. The operation of reclosing is programmable. Typically, two or three fast reclosings will occur with a few seconds delay, then a longer delay will occur followed by a final attempt to reclose. If the fault has not cleared by the final attempt to reclose, a "lockout" will occur. When this happens, the recloser essentially "gives up" and remains open, isolating the faulted section of the line.

As previously mentioned, the recloser may work with another device known as a sectionalizer. The sectionalizer is a less-expensive device and is not able to open when current is flowing. However, the sectionalizer can count the number of interruptions caused by the recloser opening and closing and determine when the recloser has locked out. When this happens, there is no current and the sectionalizer can safely open, ensuring that the faulted section of the line remains isolated.

4.2.2 Distance Relays

As mentioned briefly, distance relays have been a commonly used protective relay. They are popular because they can be configured to trip for faults that occur within a specific distance, or range of distances, from the relay. Conceptually, this makes it easier to coordinate multiple relays so that each relay protects a specific portion of the grid or serves as backup protection for a specific portion of the grid.

The concept behind distance relay operation is that power lines have a known and fairly constant impedance per kilometer. Since impedance is a function of voltage and current, the voltage and current can be sensed. Knowing what the impedance of the line should be, a determination about a distance of the fault from the relay can be deduced. Thus, the relay can discriminate between faults in a specific section of the line from faults in other sections of the line. The distance from the relay for which the relay is configured to trip for faults is known as the "reach point."

Consider that the impedance from the relay increases as the distance from the relay increases. Then for a given distance, a threshold impedance can be derived. When the relay senses the voltage and current and computes the impedance, if the computed impedance is less than the threshold impedance, then a fault can be assumed to have occurred between the relay and the configured distance or reach point.

Recall that impedance depends not only upon voltage and current, but also the phase angle between these values. This information can be captured on a resistance–impedance (R–X) diagram, as shown in Figure 4.5. Let us consider current, voltage, impedance, and resistance in more detail using this figure. First, let $|Z|$ be the ratio of the voltage magnitude difference (across the impedance) to the current magnitude. The complex impedance is then $Z = |Z|e^{j\theta}$, where θ is the difference in phase angle between the voltage and current. This can also be expressed in Cartesian coordinates as $Z = R + jX$, where R is the resistance and jX is the complex impedance. Note that j is used to represent complex numbers instead of "i" so that there will be no confusion with current, which is i. The result is that an area within the R–X diagram can be specified for which the relay will consider a fault to have occurred. Ideally, one could plot all the "normal" variation that occurs among voltage and current magnitude and phase and plot it within this R–X diagram. This can be used to help separate fault conditions, which should be outside the normal operating area, from normal operating conditions and thus help ensure that the relay only opens for fault conditions.

Consider some examples of defining fault conditions by defining areas in the R–X diagram. A circle, centered at the origin, would respond only to changes in resistance due to a fault since it would trip for all phase angles at a given resistance. It would also trip for faults behind and

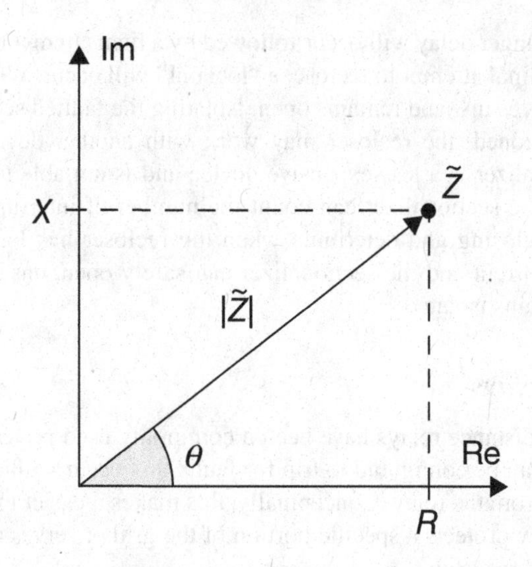

Figure 4.5 Complex impedance Z is shown comprised of two components; X is the reactance, R is the resistance, θ is the phase, and $|Z|$ is the magnitude of the impedance. Source: By Inductiveload (Own work) [Public domain], via Wikimedia Commons.

in front of the relay; that is, it would be nondirectional. If directional control were required, then a semicircular shape would be used. This can be generalized further; any complex shape can be created to correspond to any set of conditions that are dependent upon voltage and current magnitude and their relative phase. Notice that no explicit communication is required to accomplish this form of protection. Communication is inherent in the locally sensed voltage, current, and phase.

4.2.3 Pilot Protection

Pilot protection, also known as differential protection, utilizes Kirchhoff's current law to detect the presence of a fault. The conservation of electric charge implies that at any junction, or node, within an electric circuit the sum of the currents flowing into that junction is equal to the sum of the currents flowing out of that junction. In other words, the sum of currents flowing into and out of a node is zero. A pilot, or differential, protection scheme checks for precisely this condition. Current transformers can be used to measure the currents flowing into and out of any point in the power system. If their sum is not zero, then it is assumed that a power fault has occurred. As an example, the current can be measured at both ends of a power line or transmission line. If a fault occurs along the line, a significant amount of current may be drawn through ground, reducing the current measured at the downstream end of the line. A greater amount of current will be entering the line than exiting at the intended termination of the line. The result will be a nonzero current differential indicating a fault condition. Notice that, for the current summation to occur using this protection scheme, remote current (amperes) information has to be compared. This requires communication, and a typical technique that has long been utilized is to transfer the information using power line carrier communication over the same line being protected. Consideration has to be given to the fact that the power line could be severed, in which case communication would be lost if power line carrier were used.

There are two general types of protection schemes, depending upon the confidence that protection designers have in the reliability of the communication system: blocking and permissive. A blocking scheme is somewhat pessimistic about the communication system. It uses communication to "block" the protection relay from opening. If communication is lost, the relay will open by default. This implies that, if communication is faulty and no real power fault exists, there will be nuisance tripping of the line. On the other hand, a permissive scheme requires a communication signal to "permit" tripping. If the communication line is faulty, a fault may be missed and the power system will not be protected. Thus, we can already see the mindset of power system engineers who view the communication system as added complexity and another point of failure in the power system. Although communication has improved greatly since the early twentieth century when this was first applied, it is a wise precaution and still largely true. Communication adds complexity, and even with near-perfect reliability it contributes a finite degree of unreliability to the system unless used with, or as, a backup mechanism.

The pilot scheme also needs to be cognizant of a few complicating issues. When applied to sections of a power system that normally draw large amounts of current, the current differential that triggers a fault must also be larger. This is because larger normal operating currents can have correspondingly larger anomalies that are not faults. Also, the measurement system itself is not ideal. A current transformer is typically used to measure the current. However, the current transformer can saturate. This means that current is beyond the range in which the output changes proportionally with the input current.

4.2.4 *Protection and Stability*

It is important to consider the relationship between protection and stability. Stability is covered in more detail in Section 12.5. In this case, we are referring to power flow stability, the ability of the power grid to settle back into its normal operating state after power flows are perturbed. Multiple generator rotors are all moving at a speed engineered to produce the power flows required throughout the power grid. Naturally occurring magnetic coupling within the power grid balances the mechanical forces keeping the rotors spinning. A sudden change in overall load can cause individual generator rotors to experience a sudden change in acceleration. If the change is large enough, the rotors can lose synchronization with one another and begin oscillating or spinning chaotically.

A protection system that suddenly trips a large transmission line could potentially cause this kind of instability. It could cause the load as seen by the generators to change rapidly. Conversely, temporary instability in the system could cause a protection system to trip. For example, an oscillation in power flow due to transient instability could possibly exceed a power fault threshold and cause a protective relay to think it was a fault. If the relay trips, it could cause even further instability. The main point is that careful consideration of interaction among system components is required. There is always opportunity for new ideas involving communication to help mitigate these issues.

4.2.5 *Special Protection Schemes and Remedial Action Schemes*

"Special protection scheme" is the name given to a category of protection that is specially designed to respond to specific, unique, potentially catastrophic contingencies that tend to have a low probability of occurrence. This type of protection mechanism goes under many different names, including "special stability controls," "dynamic security control," "contingency arming schemes," "remedial action schemes," "adaptive protection schemes," "corrective action schemes," as well as numerous other names. It may be gleaned from all these terms that there are common features within this category of protection schemes. First, the protection mechanism may be turned on or off; it is not permanently operational. The events that the protection mechanism are designed to mitigate are relatively rare events, perhaps occurring once a year. The control mechanism is typically ad hoc in nature, having been carefully designed offline. Finally, and most importantly for communication, this type of protection mechanism, because it is typically ad hoc and built over a wide area, requires communication in order to operate.

A special protection scheme is designed as follows. Extensive offline simulations are performed accounting for all possible conditions that could impact a given power system implementation. Conditions that are found to be most harmful and most likely to occur are identified and studied in offline simulations. Measurable indicators that can be utilized to identify the onset of critical conditions are identified. These can be complex, ad hoc sets of signals. Finally, even though the system has been extensively modeled and the onset of critical conditions detected automatically, a human is typically included in the feedback loop to ensure correct operation. As mentioned, a human is required in order to decide whether to arm or disarm the special protection system.

4.2.6 Oscillographs

The oscillograph has been used for measuring alternating current and voltage. While many power grid devices are becoming more sophisticated, the oscillograph and the oscillogram that it produces represent the type of information that now tends to be digitally recorded and transmitted. Oscillographs throughout the power grid allowed measurements of current waveforms that enabled projection systems to be analyzed. Experts reading oscillograms could characterize a fault, including the type and location of fault, as well as the performance of the protection system in response to the fault. For example, an oscillogram could indicate whether a power line carrier communication system was operating correctly with regard to blocking or permissive operation in response to a fault.

4.2.7 Protection Coordination

Protection coordination is a general term for designing a protection system comprised of one or more protective mechanisms such that the mechanisms work with one another to provide the level of protection required with minimal impact upon normal operation. As a simple, specific example, consider a set of protective relays positioned along a radial distribution line emanating from a feeder. The relays should be designed to trip only when a fault occurs within the section of the line that they are designed to protect. This means that protective relays located farther from the feeder should be more sensitive; that is, they should be set to trip sooner than relays closer to the feeder. This is because all the protective relays on the same radial line will sense a fault current, but we only want the relay upstream, but closest to the fault, to trip. This way, only a minimal number of customers will be impacted when the relay opens the circuit. If a protective relay further upstream from the fault, and thus closer to the feeder, were to trip first, it would cut off power to more sections of the radial line than necessary. This would mean more customers lose power than is necessary. However, this is a relatively simple example. Recall all the different types of protection mechanisms that have been discussed, from fuses to distance relays, automatic reclosers, and current limiters. Each of these protection mechanisms can interoperate with, and overlap with, other protection mechanisms in complex ways. As another example, distance relays can be set up to provide backup protection for other protection mechanisms.

Utilities have a variety of metrics, discussed in more detail in Section 12.5.1, that measure the number of customers impacted by a power outage and the duration of the outage. Thus, there is motivation to provide adequate protection from faults while also minimizing the impact upon customers caused by unnecessary or improper tripping of protection mechanisms. There has been much work involved in designing optimization algorithms that attempt to automatically configure protection systems that minimize the number of customers impacted and the duration of outage subject to the constraint of adequately isolating a fault. A related topic is that of FDIR, which is covered in more detail in Chapter 9. The concept is to rapidly isolate a fault while impacting as few customers as possible and then finding a means of mitigating the fault and restoring power as quickly as possible. Distribution automation, or simply "DA," includes attempts to automate the FDIR process by means of communication among protective mechanisms.

From a communication perspective, protection coordination is a complex system in which much of the communication that takes place is implicit; the fault current itself is the communication signal, and local reaction to that signal accomplishes the required response to the fault. Explicit communication among protective devices can help to ensure correct and efficient operation. However, explicit communication always has a chance of failure, however small, that adds to the probability of failure for the entire system if not designed properly. Typically, explicit communication mechanisms are included on top of (that is, in parallel with) existing, proven protection techniques. This ensures that protection system reliability will improve when advanced communication is added.

4.3 Conservation Voltage Reduction

CVR dates back to the 1980s, when utilities began experimenting with reducing the voltage of their feeders. Since power is the product of current and voltage, reducing voltage reduces the amount of power consumed. Many devices are unaffected by small reductions in voltage. Not only the active power is reduced, but the reactive power required is also reduced. Typically, utilities used models of a particular distribution system as well as assumptions of user usage profiles to set lower voltage levels. The problem is that a voltage that is too low can lead to voltage collapse, causing brownout or blackout. Industrial consumers that have widely varying loads have been a particularly challenging area in which to implement conservation voltage reduction. More advanced CVR implementations were developed that could measure voltage levels in near-real time and respond to dynamic changes in load and voltage. This is, of course, the essence of the smart grid: the ability to dynamically sense and respond to changing conditions using communication.

It should be understood that CVR is old terminology that has largely been used with, or replaced by, voltage optimization (VO), volt-VAr optimization (VVO), and IVVC. All of these terms express the notion of optimizing voltage in conjunction with other closely coupled values such as reactive power. Because CVR has long been recognized as a relatively simple means of reducing power consumption, it is likely to be among the first applications of the smart grid and smart meters. The goal is to carefully monitor and control the voltage to keep it as low as possible without sinking below the American National Standards Institute (ANSI) C84.1 standard that establishes the required voltage ratings. Voltage fluctuates as power consumption varies throughout the day, so reducing the voltage without falling below the standard is not simple to control.

The impact of CVR on loads varies with the type of load. For simple resistive loads, the power varies linearly with the voltage, thus resulting in a simple reduction in active power. However, there is a nonlinear relationship between voltage and power in inductive loads, such as motors. The result of CVR on inductive loads is to reduce the need for reactive power, thus increasing the available capacity to deliver active power. Keeping the voltage relatively low for three-phase motors also has other benefits, including increasing the motor's lifetime. This is because, when such motors are driven with more voltage than needed, their iron cores saturate and eddy currents form. This results in heat buildup rather than useful work. The additional stress on the motor reduces its lifetime. On the other hand, if voltage falls too low, slip in the motor can increase. Slip is a reduction in the speed of the rotor relative to the changing magnetic field that is driving the rotor. On the other hand, sophisticated motors with variable-speed drives may draw more current to compensate for the reduction in voltage. However, these motors will be more susceptible to dips in power.

Now that we have seen many existing applications for communication in the distribution system before the smart grid came along, let us consider how communication was handled. Specifically, a very old communication system, distribution line carrier, utilized the distribution system itself for communication.

4.4 Distribution Line Carrier

Distribution line carrier, as the term implies, is a form of power line carrier designed for use in the distribution system. The power line carrier in general is discussed in more detail in Section 8.5. In examining the distribution line carrier, it is clear that the smart grid concept is far from new; a British patent from 1898 describes power line communication for transmitting meter data through the power grid. Thus, "smart grid" communication has been in use for well over a century. A basic concept recognized very early in the development of the power grid was that power lines throughout the distribution system can serve a double purpose: they are a conductive path for power and also serve as a waveguide for communication using electromagnetic waves. Without a waveguide, electromagnetic waves would travel outward from a source in every direction simultaneously. However, if carefully coupled to a conductor, most of the electromagnetic wave energy can be directed into the conductor such that the wave is confined to the geometry of the conductor, also known as a waveguide. Thus, the waveguide directs the wave along the length of the waveguide structure, allowing most of the energy to reach the opposite end of the waveguide from the source rather than propagating omnidirectionally through space. From Maxwell's equations, the electromagnetic waves are comprised of both an electric and magnetic field, or a corresponding E and H field. The E and H fields travel through space at right angles to one another. The system can be analyzed using basic wave concepts such as interference and reflection.

Some of the classic problems with using distribution line carrier include the fact that higher carrier frequencies will not pass through transformers very efficiently and that harmonic noise interferes with the signal. A lower frequency will pass through transformers more effectively;

however, 60 Hz related harmonic noise becomes an issue. Techniques have been in use since at least the 1980s that divide the information from a communication channel into multiple bands that reside in the low-noise regions of the 60 Hz spectrum. This has allowed more data to flow through distribution line carrier without being impacted by noise. Application of this technology has been used primarily for automatic time-of-use meter reading and protection mechanisms. Interestingly, researchers from the 1980s and earlier specifically mention that they are solving the problem of "DA"; thus, again, we see that "smart grid" communication applications are far from novel ideas.

4.5 Summary

This chapter introduced the pre-smart-grid era power distribution system. Individual components of the distribution system were introduced. Then the functionality of the distribution system was explored primarily through its topology and protection mechanisms. The size of the distribution area, topology, and protection mechanisms used in the distribution system will have a significant impact on the communication architecture, as will be seen in later chapters. Common applications used to manage the distribution system were also introduced with an eye toward understanding their communication requirements. Finally, distribution line carrier was introduced as an old technique for utilizing power lines within the distribution system for communication.

Chapter 5 introduces the next step in the traditional transport of power; namely, power consumption. As one may joke, the power system would be so much easier to manage if it was not for those annoying loads. The consumer side of the power grid is why the system exists in the first place. Responding to the impact of large numbers of loads, each of which exhibits dynamic power consumption, is one of the main challenges of the power grid. DSM has been an approach to solving this problem and is one of the main topics introduced in Chapter 5, along with early solutions leading toward the microgrid concept. Chapter 5 will close with some early evidence of consumers' willingness to utilize smart grid concepts and the introduction of the electric vehicle, which directly impacts the consumer.

4.6 Exercises

Exercise 4.1 Distribution Network Topology

1. What are three general types of distribution network topologies?
2. Why is power protection critical in the distribution system?

Exercise 4.2 Fault Detection, Isolation, and Restoration

1. What is FDIR?
2. How does fault detection, isolation, and recovery depend upon the topology of the power distribution network?
3. What role could communication play in automating FDIR, and what would be the most important requirements for such a communication network?
4. What factors make communication more challenging in the distribution system than in the transmission system?

Exercise 4.3 Power Protection Coordination

1. What is power protection coordination?
2. How is timing used in power protection coordination?
3. How could communication aid in power protection coordination?

Exercise 4.4 Pilot Protection

1. What is pilot protection coordination?
2. What are permissive and blocking protection schemes?
3. How has communication been used in pilot protection schemes? What are the key requirements of communication in such protection schemes?

Exercise 4.5 Overcurrent Protection

1. What is the inverse TCC curve and how is it used to implement protection?
2. Does the inverse TCC curve provide an indication of the latency requirements for a communication in support of protection? If so, how?

Exercise 4.6 Distance Relay

1. What is a distance relay, and how does it help implement power protection?
2. What are three different types of distance relays?
3. What is the "reach" of a protective relay?
4. Why is developing a protective relay for direct current more challenging than for alternating current?

Exercise 4.7 Distribution Line Carrier

1. What is distribution line carrier?
2. What are sources of noise that interfere with distribution line carrier communication?

Exercise 4.8 Digital Relay and Recloser

1. What is a digital relay and how does it operate?
2. How does a recloser differ from a digital relay?
3. What is a pulse recloser?
4. Why is there a serious potential problem if there is a DG downstream (on the opposite end from the feeder) from a recloser or relay?
5. How could communication help the above problem?

Exercise 4.9 Special Protection Scheme

1. What are special protection schemes?
2. Why do they generally require significant use of communication?

Exercise 4.10 Phasor

1. Explain how a phasor relates a periodic waveform to a vector in the complex plane. What well-known identity enables this relationship?
2. What geometric shape would the phasors form for three waves to perfectly cancel one another?
3. How do phasors convert the analysis of problems from alternating current to direct current? Note that $dq0$-transformation uses a similar concept and is described later, in Chapter 8.

5

Consumption

We will make electricity so cheap that only the rich will burn candles.

—Thomas Edison

5.1 Introduction

This chapter ends Part One of the book dealing with the power grid in the pre-smart-grid era. Consumption also ends the classical chain of power systems leading to the end user. We began with generation and the tendency toward large, centralized power plants, transmission of large amounts of power over long distances to reach populated areas, and distribution, which transforms power to become suitable for the end user and distributes it to individual users. Even though it may not be readily apparent, the end user can be considered part of the power system. In fact, some aspects of the smart grid involve finding ways to either manage loads the end users operate or manage end users themselves through pricing schemes such as DR. All of these concepts existed and have been tested or put into use long before the smart grid. The hope is that recent advances in communication will make such mechanisms more reliable, ubiquitous, and cost effective. Thus, it is important to learn from the past before jumping into smart grid promises for the consumer. It is easy to become so involved in the technology that we forget that the consumer is the most important aspect of the power grid. If the consumer cannot see value in the smart grid, it will not become reality.

This chapter begins with a discussion of electric loads and their interaction with consumers and the power grid. This includes the electric power market and its brief history and direction; the relationship among load, supply, and electric current frequency; and operation and management of the power grid. The next section focuses upon issues related to the variability and lack of control of power consumption. In particular, we look not only at the problem of peak power demand, but also various ways in which peak power has been quantified and measured. As mentioned earlier, this chapter is focused on the most important aspect of the power grid: the consumer. Thus, the next section focuses upon problems that the power grid may be able to solve for the consumer. Similar to the development of the Internet, the best new ideas and many of their solutions will come not from academia and researchers, but from practical problems faced by the average consumer along with their creative ideas for problem

Smart Grid: Communication-Enabled Intelligence for the Electric Power Grid, First Edition. Stephen F. Bush.

solving. The Internet became a useful tool only after the average programmer of the 1980s and 1990s gained interest in developing standards and protocols to solve practical problems; for example, as part of Internet Engineering Task Force working groups. Similarly, it will only be when the average electric power consumer finds smart grid services worth developing into practical solutions to practical problems that the smart grid may be considered successful. While it is nearly impossible to know which direction the creativity of consumers will go, there are some obvious practical problems that are likely to be addressed as well as hints from surveys discussed in this chapter. One anticipated problem relates to the lack of visibility into the power grid. It would be ideal if the utility had enough visibility to determine precisely when and which consumers are impacted by a power outage, rather than relying on customers to call in outages. It would also be ideal if consumers had enough visibility to determine precisely when an outage may happen and precisely when power will be restored. It would also be ideal for customers to determine precisely how their power is being used and how best to save money. Another anticipated improvement for the consumer is how best to purchase power. This also could include whether consumers are able to pick and choose their power producers, perhaps selecting aspects such as whether their power source is renewable, the power quality, the reliability, the efficiency of transport, along with, of course, cost. Then there is the issue of the consumer that can also be a producer of power. The portmanteau for this is "prosumer." This leads to a discussion of microgrids, particularly small microgrids used by consumers allowing them to sell the excess power that they produce. Finally, the issue of flexibility for the consumer or prosumer is discussed. This covers aspects related to how well the power grid can adapt to user preferences. This could include whether the consumer, or even the prosumer, can operate in a mobile environment. An example is whether the consumer or prosumer can plug into the power grid at any location and expect it to properly charge, or pay for, power consumed or generated while on the move. This chapter ends with a summary of key points and exercises allowing the reader to think further about the concepts introduced in this chapter. While this part of the book has served to provide the required background focusing upon the power grid before the smart grid era, this chapter, by examining the impact upon the consumer, begins to look at smart grid concepts, which will be picked up in more detail in Part Two of the book.

5.2 Loads

Electric power consumption is all about the characteristics of loads on the power grid, particularly the aggregate characteristics of loads. A primary characteristic of a load is its impedance. Recall that impedance has both a resistive and a reactive component. An individual load may have a fixed impedance, one of the simplest examples being the old-fashioned incandescent light bulb, which is a highly resistive load. Its impedance is fixed; the bulb is either on or off. Other loads may be more complex, such as motors that have a much larger reactive inductive component as well as a variable speed, allowing it to draw more or less power through a continuous range over time. Loads may also be capacitive, such as electronic computing devices. However, the impact of capacitive load tends to be negligible; loads are overwhelmingly inductive. From the electric power utility perspective, a primary concern is the aggregate statistical properties of these loads: namely, what does the combined load look like and how does it change over time? Until energy storage is perfected on a large scale, generation has to precisely match this aggregate load profile at all times.

As mentioned, purely resistive loads are the simplest to understand. They have no inductive or capacitive reactance, which means that their power factor is one. This includes incandescent light bulbs as well as almost all other forms of resistance-based heating devices, such as toasters, electric blankets, portable heaters and electric furnaces, and electric ovens. In each case, power is converted to heat based upon the familiar I^2R power loss. Using Ohm's law, this is equivalent to V^2/R, where we have substituted $V = IR$ into I^2R. V^2/R explicitly shows the role of resistance of the device. As the resistance decreases, power given off as heat increases. Since loads are attached to the grid in parallel, the voltage remains at a constant level unless otherwise modified at the consumer premises. Purely resistive loads are relatively tolerant to changes in power quality; they can generally operate effectively even if the voltage or the frequency changes moderately. Clearly, an excessive voltage increase, as seen in the V^2/R version of power dissipated, will cause the device to rapidly overheat as the square of the voltage and likely lead to destruction of the device. Utilities in the USA generally must keep the voltage within $\pm 5\%$ of the standard 120 V. Finally, resistive loads do not even require alternating current; they will often operate just as well with the equivalent amount of power supplied in the form of direct current.

Dimmer circuits used with lights can reduce power, not by changing the resistance, but by rapidly turning the voltage on and off, similar to the concept of pulse-width modulation. Thus, the longer the on-time, the more power is received and dissipated as light or heat by the device. The pulses are so rapid that the human eye cannot detect them. The dimmer does not need to be very accurate in the quality of its waveform output because is only sensitive to the amount of power received.

It is instructive to consider the difference between an incandescent bulb and a fluorescent lamp. A fluorescent lamp, as a load, differs considerably from the discussion regarding the incandescent bulb. The fluorescent lamp is not purely resistive as it has a ballast that acts as an inductive load. Generally, resistance has a positive value. If we examine resistance as $R = V/I$, a positive change in voltage through a resistance results in a positive increase in current through the resistance. However, for some materials, such as the resistance implemented in many types of fluorescent lights, a reduction in voltage can actually result in drawing more current. Thus, instead of a voltage drop across the resistance, or load, there is a voltage rise across the load. An increasingly large amount of current will be drawn by the device until damage occurs. Thus, a current-limiting device known as an electronic ballast must be used to keep the negative resistance from causing damage. The goal of the ballast is to make the device appear as a typical load, with normal impedance. A simple ballast can be purely resistive, but most used in fluorescent lights provide an inductive impedance. Such a device could not utilize a dimmer that operates as previously described because the inductive ballast depends upon a reasonably shaped sinusoidal waveform. It is more sensitive to power quality than the purely resistive incandescent bulb.

Instead of using a dimmer that changes the duty cycle, or the relative on-time of the voltage, one could simply attempt to change the voltage itself by inserting a rheostat, or variable resistor, in series with the device to be controlled. There would be a corresponding voltage drop across the rheostat, thus changing the voltage supplied to the light or the device being controlled. The problem with this approach is that the rheostat then becomes a load. In order to provide the voltage drop, it dissipates power according to a rate of I^2R. This is not only wasteful and inefficient, but it dissipates enough heat to be dangerous.

Let us move from the purely resistive to the inductive. Electric motors are largely comprised of conductor windings in both the stator and rotor. Thus, there is a tremendous amount of inductance in a motor. Over 60% of the power consumed in the USA is consumed by motors. This includes motors in fans and pumps to power tools and electric lawnmowers. Motor power is historically given in horsepower, which is equivalent to 0.746 kW. Recall Chapter 2.2.4 on generators; a motor is essentially the same device as a generator operated in an inverse manner. In other words, instead of supplying mechanical power to turn the rotor to create an electromotive force to generate current, current is applied to the same device to cause its rotor to turn. Similar to generators, there are three types of motors: synchronous, induction, and direct-current. As previously discussed in Chapter 2, the induction motor is simplest in construction; over two-thirds of the motors in the USA are induction motors, and they use 90% of the motor power consumed.

In order for the induction motor to cause its rotor to turn, magnetic fields must create torque on the rotor. Torque is the amount of force applied multiplied by the distance from the axis of rotation; applying force further from the axis requires a smaller force. Just like an induction generator, an induction motor induces the magnetic field that exists in its rotor, which pushes against the magnetic field in the stator. However, as described in the operation of the induction generator, for this induction to occur the stator field must be rotating, not at the same speed as the rotor, but faster. This difference in speed between the stator and rotor is known as slip.

An interesting phenomenon occurs when the motor first starts up. Because there is initially no alternating current and no field to impede current flow, the initial current entering the motor windings has very little impedance. It is almost as if a short circuit occurs for a short period of time, namely the first few alternating current cycles. This is known as inrush current, and it is not limited to motors. Any device with highly inductive coils that initially starts up will experience an inrush current, particularly transformers and capacitor banks. Inrush current makes electrical protection mechanisms difficult to implement because it can be difficult to distinguish inrush current from a true electrical fault; that is, a real short circuit. Protection devices are generally time based, meaning that they are designed to operate based upon a time–current curve such that an overcurrent is not considered a fault unless it exceeds a given duration. Usually, protection is designed such that large currents trigger operation of the protective device, such as a fuse or circuit breaker sooner, while a lower amount of overcurrent takes longer to cause the protective device to react. The topic of protection is discussed in more detail in Section 4.2.7. The bottom line is that a large inrush current may not trip a fuse or circuit breaker, but it may cause lights to dim and other devices to slow as short-duration, large-current inrush causes the voltage to drop temporarily.

After the inrush current enters the motor and establishes the required internal electromagnetic fields, the motor still draws significant power until it can "get up to speed." This is known as the starting current. Large motors contain electronics to help soften the impact on the power system as these motors first start up and achieve their operational speed. Again, motors are significant inductive loads for the power grid; their power factors range from approximately 0.6 to 0.9.

Synchronous motors, like synchronous generators, have a rotor with its own power supply for its magnetic field. Thus, induction is not required to create the rotor's magnetic field. The stator magnetic field directly "pushes" the rotor, so that the stator and rotor follow one another at the same speed. Thus, an equivalent synchronous motor tends to be less inductive than an induction motor, although it is still highly inductive.

Reluctance motors are another form of motor. They have a stator that is similar to the synchronous and induction motor; however, the rotor has no windings. It consists only of ferromagnetic material. The concept is that the rotor's ferromagnetic material will align itself so as to minimize the magnetic reluctance through the poles of the stator. Both reluctance motors and direct-current motors have much less of an impact on the power grid since there are relatively few of these devices.

Another motor distinction resides between single- and three-phase motors. Three-phase motors run more smoothly, given the naturally even and balanced separation of waves in three-phase current. Most residential homes use only single-phase current, although it is possible to convert single-phase current into three-phase current by simply splitting the voltage three ways and adjusting their phases.

Because electric motors consume the majority of power that is produced, their operating efficiency becomes critical in the aggregate. Motors have efficiencies that range from 65% to 95%. Part of increasing motor efficiency can come from efficient monitoring and sensing of motor health and motor controls. Adjustable speed drives work by converting alternating current to direct current, adjusting power, and converting back to alternating current. There is power loss in this conversion process; however, the gains in motor efficiency by having variable-speed control generally outweigh the loss due to conversion. The cost of power to run a motor can exceed the original cost of a motor by 10 times or more; thus, increased motor cost to reduce energy consumption can be a net gain many times over. In fact, owners of expensive motors in an industrial plant will purchase additional power system hardware to improve the utility power quality in order to improve the lifetime of their motors.

Finally, we come to consumer electronics, the semiconductor, computer, and communication devices upon which we have all come to depend. These consume very low direct-current power; thus, they would seem to have very little to do with the power grid. However, almost all of these devices operate, or recharge, through a power supply that plugs into the power grid. These devices are focused largely on moving and manipulating information, something that sounds minuscule compared with the real work done by heavy-duty industrial motors.

Clearly there is a small amount of power dissipation in computational electronic devices through resistive elements of the electronics as well as perhaps small inductive and capacitive reactances, being small relative to industrial machinery. However, there is another small, but interesting phenomenon that may dissipate energy as well. This phenomenon is interesting because it ties energy to information, something that is a significant goal of the smart grid. This phenomenon is known as Landauer's principle. Conceptually, the idea is rather simple; the second law of thermodynamics states that entropy within a closed system can never decrease. The concept is that, as an electronic computation takes place, if the number of logical states of the computation were to decrease – that is, become irreversible – without any compensating effect, this would result in a violation of the second law of thermodynamics. Thus, the way nature compensates for loss of computational entropy is by a corresponding increase in physical entropy, in the form of heat. Simply put, for temperature T, the energy emitted given entropy S is $E = ST$. If one bit of logical information is irreversibly lost, then the amount of entropy generated is $k \ln 2$. The energy dissipated into the environment for one bit is thus $E = kT \ln 2$, a very small amount.

The goal of semiconductor electronics is to fit more transistors into a smaller space while using less current so as not to overheat the device due to the dense circuitry. Even when electronic devices are turned off, they may continue to consume small amounts of power. For example, remote-controlled devices need to have enough system components running in order to detect when the remote turns the device on. This is known as standby power. Also, power converters and chargers left plugged into outlets continue to run current through their coils. While the dissipated energy is small for a single device over a short period of time, the energy dissipated by many such devices adds up over the course of a year. The heat given off by these devices is often accompanied by energy required to cool the surrounding environment. This may be done explicitly, as an example, for a room full of large computers that needs its own cooling system. However, this may also occur unconsciously, as people turn up air conditioning in order to counter the impact of heat contributed from electronic devices in their homes. The bottom line is that reducing heating in devices not only reduces energy wasted by the device itself, but it also reduces energy required for additional cooling.

From the utility perspective, the combination of electronic devices looks like a resistive and inductive load. The primary difference from other inductive loads, such as motors, is the significant sensitivity of electronic devices to power quality. One simple example of sensitivity is the relationship between stable alternating current frequency and clocks. As a personal example, consider the digital clock; we have become acutely aware of power outages by the blinking digital clocks and stereos that need to be reset after every momentary power outage. It was not until 1926 that electric clocks, driven by a synchronous motor, were developed. This means, of course, that the clock motor's stator requires a precise 60 Hz waveform otherwise the clock's time will speed up or slow down along with the alternating current frequency. Thus, clocks became dependent upon power grid operators to ensure that the frequency was maintained. In fact, power grid operators regulate the power grid frequency so that synchronous motor clocks will stay within a few seconds of the correct time. Whenever clock error exceeds a given amount in North America, which varies among North American interconnect systems, a correction of ±0.02 Hz may be applied. This occurs when the clock error exceeds 10 s in the East, 3 s in Texas, or 2 s in the West. Time error correction starts and ends on either the hour or half-hour. In continental Europe, the deviation between the alternating current phase and the standard 50 Hz frequency is computed each day at 8 a.m. in a control center in Switzerland. The frequency is adjusted by up to ±0.01 Hz away from the standard 50 Hz as required in order to maintain a long-term average frequency of $24 \times 3600 \times 50$ cycles per day.

From the perspective of the power grid, an individual load is an infinitesimal part of the entire load. The power grid sees one aggregate load comprised of many individual loads. However, each individual load contributes a small portion depending upon its individual usage; that is, when it is turned on and the magnitude of its load over time. Remember that the power supplied needs to balance the power demanded; thus, the grid operator is concerned with instantaneous power (that is, power supply and power demand) at any instant in time. The typical user, on the other hand, is usually primarily concerned about their electricity bill; that is, how much energy, which is power multiplied by time, they consume.

Until recently, it has always been assumed that it is the job of the utility to meet the load requirements of consumers, regardless of cost or efficiency. In other words, power systems engineers would treat load as an external parameter that is outside their control; the consumers' demand must be met at all times at all costs. However, there has been a steady and growing trend to balance this policy; progressively more pervasive mechanisms have been developed in an attempt to allow the utilities to exercise some control over the loads themselves. This has ranged from simple suggestions and requests, to pricing schemes that encourage desired change in load usage, to direct control of industrial and residential appliances by utilities. The pricing mechanisms to control loads have resulted in moving from a service-oriented power grid to a market-oriented power grid. This falls under the DR schemes of the smart grid and is discussed in Section 7.2. A market-oriented power grid provides great incentive to predict loads for pricing purposes. Load forecasting uses many types of historical data, ranging from user profiles to weather information, to attempt to predict load.

Now we discuss in more detail the timing issues involved in power supply and demand. The term "coincident demand" is used to classify the amount of energy required by a customer or set of customers during a given time period. It is determined by the number of devices that the customer group of interest has operating at the same time during the time period of interest. If we think of each device's operating time as a line segment on the real number line, it is the number of such line segments that overlap with the time period of interest. The term coincident *peak* demand is the demand by the same group of customers during peak system demand. A customer's coincident peak demand is typically taken from meter readings at the time when that customer's demand is likely to be highest. Noncoincident demand is the demand by a given customer or set of customers assuming *all* their loads are operating. Typically, a customer or set of customers would not have all their devices turned on at the same time; thus, their operation would not normally "coincide" with one another.

Thus, determining total energy demand is often viewed as the problem of determining, from a statistical standpoint, how much coincident load will exist at every point in time. As many network communication engineers see almost immediately, coincident load is similar to transmission requests at the media-access layer of a communication network. Coincident demand is similar to the expected number of transmissions that would overlap in a given number of time slots. If we consider a hot, humid day, many residences have running air-conditioners. However, air-conditioners have thermostats that limit their lower temperature; they do not have to run continuously, only enough to maintain a given temperature. Assuming the air-conditioner of each residence is set to roughly the same temperature, there are differences between when the compressors of the air-conditioners are actually operating. Only a subset of all the compressors is actually running simultaneously; this variance helps to minimize the peak load.

Another way to think of the problem is from the utility and supply perspective. If the utility had to supply all customer noncoincident load – that is, if all possible loads were operating simultaneously – it is unlikely that any utility could meet the demand. It is diversity in load usage that allows the utility to meet the demand; generally speaking, the more diversity, the better, since the load is then spread as evenly as possible over time. In fact, there is a measure called the "diversity factor" that measures the ratio of the total noncoincident load for a set of customers to the coincident load for the same set of customers. Thus, the diversity factor is a number that is typically larger than one that provides an indication of how spread out load usage is over time.

While it may seem as if noncoincident demand plays a small role, since all loads are rarely on at the same time, it is possible for this to happen. As a simple example, consider what happens after a power outage of moderate duration. When power is restored, it likely that nearly all devices, including air-conditioner and refrigerator compressors, will be activated immediately when power is restored in order to compensate for the increase in temperature that occurred during the outage. A larger problem is the inrush current, as previously described. This larger-than-normal current that occurs due to the initial low impedance of inductive loads can cause a current spike capable of destroying power system equipment, such as transformers, when power is turned on after an outage. Utilities prefer that consumers turn off devices when power is initially restored in order to avoid this problem.

There are several different ways that utilities look at load statistics. One of the simplest is a load profile. A load profile is simply the instantaneous power used versus time by a consumer or aggregate of consumers. A chart similar to this is typically included in one's electricity bill, sometimes along with a comparison with the load profile from the previous year in order for the customer to see the difference in usage from the previous year. From the utilities' perspective, one of the key attributes of the load profile is the characteristic of its maximum value, or peak. A sharp peak could indicate that additional generators are required in order to meet demand over that peak period. These generators are expensive to start up and shut down, particularly when run for a short duration – that is, just to meet the peak load requirements.

Another way to view consumption statistics is called a duration curve. This is similar to a load profile, except that the values are sorted by the value of the instantaneous power demand itself, rather than time. Sorting is done in ascending order, so that the curve will start from higher values and is monotonically decreasing. This curve is useful for finding the probability that a given load power demand will be exceeded.

The ratio between the average and peak power demand is known as the load factor, which is not to be confused with the power factor. From the utility standpoint, a high load profile with a high power factor is the ideal case. This is because the utility is paid by consumers based upon consumption rate, but with low peaks in consumption there is no need to pay the expense of starting up a new generator to meet peak demand.

The growth of interconnectivity in the power system allows many advantages related to cost saving. This is a prime example of the power of networking in general. For example, a highly interconnected power system allows power generation, and the generators themselves, to be shared efficiently. Also, interconnectivity allows utilities to share load in such a manner as to increase diversity.

A typical residential consumer does not have full, three-phase power provided by a generator, only a single phase; that is, one of the three phases. In fact, what the typical residential customer sees is a "hot" black wire that connects to the smaller slot in a power plug. Such a polarized plug has different slot sizes to ensure that the "hot" wire and neutral wire are connected in the correct manner. Of course, current is changing direction, so this is rarely important for the devices themselves; however, devices are often designed to improve safety. The hot wire is placed in a harder-to-reach location so that it is not inadvertently touched. For example, it is much safer to have the hot contact located on the inside back of a light fixture rather than the larger, metal-threaded cylinder that may be touched. The neutral wire is white and completes the circuit. The ground wire is not part of the circuit unless there were an electric fault. In this case, the ground dissipates excess current. Many residences also have an additional, higher voltage line entering the home. This is a 240 or 208 V line for larger appliances that require higher voltage. It is important to note that while this involves three wires entering the home – namely, a neutral line, a 120 V line, and a 240 V line – this is not three-phase current. These currents come from a transformer that is on a single phase; the different voltages come from different points, or taps, in the secondary transformer winding.

In Section 5.2.1, we move from loads and their impact on the power grid to paying for the power that runs the loads. Part of the smart grid includes a closer coupling between operation of the power grid and the purchase and price of power. More specifically, part of DSM includes utilizing the price of power as a signal to influence demand. To understand this, we need to review the power market as it existed prior to the smart grid and where it is evolving.

5.2.1 *The Power Market*

Let us begin with a simplified view of the power market. There are three main entities: (1) the ISO, who runs the system but has no financial stake in the business components; (2) the LSEs, who are essentially the utilities or retail power providers; and (3) the GenCos, who generate the power. There are both day-ahead and a real-time markets. We can assume that the ISO has the goal of operating an efficient and reliable market. The ISO operates the day-ahead market using location marginal pricing (LMP), which results in a marginal price for each unit of power at any location within the grid while taking possible congestion into account. Both the LSEs and the GenCos have the main goal of making a profit. The LSEs want to buy power as cheaply as possible and sell it at a higher price to end customers. Thus, the LSEs need to have some knowledge of the demand for power for the following day. An LSE places a bid for power at both a single price for the next day as well as 24 hour price-sensitive bid requests. The GenCos want to produce power as cheaply as possible and sell it to the LSEs at a profit. Each GenCo will submit an offer for a price to sell power for each of 24 hours for the next day. The offer is comprised of the marginal cost of the power over an operating capacity interval. The ISO then publicly reports both the hourly supply and demand commitments and the LMPs for the next day for a bid-/offer-based, direct-current optimal power flow problem. Recall that power grid congestion is accounted for in the LMP, so the goal is for the market to take congestion into account. Any differences that arise between the amounts committed to be bought and sold and the amounts actually used are resolved in the real-time market at the end of the day on which the power was used. Thus, the day-ahead market is always trading ahead for the following day and the real-time market is clearing any differences in purchased and actually used power for the current day.

In the past, each utility had its own system operator to regulate power generation within its region of operation, keeping in mind that a particular region can span many states. However, the system has changed such that an ISO controls power generation within a given region for all the utilities in that region. The system operator has the job of maintaining the balance of power generation and demand in their area. The real-time error between actual and scheduled power imports and exports in a region controlled by the operator is called the area control error (ACE). A positive value indicates more imported power, generation, or less load or exported power than expected; in this case, generation capacity can be reduced. A negative value indicates that generation within the area must be increased because more power than expected is leaving the system. Typically, the command to increase power generation required human intervention; however, generators implemented with automatic generation control can receive a signal directly from the ISO to adjust generation capacity. When load becomes too high, another real-time control operation is to manage load shedding, deciding when and whom to disconnect from the grid until the system is able to handle the load again. This can involve disconnecting customers that agree to have interruptible loads, or it could mean forced outages. These outages can be done through rotating blocks of customers via a rotating outage. This form of load shedding takes place within the distribution system. Finally, real-time control has been required for both reconfiguring the system for maintenance and restoring the system after a fault, particularly after a large or complete blackout. In these cases, very careful control must be exercised in order to either reconfigure or reconnect the system so as not to introduce instability, as discussed in an earlier section.

The next level of time scale is the level of scheduling, which begins to impact more closely upon economic concerns. Up until now we have been discussing purely technical aspects. In the past, both economic and technical concerns were handled by the same operational organization. In order to prevent fraud, abuse, or simply unconscious bias, power system operation has been restructured so that economic and technical aspects are handled separately, by different people with different interests. When technical and economic goals are in the hands of the same operator or operators, they can potentially operate the system so as to gain an economic advantage. As a simple example, they can schedule particular generators to operate for nontechnical, purely profit-driven, reasons. Perhaps they could route power or cause congestion in a preferential manner. Thus, the term "independent" in ISO is meant to indicate a certain detachment from potentially abusive economic motivations when operating the power system.

The concept of changing DG from being considered a negative load toward explicit incorporation into economic dispatch is taking place. The challenge with wind, as with other forms of DG, is that future generation capacity is unknown. In other words, if one owns the wind turbine or wind farm, there is little operational cost; however, the duration and strength of wind gusts is, of course, impossible to control and nearly impossible to predict with any significant level of accuracy. In a conventional generator, such as a steam generator, the operator has precise control over the generation capacity and the corresponding amount of fuel used. It is true that it is difficult to predict the price of fuel; however, predicting wind is a much more difficult task given the larger volatility of wind. Thus, we have unpredictable renewable power sources and unpredictable load behavior; finding ways to allow these unpredictable systems to cancel one another is one solution.

The goal, as in all economic dispatch algorithms, is to solve an optimization problem, to determine how to run the generators so as to minimize the cost to produce precisely enough power to balance anticipated load. Since the instantaneous wind power available at any given time is difficult to predict, simplifying assumptions have been made. For example, prior research has shown that the wind-speed profile at a given location tends to follow a Weibull distribution. Assuming this to be true allows for a great deal of simplification in setting up the optimization problem. The basic idea is that factors must be included in the optimization model for either over- or under-estimating available wind power. If wind power is overestimated, then that means, in actuality, there will not be enough power to meet demand. There needs to be a penalty for having to purchase the power from an alternate source to make up for the shortfall. On the other hand, if the available wind power is underestimated, and more is produced than was planned, the power will have to be exported, causing other generators to change their schedules, or maybe wasted by being forced through breaking resistors.

Economic decisions are made by a scheduling coordinator. Scheduling generators to meet load requirements is known as *unit commitment*. The scheduling algorithm is known as an *economic dispatch algorithm*. The goal of the economic dispatch is to fill in the area under the load–duration curve, discussed previously, with generation while also meeting all necessary constraints. The cost of generation takes into account the marginal cost of each generator's output in both fuel and operational cost. The algorithm must also take into account power flow analysis and possible line losses that would occur when using each generator. This is expressed as a penalty factor that is used to adjust each generator's cost. There are three general categories of generation: *baseload*, *load-following*, and *peaking* units. Baseload units are the cheapest to operate and are best when used on a continuous basis; examples would be coal-fired or nuclear plants. Load-following units are ideal for responding to changes in demand; examples are hydroelectric and some steam plants. Peaking units are expensive to operate and are best held in reserve to meet demand peaks; an example is gas turbines. DG, such as solar and wind power, has often not been explicitly included in the economic dispatch algorithm. Instead, such smaller and distributed generating units have been considered "negative load," in that their relatively random nature of generation has been aggregated with demand to more or less cancel the demand.

Note that, in the past, the economic dispatch algorithm was implemented in a centralized manner; a single organization would predict the load and schedule generators accordingly. In later changes to regulation, a more competitive, free-market approach has been utilized. Here, there are bilateral contracts between consumers and power generators with a "power pool" that serves as a clearing house for buying and selling power. An entity serves as the scheduling coordinator to keep track of the power being bought and sold; the process can operate similar to an auction in which generation is awarded to the lowest bidder on an hour- or day-ahead basis. An item that has to be accounted for is known as *spinning reserve*. These are generators that are actually up and running in synchrony with the grid, but not connected or generating power. They can be brought online quickly to meet any unanticipated changes in demand. Another complication is control and operation of reactive power. Reactive power is handled in a manner similar to active power; its demand is predicted and generators bid to produce it.

Even though modern economic dispatch is theoretically a nice, neat, orderly, free-market approach to the problem of running the power grid, problems can occur; what should have been planned ahead of time can be a truly real-time event if things do not go as planned. Operation of the grid will revert to simply keeping the grid up and running at all costs when extreme situations occur.

The next, and highest level, time scale is that of planning. System operation, as described to this point, has the constraint of using equipment that is existing and functional at the time of operation. The planning stage is about redesigning and upgrading the system; this is typically performed on the time scale of years. The planning process generally assumes that demand is an independent variable driven by population size and the types of loads that can be anticipated. However, the utility itself can also influence the market by advertising and actively attempting to acquire new customers. However, this point aside, the trend in the past has been to continue to leverage economies-of-scale discussed earlier. A growing customer base that had a growing appetite for power consumption and is used to having power always available was the norm. Thus, it made sense to anticipate growth far into the future and build much larger capacity than was immediately necessary. This meant that planning cycles tended to be very long. It was much cheaper to build components larger than immediately necessary and then operate for many years or decades before considering major upgrades. From a technology standpoint, this is an interesting contrast with communication, where, for example, a cell phone becomes obsolete the day after it is purchased or telecommunication carriers advertise their next greatest technology on what sometimes seems like a monthly basis. The point is that technological advances in power systems had to wait relatively long periods of time to be deployed, and this has almost certainly impacted the rate of advance of power system technology. Over the same time periods, many generations of communication technology have had the opportunity to be deployed and lessons had been learned from actual field experience with large numbers of customers.

5.2.2 *Implications for Restructuring*

In the past, power production was a vertically integrated business: a single organization, the utility, managed the entire power system, from generation to transmission, to the ultimate delivery and meter reading at the customer site. The utility, in turn, was tightly regulated with stringent local and national oversight. The 1990s saw a shift away from the vertically integrated business, throughout the world, through deregulation or liberalization. This process involved the unbundling of the utility business into many different and competing components. This has occurred in different ways throughout the world; there is no standard model for how the energy market is to operate. However, some generalizations can be made. There are two main types of markets: centralized markets or power pools, and decentralized markets or the bilateral contract model.

In a power pool, all power suppliers offer a set of price–quantity points that form a supply curve. System operators can then use the demand forecast along with the supply curve to allocate generation. This is called a one-sided pool. In a more complex two-sided pool, a price–quantity curve can be derived for demand and that can be paired with the price–quantity curve to allocate generation. Markets can operate from day-ahead up to near-real time; that is, 5 minutes ahead. Some real-time markets allow non-firm hour- or day-ahead markets to operate in parallel in order to allow operators to forecast future supply and demand. Negotiation can be a several-step process. Initially, the prices are determined assuming the power network is a "copper plate"; that is, there are no constraints in delivery of power to any location. In the next stage, the feasibility of power delivery is considered. At this stage, any constraints on delivery may have to be alleviated by using different, perhaps higher, cost generators. This is added to the cost and is known as an "uplift" charge.

In a pooled market, it is more common to include transmission constraints along with generation cost. In this case, it allows for LMP. LMP allows the marginal cost – that is, the cost for the next unit of energy – at any location in the network to be determined taking into account both generation and power transport constraints. This allows for clear price differentiation that includes the cost of congestion and is more accurately applied than the more general uplift charge. The location for which an LMP is computed is known as a "node" and is sometimes referred to as a nodal charge.

The alternative to the power pool is direct bilateral contracts between energy buyers, usually distribution operators, and sellers, usually generators. However, a power generator can become a buyer if a shortfall in production is anticipated; likewise, a distribution operator can become a seller if excess capacity is anticipated. Brokers can intermediate in this process and over-the-counter trading can take place. Of course, forecasts are never perfect, and there will always be some discrepancy between bulk power purchased and the actual power consumed. These discrepancies have to be accounted for and settled by system operators. This can take place in a separate market, known as a balancing market, in which a market price is established for the cost of these discrepancies.

The details of the energy market are outside the scope of this book. However, because the business of operating the power grid impacts the way it is designed, it is necessary to have a basic understanding of the business and market aspects and how they interface with the technical components of the system.

On first blush, it seems that electricity could be treated as a commodity similar to oil or wheat. There would be a given supply and corresponding demand; an equilibrium would be established when the correct price is reached. In effect, price can be a mediating factor that serves to balance supply and demand. However, a significant difference between the electricity market and the commodities market is the fact, that unlike other commodities, electricity cannot yet be created ahead of time and held in inventory; as is often repeated, supply of electricity must instantaneously balance the demand as a standard part of operation. With commodities, supply and demand can vary significantly; this cannot literally be the case in the electricity market – recall the impact that overgeneration has on the system.

Another challenge has to do with the price elasticity of electricity. Electricity has always been viewed as an essential service, one that is *assumed* to exist. Consumers are used to paying whatever fixed price the utilities charge. It would be a significant change in the consumer mindset to consider electricity as something that can either be bought cheaper at a lower quality or simply do without. Because in the past utilities worked with economies-of-scale, very large systems were constructed with capacity to support growth far into the future. Large amounts of cheap power were the norm and increases in price were relatively few and small.

On the supply side, another challenge to a free-market system is the inability for new suppliers to enter the market. Building a centralized generation plant or a transmission and distribution system is a huge undertaking requiring massive amounts of capital. It is not a market where suppliers can easily enter to meet excess demand and leave when demand is low. In theory, it would be ideal for any supplier, however small, to enter the market to help meet demand. Making this ideal a practical reality is something that the smart grid should help to make more practical from a technical standpoint. The ability for the individual consumer to become a power producer and inject power into the grid for sale would be an ideal, but radical, change in the way the market operates. This can be taken further to nanogrids and nanogenerators, discussed in Chapter 15.

Another difference between a typical commodity market and the electricity market is the transport of power. The cost of transporting power is not insignificant. Congestion in transmission and distribution lines is becoming a more frequent issue. Thus, it is not simply a case purely of supply and demand, because an oversupply of potential generation capacity in one region may simply not be feasibly transported in a cost-effective enough manner to another region where demand is high. In other words, transport becomes a much more significant and complicating issue than in other commodities. However, there is one other commodity that is similar in nature, and that is communication and networking. The communication and networking industry has a primary focus on efficiently routing and transporting data through potentially congested links.

One of the problems with designing a free electricity market has been the potential for suppliers to "game" the system. Because of the price inelasticity of electric power and because the power system is approaching its limit for supply and transportation capacity in some regions, the sensitivity of the market to artificial drops in supply – that is, finding reasons to withhold power – can reap large rewards for suppliers. It is not hard to see how the complexity of determining and predicting power flow can be subtly combined with profit-making intentions in order to create the supply and demand required for huge profits.

Another issue regarding a free-market approach involves transparency. Public knowledge of the costs and operational aspects of the system would enable consumers to have information required to converge on the right choices. However, there are competing reasons to keep information hidden; namely, security, privacy, and, counter-intuitively, to allow competitive advantage among suppliers. For national security reasons, revealing the location and capacity of all generators and transmission lines would allow an adversary to know precisely where to strike with minimal effort to do the most damage. From a privacy standpoint, particularly as advanced metering and DR mechanisms become widespread, consumers may not wish to reveal the details of their personal usage information. Finally, a supplier would not want their competition to know any potential unique limitations of their power production capacity. When suppliers have detailed knowledge of competitors' vulnerabilities or unique challenges in meeting demand, this could become an "unfair" advantage in some situations.

Another issue, somewhat related to the one above, is the unique impact to the environment of particular power generation methods. Like many of these business and market issues, this can be a sensitive and highly charged subject. The bottom line is that policies such as the carbon tax or tax benefits for using "cleaner" energy production have been part of the mix in order to attempt to include the cost of change to the environment into the cost of producing power. Whether it is really possible for humanity to have any significant impact on the environment or even measure it with the precision necessary to assign a meaningful cost seems be more of a political, rather than a scientific or engineering, debate.

Finally, there is the issue of "fairness" in providing quality power to everyone. Clearly, from a purely technical standpoint, it is more cost effective to supply power to large metropolitan areas; these areas have large numbers of relatively wealthy consumers in a single, densely populated location. Long and costly transmission and distribution lines can be minimized. However, this means that rural customers, those located in areas sparsely populated and who are relatively poorer, demand power to be supplied over long and potentially difficult terrain and in areas with potentially bad weather conditions. In a purely free market, they would have to pay significantly more in order to entice suppliers to cover the expense of reaching them. However, in the past, such customers have received power at similar rates to everyone else; the costs had been split roughly equally, with urban centers essentially helping to subsidize outlying rural areas. It is possible that remote rural areas may be able to more cost effectively generate their own power, particularly with advances in DG. However, this illustrates one of the many complications that need to be taken into account when considering business and market aspects of a free power market. The last point above highlights the expression of either being "on the grid" or living "off the grid." The former expresses the shared interconnection of a united and advanced society, while the later indicates self-sufficiency, individualism, and perhaps a sense of being more in tune with nature, or it may just mean operating your own microgrid.

5.2.3 *Frequency*

If the market fails, frequency is one of the first indicators that supply and demand are not in balance. When power demand exceeds the generator's ability to provide it, energy is taken from the kinetic, rotational energy of the rotor, thus slowing it down. The primary danger of changes in frequency are to synchronous motors and generators; their windings will experience irregular current flows and be subjected to a risk of current overload. Such frequency-sensitive devices generally have their own frequency monitors with the ability to disconnect from the grid if the frequency strays too far from its nominal value.

Some sections of the transmission and distribution system will utilize frequency as a means to trip a relay to release a line from the grid if frequency exceeds a given range from its nominal value. The idea is that a relatively large change in frequency indicates a potential inability of the generators to meet the given demand; in that case it can be better to release all the loads on a line in order to reduce the demand seen by the generator, rather than remain connected and have the loads take the system down.

The range of acceptable requirements for frequency vary with location. In areas where generation generally has trouble meeting demand, the acceptable frequency range may be larger than in areas where there is an abundance of available generation capacity. As mentioned earlier, some electric clocks rely on a constant grid frequency to keep time. In this case, the gain or loss of frequency cycles becomes important; the utility will purposely add or subtract the number of cycles necessary to keep such clocks on time, usually on evenings or weekends. Clearly, precise timing applications such as GPS or communications, will not rely on the power grid, but implement their own internal clocks with the precision required. Thus, the impact of keeping grid frequency constant for time-keeping purposes is really only a convenience.

The next, and perhaps most interesting, aspect of power quality is maintaining clean, sinusoidal waveforms. Ultimately, it is the shape of the windings and the trajectory of the rotor through the windings that creates the electromotive force that causes current to flow. Just as in communications, noise can be injected into the system. Typically, distortion in voltage comes from generators and distortion in current comes from loads. A typical form of waveform distortion is caused by harmonics, integer multiples of the fundamental frequency. Harmonic distortion causes a nice smooth sinusoidal waveform to appear "scribbled" with additional bumps instead of a smooth curve. Because the poor waveform is composed of the fundamental frequency and its harmonics, these frequencies can be decomposed via simple Fourier transforms. The degree to which the higher harmonics dominate the main fundamental frequency determines the amount of harmonic distortion. Total harmonic distortion (THD) quantifies the amount of power contained in harmonic frequencies relative to the power in the main, fundamental frequency.

Purely resistive loads are not impacted by harmonic distortion; however, in motors, transformers, and other electronic devices, harmonic distortion can cause buzzing sounds as well as add power loss and overheating. Loads with high reactance are particularly affected. A transformer, for example, has a high inductance and resists frequent changes in current. High harmonic frequencies cannot easily pass through; the energy is lost and is converted to heat. Thus, such a transformer operates at a lower efficiency and will have a shorter life.

Now consider a delta transformer connection. Recall that the A–B, B–C, C–A, 120°, relative-voltage phase difference is what creates the current flow. The third harmonic of a 60 Hz system is a 180 Hz sinusoidal signal. This third harmonic on phase A turns out to be indistinguishable from phases B or C, and this is true for every multiple of the third harmonic. This means that there is no voltage difference in these third harmonics, causing no current or power flow to occur. Since this occurs for all multiples of three of the base frequency, this means that one-third of all the harmonic frequencies "block," or do not add to, power flow. For example, if the total harmonic distortion were 3%, then this implies there will be, at minimum, a 1% power loss. Also, with tremendous increase in DG comes a corresponding increase in harmonic distortion, particularly from inverters required to convert direct current to alternating current, which was discussed in Chapter 3.

5.2.4 System Operation and Management

This section begins to touch on the topic of the power grid as a complex system. It is often repeated that generation and demand must always be perfectly balanced in the power grid – at least until energy storage systems become widely deployed. However, the statement that generation and demand must always be perfectly and instantaneously balanced has never been perfectly accurate. If this were true, the power grid would be impossible to control. Instead, there are relatively small amounts of energy stored within elements throughout the grid; for example, within the generator and transformer windings and the inertial energy of rotors. This allows time, however short, for control actions to be taken. Thus, operation and control actions depend upon the time scale allowable for actions to take place.

As discussed earlier, as long as changes are not too drastic, generators are able to adapt to change in load by absorbing or releasing kinetic energy. If demand is not matched by generation, frequency will change and there is a small, but acceptable change in frequency that can act as a buffer and allow time to adapt to changes. Voltage magnitude changes also have some buffering action. Namely, when load requirements exceed power generation capability, voltage begins to sag. The sag in voltage actually causes loads to draw less current, and thus less power, allowing time to adjust to the situation. This might be considered unintentional conservation voltage reduction. An opposite, and equally beneficial reaction occurs when generation exceeds demand; that is, voltage will increase, causing loads to draw more current, and thus more power, helping the generator to slow down. The purpose of this discussion is to illustrate that, although the time scales are relatively small, there is at least some time to adapt to changes in the system.

Thus, we can view operation and control of the grid as being categorized by the time scale of events. The smallest scale events are those that happen within a fraction of second; that is, those that happen within a matter of cycles. For 60 Hz systems, this is 1/60 s or 16 ms. Frequency regulation occurs on this time scale. The first level of response is the natural, passive response of the system to a generator that speeds up or slows down relative to others. As discussed previously, there is a natural magnetic restoring force that tends to bring the generator back to the proper frequency. The term "passive" is used here to indicate that no human intervention is required.

The second level of frequency control is active – although, of course, it can be automated. This is simply the control of mechanical power to the rotor in order to actively increase or decrease the speed of the rotor to restore the standard frequency. The implementation depends upon the type of generator. It may require controlling the amount of steam in a steam turbine, the amount of water in a hydroelectric generator, inserting or removing control rods from a reactor, and so on. Power system protection mechanisms can also operate on the scale of a single cycle. It is important to distinguish a fault from an abnormally high load and isolate the fault as quickly as possible.

The next-fastest scale of operation is the scale of human, real-time operation. Here, operational control and response times are on the order of the speed with which a human can perceive an event and take action, from seconds to minutes. Such control is typically required in abnormal circumstances, such as generator start-up or shutdown and some switching events.

Large, centralized generation start-up is a complex process that requires coordination of pumps, valves, flow rates, pressures, and temperatures throughout the power plant. It can take hours to bring a generator up to equilibrium at its nominal speed. Once it is running at the proper speed in parallel with the power grid and at the same frequency as the power grid, it is connected to the grid. A skilled operator can even manually control the generator speed to match the load when the load has changed dramatically from normal. An interesting example comes from operators at the Pittsburgh Power Plant in California that kept their plant online after the 1989 Loma Prieta earthquake caused a complete communication shutdown. The only information the operators had was their local frequency and voltage measurements without any knowledge of how much of the grid was actually still connected.

5.2.5 *Automation: Motivation for the "Smart Grid"*

The main components of the power grid are still comprised of the same simple devices used since the grid's early years, a century ago. The fundamentals of electrical power generation and control devices have not changed; the synchronous generator, transformer, and capacitor are still used much as they always were. However, the relative simplicity of the devices stands in contrast to the complexity of the power grid as a whole; each of these simple devices, when connected and operated properly, results in the largest, most complex machine on Earth. There is plenty of opportunity not only for more advanced components, but also simply to understand and improve this large-scale, complex system. One of the first and perhaps most obvious improvements has been the introduction of SCADA systems allowing remote monitoring and control of the system. Implementing communication for SCADA systems has been one of the fundamental advances of communications in the power grid. Wide-area sensing and control can be implemented via dedicated phone lines, microwave, (now WiMAX perhaps), or power line carrier. Without additional communication, there is no way for a transmission or distribution operator to know the current state of the system – that is, until it either reaches a level bad enough to influence power throughout the system or, more likely, until a customer calls to complain. Utilities have installed outage reporting systems that correlate customers' phone numbers with locations in the grid.

Expert systems can be utilized for automation. In open-loop mode, they provide input to a human who makes the ultimate decisions. In closed-loop mode, the expert system directly controls the system. Automatic machine-learning systems could be applied throughout the grid; for example, to quickly reconfigure the distribution system for load balancing as loads vary – recall that this involves reducing line losses by allocating loads evenly among feeders and phases. Automated service restoration is another example of a potential switching operation that may be performed more quickly in an automated fashion. However, these are also highly critical operations with potentially dangerous consequences if there is any error or unexpected event. It is difficult to make the case for such tasks to be fully automated.

On the other hand, we should always keep in mind that the consumer will ultimately pay for the smart grid, including both initial development and installation costs as well as ongoing operational costs. Smart-grid communication makes more sense as the cost of communication decreases relative to the cost of energy. As DG continues to increase, there may be a point where massive, centralized data collection and control may be necessary, although the ideal case is always more autonomous and distributed control to achieve common goals if possible.

5.2.6 *Human Factors*

Engineers and operators have had a different mindset with regard to the power grid. Engineers view the grid as a system that can be understood through mathematical modeling and controlled via careful design. Operators, who have much more hands-on experience with the power grid, tend to view the grid more as a complex living organism, one that can sometimes defy our ability to succumb to analysis and design. An operator or person in the field can use all of their senses to detect and fix problems. They can hear the hum of a transformer or power line and, from vast experience, immediately recognize a problem. They know which transformers tend to overheat faster and under which conditions. They also know from experience how to correct abnormal conditions – they also know that the system does not always respond the way it should according to the manual. It is not always clear which approach turns out to be better, the analytical, engineering approach or the operator's phenomenological understanding, particularly in critical and high-pressure situations. This difference in thinking becomes extremely important when thinking about how to present and use information and apply communication in the power grid in a manner that will add value to the operators that need it.

The engineer focuses upon efficiency, while the operator focuses upon safety. The possibility of injury or death is very real for the operator, and this, of course, takes first priority. The difference between an engineer thinking about safety and an operator who actually lives it can be a significant difference. The engineer will tend to focus upon speed and precision in pursuit of efficiency. However, speed and precision are often not the operator's main priorities and can in fact work against the operator. A system that operates too quickly and requires too much precision to operate makes things more difficult for the operator. What the operator wants is stability. The operator wants a system that is predictable and moves from equilibrium slowly enough to allow human intervention to bring the system back to its proper state.

Of course, as in all systems, too much information can be worse than no information. While the communications or computer expert may work to make all possible sensor and control information available to the operator, the operator wants only the important information, and more importantly the operator must be able to feel that the information can be trusted at the time it is displayed. Also, providing more control than is necessary can do more harm than good to the operator. Generally, the operator knows the system and how to coax it back toward equilibrium when problems occur. Adding more control and complexity will simply offer more opportunity for confusion and mistakes. The operator needs a simple, robust system, while the engineer is focused upon efficiency, which often means adding complexity and adding pressure on the operator. The important point is that operators must be included in any proposed "advancements," particularly, in our case, related to communication. Now consider how much of this operational complexity can be shifted to users, how much should be shifted to users, or how much would users even want to have? That is the topic of the remainder of this chapter.

5.3 Variability in Consumption

One of the most significant selling points for the smart grid concept has been the notion that providing more information to the consumer and allowing consumers to have more visibility into their relationship with the power grid will motivate them in a manner that benefits the power grid as a whole. The classic example is the old problem of reducing peak power consumption. Variability in power consumption was discussed earlier in this chapter, and it causes the primary problem of requiring costly generation sources to be brought online for short periods of time, which is very cost inefficient.

Large swings in power consumption not only impact generation; they also exacerbate problems in the power grid that impact all grid systems. Both transmission and distribution systems are stressed. Active and reactive power have to be adjusted, and voltage regulation mechanisms, capacitor banks, and almost all other supporting grid components have to make large adjustments in their settings; this reduces the lifetime of these components. The ultimate impact will eventually be felt by the consumer if load shedding must be implemented or frequency or voltage instability results in damage to sensitive consumer electronics. Finally, large swings in power demand could lead to angular instability in generators and loss of control of the power grid or tripping of protection mechanisms on large transmission lines. Thus, maintaining consumption as constant as possible simplifies many aspects of grid operation, leading to lower cost, better reliability, and smoother operation.

It is worth injecting at this point, and also will become apparent in the remainder of this book, that it is not entirely clear whether the many and diverse smart-grid activities are leading toward a reduction in peak load. For example, renewable energy sources enabled by the smart grid are a form of intermittent energy source for which it is difficult to predict when they will be operational and what their power output will be. Smart meters may increase demand uncertainty rather than reduce it. The increased reliance upon communication may reduce the overall reliability of the system; communication faults and cyberattacks could increase the likelihood of intermittent outages. Consumers that are also producing electric power, or storing it for profit in new forms of energy storage systems, could increase the electric power market volatility rather than reduce it. Finally, the ability for the power grid to operate close to its limits given the ability for increased sensing and control enabled by the smart grid could drive the system toward becoming more brittle, requiring sudden, random mitigation techniques to avoid problems. In other words, the smart grid is increasing the features and dimensions over which the power grid operates, which can reduce the certainty over which state trajectory will be taken, leading to greater entropy in the power grid. Now that we have considered the power market and an operations perspective, let us consider the consumer perspective.

5.4 The Consumer Perspective

The smart grid will only be successful to the extent that it satisfies consumer needs and desires. Unfortunately, understanding consumer demands is often an overlooked aspect of technology development, and the same may be true for the smart grid. Certainly, there are many features that can be developed for consumers, ranging from detailed visibility and control of their electronic devices by the utility to consumer visibility and new types of control of electronic devices via automated methods under consumer control. One of the biggest questions upon which significant portions of the smart grid depends is what the consumer really wants, will actually use, and be willing to purchase. Another significant question involves to what degree the consumer will be motivated to change their behavior even if they are willing to purchase and utilize smart-grid features. These are the types of questions considered in this section.

There have been early studies that attempted to identify consumer acceptance of smart-grid technologies. Let us consider the results of a few such studies. Some studies utilize the technology acceptance model (TAM) (Stragier *et al.*, 2010). This model has been developed and used for predicting the adoption of new information technologies. It hypothesizes that "perceived usefulness" and "perceived ease-of-use" are the two most important factors in determining whether a new technology will be adopted by consumers. It also includes an "attitude toward using" the new technology and the "behavioral intention to use" the new technology. The model places each of these factors in relation to each other and examines how these factors influence one another. Many of the results have been more-or-less common sense. For example, perceived ease-of-use was found to have a significant influence on perceived usefulness. Thus, smart appliances would only be considered as useful and adopted if they were easy to use. Clearly, ease of use must be taken into account when considering consumer adoption. Both perceived ease-of-use and perceived usefulness had a strong influence on "attitude toward using," with "perceived usefulness" having the strongest effect. Thus, consumers must be convinced of the usefulness of smart-grid features in order to adopt them. On the other hand, if new smart-grid capability is as easy to use as existing devices, then consumer attitudes toward adopting the new technology will also be positive. Interestingly, there was no correlation between "perceived usefulness" and "intention to use," indicating that there was little knowledge by the user of the "perceived usefulness" of smart-grid devices. Another interesting result was that "attitude toward using" did have a positive impact on "intention to use." To put it crudely, it appears as if consumers had a feeling that there was something good that could drive their intention to use it, but they were not sure why, or how, it would be useful. It should also be noted that another part of the survey, not directly related to the TAM, indicated that consumers had concerns regarding the safety of DSM enabled by smart-grid appliances and the potential loss of control that they might experience over their appliances.

The results of a survey by Silva *et al.* (2012) also provide some indication of consumer attitudes with regard to the emerging smart-grid technologies. This survey did not implement a formal model, but rather attempted to gauge consumer thinking in five categories: (1) willingness to change, (2) monitoring and understanding energy usage, (3) automated control and energy management, (4) value-added services, and (5) privacy issues. "Willingness to change" tests the concept behind DR; namely, whether consumers would be willing to adjust their behavior based upon information provided to them from smart-grid technologies. The question involves not only whether they would adjust their behavior, but also to what degree and under what conditions such as changing power use in response to price signals, participating in energy trading, or paying more for power obtained from renewable sources. The overall result was that 80% of consumers in this survey were willing to adjust their behavior in all these situations. A majority of those surveyed were also willing to share unused energy, particularly for monetary reward. A large majority were willing to share information about energy usage with a third party, such as an energy retailer, if it would result in reducing energy cost. However, when it came to revealing detailed usage information, such as indicating when consumers were on vacation, the number willing to do this dropped from a large majority to half. Thus, a means of conveying necessary energy usage information without revealing details of individual consumers appears to be a requirement for success.

Over 90% of those surveyed desired more information regarding an overview of their energy consumption and would also like to understand the details of the energy use of individual devices. This implies that there are not only smart meters available in the home, but that every home appliance can monitor and correlate its energy consumption with the manner in which the consumer uses it. Some consumers also voiced a desire for more visibility regarding how their energy use impacts larger energy management and environmental efforts. Additionally, over 70% of the consumers expressed a desire to compare their energy-use profile with other consumers. This appears to indicate that consumers in this survey would like to be viewed within the context of a smart neighborhood or city. Over 80% of the consumers wanted to have the capability of viewing their energy use in real time, or near-real time, on mobile devices such as cell phones or laptop computers. The inference drawn from this result was that consumers wanted to view their energy usage on existing devices rather than being required to purchase new, dedicated devices.

Interestingly, over 90% of the consumers in this survey would tolerate devices with automated control that reduced energy consumption without impacting consumer comfort. However, the result dropped to less than 50% when similar control was give to a third party. This indicated that consumers were more willing to accept appliances that automatically respond to price signals than having external entities manage the appliance for the consumer, even if the results were the same. This seems to indicate that consumers are more willing to adopt a DR system in which they feel that they have some control and does not degrade their life style. However, the same consumers, when asked if they would allow a third party to manage the sale of the excess power produced by their renewable energy systems, overwhelmingly said yes. Thus, consumers appeared to want control over power consumption but were not as concerned with control of their power production. While this could imply a business opportunity for neighborhood energy management services, it may also be a temporary symptom of the fact that consumers are less experienced in energy production. Value-added services are information analysis types of service that include ideas such as comparison with similar consumers, suggestions and recommendations for saving energy, as well as predicting energy bills and detecting anomalous usage, such as a larger-than-normal bill. While the majority of consumers in the survey wanted these features, it was not clear whether, or how much, they would pay for such services. Finally, regarding privacy issues, more than half the consumers in the survey would trade some privacy for reduction in cost. More specifically, if they had a means of accurately controlling the amount and type of information shared with a utility, more than half of the consumers would do so in order to reduce their energy bill. This implies that the more consumer information that can be kept anonymous or private while still revealing useful usage information, the more information consumers would be willing to negotiate for a reduced energy bill.

5.5 Visibility

Visibility at the consumption level can occur in two directions: (1) visibility by the utility through the power grid toward the consumer; (2) visibility by the consumer through the power grid toward the utility. The utility could use such visibility to ensure that requirements are being met with regard to delivery of power to the consumer. Additionally, the utility could utilize greater visibility and control to help guide the consumer toward actions that increase the efficiency of power delivery to all consumers, with reducing peak load being a prime example.

Improved visibility by the consumer could allow the consumer to make decisions that minimize their individual cost of power, better anticipate and respond to outages and restoration, and even participate in the power generation and distribution process by selling or storing excess power. The idea of virtual power plants allows consumers to make more detailed choices about how their power is generated and where their power originates.

A virtual power plant is a collection of distributed generators that are operated by a central authority. The distributed generators in a virtual power plant are often assumed to be renewable energy sources, but they need not be. For example, they may be simply gensets – engine–generator pairs in which the engine powers the generator to provide power. The main point of a virtual power plant is its flexibility; collectively, it should appear no different than a typical centralized generator, but the many small units of the virtual power plant allow it to more cost effectively ramp-up to handle peak loads. The trade-off for this degree of flexibility is the added complexity of running a large collection of often heterogeneous generators. It requires more complicated optimization and control, and an often-made assumption is that a corresponding increase in communication is required as well. Just as the "Internet of Things" (IoT) has been an academic buzzword, the "Internet of Energy" is related to collections of smaller scale power sources that exhibit great flexibility.

Before the "smart grid" era, consumers could lose power without the utility being aware of it. It often took a phone call to alert the utility of an outage, and then a repair truck would have to be sent to diagnose and correct the problem. Instead, outage management systems should provide automatic detection and location of faults, as well as the consumers impacted, and ultimately an automated restoration process to the extent possible. Meter reading used to be largely manual, requiring personnel to physically drive to consumer locations to read meters; however, this process is now becoming automated. So-called "smart meters" could automatically collect information, not only total power used, but more information about how the power was used as well as aspects related to power quality. If improved visibility comes with more control, utilities could make subtle changes to improve the overall power efficiency at the consumer level.

However, some overarching questions regarding all this improved visibility and control are: (1) Is all this added visibility at the utility level really necessary or helpful? (2) How much privacy and control will consumers be willing to give to a third party? Regarding the first question, basic information such as that already provided today, such as total energy used, along with improved outage information would certainly be welcome. However, the utility probably does not need to know when someone turns on a hair drier or opens a refrigerator door. This level of detail and control could be automated under the control of the consumer. The second question, regarding privacy and control, is outside the scope of this book. However, recent surveys discussed in Section 5.4 begin to provide insight into this question.

5.5.1 Microgrids

Microgrids originated from the need to reduce the complexity created by DG. Rather than add distributed generators to the power grid in an ad hoc manner, a new system-level approach was suggested in which the global power grid is, in essence, apportioned into smaller power grids, known as microgrids. Each microgrid can then be theoretically managed independently from the larger grid. The goal is to provide a simpler, modularized approach to designing the power grid with DG. In the ultimate implementation, the entire power grid would be decomposed into microgrid modules. However, until (or if) that time arrives, microgrids need to be interoperable with the main power grid.

One of the interoperability concerns with microgrids, particularly those injecting power into the main power grid, has been how to ensure that their frequency remains synchronized with

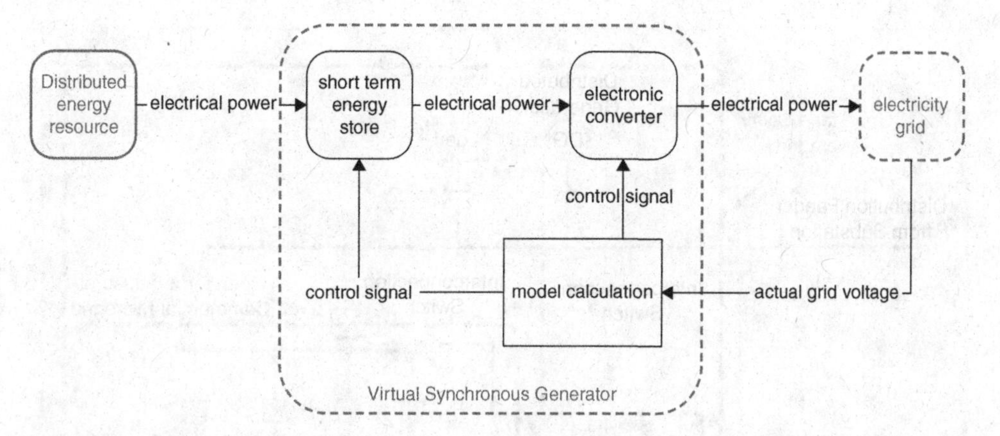

Figure 5.1 The concept of using virtual inertia is illustrated. The electronic converter is controlled in such a manner as to reproduce the inertia of a rotor in a large generator. Source: Kroposki *et al.*, 2008. Reproduced with permission of IEEE.

the main power grid even when the loads that they supply have significant variation. Numerous approaches have been proposed to this problem. One of them is shown in Figure 5.1, which illustrates a version of the concept for a virtual prime mover. In this approach, an energy store is used as a buffer; the main power grid frequency is sensed and computation is performed to control a power electronic converter such that power output is synchronized with the main power grid. Power electronic converters are discussed in detail in Section 14.3. The dashed box illustrates the hypothetical virtual prime mover system with control signals illustrating where communication may be required.

An example of a large microgrid used within the distribution system is shown in Figure 5.2. Power arrives from the distribution feeder on the left side of the figure. Assume that the distribution feeder line continues across the figure to the right with additional consumer loads attached to the feeder. The key components of this figure are the two switches: the interconnection switch labeled "open for a microgrid" and the interconnection switch labeled "open for a industrial/commercial microgrid." If the utility microgrid switch is open, then all power downstream of the switch must be supplied by DG connected to the feeder. This comprises the larger, dashed box in the figure, which outlines a utility microgrid. If the industrial/commercial microgrid switch is open, then the microgrid outlined by the dashed, smaller box in the lower right becomes a stand-alone microgrid feeding only its own local loads. The control system is also illustrated showing roughly where communication could be required.

A more detailed view of the connection between a microgrid and the main power grid is shown in Figure 5.3. The key point of this figure is that the current and voltage of the main power grid (on the left side of the figure) and the microgrid (on the right side of the figure) are sensed via a current transformer (CT) and potential transformer (PT) and the results sent to a centralized, digital signal processor (DSP) where monitoring, protection, and diagnostics take place. Note that communication in this figure is not for high-bandwidth, dynamic control of the system; rather, it is for general configuration and reporting. Dynamic control does not necessarily require explicit communication. Droop control, which is explained next, is an example of a common control mechanism that does not require explicit communication.

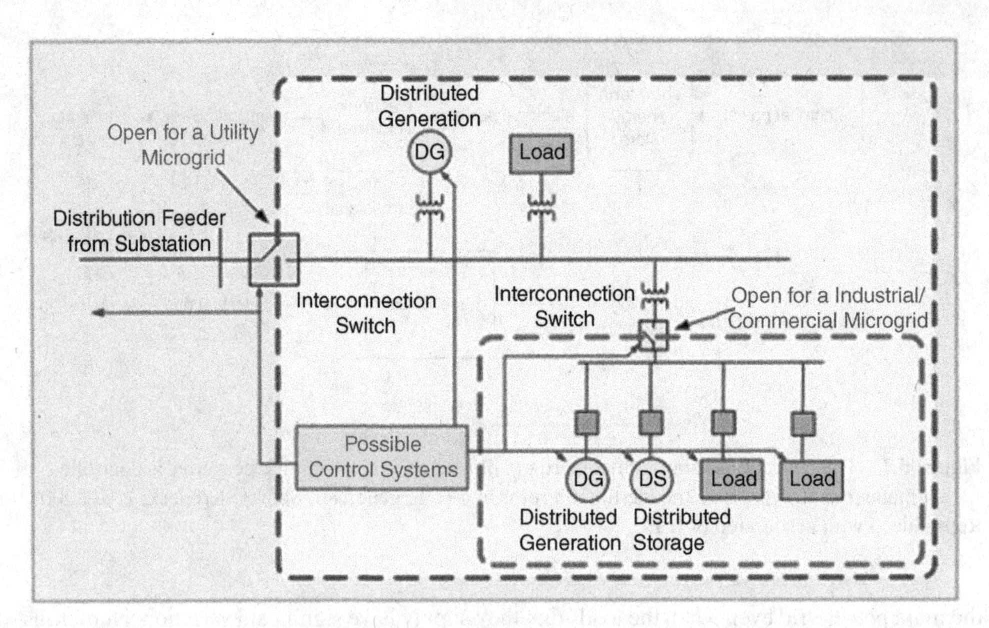

Figure 5.2 The layout of a utility and industrial microgrid. Notice that both types of microgrids are shown attached to a radial substation feeder as described in Section 4.1.4. A FAN communication system may be required to control the microgrids. Source: Kroposki *et al.*, 2008. Reproduced with permission of IEEE.

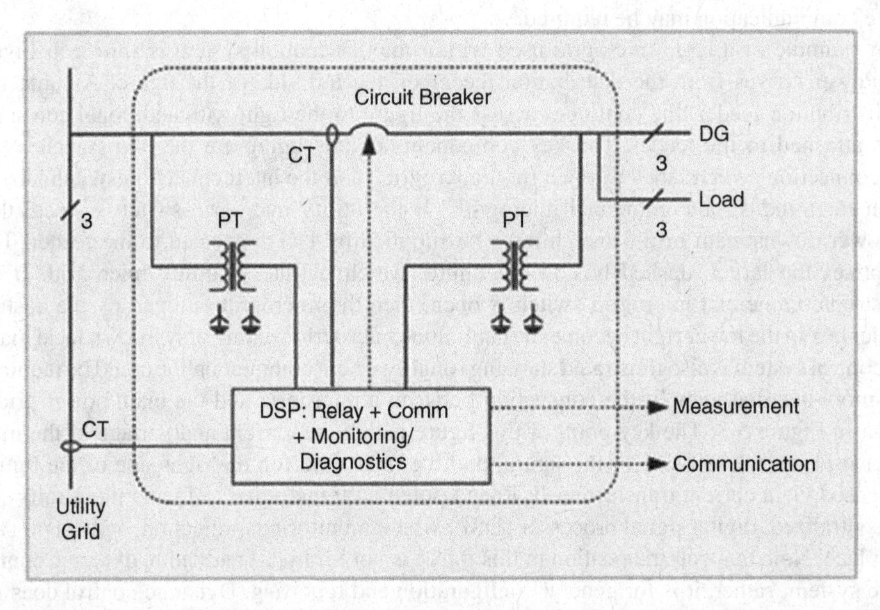

Figure 5.3 A microgrid connection to the main power grid is shown using a circuit-breaker-based interconnection switch. Source: Kroposki *et al.*, 2008. Reproduced with permission of IEEE.

Droop control is a common control method used in the power grid that avoids the need for explicit communication. Thus, it is important for communication experts to be aware of it and understand its advantages and disadvantages. In general, droop control involves locally measuring a value that may decrease, sag, or droop over time and using the drooping value to control the local system to compensate for droop in the measured value. Thus, many components in the power grid can each be simultaneously measuring a value locally, taking local action to raise the value back within its nominal range. A simplified example to quickly illustrate the concept is to consider a set of generators, each locally measuring frequency. If a generator goes offline or additional load is added, each generator will individually sense a drop in frequency. The droop control system at each remaining operating generator will increase its frequency to compensate for the drooping frequency. This simple technique, which avoids the need for explicit communication, can be used for many other power grid control parameters.

Let us consider an example in more detail. It should be very familiar by this point in the book that the active power across an ideal transmission line is shown by

$$P = \frac{V_1 V_2}{X} \sin \delta, \tag{5.1}$$

where δ is the difference in phase between the voltages on each side of the line, V_1 and V_2 are the magnitudes of the voltages on each side of the line, and X is the impedance of the line. The reactive power across the ideal transmission line is

$$Q = \frac{V_2}{X}(V_2 - V_1 \cos \delta). \tag{5.2}$$

As long as the power angle δ is small, a simplifying approximation can be made by assuming that $\sin \delta \propto \delta$ and $\cos \delta = 1$. Using this approximation in Equation 5.1 yields

$$\delta \propto \frac{PX}{V_1 V_2}. \tag{5.3}$$

Also, using the above approximation that $\cos \delta = 1$ in Equation 5.2 yields

$$V_1 - V_2 \propto \frac{QX}{V_2}. \tag{5.4}$$

Equations 5.3 and 5.4 show how active power has a significant impact on the power angle and reactive power has a significant impact on the voltage difference across the line.

Another key piece of information to recall is that frequency is related to power angle via the swing equation. The swing equation shows the relationship among the generator rotor position, frequency, and the power angle. Thus, from Equation 5.3, we can see that active power has a key impact on frequency by, in a sense, tightening or loosening the power angle.

We can use this information to construct a frequency droop control system. First, let f be the system frequency and f_0 be the nominal or correct operating frequency. Similarly, P is the active power output of the generator, P_0 is the nominal active power, V is the local voltage measurement at the generator, and V_0 is the nominal voltage output of the generator. Also, Q is the reactive power of the generator and Q_0 is the nominal reactive power for the generator.

Now it is possible to construct the droop control for both the generator's active and reactive power outputs. The concept is to *locally* measure change in frequency and adjust the active power and *locally* measure the voltage and adjust the reactive power. A drooping or falling frequency indicates the need for more active power, and a drooping voltage indicates the need for more reactive power. This is shown in Equations 5.5 and 5.6, where k_p is the frequency droop control setting k_q is the voltage droop control setting:

$$f = f_0 - k_p(P - P_0), \tag{5.5}$$

$$V = V_0 - k_q(Q - Q_0). \tag{5.6}$$

In this example, droop control allows multiple generators to react in such a manner that they fall into an equilibrium state such they share the load. Without droop control, there would be no control over how generators supplied the load and they could end up oscillating wildly or even becoming unstable.

Droop control settings are indicated by percentage droop. This is the amount by which the measured value changes in order to cause a 100% change in the controlled value. For example a 3% voltage droop means that for a 3% change in voltage the generator will require a 100% change in reactive power output. This is assumed to be a linear relationship; so once the percentage droop is known, an intermediate value can be easily interpolated.

As mentioned, a droop control system allows distributed control of a large number of dynamically changing components without requiring explicit communication. However, if a significant change occurs, such as loss of a large generator, the remaining generators may reach equilibrium at a lower-than-nominal frequency. The basic droop control system as described here will not automatically return the system to the normal operating frequency when the lost generator comes back online. In other words, droop control, as its name implies, compensates only for downward or drooping changes in measured value. Additional control must be used to sense the new power availability and bring the values up toward their nominal values.

The reason for this detour into the details of droop control is that the question of when communication really needs to be used in the power grid and its benefits are still open in general. For example, under what general, theoretical conditions can it be proven that communication is better than local-sensing techniques such as droop control? Remember that communication adds extra cost and increases complexity of the system, which increases the likelihood of component failure and thus potentially decreases overall reliability.

5.6 Flexibility for the Consumer

This section covers the notion of flexibility in electric power consumption for the consumer. There used to be very little choice or flexibility for the consumer with regard to electric power. The consumer paid the monthly electricity bill and the only commonly available information for the consumer was the monthly bill and the only form of control was to manually turn devices on or off. Varying the price of electricity by time of use was an early attempt by utilities to control consumption in order to reduce peak demand. More recently, devices have become more cognizant of how they utilize energy; they can modify their operation to control energy consumption, either autonomously or with user feedback. Also, the consumer is gaining more choice regarding the type and location of electric power that they purchase. The result is ever-increasing information, flexibility, and choice for the consumer regarding electric power purchase and use. Consumer energy management, also known as DSM, is discussed, followed by an introduction to plug-in electric vehicles (PEVs).

While the smart grid can increase flexibility for the consumer, there is an element contradictory to flexibility: the utility is trying to motivate consumers' behavior; that is, to shape the aggregate consumer energy-usage profile in order to minimize peak load. This is known as DSM. Of course, using price to control demand is not new, charging phone customers more during peak hours of telecommunication traffic has a long history. More detail on DSM, and DR in particular, will be covered in Section 7.2.

DSM is broader than DR and includes techniques that increase energy efficiency by using less power to perform the same tasks as well as dynamic demand, in which devices autonomously adjust their operating cycles on the order of seconds in order to increase the diversity factor for a set of loads. The latter concept is an old one and involves devices that not only monitor power frequency and their own operating profile, but also the power factor, resulting in many, small, local actions, unobservable to the consumer, that flatten overall power demand and improve the operation of the power grid.

5.6.1 Consumer Energy Management

Energy management at the consumer level was available long before the term smart grid was coined. This includes techniques that range from building EMSs to consumer electronics that autonomously manage their own energy, often without the consumer realizing it. Here, we take a quick look at two specific EMSs: building EMSs and electronics energy management. Both existed long before the smart grid. In fact, concepts derived from electronics energy management are quite sophisticated and foreshadow technology that might be used in the future for nanogrids, discussed in Section 15.3.3.

Building EMSs (McCann *et al.*, 2008; Virk *et al.*, 1990) are interesting because they have typically utilized model-based control and have focused on the development of thermal models at the human scale. Much of the electric power that we utilize in buildings goes toward heating or cooling: heating or cooling water and air for a variety of purposes, including ovens, hot-water heaters, washing machines, clothes dryers, and of course home heating, as well as air-conditioning, refrigerators, and freezers. It becomes important to model the thermal mass, or thermal inertia, within a building in order to understand how long it can retain its current temperature before being influenced by external fluctuations in temperature. Thermal mass is also known as thermal capacitance or heat capacity, since matter can store and release heat at different rates under different conditions. Much like an electric circuit, components of a building can store and release thermal energy in ways that have an overall benefit to the energy efficiency of the building. Understanding the "thermal circuit" results in a more energy-efficient control system that can be integrated with evolving DSM techniques.

As building EMSs have optimized energy usage, so too have many consumer electronics been designed to automatically minimize their energy consumption with minimal impact to the user. As the density of switching elements has increased within integrated circuits, the need to manage heat buildup becomes more important. Reducing power consumption is one way to reduce heat buildup as well as save energy for the user. Note that while power required by a single logic gate has steadily decreased, the increasing density of gates has had the overall effect of intensifying heat production and increasing overall power consumption. There is a deep relationship between energy consumption and information that will be discussed in Section 6.3.

There are many techniques used to optimize integrated-circuit power consumption. Some integrated circuits require a minimum clock rate to be maintained, even when they are not performing computation. Since this wastes power, some integrated circuits can either reduce their clock speed or turn off their clock and hold information in a stable state without a clock. However, some leakage current is still wasted. Another way to reduce leakage current is to reduce voltage levels required to operate the circuit. Unfortunately, this causes the circuit to operate more slowly. However, it is possible to design circuits to use lower voltages for logic that can operate more slowly without impacting the performance experienced by the user. One might relate this concept to conservation voltage reduction in the power grid.

An obvious way to reduce power consumption is to turn off entire sections of the integrated circuit when not needed. The powered-down sections of the integrated circuit can be awakened or periodically awaken to check when they need to operate. Another approach, related to superconductivity discussed in Section 14.5 is to reduce the temperature of the circuit. A lower temperature reduces thermal noise that the voltage needs to overcome; if thermal noise is reduced, voltage can also be reduced. Adiabatic, also known as reversible, or isentropic computing utilizes logic gates that are completely reversible. Thus, there is no change in entropy as the circuit operates and no corresponding loss of power. Unfortunately, this is difficult to implement and is discussed further in Section 6.3.

Communication itself is undergoing development toward more energy efficiency. Since communication requires computation, all the previously discussed techniques apply to reduce power consumption in communication. In addition, since communication involves the transmission of power over a distance, there are numerous techniques that address the issue of how to minimize transmission and reception power. For example, there is an optimal distance for power transmission to nodes in a network that minimizes transmission energy. Transmitting too far requires more energy than necessary from a single node that could have been shared among intermediate relay nodes. Transmitting over too short a distance causes more intermediate relay nodes to operate than necessary, which also wastes energy. Thus, there is an optimal "sweet spot" with regard to the distance of transmission, known as the characteristic distance. Also, receiver energy is often wasted waiting for transmissions. Techniques that allow the receiver to sleep as long as possible help to minimize overall power consumption.

In summary, consumer electronics have long implemented a variety of techniques to optimize and manage energy consumption. Emerging smart-grid communication and applications should be cognizant of these techniques already implemented on the demand side and consider ways to leverage, or work with, these techniques.

5.6.2 *Plug-in Electric Vehicles*

We will define the PEV to include any motor vehicle that is designed to draw electric power from the power grid in order to provide mechanical power to the vehicle's motor. This can include a wide variety of vehicles, including battery electric, plug-in hybrid electric, and electric conversion of existing combustion engine vehicles. There are numerous other terms for such vehicles, including "grid-enabled vehicle."

Battery technology has been a challenge in producing a cost-efficient, electric vehicle. Plug-in hybrid vehicles allow traditional combustion engine operation to help overcome battery problems. Not all hybrid vehicles are "plug-in." Some hybrid vehicles extract electrical power from the internal combustion engine and regenerative braking; however, they cannot be recharged from a source external to the vehicle, such as the power grid. A battery electric vehicle is driven via the chemical energy stored in batteries. The engine is an electric motor instead of an internal combustion engine. A plug-in hybrid electric vehicle operates using an electric motor in charge-depletion mode. It can switch to a charge-sustaining mode when the battery has reached its minimum state of charge.

Clearly, PEVs will require an adequate infrastructure for recharging from the power grid. Charging stations, battery swapping, and trickle charging from vehicle roof-mounted solar panels are a few of the numerous ideas. Charging stations are specialized connections to the power grid specifically designed for vehicle recharging. They maintain the consumer psychological momentum of a gas pump. Battery swapping is the notion of simply swapping a depleted battery pack for a fully charged pack. The depleted battery can then be recharged without the need for the consumer's car to be parked and waiting for the charging process to complete. Trickle charging is the notion of utilizing low-power sources, such as roof-mounted photovoltaics (PVs), to gradually recharge the battery whenever possible. Other more innovative approaches involve wireless power transmission to the vehicles.

The topic of electric vehicles is a broad one; our concern is limited to the relationship between electric vehicles and the power grid. Of course, there has been concern about whether the power grid can handle the increased load from large numbers of PEVs. Some estimates are that PEVs could require three times the amount of power as the average home. However, electric vehicles can be considered mobile energy storage devices that could potentially provide power as well as extract power from the grid. Many complex schemes for bidirectional energy flow among the power grid and PEVs have been explored.

5.7 Summary

This chapter focused upon electric power consumption; namely, power loads and the consumer. It became apparent early in the history of the power grid that loads may be viewed as managed components of the electric power grid, either remotely managed by the utility or by inducing consumers to operate their loads in a manner that is beneficial to the power grid using price as a controlling signal. A brief history of the power market and its current trends were reviewed. The relationship between power supply and load was examined with respect to alternating current frequency as well as operation and management of the power grid. Reducing variability of power demand has long been one of the goals related to loads and consumption. Various measures related to peak demand were derived.

Value for the consumer is *sine quo non* for the smart grid; success of the smart grid is dependent upon the support of the consumer. Thus, it becomes paramount to look at the smart grid not only from the utility perspective, but also from the consumer perspective. Consumers will drive innovation behind smart-grid applications if they are given the proper building blocks, just as the most useful and innovative features of the Internet were developed once access and tools were give to the average programmer of the 1980s and 1990s.

We also examined the power grid from several consumer perspectives; namely, visibility into the grid, capability to purchase power based upon attributes of the provenance and quality of power, the ability to generate and sell power, and the flexibility of the power grid to work for the consumer.

Now that we have completed an introductory tour of the pre-smart-grid power grid, Part Two of the book considers the "smart" aspects of the evolving power grid. We begin Chapter 6 with an overview of communication and its relationship to the power grid. Communication has been used in the power grid for over a century; new concepts addressed by smart grid communication need to be clearly articulated. Fundamental physics has shown the relationship between energy and information; this relationship quantifies the unique aspects of communication in the power grid and how it improves energy efficiency. This forms the core of a new field known as power system information theory, explained for the first time in the upcoming chapter. The energy-information relationship leads to a fundamental understanding of the minimal amount of processing and communication required to compensate for energy transmission inefficiency. As information theory has done for communication, power system information theory provides a fundamental understanding and theoretical optimum limits to what smart grid communication can achieve for power system efficiency. The energy-information relationship is also applied to fundamental physical aspects of communication through wireless and wave guide media. In a nutshell, Chapter 6 provides fundamental concepts related to communication and electric power that place communication discussions in later chapters into better context.

5.8 Exercises

Exercise 5.1 Microgrids

1. What is a microgrid?
2. What are the advantages of a microgrid?
3. How does a microgrid differ from the concept of DG?

Exercise 5.2 Peak Power Demand

1. What is peak power demand? How has it been defined?
2. Why is peak demand a problem?
3. What techniques have been used to reduce peak demand?
4. How could adding communication help?

Exercise 5.3 Energy Storage

1. How could energy storage help reduce peak power demand?
2. What characteristics of energy storage are required to be effective?
3. What characteristics for communication will be required to control such energy storage systems?

Exercise 5.4 Automatic Meter Reading

1. What is the difference between automated meter reading (AMR) and AMI?
2. If AMR has been in operation since the early 1900s, what are the relatively recent concerns over privacy and cybersecurity related to "smart meters?"
3. What are the communication requirements with respect to AMI?

Exercise 5.5 Building Energy Management

1. What is a zero-energy building?
2. What are some of the issues regarding communication and networking within such a building?

Exercise 5.6 Power Demand Entropy

1. What is power demand entropy?
2. Predictability and the ability to compress information are intimately correlated. What does this tell us about electricity usage and communication of usage information?

Exercise 5.7 Power Demand Modeling

DR mechanisms have tended to ignore the cost of communication. The assumption is that load and price information are always immediately available.

1. What happens when the cost of communication for DR is taken into account?
2. How might DR algorithms change if realistic costs and constraints on communication are taken into account?

Exercise 5.8 Questions of Efficiency

1. Is maximizing power transfer equivalent to maximizing electrical efficiency?

Exercise 5.9 Prosumers

1. One of the proposed features of the smart grid is the ability for consumers to be able to also generate and sell their excess power. What changes would have to occur to the power grid to support this capability?
2. What communication requirements would this entail?

Part Two

Communication and Networking: The Enabler

6

What is Smart Grid Communication?

Imagine the fundamental advances that would have taken place if Maxwell, Tesla, and Shannon could have met.

—Stephen F Bush

6.1 Introduction

In Part One, *Electric Power Systems: The Main Component*, we reviewed the power grid and its fundamental operation as it existed before the term "smart grid" was fabricated. Part One is an important prerequisite for the remainder of the book. It sets the historical context from which the "smart grid" is evolving. Without this context, much of the smart grid may seem a *non sequitur*. It also explains underlying physics behind power grid operation, crucial to understand when applying communication and networking. It also explains the mindset behind power system design, which differs considerably from communication and networking.

This chapter opens Part Two, *Communication and Networking: The Enabler*, which focuses upon communication and networking for the power grid. The term "enabler" in the title is important; communication must be viewed as supporting the power grid, not as an end in itself. The question, of course, is what precisely is being enabled? Communication has been used in the power grid almost since its inception, and its use is naturally increasing; a question is whether communication is enabling anything fundamentally new or simply enhancing ideas that have been around for a long time. The "elephant in the room," so to speak, is why more communication is needed. Why cannot most grid operation be done locally, as has typically been the case? In other words, communication has occurred implicitly, through local detection of physical phenomena in the grid without the need for a complex communication infrastructure that could serve as another point of failure? What benefit does explicit communication add? Typically, communication provides advanced warning of an impending event or it enables remote control; that is, networked control. So the real question is: What does communication enable that is better than local control? An exhaustive list would be tedious and incomplete; however, an abstract model of the situation could help answer this question. This involves a

Smart Grid: Communication-Enabled Intelligence for the Electric Power Grid, First Edition. Stephen F. Bush.

"power system information theory" that explicitly combines power systems and information theory. While such a field does not yet exist, we explore the concept in this chapter. Part Three, *Embedded and Distributed Intelligence for a Smarter Grid: The Ultimate Goal*, dives deeper into specific applications and sheds more light upon the answer to this question.

This part of the book examines each of the classical divisions of the power grid – namely, generation, transmission, distribution, and consumption – covered in Part One and examines the use of communication in those systems in a smart-grid context. As more communication and networking are being utilized within the power grid, we must keep in mind that the power grid itself is evolving. More DG and energy storage systems are being incorporated. Power electronics is advancing, and new interactions with the consumer are being contemplated in the form of DR and so-called "smart meters." We begin this part of the book with a brief overview of communication and networking including communication architectures relevant to the smart grid. Then we look at consumption, namely, the advanced metering infrastructure and DR. Then we explore communication and networking as it relates to DG and transmission. Note that as DG becomes more ubiquitous, the need for transmission will tend to decrease. Next, communication within the distribution system is examined, particularly with regard to fault detection, isolation, and restoration. Finally, we review some key standards and the role they are playing in the smart grid. Success of the smart grid will depend upon having communication and power system components that interoperate with one another allowing the power grid to be easily constructed and evolve in a modular manner.

The goal of this chapter, *What is Smart-Grid Communication?*, is to provide an introduction to fundamental aspects of communication and networking. The answer to the title of this chapter is that smart grid communication is leveraging the fundamental relationship between power and communication. To accomplish this, that fundamental relationship must be understood. Communication and networking are broad topics and cannot be covered exhaustively in a single chapter. Instead, this chapter focuses upon topics that appear most relevant to power systems; an attempt is made to explain material in a manner that compares and contrasts communication with power systems when such comparisons are particularly insightful. The chapter begins with a fundamental comparison between energy and information. There are direct physical links among information, communication, and energy, which provide inspiration for the notion of a power system information theory. The next section shifts from a physics-based relationship among communication and energy to a system-level view of the power grid. Here, we examine the power grid from the perspective of complexity theory. Next, graph theory and network science are introduced. These topics are used not only in communication and networking, but in a wide range of disciplines, including the power grid. Then a section devoted to classical information theory and the power grid follows. Information theory deals with quantifiable characteristics of information, including its compression and protection, which includes both protection of information from corruption during transmission and from cybersecurity attack. Then in the next section, a communication architecture for the power grid is discussed. A communication network architecture can be as loose as simply specifying the framework for a network's physical components and their functional organization and configuration and perhaps incorporating its operational principles and procedures. An architecture may also be as specific as covering data formats used during operation. There are a wide variety of communication technologies and applications within the power grid; thus, it has been extremely challenging to define a detailed architecture. There are a myriad of detailed communication protocol standards, as will be seen in Chapter 10. This section is

Table 6.1 Some specific smart grid communication technologies discussed later.

Technology	Chapter	Location
WiMAX	DA	Section 9.3.6, p. 332
LTE	DA	Section 9.3.5, p. 328
802.11	DR and AMI	Section 7.5, p. 261
Fiber optics	DA	Section 8.5, p. 295
Power line carrier	DR and AMI	Section 8.5, p. 289
Cognitive radio	Machine intelligence	Section 11.5.2, p. 390
802.15.4	DR and AMI	Section 7.4, p. 250

focused upon higher level, generalized frameworks, or modes, of thinking about the relationship among classes of communication technologies and the power grid. The next section dives into fundamental physical aspects of communication that apply to both wireless and waveguide communication. These are comprised of a few key concepts, such as electromagnetic radiation and the wave equation, that are useful for understanding specific communication technologies discussed later in this part of the book. Thus, this chapter is an overview of communication fundamentals; more detailed information on specific communication and networking architectures and protocols appears in the following chapters as communication is applied to specific aspects of the power grid. Exercises are available at the end of the chapter that encourage the reader to explore further. Table 6.1 indicates where specific communication technologies are explained in more detail.

Communication has been used in the power grid since its inception. Power grid communication is currently comprised of a variety of communication technologies and network protocols. Understanding and insight into smart grid communication will come only when viewed at the common theoretical level of communication and information theory. Communication and networking often appear to be a rather confusing mixture of meaningless protocol acronyms. This chapter is about taking a step back to view communication from different fundamental perspectives that may help in making sense of communication and networking as it relates to the power grid and perhaps inspire new, fundamental ways of combining communication and power systems.

Underlying any communication network is a common theory of information, the communication channel, and channel noise, known collectively as either information theory or communication theory. Tremendous insight is gained when one views both the power grid and its communication and networks at this fundamental level. This enables designers and engineers to transition from adding communication in an ad hoc manner toward understanding and optimizing a holistic system. For example, different components will operate and generate data with different information entropy; knowing this allows optimization of different compression rates, enabling communication channels to be better utilized.

While individual communication channels can be optimized, network analysis can used to optimize the topology formed by the interconnection of components within the power grid. Network analysis can impact not only how information is routed, but it also enables efficient analysis of what happens when different parts of a system are allowed to directly interact, often seen as the effect that a perturbation has upon the stability of a system. Eigenvalues and eigenvectors play a significant role because the network can be characterized as a graph comprised of nodes and edges, typically in the form of a matrix.

In a broad sense, there are two network topologies: the topology of the power system, comprised of generators, power lines, and loads, and the topology of the communication network, comprised of transmitters, communication channels, and receivers. Some communication technologies are constrained to follow the power grid topology (for example, power line carrier), while many communication technologies have more freedom to deviate from the power grid topology (for example, short-range wireless systems). Finally, some communication technologies are completely independent, such as those utilizing a telecommunication company or common carrier to implement communication. Protection mechanisms are a nice illustration of the impact of topology; a segment of the power system may have to be isolated due to an electrical fault while minimizing the impact of the fault on the flow of power to consumers. Changes in the power system topology become critical to achieve the best result; communication latency must be low in order to operate as quickly as possible. Of course, from a communication perspective, topology plays a critical role in how the communication media are shared and the efficiency of routing through the network.

Communication theory addresses the fundamental nature of communication; namely, transporting a packet across a potentially noisy communication channel given the power of the signal and amount of extraneous noise. Information theory addresses the efficiency of the encoding of the information over that channel, both how to compress the information to fit as much as possible into the channel (known as source coding), and how to protect the information (known as channel coding), where some of the information may be received in error due to noise on the channel. Channel coding enables error correction, a form of self-healing for information. Power grid communication should be both efficient, transmitted quickly, and use as little of the channel as possible to save room for other communication, and reliable – that is, error-free. It turns out that information theory and network topology are related; graphs have been used to analyze information coding, and the properties of a network can be inferred using information-theoretic techniques upon graphs.

Theoretical approaches are useful not only because they provide deep insight, but also because they provide ideal limits. It is possible to know how close the design of a system is toward reaching optimality. Both information theory and analysis of networks apply also to machine learning, a natural concept for use in the smart grid once the smart grid communication infrastructure is sufficiently implemented. One line of reasoning is that smart grid communication will be much like the Internet: new, sophisticated applications will develop via communication and innovation in the power grid in a manner similar to the way the Internet has enabled innovation.

Consider the interaction of communication and power system applications such as stability, load balancing, DR, switching, IVVC, automatic gain control, protection, FACTSs, state estimation, and a myriad of other power system applications. Then consider what is common and fundamental with regard to communication and power systems. Is there a power system information theory that is yet to be discovered? What would it look like and how would it change the nature of the power grid? Power systems technology has traditionally been fundamentally concerned with the dynamics of electromagnetic fields found in such components as generators, inductors, capacitor banks, transformers, and loads. The dynamics are nicely captured in Maxwell's equations, as well as in higher level simplifications such as Kirchhoff's laws for example. As an aside, network analysis, long used in power systems in the form of admittance matrices and Laplacian matrices for example, are continually being rediscovered today for use in network science. Power systems and communication parted ways long ago, one to focus on optimizing power transmission and the other to optimize information transmission. Can they be united again at a fundamental level, perhaps in the form of Shannon information theory and Maxwell's equations? If so, the unified field of power system information theory would be at the core of smart grid communications.

Just as power plays a role in communications based upon the signal-to-noise ratio, there should, in essence, be the inverse consideration; namely, how much electrical power a bit of information influences or controls. Then the question becomes where that bit is transmitted and received, both within the power system and the communication system; namely, the topology of the both the power and communication networks. The key elements of communication are a triumvirate of information (information theory), control (networked control), and network topology (network science). Note that while a unification of Maxwell's equations and Shannon information theory is suggested for a power system information theory, the unification could perhaps more easily take place at a simpler level, such as Kirchhoff's laws or even at the level of individual power system components. We explore the fundamentals of communication with this unification in mind.

Another common aspect between electric power and communication within the power grid is the notion of changing the power grid from a relatively uniform and static conduit of power to a more dynamic system, one that is highly flexible and customized for the consumer. In this aspect, the power grid is analogous to an active communication network, one in which messages contain executable code to modify the communication network as they flow through it. Control is highly distributed in such networks, and it is instructive to see how they relate to smart grid concepts. Finally, this chapter is meant to be clear and easily accessible. Some equations are introduced along with simple background information; however, a mathematical background is not required to follow the chapter.

6.1.1 Maxwell's Equations

Maxwell's equations form a theoretical basis for electricity and magnetism, applied to power systems and communications. These equations govern the electric power from within an electric generator, through transmission and distribution to a consumer's load. Any electromagnetic-based communication will also be governed by these equations. A "power system information theory" could reach down to this level to unite power systems and communications in a fundamental, useful manner. The equations are summarized as follows:

- **Coulomb's law**

$$\nu \cdot \mathbf{D} = 4\pi\rho. \tag{6.1}$$

Coulomb's law equation deals with the magnitude of the electrostatic force of interaction between two point charges. The force is directly proportional to the scalar multiplication of the magnitude of the charges and inversely proportional to the square of the distance between them.

- **Ampère's law**

$$\nu \times \mathbf{H} = \frac{4\pi}{c}\mathbf{J}. \tag{6.2}$$

Ampère's law equation relates the integrated magnetic field around a closed loop to the electric current passing through the loop.

- **Faraday's law**

$$\nu \times \mathbf{E} + \frac{1}{c}\frac{\partial \mathbf{B}}{\partial t} = 0. \tag{6.3}$$

Faraday's law equation characterizes how a changing magnetic field creates an electric field.

- **No magnetic monopoles**

$$\nu \cdot \mathbf{B} = 0. \tag{6.4}$$

The "no magnetic monopoles" equation describes that, for each volume element in space, there is exactly the same number of "magnetic field lines" entering and exiting the volume. No total "magnetic charge" can build up within any point of space. For example, a magnet's north and south poles are equally strong. Free-floating south poles, without accompanying north poles (magnetic monopoles), are not allowed.

In summary, information theory allows for (1) comparison of design and implementation to optimum theoretical results allowing one to know when to stop trying to optimize design of the smart grid, (2) elimination of redundancy in information, which is critical for the potentially massive amount of data in the smart grid, and (3) minimization of the rate, and thus bandwidth required, for polling or transmitting information throughout the smart grid.

Network science is a broad, interdisciplinary field that studies complex networks arising in a variety of other fields. It tends to draw heavily from graph theory and statistical mechanics, among other fields, and focuses upon network properties arising from the interconnections among nodes within the network. For example, network science has been used to find vulnerabilities in power grid networks. The power grid is routinely analyzed as a network to estimate and control power flow. The stability of the power grid is another property that is greatly influenced by the topology of the interconnected network nodes. Network science is an emerging science that will provide fundamental theories uniting both electric power and its supporting communication infrastructure.

Information theory addresses the study of efficient information representation and communication. Regardless of the specific communication technology, link, or protocol in a smart grid application, information theory will guide the choice of the best type of source and channel coding, which may include a variety of approaches, including compressive sensing, network coding, or joint source–channel coding (Hekland, 2004) to name a few. Applications may range from customer meter reading in the AMI to FDIR within the distribution network. Each of these communication applications has different requirements in terms of message size, latency, and reliability, ranging from short, life-critical messages with low latency requirements to longer, less-critical messages that can tolerate longer latency. Classical information theory enables computational techniques to optimize the communication system to meet these requirements.

Network science and information theory did not develop in complete isolation; notions such as the "maximum capacity of a graph" and "graph entropy" used graph theory, what later became network science, as a tool to reason about and compute the best coding algorithm for a noisy channel. Here, a graph is constructed from the symbols being transmitted, which are the nodes of the graph. The likelihood of confusion, or inability to distinguish the symbols at the receiver, is represented by edges of the graph. Thus, one desires to transmit the set of codes that conveys the most information but are subsets of the graph that have no connection to one another, known as a stable or independent set. Thus, instead of a channel being an edge of the graph, the entire graph represents a channel.

Another fundamental relationship between network science and energy comes from the notion of "graph energy." The mathematics of the concept will be avoided in order to keep the explanation relatively easy to understand. The basic idea comes from physics and chemistry; namely, the low-level behavior of atoms and quantum mechanics. In the quantum realm, energy only exists at discrete values. These values can be determined by the solution of the wave equation, which makes sense only when the value for energy takes on discrete values. These values happen to be eigenvalues of the wave equation. Any graph can be represented as a square matrix, an adjacency matrix, or a Laplacian matrix, for example, whose eigenvalues and eigenvectors can be computed. The graph represents a system and the eigenvalues represent the allowed energy states of the graph. Thus, the term "graph energy" was derived. An adjacency matrix is a square matrix with elements that indicate the connection between each node in the network. An incidence matrix has rows equal to the number of vertices, and columns equal to the number of edges. A "1" can be placed in an edge leaving a corresponding node and a "0" in an edge entering the corresponding node. Thus, the incidence matrix translates edges into a "differential" between connected nodes. If one considers the incidence matrix as the first derivative of a graph because it captures the differences between connected nodes, then the Laplacian matrix (also known as the admittance matrix or Kirchhoff matrix) represents the second derivative; namely, it is the incidence matrix squared. The Kirchhoff matrix plays a significant role in power flow studies. A related field is random matrix theory (RMT), which also examines random matrix properties, often as they apply to network structures.

6.1.2 Eigensystems and Graph Spectra

The eigenvectors of a matrix, in which the matrix is square, are vectors that have a special property. Recall that vectors are simply entities with a magnitude and direction in an n-dimensional space. When multiplied by the matrix, the vectors maintain the same direction; this is the special property mentioned above. The magnitude of the vector may change; however, the magnitude is captured by the eigenvalue, a scalar value as shown in Equation 6.5, where λ is the eigenvalue, v is the eigenvector, and A is a matrix:

$$\lambda v = Av. \tag{6.5}$$

There are as many eigenvectors and eigenvalues as there are rows in the matrix. The term spectra comes from studying the eigenvalues of matrices related to fundamental concepts in physics; namely, waves and energy.

6.2 Energy and Information

Power is the rate at which energy flows and is measured in joules per second or in watts. The purpose of the power grid is to support the flow, or spatial displacement, of power from source to consumer. The integration of power over time yields work where the result for work is dependent upon the path of the integral and it is in units of joules. On the other hand, information is measured in bits, yet information and energy are related at a fundamental level. Since this is true, then information and power are related at a fundamental level as well. If information, power, and control are related in a coherent manner, then there is hope for a theory that better unites communication and power systems, as we will see.

The story, in a nutshell, begins with Maxwell's demon (Plenio and Vitelli, 2001). This is a thought experiment used to test the second law of thermodynamics. Recall that this law states that, in an isolated system, entropy never decreases. Maxwell proposed a hypothetical demon that would sit between two masses that are at different temperatures; assume the gases are in a box separated by a wall in the middle. The demon knows precisely when to open a door in the wall allowing molecules to pass through, so that, for example, a higher energy gas molecule can leave the cooler gas enclosure and enter the warmer gas enclosure. Thus, instead of the gases coming to an equilibrium temperature, the cooler gas continues to become cooler, entropy decreases, and the second law appears to be violated. Counter-arguments were made to save the second law of thermodynamics from Maxwell's demon by stating that the demon's operation or measurements would require energy and thus release entropy into the environment to make up for the difference. However, these counter-arguments were generally shown to be incorrect until Bennett and Landauer theorized that the key lies within the information processing of the demon's mind.

Maxwell's demon's mind is a processing system; it may be biological or implemented as an electronic device, but either way it is a physical computational system that has its own physical properties, including entropy. If the number of logical states of a computation were to decrease, then there must be a corresponding physical increase in entropy in order to conserve the entropy of the system. In other words, Landauer's principle (Bennett, 2002; Landauer, 1961) proposes that any irreversible logical operation (loss of information entropy) must result in a corresponding increase in the physical entropy of the immediate environment. More specifically, loss of a bit of information results in the release of $\ln 2kT$ joules of heat, where k is Boltzmann's constant, and T is absolute temperature. This has direct application to the use of electric power in computation. By developing irreversible computing elements, power consumption should be greatly reduced. But of course this is looking at the problem of a large amount of computation with a small amount of power, which is the reverse of power system information theory – namely, minimizing computational requirements to efficiently control large amounts of power. But nonetheless, this relationship between energy and information clearly exists and has been verified.

6.2.1 Back to the Physics of Information

Because digital information has been defined via the notion of entropy, borrowing from the notion of Boltzmann's entropy, there has a been a myriad of relationships and similarities drawn between information and communication with physics and thermodynamics. It will be instructive to visit one of these relationships in detail. The benefits will be twofold. First, it will provide background for the reader who may have expertise in power systems, but less familiarity with communications. Second, it will hopefully reinforce the notion that there is a deeper, yet-to-be-explored, fundamental relationship between power and communication.

Let us first derive the Shannon–Hartley equation; then we can proceed to its relationship with thermodynamics. In communication, the goal is to successfully transmit information in the presence of noise. In other words, the signal must be distinguishable from the noise. While the goal of the power grid is to maximize power transfer without noise or information, the goal of communication is to transmit information, typically while minimizing power. Consider a transmitter, channel, and receiver. The total power received is shown by

$$P = S + N, \tag{6.6}$$

where S is the signal power and N is the noise power. An assumption in this quation is that the signal and noise are uncorrelated with one another.

The RMS voltages are related thus:

$$V^2 = V_S^2 + V_N^2. \tag{6.7}$$

Here, we assume that information is being encoded by voltage level. Thus, the total range of voltage can be divided into a number of voltage bands, 2^b, each of equal size. A large number of bands allows more information to be encoded, but then individual bands must be smaller. A smaller band results in a greater probability of noise interference causing the voltage level to fall outside the intended band; this creates an error. Let the width of each band be ΔV.

Since we know the noise voltage is V_N, Equation 6.8 shows the maximum number of bands:

$$2^b = \frac{V}{V_N}. \tag{6.8}$$

Next, some algebraic manipulation and substitution are used to obtain a result in terms of the signal and noise power. Equation 6.9 is simply the square root of Equation 6.8 squared:

$$2^b = \sqrt{\frac{V^2}{V_N^2}}. \tag{6.9}$$

Equation 6.10 replaces V using Equation 6.7:

$$\sqrt{\frac{V_N^2}{V_N^2} + \frac{V^2}{V_N^2}}. \tag{6.10}$$

Finally, Equation 6.11 simplifies Equation 6.10:

$$\sqrt{1 + \frac{S}{N}}. \tag{6.11}$$

From Equations 6.9–6.11, we know that $2^b = \sqrt{1 + (S/N)}$. Solving for b yields

$$b = \log_2\left(\sqrt{1 + \frac{S}{N}}\right). \tag{6.12}$$

Now assume that we can make M measurements of b bits in time period T. The number of bits in one time period is shown by

$$Mb = M \log_2\left(\sqrt{1 + \frac{S}{N}}\right). \tag{6.13}$$

The information transmission rate is shown by

$$I = \frac{M}{T} \log_2\left(\sqrt{1 + \frac{S}{N}}\right). \tag{6.14}$$

Just as transmission of power is not perfectly efficient, the transmission of information is also not noise free. Even at the lowest physical level there is thermal noise in the communication channel, and effort is required to compensate for the noise.

From the Nyquist theorem, we know that if B is the highest frequency component of a signal, then the signal can be perfectly reconstructed if the sampling rate is at least $2B$. This is reflected in

$$\frac{M}{T} = 2B, \tag{6.15}$$

where M is the number of measurements as previously defined and T is the time period. Their ratio is the sampling rate.

Notice the M/T in both Equations 6.14 and 6.15. Using this to combine the two equations yields

$$C = 2B \log_2\left(\sqrt{1 + \frac{S}{N}}\right) = B \log_2\left(1 + \frac{S}{N}\right). \tag{6.16}$$

This equation is useful because it defines a general bound on the maximum rate at which information can flow through a channel given the bandwidth, signal, and noise levels.

Following the explanation above, the additive white Gaussian noise channel is described as shown

$$y = x + n, \tag{6.17}$$

where x is the transmitted signal, n is the noise, and y is the received signal.

The noise is assumed to follow a complex Gaussian distribution $CN(0, N_0)$ where N_0 is the noise variance $N_0 = E[nn^*]$. Note that the star superscript is the complex conjugate, but do not worry about that here. Our goal is to focus upon the larger picture by showing the analogy between compression of an ideal gas and the energy required to communicate over a channel. Similar to N_0, $E_S = E[xx^*]$ represents the energy of the signal. Ideally, the signal needs to be strong enough to overcome the noise for the receiver to be capable of decoding the signal.

Following the explanation above in the derivation of Equation 6.16, the maximum channel capacity for the additive white Gaussian noise channel is

$$C = \log_2\left(1 + \frac{E_S}{N_0}\right). \tag{6.18}$$

Notice that the bandwidth B is missing in this equation because we are assuming infinite bandwidth. In other words, we are allowing for an infinite amount of time for transmission to occur.

With a little algebraic manipulation, the total signal energy is shown by

$$E_S = N_0(2^C - 1). \tag{6.19}$$

Now we can obtain the energy per bit by simply dividing the total energy by the channel capacity, in bits:

$$E_\mathrm{b} = \frac{E_S}{C} = N_0\frac{2^C - 1}{C}. \tag{6.20}$$

Consider the minimum energy required to transmit one bit given an infinite amount of time. This is shown by

$$E_\mathrm{b}^\mathrm{min} = N_0 \lim_{C \to 0} \frac{2^C - 1}{C} = N_0 \ln 2, \tag{6.21}$$

where the channel capacity goes to zero.

Now consider an ideal gas. This is particularly relevant to the power grid because electric power is often generated by compressing and expanding fluids through a turbine; for example, steam generation, hydroelectric generation, compressed-air energy storage, and even wind generation are examples. Let us see how this relates to communication (Samardzija, 2007).

The differential change in the internal energy of a gas is

$$\mathrm{d}U = \mathrm{d}Q - \mathrm{d}W, \tag{6.22}$$

where Q is thermal energy and W is energy used to do mechanical work. This equation is concerned with differential changes in value, where the energy used to do work is removed from the total energy.

Now consider each of the terms in Equation 6.22 in more detail. The thermal energy

$$\mathrm{d}Q = T\,\mathrm{d}S \tag{6.23}$$

is the product of the temperature T and $\mathrm{d}S$, the thermodynamic entropy. The term "entropy" should raise some interest as we know that information entropy borrows a similar notion. The thermodynamic entropy might be considered as characterizing how the internal energy U is distributed among the components, in this case the individual gas molecules.

The differential work is made more precise in

$$\mathrm{d}W = p\,\mathrm{d}V, \tag{6.24}$$

where p is pressure (the pressure of the gas on the sides of its container for example), V is volume, and $\mathrm{d}V$ is the differential change in volume. Thus, a large pressure or large change in volume increases the work that is done.

The average energy of a particle is $kT/2$ per degree of freedom, where k is Boltzmann's constant. Building upon this, the change in average internal energy, mentioned in Equation 6.22 is,

$$\mathrm{d}U = \frac{LMk}{2}\,\mathrm{d}T, \tag{6.25}$$

where L is the number of particles and M is the degree of freedom.

Equation 6.26 recalls for us the well-known ideal gas law:

$$pV = LkT. \tag{6.26}$$

Without loss of generality, the rest of the discussion implicitly assumes $L = 1$ reflecting a single particle and $M = 2$ representing two degrees of freedom. This corresponds to a single particle moving along a plane. In order to develop the relationship, consider the gas in a container and reducing its volume by half. This process is assumed to be done isothermally; that is, the temperature is held constant throughout the process. Thus, $\mathrm{d}T = 0$. Using Equation 6.22 and substituting the appropriate values yields

$$0 = T\,\mathrm{d}S - p\,\mathrm{d}V = T\,\mathrm{d}S - \frac{kT\,\mathrm{d}V}{V}. \tag{6.27}$$

Using Equation 6.27, the amount work required to compress a volume of gas by half is

$$\Delta W = -\int_{V}^{V/2} \frac{kT\,\mathrm{d}V}{V} = kT\ln 2. \tag{6.28}$$

The integral is taken over the volume that is being removed due to compression.

Of course, choosing to reduce the volume by half was not arbitrary; it is related to binary information by the fact the amount of information required to track each particle is also reduced by half. This is equivalent to reducing the number of binary information bits by one. Conversely, instead of tracking gas particles, the fact that gas particles are confined within a specific volume can be thought of as storing information. Compressing the gas to half of its original volume can be considered as representing either a "1" or a "0." This is also related to Landauer's principle of erasure, which theoretically characterizes the minimum amount of information to perform a computation.

Recall Equation 6.21 and that N_0 is the noise variance in a channel. If $N_0 = kT$, then Equation 6.28 and Equation 6.21 become equal, as shown by

$$\Delta W = E_b^{min} \quad \text{for} \quad N_0 = kT. \tag{6.29}$$

This equivalence implies that the same amount of energy is required to transmit one bit of information over an additive white Gaussian noise channel, where $N_0 = kT$, as that needed to store one information bit via isothermal compression of an ideal gas, which requires $kT \ln 2$ joules of energy. Of course, the analogy between an ideal gas and communication can be taken further ad nauseam; the goal here is simply to point to the notion that communication, information, and energy have a fundamental relationship and that there is much more for the reader to discover and innovate regarding the relationship between communication and the power grid.

Figure 6.1 illustrates relationships among energy, information, and communication. Figure 6.1a illustrates Maxwell's demon opening and closing a door separating two volumes such that one becomes warmer and the other cooler in violation of the second law of thermodynamics, related to Landauer's principle. Figure 6.1b illustrates the equivalence between ideal gas compression and information previously discussed. Figure 6.1c shows the small size of components considered for nanoscale communication networks, discussed in more detail

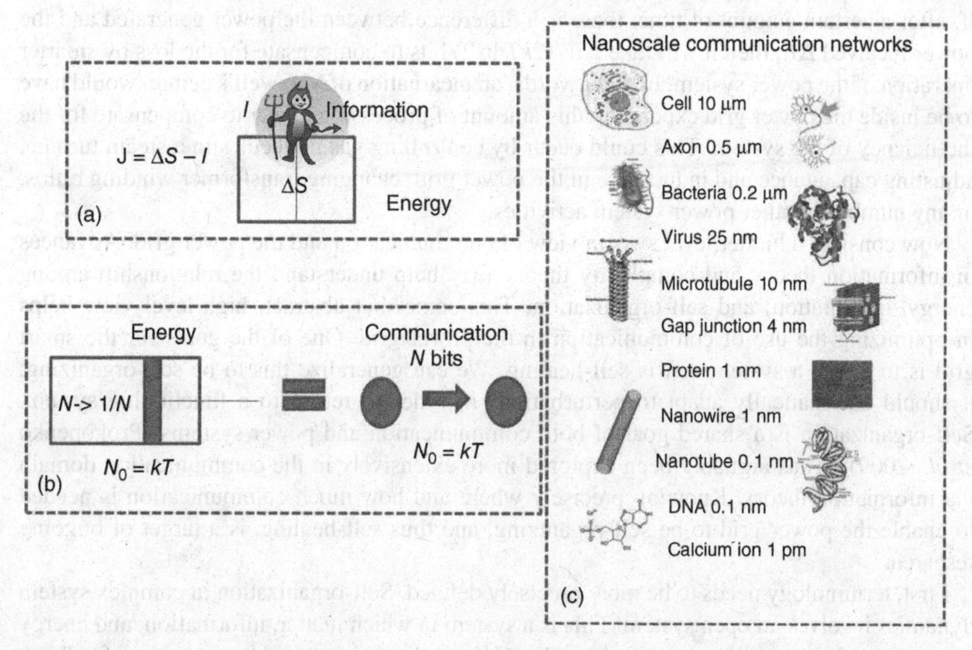

Figure 6.1 The relationship among energy, information, and communication: (a) energy and information are related; (b) communication and gas laws (energy) are related; (c) nanoscale communication networks operate at the lowest levels at which these relationships come about. Source: Bush, 2010a. Reproduced with permission of IEEE.

in Section 15.4. Thus, Figure 6.1 illustrates that (a) energy and information are related, (b) communication and gas laws (energy) are related, and (c) nanoscale communication networks operate at the lowest levels at which these relationships come about.

6.3 System View

The astute reader can probably see where this is leading us; however, let us review the situation before making the final leap. The power grid's main purpose is transferring power efficiently. Transmission can be implemented in the form of wires, wireless power transmission, quantum teleportation of power, or many other possible forms. Power is the flow of energy. If the energy transmitted does not equal the energy received, then some of that energy is lost due to inefficiency in the system. The role of smart grid communication is to enable computational processes to interact in a manner that can hopefully minimize inefficiency. The energy–information relationships lead to a fundamental understanding of the minimal amount of processing and communication required to compensate for energy transmission inefficiencies. Of course, this deals with theoretical optimum values but, as information theory has done for communication, it provides a fundamental understanding and theoretical optimum limits to what can be achieved. The results should also be independent of scale; that is, it should be independent of whether it relates to the small amounts of energy of an ideal gas or large amounts of energy transmitted over a transmission line.

Consider a simple case: power transmitted over a period of time from source to destination. If, after a certain amount of time, there is a difference between the power generated and the power received ΔE, then it will take $\Delta E/(2kT \ln 2)$ bits to compensate for the loss by smarter operation of the power system. In other words, an incarnation of Maxwell's demon would have to be inside the power grid expending this amount of processing power to compensate for the inefficiency of the system. This could occur by controlling gas molecules in a steam turbine, adjusting capacitance and inductance in the power grid, changing transformer winding ratios, or any number of other power system activities.

Now consider a higher level system view of communication and the power grid. Advances in information theory and complexity theory may help understand the relationship among energy, information, and self-organization. This somewhat abstract, high-level view helps in optimizing the use of communication in the power grid. One of the goals for the smart grid is to create a system that is self-healing. We can generalize this to be self-organizing; it should automatically adapt to perturbations in order to return to a functioning system. Self-organization is a shared goal of both communication and power systems (Prokopenko *et al.*, 2009). It has arguably been explored more extensively in the communication domain via information theory. Knowing precisely where and how much communication is needed to enable the power grid to be self-organizing, and thus self-healing, is a target of ongoing research.

First, terminology needs to be more precisely defined. Self-organization in complex system dynamics involves an open system. This is a system in which matter, information, and energy are entering the system from the environment. It is explicitly not a highly engineered feedback control system. That is dealt with using classical control theory in Section 8.6. A complex dynamic system is more challenging. For example, the smart grid is receiving price signals from consumers, loads are constantly changing, supply may be constantly changing, power lines or power flow may be disrupted, any number of inputs and events exogenous to the system are occurring, and the smart grid should self-organize to handle the events. Self-organization is taken to mean the ability of the system to organize without central or hierarchical control or explicit instructions indicating what each component must do. The notion of emergence, which is the idea that many small components with relatively simple rules interact to create a global, adaptive, "living" system, is a highly seductive form of self-organization that researchers have sought to understand and emulate. The means by which the global system self-organizes may not be apparent or readily understood. There have been many different approaches in attempting to understand self-organization, and some of them involve the notion of thermodynamics and the use of energy by a system. However, it should be kept in mind that, in these analyses, energy refers to energy used by the system to maintain its organization, not necessarily energy being transported.

6.4 Power System Information Theory

The electric power grid currently utilizes concepts from information theory and network analysis. It is beyond the scope of this chapter to list every possible application; however, a subset of selected applications is shown in Table 6.2. Application of classical information theory effectively increases communication network capacity. AMI may further benefit from decreased traffic using compressed sensing. Compression techniques can be applied to synchrophasors to reduce load. FDIR benefits network analysis; electrical faults require switches

Table 6.2 Information theory and network analysis mapped to selected areas within the smart grid.

Concept	Smart grid application	Benefit
Classical channel capacity	Throughout the smart grid	Decrease load
Classical compression	Synchrophasors	Phasor data compression
Compressive sensing	AMI	Decreases meter load
Entropy	Security	Increase encryption strength
Entropy & prediction	DR	Predict power output
Entropy & prediction	DG	Smooth peak demand
Entropy & prediction	Energy storage	Reduce variance
Entropy & prediction	stability	Channel coding
Inference	State estimation	Inferring state with less data
Nanoscale communication	Nanogeneration, nanogrids	Control power grids comprised of large number of very small components
Network coding	AMI	Efficient transmission
Quantum information theory	Security	Quantum key distribution
Spectral graph theory	FDIR	Distribution network analysis
Spectral graph theory	Power grid and communication networks	Network structures
Spectral graph theory	Grid-communication networks	Network structures
Entropy	Security	Encryption strength
…	…	…

to reconfigure in such a manner that the impact to customers is minimized. State estimation will benefit from less data required to infer the state of the power grid. Information theoretic techniques will lead to improved power demand prediction and better response. DG will benefit from better stability control derived from network analysis. Stability in general will improve from application of information theory and network analysis. The concept of network coding, which combines source coding with network analysis, may someday further reduce traffic load for AMI as well as other applications within the smart grid. Spectral graph theory applies toward improving both power grid topology and communication network topology. Entropy and quantum information theory apply to cybersecurity within the smart grid (for example, quantum key distribution discussed in Chapter 15). Also, entropy and prediction will apply toward optimizing the use of energy storage within the network. This is because energy storage is used to help reduce peak demand for power; smoothing power demand is equivalent to reducing its entropy. Finally, one of the most challenging advances to consider is a power grid that reaches down to the nanoscale – one in which every joule of energy is efficiently harvested, down to the molecular level. This will require new forms of communication capable of operating at the molecular level (Bush, 2011a).

This chapter posits that advances in smart grid communication will remain both superficial and ad hoc until a fundamental extension of information theory is developed that unites our understanding of communications and power systems. The need and requirements for such a new power systems information theory are laid out with respect to a vision for advances in power grid operation anticipated far into the future. Highlights of this vision include the trend toward increased dynamization of the power system, including greater use of physical fields and mobile components, a trend toward microscale and nanoscale distributed power generation, and a significant increase in controllability and complexity. All components of this vision increase information entropy, and motivate the requirement to determine the fundamental limits of communication given the underlying physics of future power systems.

Industry, utilities, and standards working groups would benefit from knowing the optimal communication architecture for the power grid as a whole. The longer the grid continues to develop in a piecemeal, non-holistic fashion, the lower the chance that it will be done in an efficient, cost-effective manner. The overarching question is what is the impact of communication on electric power transmission efficiency? It is relatively easy, given a specific grid application, to analyze the impact of noise, bandwidth, latency and jitter upon operation of such an application. However, there are many such grid applications – for example, stability control, FDIR, AMI, and DR. Each of these applications is being designed as a separate, individual control system to which communication is often assumed to work cheaply, reliably, and with low latency. Each grid application often assumes its own control system supported by an overprovisioned communication infrastructure. Researchers need to step back and take a holistic view in order to ensure that the entire system is being designed optimally, and to avoid inefficiency, redundancy, and expensive overprovisioning of communications. In order to accomplish this, we should turn to fundamentals. We can draw inspiration from communications itself, which was modeled and analyzed in a holistic manner in the early days of digital communications.

Among the high-level goals is the desire to minimize cost and pollution by maximizing efficiency of power transmission from centralized or renewable power generators to individual consumers. We need to know (1) if our standards and architectures are leading toward an optimal solution, (2) the theoretical optimum that can be achieved, and (3) when we have achieved the desired theoretical, low-cost, optimal-efficiency trade-off within the space of architectural solutions.

For example, it would be ideal to know the theoretical value of metrics such as the minimum number of links, interfaces and communication cable or radio frequency (RF) energy required to achieve particular physical-layer communication solutions. It is necessary to know, based upon the characteristics of the physical layer, what the impact of those solutions will be on the overall operation of power system efficiency (and not just on a particular application implemented in a particular manner), which may be designed inefficiently or impact the rest of the system and other applications in a nonoptimal manner. As an example, a few fundamental communication metrics are:

- minimum bits (of communication) per kilowatt (of active power delivered);
- maximum transmission efficiency (power delivered/generated) per bit (of communication);
- maximum stability margin (voltage angle) per bit (of communication);
- minimum protection and restoration system average interruption duration index (SAIDI) per bit (of communication);
- maximum load control (percentage of optimal scheduling) per bit (of communication);
- maximum power factor (dimensionless) per bit (of communication);
- maximum distribution load balancing per bit (of communication);
- all of the above in terms of minimum RF power or optimal spatial reuse if communication is implemented wirelessly.

We also want these metrics as functions of communication latency, jitter, and ability to tolerate packet loss. These values are easy to obtain for a *particular* implementation, but that is not the goal. The goal is to determine the theoretical optimum for these values, assuming perfect sensing, actuation, and control.

Figure 6.2 illustrates an abstract model for our discussion of smart grid communication. On the left side of the figure, electric power flows vertically. On the right side of the figure, a control system senses a subset of power parameters and makes the decisions necessary to send control signals to the actuator. An application on the top right of the figure has high-level management of the system. In our model, we assume "perfect" sensing, actuation and control. In addition, since we want theoretical optimum values, we assume the most fundamental model in which all parameters can be sensed and any actuation can take place. The fact that some parameters may have limited detectability today should not be considered at this point; we want to know the best that can physically be achieved between communication and electric power transport. The control system has a perfect model of the power flow and makes perfect decisions with no computational latency. The only potentially imperfect component that we consider communications, so that we can focus on the impact of communication upon the system.

It has been shown that there is a correspondence between energy in the form of thermodynamic laws and Shannon information theory (Samardzija, 2007). The average energy required for adiabatic compression of an ideal gas to $1/N$ of its initial volume is the same as the average energy needed to achieve the capacity $C = \log_2 N$ of the equivalent additive white Gaussian noise channel. Power generators, particularly turbine generators, extract energy from such thermodynamic processes, having a direct analogy with communication. There is also a correspondence between ideal gas volume and the minimum square code-word distance. Also,

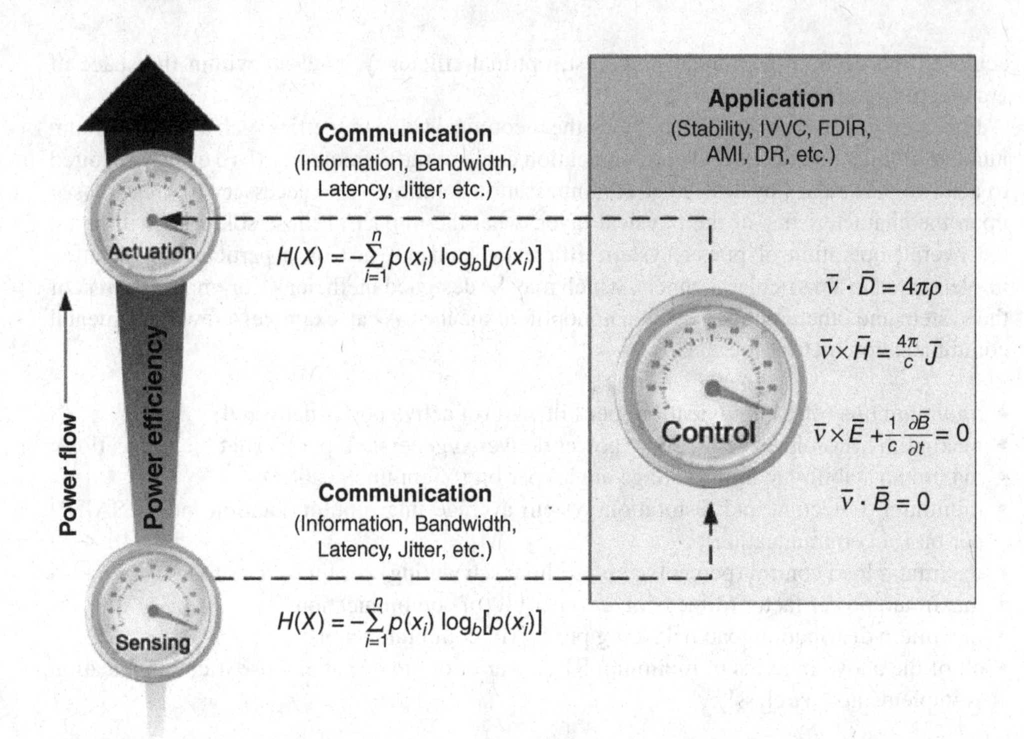

Figure 6.2 An abstract communication model for smart grid communication. Electric power flows vertically along the left side of the figure; power parameters are sensed and then modified via an actuator. On the right side of the figure, a control system makes the decisions necessary to act upon the sensed parameters and sends control signals to the actuator. An application on the top right maintains high-level management of the system. Performance of communication impacts the ability of the application to have perfect control over the efficiency of power transport. Source: Bush, 2013a. Reproduced with permission of IEEE.

Landauer's principle shows a direct connection between information and energy via information and physical entropy. These relationships between communication, information, and energy hint at the possibility that there is a deeper relationship between communication theory and power systems that has yet to be explored for the benefit of smart grid communication, as illustrated in Figure 6.2.

Communication is utilized in support of control systems that reduce power loss and improve power quality within the power grid. A notional illustration is shown in Figure 6.3, in which the thickness of the black bar represents power loss in the grid and the thickness of the gray bar represents the amount of communication channel capacity utilized. This figure illustrates the relation between power grid efficiency and information flow through communication channels. A fundamental understanding of this relationship is lacking. Instead, communication links tend to be added to the power grid in a haphazard, ad hoc manner. Only a small subset of power applications is shown in the illustration, namely stability control, IVVC, FDIR, and AMI. Each application implements independent control mechanisms to optimize grid operation. IVVC was explained in detail in Section 3.4.1.

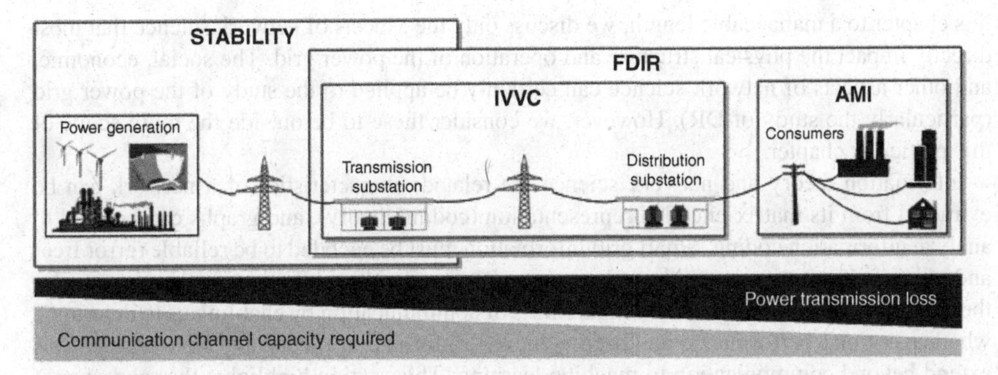

Figure 6.3 Revisiting Figure 3.7, which was a conceptual illustration of communication bandwidth (gray) versus power loss (black) during power transport. More communication bandwidth can enable better control and a reduction in power loss. The author posits that there is a fundamental trade-off among communication power, channel capacity requirements, and reduction in power loss within the power grid. Source: Bush, 2013a. Reproduced with permission of IEEE.

The closest similar theory to that proposed here, in terms of power system information theory, is the theory of networked control systems (Figueredo *et al.*, 2009). However, networked control theory is still in its infancy and does not directly address the more relevant issue of the relationship among power, information, and communication. Concepts from information theory and those related to network science play a significant and growing role in the smart grid, since both the power grid and communication networks are themselves complex networks. The information carried by the communication network must be encoded in the most efficient format possible in order to reduce required bandwidth (source coding) and to possess the ability to "self-heal" (for example, error correction or channel coding).

Network science encompasses terms that are used interchangeably in this chapter, such as network analysis and network topology. Because network science is an emerging area of study, it lacks an official, uniformly accepted definition. The National Research Council defines network science as the "study of network representations of physical, biological, and social phenomena leading to predictive models of these phenomena … A working definition of network science is the study of network representations of physical, biological, and social phenomena leading to predictive models of these phenomena. Initiation of a field of network science would be appropriate to provide a body of rigorous results that would improve the predictability of the engineering design of complex networks and also speed up basic research in a variety of application areas." (Committee on Network Science for Future Army Applications, 2006). Thus, network science focuses on deriving the impact upon a system due to its network structure and not the details of a particular application. In other words, network science attempts to isolate the "network" from the "application." Certainly, there is no question that the performance of both the power grid and communication networks are intimately involved with their network structures. When both of these network structures are combined in a holistic manner within the smart grid, there is the potential for complex network effects. Network science has broad application to many disciplines, including social networks, economics, and biology – all of which play a role in the smart grid. However, in order to keep this chapter to a manageable length, we discuss only the aspects of network science that most directly impact the physical structure and operation of the power grid. The social, economic, and other aspects of network science can certainly be applied to the study of the power grid (particularly the study of DR). However, we consider these to be outside the main scope of this particular chapter.

Information theory and network science are related; characteristics of a network can be extracted from its matrix-encoded representation (coding theory), and graphs can be used to analyze information coding. Smart grid information must be encoded to be reliable (error free) and efficient (transmitted quickly and consume less bandwidth). Information theory provides the analytical tools to find the theoretical limits of communication by enabling us to determine whether optimal performance has been achieved. Information theory and network analysis extend beyond communication to machine learning. This section highlights the relevance of these fundamental theories to the electric power grid and discusses the current state of these theories in a broad context within the power grid. For example, consider Figure 6.3 again. Engineers would like to know – with minimal communication cost – when optimal power transport has been achieved. Can we develop a theory to tell them how close they are to achieving the optimal communication architecture? Certainly, in the communication domain, the optimal channel capacity and optimal compression rate can be determined. But the relation between optimal power transmission efficiency and communication has not yet been achieved for power systems.

Power systems are fundamentally concerned with the dynamics of electromagnetic fields found in generators, power lines, capacitor banks, transformers, and highly inductive motor loads. These dynamics are elegantly defined by Maxwell's equations, discussed briefly in Section 6.1.1. A layer above electromagnetic fields is Kirchhoff's laws, which define how current and voltage behave through an electrical network. Kirchhoff's laws, admittance matrices, and Laplacian matrices have a long history of use in network analysis, and are described in this chapter. Unfortunately, power systems and communication have developed along independent theoretical paths. The smart grid provides an opportunity to unite these disparate fields, and this chapter proposes precisely this vision: the need for unification, perhaps in the form of Shannon information theory in concert with Maxwell's equations.

As we look at communication and power system applications such as stability, load balancing, DR, switching, IVVC, automatic gain control, protection, FACTSs, state estimation, and a myriad of other power system applications, we need to consider what is common and fundamental with regard to communication. This chapter contends that the common and most fundamental disciplines are a triumvirate of information (information theory), control (networked control), and network topology (network science). While we suggest a unification of Maxwell's equations and Shannon information theory to create a "power system information theory," the unification could perhaps more easily take place at the level of the simpler Kirchhoff's laws, or even at the level of individual power system components. However, just as power plays a role in communication based upon the signal, there should, in essence, be the inverse consideration: how much electrical power one bit of information influences or controls within the power grid. Then the question becomes where that bit is transmitted and received, both within the power system and within the communication system; namely, the network topology of both the power and communication networks. The smart grid is also turning the power grid from a rather passive conduit of power to one that is more autonomous and active. Active power distribution networks are a prime example. Thus, we examine active communication networks from an information theoretic perspective as a means of understanding and implementing self-healing and as a form of machine-to-machine (M2M) communication.

A well-designed electric power grid requires application of fundamental theory in order to examine the theoretical tools and limits that help design the technologies, and enable a comparison of those with actual results. Complete control of the power grid would require real-time collection of information to implement closed-loop dynamic control. Ideally, the complete set of data (measurements from all points in the power grid – substations, power line devices and consumers) should be continuously available. Since the size of the grid makes this unlikely, real-time requirement may be relaxed in order to make the problem tractable. This is done today by means of state estimation (Schweppe and Wildes, 1970) within the power grid. State estimation is used in control theory and in the more relevant field of networked control. Networked control attempts to understand the impact of communication on control system behavior – for example, how latency will impact the performance and stability of a control system. State estimation has been thoroughly researched and developed by industry since its initial introduction; decentralized and distributed techniques for state estimation have been developed, as well as hiding, manipulating, and obfuscating information within the associated state space. An obvious path of development has been to combine state estimation with all of the techniques described in this chapter. Simply put, there is redundancy in power grid data; this means that there is data that may be omitted to reduce bandwidth. Determining to what degree the "real-time" requirement can be relaxed and the conditions under which measurements can be omitted due to redundancy are two of the problems effectively addressed by information theory. In addition, analyzing the power grid using network science provides insight into possible strengths, weaknesses, and even the dynamic range of the grid; that is, how rapidly disturbances flow through the grid and either resonate or become dampened. From an information theory perspective, the power grid is a complex system in which electric power is carried over a network of conductive lines. It can be modeled as a graph (unidirectional or directional, depending on the point of view) with electrical devices as the nodes of a graph and power flows as edges of the graph. This is routinely done by power systems engineers in power flow analysis.

The study of efficient information representation and communication is addressed by information theory. Application to the electric grid includes the classical uses of information theory for determining the best form of source and channel coding, including such topics as compressive sensing, network coding and joint source–channel coding (Hekland, 2004), ranging from coding for customer meter reading in the AMI to FDIR within the distribution system. Each of these communication applications has different requirements in terms of message size, latency, and reliability, ranging from short, life-critical messages with low latency requirements to longer, less-critical messages that can tolerate longer delay. Classical information theory enables computational techniques to optimize the communication system to meet these requirements. For the power grid, it would be ideal to know the theoretical optimal bits per kilowatt wherein each kilowatt is delivered as efficiently as possible, or perhaps bits per efficiency of a kilowatt of electric power. This concept can be taken down to the physical communication layer, such that the power grid efficiency can be derived as a function of the optimal communication transmission power – for example, RF transmission power to achieve the channel capacity to provide the necessary power grid control. In fact, at the physical communication layer, the beauty of the unity between communication power (typically measured in dBm) and grid power (measured in kilowatts) becomes apparent. It should be possible to treat grid power similar to communication power and vice versa. For example, can grid power be wireless? Can grid power be transmitted in a store-and-forward manner? Can communication be embedded, or more integrated, with grid power? Of course, the answer to all these questions is yes, as we will see.

Information theory, network science, and graph theory are related via such topics as the maximum capacity of a graph and graph entropy. In these applications, a graph is used to analyze the efficiency of channel coding. The maximum capacity of a graph uses a graph to represent the potential confusion (corruption) of symbols being transmitted, where symbols are nodes and links represent confusion between symbols. The capacity of the graph is the maximum number of messages that can be transmitted without corruption using the given symbols. Graph entropy is related to the maximum capacity of a graph; given the probability of occurrence of the symbols in the graph, it returns the information entropy of the graph (that is, its information carrying capacity). A more general study of graph and network structure is found in spectral graph theory, which examines the characteristic polynomial, eigenvalues, and eigenvector of matrices associated with a graph, such as its adjacency matrix or Laplacian matrix. An adjacency matrix is a square matrix with elements indicating the connection between each node in the network. An incidence matrix has rows equal to the number of vertices and columns equal to the number of edges. A "1" can be placed in an edge leaving a corresponding node and a "0" in an edge entering the corresponding node. Thus, the incidence matrix translates edges into a "differential" between connected nodes. If one considers the incidence matrix as the first derivative of a graph, because it captures the difference between connected nodes, then the Laplacian matrix (also known as the admittance matrix or Kirchhoff matrix) represents the second derivative; namely, it is the incidence matrix squared. It plays a significant role in electric power flow studies. A related field is RMT, discussed in Section 6.4.3, which also examines random matrix properties, often as they apply to network structures.

Relatively recent advances in information theory, namely algorithmic information theory, focus on the relationship between computation and information; namely, the minimum characteristics (minimum size, minimum execution time) of a program that produces a given piece of information. This is seen in the minimum description length (MDL) principle and Kolmogorov complexity (Barron *et al.*, 1998). These can be used for classification of normal and abnormal operating states with application to cybersecurity and fault detection for the smart grid (Bush, 2002). It also plays a significant role in active networking, as explained later in Section 6.5.2. Information theory also has strong ties with complexity theory and understanding complex system behavior. Complexity arises because power must be delivered in a controlled manner; the system must adapt to exogenous events in order to maintain power quality. The efficient exchange of information among components within the power grid is required in order to maintain the required power quality to consumers. Because the power grid is a complex system, complexity theory (e.g. self-organized criticality) can provide understanding regarding the rate at which information may be generated, and thus the corresponding communication requirements, cybersecurity vulnerabilities, and operational efficiency.

6.4.1 *Complexity Theory*

There is little doubt the power grid is a complex system capable of maintaining the balance between supply and demand through many changing conditions. Figure 6.4 illustrates that the power grid, as it becomes the "smart grid," begins to interact more between two complex

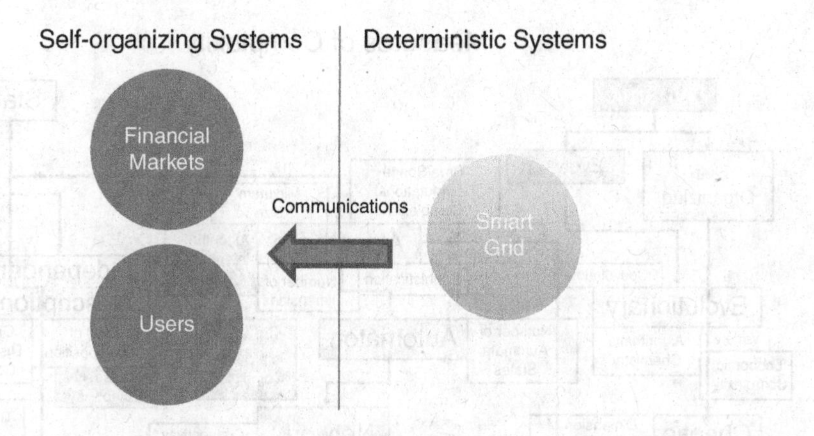

Figure 6.4 Financial markets are a form of self-organizing system. The power grid has been primarily analyzed and designed as a deterministic system. One of the smart grid components, namely DR, is using communication to couple the self-organizing financial market with the power grid, one of the largest, most complex machines built by man. The result will be a highly complex system.

systems: the market and consumers. It is becoming a bidirectional power market mediator. Thus, complexity theory becomes an enticing area in which to look for theories related to smart grid behavior.

Kolmogorov complexity is a quantification of information based upon the size of the smallest program capable of reproducing the information and is discussed in more detail in Section 11.2.1. The MDL principle is a formalization of Occam's razor, which is that the most likely hypothesis that explains a set of data is the one that leads to maximum compression of the information. There are numerous other definitions of complexity, as illustrated in Figure 6.5, which attempts to classify the various forms of complexity.

The electric power grid currently uses concepts from information theory and network analysis. It would be beyond the scope of this chapter to list every possible application; however, a subset of selected applications is shown in Table 6.2. Application of classical information theory effectively increases communication network capacity. AMI may further benefit from decreased traffic using compressive sensing. Compression techniques can be applied to synchrophasors to reduce load. FDIR benefits from network analysis; electrical faults require switches to reconfigure in such a manner that the impact to customers is minimized. State estimation will benefit from less data required to infer the state of the power grid. Information theoretic techniques will lead to improved power demand prediction and better response. DG will benefit from better stability control derived from network analysis. Stability in general will improve from application of information theory and network analysis. The concept of network coding, which combines source coding with network analysis, may someday further reduce traffic load for AMI, as well as other applications within the smart grid. Spectral graph theory applies toward improving both the power grid topology and communications network topology. Entropy and quantum information theory apply to cybersecurity within the smart grid (for example, quantum key distribution). Also, entropy and prediction will apply toward optimizing the use of energy storage within the network. This is because energy storage is used to help reduce peak demand for power; smoothing power demand is equivalent to reducing its

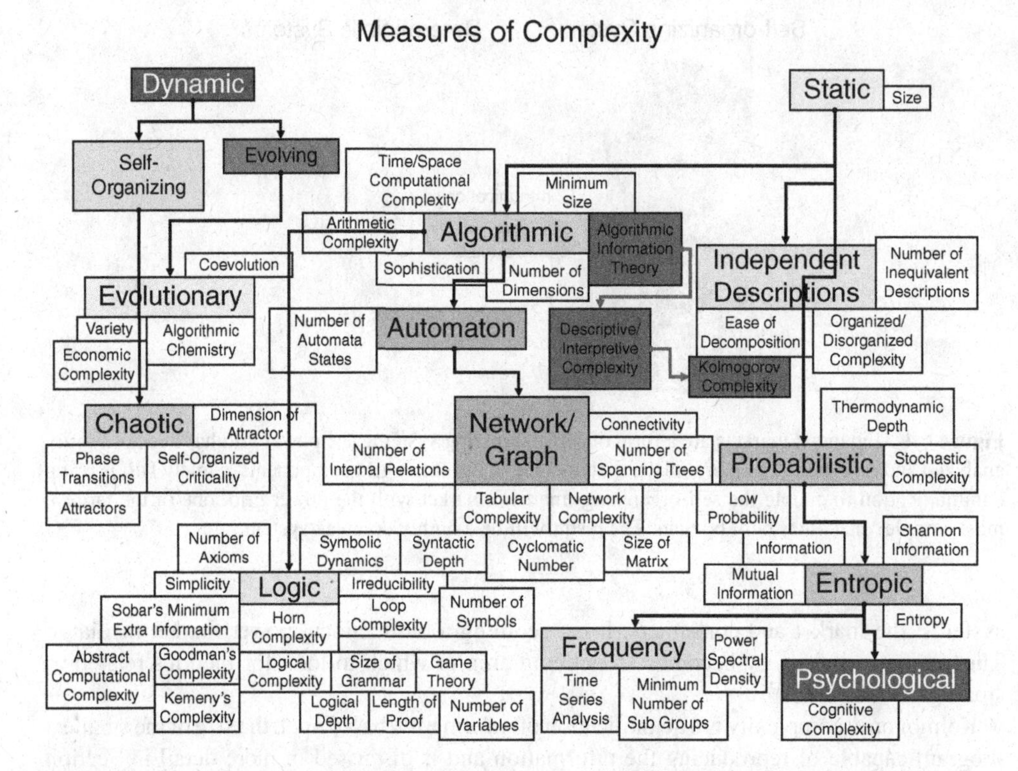

Figure 6.5 There are a myriad definitions of complexity. This illustration attempts to classify a few definitions of complexity that stem from dynamic and static techniques. Dynamic techniques involve measuring how a system transitions or evolves, whereas static techniques attempt to estimate complexity from a static snapshot of the system.

entropy. Finally, one of the most challenging advances to consider is a power grid that reaches down to the nanoscale – one in which every joule of energy is efficiently harvested, down to the molecular level. This will require new forms of communications capable of operating at the molecular level (Bush, 2011a). More details of this aspect will be explained later in Section 15.3.

6.4.2 Network Coding

Network coding is one example of information theory that could be applied for source coding in power grid communication. The concept is quite simple to illustrate. As shown in Figure 6.6, nodes S1 and S2 along the top of the figure are able to fully utilize the network in order for their information to reach the destination nodes R1 and R2 at the bottom of the figure. In particular, notice that the middle links, labeled $a_i \oplus b_i$ (which represents a_i exclusive-or b_i), carry useful information, that it otherwise would not, for the destination nodes. Both destination nodes receive all information from both transmitter nodes. Note that without network coding the central links would only be able to carry a packet from S1 or from S2, but not both simultaneously as they do in this example.

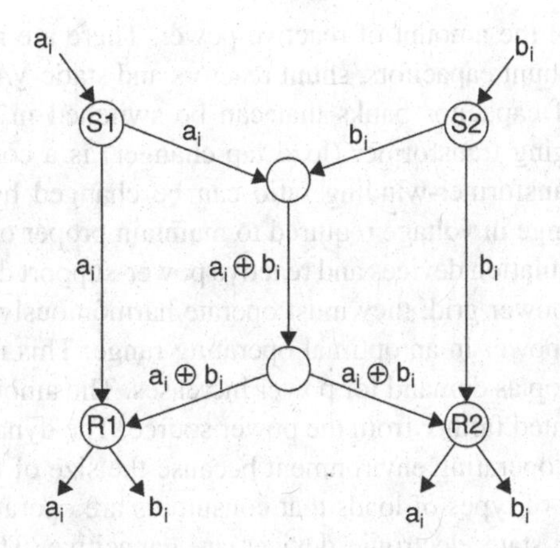

Figure 6.6 Network coding allows multiple communication channels to simultaneously carry useful information to multiple receiving nodes, allowing fuller utilization of the network. Source: Katti *et al.*, 2008. Reproduced with permission of IEEE.

This concept can be generalized in random networks by allowing nodes to broadcast random linear combinations of the messages that they receive, similar to the way the example above enabled the central links carry a linear combination of messages. The coefficients used in the random network code come from a Galois field; if the field is large enough, receiving nodes will have a high probability of receiving enough linear independent combinations to reconstruct the original message. The problem is that if the receiver does not obtain the requisite number of linearly independent combination of messages, the receiver will not be able to decode any useful information. A typical solution to insure this does not happen is for nodes to continue to transmit additional random linear combinations of messages to overcome anticipated error in the channel. Of course, this also adds additional overhead. However, the hope is that the efficiency gain of more fully utilizing the network will outweigh the additional messages such that there would still be an overall increase in network throughput.

While space limitation does not allow for explanation of all smart grid applications listed in Table 6.2, let us consider the IVVC application in more detail (Borozan *et al.*, 2001). This application serves as an example that provides insight into how communication can apply in a specific power system scenario. As the term implies, IVVC is the joint control of both voltage and VAr, where VAr denotes volt-ampere reactive, also known as reactive power. Reactive power exists when a power system contains a reactive component, such as an inductance or capacitance, that causes current to lag or lead voltage. When considered as a phasor, power can be represented in the complex plane with both real and imaginary components (reactive power is the imaginary component of complex power). In physical terms, reactive power is power that flows or pulses back and forth among reactive components within the power grid. While reactive power does not result in energy that can do useful work for the consumer, it is required in order to support the flow of real power. In fact, reactive power is strongly correlated with voltage level throughout the power grid. Reactive power flow and voltage level must be carefully controlled to allow a power system to operate within acceptable limits. There are devices within the power grid to regulate voltage and there are separate devices to control the amount of reactive power. There are many forms of reactive power control, including shunt capacitors, shunt reactors and static VAr compensators. Shunt capacitors, in the form of capacitor banks that can be switched in and out of the circuit, are common. A tap-changing transformer (load tap changer) is a common form of voltage regulator in which the transformer-winding ratio can be changed by a simple mechanical switch. The result is a change in voltage required to maintain proper operating voltage. Thus, there are many voltage-regulation devices and reactive power-support devices that are spatially separated throughout the power grid; they must operate harmoniously together in order keep both voltage and reactive power in an optimal operating range. This is a complex task given that voltage will tend to drop as demand for power increases. The amount of voltage drop will also increase for loads located farther from the power source. The dynamics of reactive power are also complex in a real operating environment because the size of the reactive component depends upon the mixture of types of loads that consumers are operating. Motors are highly inductive, and many solid-state electronic devices are capacitive. The result is a complex, dynamically changing voltage and reactive power profile that requires constant monitoring and control of the jointly interacting voltage and reactive power compensation devices. An additional consideration is that reducing voltage (within safe limits) reduces the amount of power that devices consume. Conservation voltage reduction is a technique in which voltage is purposely reduced for the power consumed by loads; however, this brings the voltage level closer to its acceptable lower limit, reducing the safety margin for handling fluctuations in voltage. Similarly, reducing reactive power – when possible and within safe limits – also frees resources for real power, which is the product for which utilities are actually paid. The important point from a communication standpoint is that voltage and VAr compensation devices must be located at specific points within the power grid that are spatially separate. In addition, monitoring and control information must be continuously and reliably exchanged among the devices in order to keep the power grid operating safely; that is, avoiding a voltage collapse or causing an electrical fault in extreme cases.

Because there are a variety of distinct metrics that IVVC can optimize, IVVC can be viewed as a multiobjective optimization problem. For example, IVVC can attempt to minimize power loss through the transmission and distribution system, maximize the power factor, which is the ratio of real to apparent power (the magnitude of the complex power mentioned earlier), or maintain the voltage profile, which is the voltage level as a function of time and distance from the power source. When applied in the transmission system, stability is another objective function to be maximized, where stability is the difference in voltage angle across a transmission line that, if too large, could cause generators to lose synchronization. As is typical in multiobjective optimization, there is no single point at which all of these objectives are simultaneously maximized. Instead, there is a Pareto-optimal front, which is a set of solutions such that it is impossible to improve one objective without degrading the others. Thus, there is a significant amount of computation required to perform IVVC, including solving the optimal power flow equation, which is highly-dependent upon the network structure of the power grid. IVVC requires the communication of large amounts of up-to-date state information as well as fast, reliable communication for control. Synchrophasors are becoming widely deployed throughout the power grid in order to supply the required information. Keep in mind that this is only one relatively simple example of a power system application that requires communication networking. Hopefully, this one simple example provides the reader without a power system background some insight into the potential interaction among power system, communication channel capacity, information theory, and network science. In this example, a poor communication channel could either: (1) slow the rate at which IVVC operation receives data or issues commands, thus reducing its reaction time and causing incorrect operation, or (2) drop packets, causing an incorrect state to be inferred and leading to incorrect operation. The ability to perform the multiobjective optimization efficiently over a large network structure requires advances in network science. Also, the ability during planning to optimize placement of sensors and actuators throughout the power network would benefit from network science.

The remainder of this chapter focuses upon the question of whether or not there is a deeper relationship among power systems, information theory, and network science that, when viewed at the proper depth and perspective, makes power grid applications, such as IVVC, a relatively simple problem rather than a complicated set of heuristics and estimates, or whether the so-called smart grid will simply carry forward old thinking utilizing new technology. The next section introduces information theory and network analysis in more detail. Specifically, there is a brief introduction to the fundamentals of information theory followed by RMT. RMT leads naturally to analysis of network topologies (that is, networks described by a random matrix). This is followed by a discussion of compressive sensing, a relatively recent fad in traditional information theory. The topics of self-healing and M2M communication, both integral to the implementation of the smart grid, are then discussed from an information theoretic viewpoint. This is done within the context of active networks, a highly flexible form of communication networking.

6.4.3 *Information Theory and Network Science*

Information theory and network science are discussed within the context of how the technologies have evolved, how we expect them to evolve, and barriers to their envisioned usefulness within the smart grid. Information theory is a broad subject encompassing classical information quantification based upon entropy developed by Claude Shannon (Shannon, 2001), as well as an alternative quantification in algorithmic information theory developed by Ray Solomonoff and Andrey Kolmogorov (Kolmogorov, 1965). Information theory is related to graph theory and network topology via such topics as the maximum capacity of a graph and graph entropy (Shannon, 1956); a more recent reference is Solé *et al.* (2004). A more general study of graph and network structure is found in spectral graph theory, which examines the characteristic polynomial, eigenvalues, and eigenvectors of matrices associated with a graph, such as its adjacency matrix or Laplacian matrix. Graph spectra and RMT have a long history, going back to at least the 1950s, and have often been applied toward understanding the topology of the electric power grid. An alternate term for the Laplacian matrix is the Kirchhoff matrix, clearly indicating that matrix and graph spectra approaches have a very long history of being used to study networks of current flow. Application of the graph Laplacian to nanoscale current flow can be found in Bush and Li (2006) and Bush and Kulkarni (2007), which is analogous to power flow analysis on a nanometer scale using carbon nanotube power lines.

The electric power grid is a complex system and can be examined through the field of complexity theory. Complexity theory can be traced back to at least 1948 (Weaver and Wirth,

2004), in which complexity is considered the degree of difficulty in predicting the properties of a system. Algorithmic information theory was developed independently by Andrey Kolmogorov in 1965 and Gregory Chaitin circa 1966. One of the important ideas from algorithmic information theory is that the compression rate of information can provide useful information about the nature of the source that generated the information. This is seen in the minimum description length principle and Kolmogorov complexity (Barron *et al.*, 1998). This can be used for classification of normal and abnormal operating states with application to cybersecurity and fault detection (Bush, 2002) within the smart grid.

RMT (Anderson *et al.*, 2011) has been applied and developed in the fields of physics, finance, economics, and wireless communication. The goal of RMT is to understand the properties of matrices whose elements come from specified probability distributions. If the matrices represent graphs, such as adjacency matrices, then RMT provides insight to the corresponding network structure. This approach is particularly useful for understanding how perturbation and failure in a network impact the global system (for example, electric power grids). RMT can also help provide the theoretical foundation for how the electric power grid can become self-healing and resilient to failure, whether accidental or malicious, as well as operate with more stability given variable renewable DG sources (Marvel and Agvaanluvsan, 2010) – it is well known that the resistance between any two points in an arbitrarily connected network of impedances can be computed using the eigenvalues of the Kirchhoff matrix (Bush, 2010b; Wu, 2004).

Consider the adjacency matrix A with elements $A_{ii} = 0$ and $A_{ij} = n$, if vertices i and j are connected by n edges. In graph theory, $n = 1$ always, but in pseudo-graphs representing realistic grids n may take any positive integer value; that is, the edges will be weighted. The eigenvalues of the A matrix provide information about the network's topology: zero eigenvalues indicate the presence of star clusters, with many peripheral vertices connected to a single, central node, while a high density of eigenvalues around a certain value indicates the presence of disconnected node pairs. But this matrix is still deterministic. We need to apply these techniques to the electric grid, which is becoming inherently nondeterministic, considering renewable energy sources (for example, wind and solar) and other DG sources that can be added and removed in a stochastic manner.

Information theory has its foundation based upon a definition of information entropy. If there are n possible outcomes, then entropy can be thought of as the expected value of the surprise (also known by the term suprisal) in obtaining the outcome. The base of the log in the following definition of entropy is typically 2:

$$H(X) = -\sum_{1}^{n} p(x_i) \log_b[p(x_i)]. \tag{6.30}$$

6.4.4 *Network Science and Routing*

Graph spectral techniques have been applied to a wide range of routing-related problems in communications – from understanding the structure of the Internet, to improving routing for wireless sensor networks (Subedi and Trajković, 2010; Wijetunge *et al.*, 2011). Node-clustering information is encoded in the eigenvalues and eigenvectors of both network adjacency matrices and Laplacian matrices. For example, eigenvectors have been used to identify clusters of connected autonomous system nodes (that is, nodes with connected Internet Protocol routing prefixes under a common network operator or provider). With regard to routing in sensor networks, which will be highly apropos within the smart grid, it will be important to have efficient routing techniques as the number of sensor nodes increases. The traditional approach of forwarding routing table information adds significant overhead, particularly to a large-scale deployment of sensors. Researchers are discovering the use of RMT and spectral graph theory for network routing (Wijetunge *et al.*, 2011).

The graph Laplacian is the difference between the degree matrix, which is a diagonal matrix indicating the number of connections to each node, and the adjacency matrix of the graph. The graph Laplacian is a fundamental attribute of a graph, and can be derived in many different, but equivalent, ways.

It has been found that the global behavior of the interconnection, or coupling, of interacting components such as oscillators is sensitive to the network structure that they form. A favorite toy example of researchers is the firefly or cricket demonstration, in which, given the proper density and interconnection of devices that are randomly set, they begin to sense each other's output and converge toward synchronized blinking or chirping. Ultimately, the entire system begins to chirp or blink in unison. This concept plays a fundamental role in the network structure of the power grid. FDIR is dependent upon the power grid's topology within the distribution system. Relays must coordinate with one another to "self-heal" a fault in the power grid (Bush *et al.*, 2011a,b).

Synchronization and stability of the power grid can be viewed at the generation- and the transmission-level by considering perturbations in power flow through the transmission network as a function of the power grid network topology. Consider one brief intuitive example. First, the swing equation describes the power angle and acceleration of the generator rotor as a function of the mechanical force applied and the load experienced by the generator; this characterization of the generator represents a node in a graph. The transmission lines can be modeled ideally by impedance and represent a graph edge. The generator interconnections with each other and with loads are described by a graph Laplacian matrix. Graph spectral techniques are applied to make general assumptions about whether disturbances in power flow will be amplified or dampened simply by the topology of the power network. The result can be reduced to examining the eigenvalues of the power grid network's graph Laplacian matrix. This can be taken further to examine the impact of adding explicit communication mechanisms versus the direct electromechanical generator-to-generator coupling (Li and Han, 2011), thus combining network science, information theory, and networked control.

6.4.5 Compressive Sensing

Compressive sensing (Candès and Wakin, 2008) was originally developed for imaging to reduce the exposure of patients to radiation during computer-aided tomography scans, and has since been applied in different fields. It can be applied whenever we have a signal that is somehow "redundant." In short, we can reduce the number of samples used to represent the signal without losing information, or at most losing only a "small" amount of information. In the case of the electric power grid, there are different cases in which compressive sensing can be usefully employed. Significant cases include measurements taken from a phasor measurement unit (PMU) in close electrical proximity and meter readers (Li *et al.*, 2010).

The terms signal and vector are synonymous in the following, and discrete time signals are assumed. A signal of length N is k-sparse if it is a linear combination of only k basis vectors with $k \ll N$. Recall that basis vectors form a set of linearly independent vectors that, through linear operations, are capable of representing any vector in the given space. In that case, typically, transform coding techniques are used to project the signal onto an N-dimensional basis (possibly different form the k-dimensional vectors mentioned above) and then the most significant components of the projection are selected. The cardinality of the most significant components is usually much smaller than N. By transmitting them, bandwidth is saved while allowing an accurate reconstruction of the signal.

The rationale behind compressive sensing is to attempt to know in advance which samples of a signal one can avoid sampling because they will have little influence on the reconstruction of the signal. It was motivated by the desire to reduce radiation exposure in computer-aided tomography scans. We avoid the sequence of full sampling, applying a transform, transmitting, and then applying an inverse transform. Instead, we sample and transmit only a sparse vector and reconstruct using our knowledge of sparseness to help in the reconstruction of the signal, as explained next.

Let A be a measurement matrix that defines the values we actually sample (this could even be a random matrix and achieve good performance), then $y = Ax$ defines the samples y. To reconstruct the signal, we look for the sparsest vector ($\min \|z\|_0$), where $\|z\|_p = (\sum_{j=1}^{N} |x_j|^p)^{1/p}$ is the p-norm (or size) of the vector that reconstructs our received samples. The optimization problem to be solved is thus: $\min \|z\|_0$ subject to $Az = y$. Since this is difficult to solve, the following is used in practice: $\min \|z\|_1$ subject to $Az = y$.

An increase in the application of machine learning and artificial intelligence is a likely outcome enabled by the smart grid. There are many applications utilizing machine learning and artificial intelligence being explored for the power grid (Pipattanasomporn *et al.*, 2009; Saleem *et al.*, 2010), and these are discussed in more detail in Chapter 11. Information theory and network analysis play a role in machine learning and artificial intelligence; however, this chapter remains focused upon the role of communications. Mobile agent technology and dynamically reprogrammable devices within the power grid will require the communication network to transport executable code. There is much to learn from a form of communication that facilitates a highly dynamic, reprogrammable communication network known as active networks, described next.

We expect that information theory and complexity theory will evolve to provide a better understanding of the power grid as a complex system. In addition, information theory will evolve to enable an understanding of active networking and of power networks that are extremely flexible and reconfigurable. Information theoretic topics such as compressive sensing and network coding may find application with the power grid for AMI and state estimation. More importantly, information theory coupled with physics may serve to unify power systems management and communication (that is, to create new properties and metrics that integrate, or blend, power systems and communication in a fundamental way).

Interesting theoretical work on the relationship between thermodynamics and information, as well as energy and information, has yielded results as described throughout this chapter. Information theoretic measures of power analysis have already been considered in the context of information entropy on chip-level power consumption (Marculescu *et al.*, 1996). The vision posited in this chapter is the extension of such analysis toward the integration of information theory and network analysis with the power grid. While the prime mover for electric power is often thermodynamically driven, electric power itself follows Maxwell's equations. It is obvious to us, in hindsight after Maxwell, that an electromagnetic wave used to form a communication channel can be directly analyzed from an information theoretic standpoint (Gruber and Marengo, 2008; Loyka, 2005). This analysis may also be used to combine power systems and communication fundamentally (that is, at the level of charges and fields). This will enable us to understand (1) precisely where to place communication, since the entropy of all power systems components would be known at a low level and with high resolution (today, communication is implemented in an ad hoc manner in the power grid); (2) the theoretical communication limits of waveguides (for example, power line carrier), whereas today they are only estimated; and (3) the precise minimum "bits per kVA" to transfer electric power. Because of (1) and (2), this can be known precisely. These concepts are discussed in more detail in Bush (2013b).

As anyone who has worked in the power industry knows, one must consider competing objectives when optimizing power systems (for example, protection versus availability). Knowing the aforementioned theoretical limits will aid in this optimization process. In addition, combining the above theoretical limits with network analysis will evolve to play a significant role in routing within sensor networks. It may serve as a theory to unify the fields of power flow and communication routing at a more fundamental level.

Information theory and network analysis already have a long history in improving communication; source-coding-, channel-coding- and networking-routing-related advances naturally continue given the existing momentum. The use of these theories to study complex systems also has a long history and will continue.

The barriers to the use of information theory and network analysis in the smart grid stem, counter-intuitively, from the momentum mentioned in the previous paragraph. Researchers in academia tend to be less innovative and slower in addressing new problems than in industry, perhaps for cultural reasons found in academic environments, such as peer pressure or fear of being isolated from publication venues. Whereas the bulk of academic papers tend to be endless reformulations of the same concept, real innovation will be required to make the fundamental advances needed for the vision presented in this chapter. It will remain easier for the community to continue publications on safe topics that are comfortable than to explore new areas. For example, the application of information theory to active networks requires thinking "outside the box" created by Shannon, beyond which only a few brave researchers dare to explore. The theoretical unification of communications and power systems engineering will require thinking about information and energy in new ways, such as those approached by researchers like Landauer (Landauer, 1961). At the same time, researchers need to respect and listen carefully to power system experts in the field. These people know the grid intimately; it would be very unwise to ignore their input and concerns. Research and its benefits must be communicated clearly, concisely, and in the language of the power systems engineer, not hidden in academic jargon.

The primary barrier to the use of information-theoretic techniques and network analysis for use in the smart grid is the efficiency of implementation, specifically distributed and decentralized implementation. Matrix and eigensystem analysis of complex networks have been around for a long time. However, they require having a complete matrix upon which to operate. In real applications of the theory, the information for the complete matrix will likely be unavailable, or it will be too inefficient or costly to implement a centralized solution with the omniscience required to construct the matrix.

For example, communications techniques such as network coding are not widely used because, while they are theoretically interesting to academics when used in ideal conditions, they are difficult to configure and coordinate in a realistic environment. Similarly, network topological techniques that rely upon computing the eigenvalues of large graph matrices cannot be done timely or efficiently in a centralized manner. The smart grid must be a fast, decentralized, adaptive system. Until efficient decentralized approaches are developed that show clear benefit over existing methods of operation, new techniques cannot, and should not, be adopted. Thus, it will be extremely important for academia not only to have strong ties with industry, but also to understand and follow the guidance of industry in order to contribute more than meaningless paper citation chains.

In a more futuristic vision, the trend toward flexibility and the evolution toward smaller scale generation can be illustrated working together via advances in both nanoscale power generation and wireless power transmission. Advances in wireless power transmission will enable it to become more accepted and pervasive. The power grid becomes more efficient and standardized in its ability to accept and aggregate large numbers of very small scale (nanogrid) sources of power. This encourages consumers to become more active in power production, and the aggregated power from a region is transmitted using wireless power transfer. Finally, individual consumer devices receive wireless power transmission and operate continuously without the need for batteries or power cords (for example, wireless recharging of electric vehicles). The world ultimately becomes tetherless from a power perspective.

Clearly, this vision of power systems requires advances on many technological fronts and moves well beyond the conventional view of the smart grid. However, this vision is a natural extension of the smart grid if one extrapolates the technological trend for microgrids to become smaller and more ubiquitous, and for the grid to become more integrated with communications and its environment. In this case, wireless communication is extrapolated to the transmission of power. In the long-term information theory and network science related to wireless communications and ad hoc networking will be paramount.

6.5 Communication Architecture

In this section, a communication architecture for the power grid is discussed. A communication network architecture can be as loose as simply specifying the framework for a network's physical components and their functional organization and configuration and perhaps including its operational principles and procedures. An architecture may also be as specific as covering data formats used during its operation. There are a wide variety of communication technologies and applications within the power grid; thus, it has been extremely challenging to define a detailed architecture. There are a myriad of detailed communication protocol standards, as will be seen in Chapter 10. This section is focused upon higher level, generalized frameworks, or modes of thinking about the relationship between classes of communication technologies and the power grid.

There are two general categories of networks: (1) active and programmable networks and (2) passive networks, whose differences are illustrated in Figure 6.7. Active and programmable networks incorporate networking technology that is highly dynamic and programmable, typically allowing the network to be programmed while it is in operation. In active networks,

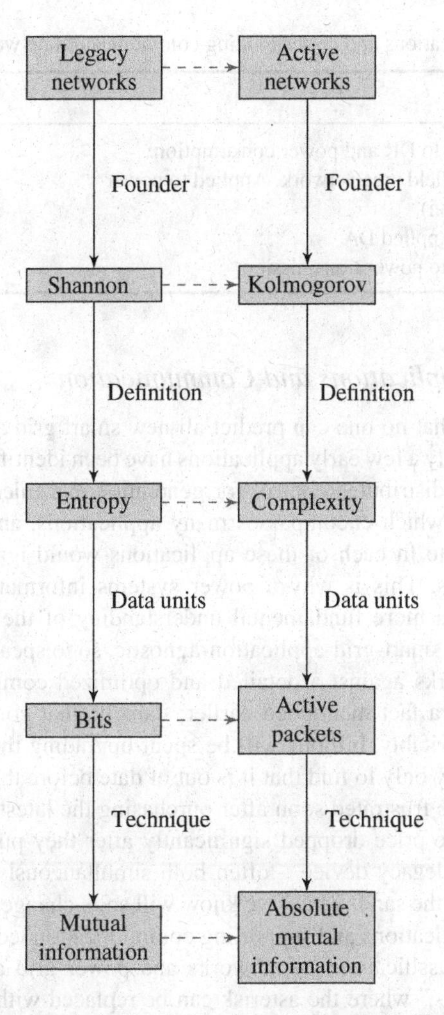

Figure 6.7 Passive, legacy communication networks are compared with active networks. Active networks intimately couple computation and communication, which accounts for many of the differences.

packets contain executable code that changes the operation of the network in a controlled manner for the packet's application as the packet flows through the network. Each packet, in essence, carries its own protocol, enabling an extremely dynamic and flexible communication network. In programmable networks, which also includes software-defined networks (SDN), communication switch interfaces are accessible in a standard, programmatic manner for developers. The active and programmable approaches differ greatly from passive networks, in which a single standard is implemented to which everyone must adhere. The Internet Protocol is a prime example of a passive network. As Figure 6.7 illustrates, active and programmable networks require a more advanced theory of operation and have proven to be more challenging and complex to understand theoretically. SDNs are somewhat of a compromise between passive and active networks.

Table 6.3 Smart grid applications and corresponding communication network types.

Application	Communication type
Home area network. Applied to DR and power consumption.	HAN
Neighborhood area network/field area network. Applied to power generation and consumption).	NAN/FAN
Metropolitan area network. Applied DA.	MAN
Wide area network. Applied to power transmission.	WAN

6.5.1 Smart Grid Applications and Communication

It should be emphasized that no one can predict all new smart grid applications that will be developed in the future; only a few early applications have been identified so far. These include applications such as DR, distributed control for generation and microgrids, synchrophasor-related applications, DA, which encompasses many applications, and so on. Attempting to optimize communication to fit each of these applications would ignore many other, as yet undiscovered, applications. This is why a power systems information theory is required; it would ideally provide a more fundamental understanding of the communication–power relationship that would be smart-grid-application-agnostic, so to speak.

Another factor that works against a detailed and optimized communication architecture for today's power grid is a fact mentioned earlier; namely, that communication evolves at a relatively fast pace. Inevitably, billions will be spent upgrading the power grid to today's communication technology only to find that it is out of date before the upgrade is completed. I often see friends who are frustrated soon after purchasing the latest communication gadget only to find that either the price dropped significantly after they purchased it or it quickly became an old-fashioned legacy device – often both simultaneously. With these caveats in mind, let us draw a line in the sand, which we know will soon change, and delve into some of the current smart grid applications and supporting communication technologies.

One of the simplest classifications of networks and power grid applications utilizes the notion of "*-area networks," where the asterisk can be replaced with any area specification. Common examples include HAN, neighborhood-area network (NAN), FAN, MAN, and WAN, along with corresponding smart grid applications, as shown in Table 6.3. One problem with this approach is the "*-area network" classifications are not rigorously defined.

Instead of the above "*-area networks," consider network classification from the perspective of the type of architecture and functionality. The Internet succeeded in part because applications could interface simply with the network. Internet applications have no knowledge of the state of the network, they simply have the ability to send packets over their respective interfaces. The network succeeds as a dumb pipe and applications blindly trust that the network will handle packet transport; applications accept whatever service the network happens to provide. Both network control and packet flow have been tightly coupled in an integrated system. Researchers have long thought about how to create a better interface between applications and networks such that applications can have knowledge of the network along with better control of network transport. SDNs (Skowyra *et al.*, 2013; Vaughan-Nichols, 2011; Yap *et al.*, 2010; Yeganeh, 2013) separate control of the network from the data path taken by packets through the network. Application program interfaces (APIs) allow control of the network to be constructed programmatically. It becomes much easier to build new network services, functions, and features when control is separate from data flow and exposed via an open API. The dumb pipes of the Internet can be seen as the classical power grid, in which both data and power flow blindly, without knowledge of their environment and without the ability for application-specific optimization. Just as the smart grid is providing new dimensions and control for power, an SDN provides application-specific optimization and control of the communication network. A step beyond the SDN involves placing control, which has been separated from the data flow as previously mentioned, into each packet, resulting in an active packet. Let us consider the smart grid analogy with advanced networking in more detail using active networks.

6.5.2 *Active Network*

One of the defining properties of the smart grid is its ability to exhibit self-healing. That is, the grid must be able to dynamically reconfigure in order to optimize its performance, particularly under adverse conditions. Since the smart grid is a collection of intelligent electronic devices, the communication for this reconfiguration takes place using M2M communications.

The electric power distribution system is a particularly interesting portion of the power grid because it resides at the front line of many different advances. Power is more "actively" managed through the distribution system; more information (and control) is required for each unit of energy. Smart meters and DR mechanisms must tie into the distribution system, DG feeds into the distribution system, and new protection mechanisms are being developed for the distribution system. All of these developments are taking place simultaneously and are known generally as DA. Another term that has been applied is an active distribution network, as shown in Figure 6.8, and its operation is known as active network management (Figueredo *et al.*, 2009; McDonald, 2008; Nguyen *et al.*, 2008, 2009; Samuelsson *et al.*, 2010). This is because the distribution system will no longer be a relatively passive electrical system that customers tap into for power. Instead, this portion of the power grid is becoming highly active in terms of supporting generation, controlling power flow, monitoring usage, and becoming more efficient at detecting, isolating, and restoring faults (that is, self-healing).

Efficient communication will be required to support the entire power grid; however, the active distribution network provides a challenging test case for communication. Experience can be drawn from another form of active network, namely active communication networks. Just like active distribution networks, active communication networking takes the view that the conduits of computer communication should no longer be passive, but highly flexible and self-healing. Active packets allow communication packets to carry not only passive data, but also executable code that can be deposited within the network to control and modify its behavior. It is also a form of M2M communication in which packets flow through the network and are automatically deposited into devices within the network; namely, execution environments that reside on intermediate network nodes within the network. Just as reactance may change to control power flow in an active distribution network, active network code may change the source or channel coding on a particular link to match changing conditions, as one example. Active networking is a highly flexible and self-healing communication networking paradigm that pushes information theory to its theoretical limits and provides lessons for the smart grid. Communication networks that are capable of transporting their own code and modifying their

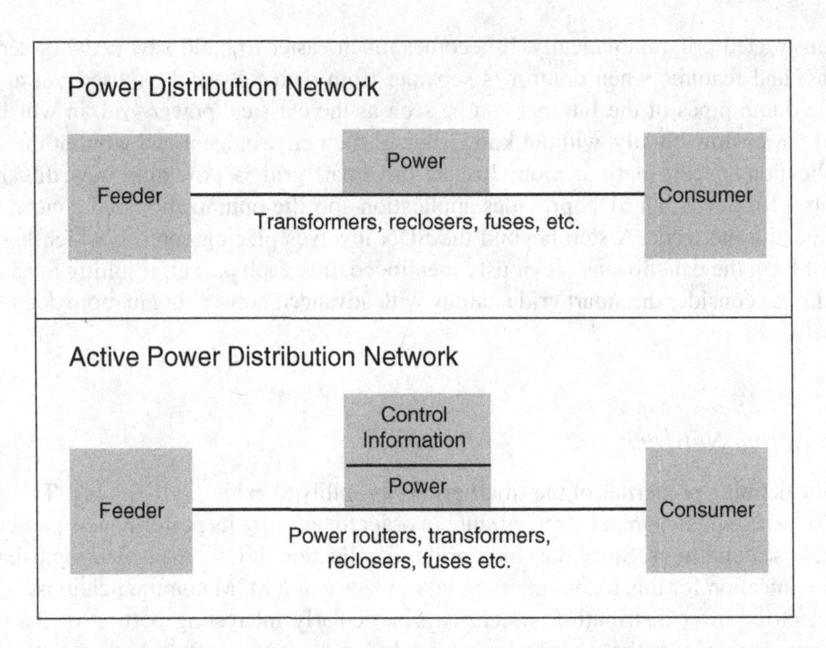

Figure 6.8 An active electric power distribution architecture is illustrated in a simplified manner as an evolution from the traditional, relatively static, distribution system. Just as reactance may change to control power flow in an active distribution network, active communication network code may change the source or channel coding algorithm on-the-fly in a particular communication link to match changing conditions, as shown in Figure 6.9. Source: Bush, 2013a. Reproduced with permission of IEEE.

own operation with their transported code have been considered in the past, as far back as the 1980s (Wall, 1982; Zander and Forchheimer, 1980, 1983). The notion of active networking as a serious communication paradigm did not occur until the mid to late 1990s (Bush and Kulkarni, 2007).

Figure 6.9c shows the active network architectural model as a natural evolution of circuit-switched (Figure 6.9a) and packet-switched (Figure 6.9b) architectures. The circuit-switched network (Figure 6.9a) pre-establishes connections. Thus, while data may be packetized, it plays a passive role, having little interaction with the network. In the packet-switched architecture (Figure 6.9b), packets play a slightly more active role, in that a header is required that interacts with the network in order to properly route the packet to its destination. In an active network (Figure 6.9c), the packet becomes more active and a distinction can be made between the packet header, code, and data. The packets, or PDUs, may carry code as well as data. The code within the packet can change the operation of a channel.

Active networking requires analyses that go beyond the limits of classical Shannon information theory. In active networking, a distinction is made in channel input between static or "passive" data X and "active" executable code X', as shown in Figure 6.10. Executable code is included within each packet; a packet carries its own transport protocol. An active network can send information as a combination of passive and active information, represented by (X, X'). The active portion of the information can be dynamically inserted into the communication channel, changing its operation to become a function of $X', f(X')$ (Bush *et al.*, 2011a; Bush, 2011a).

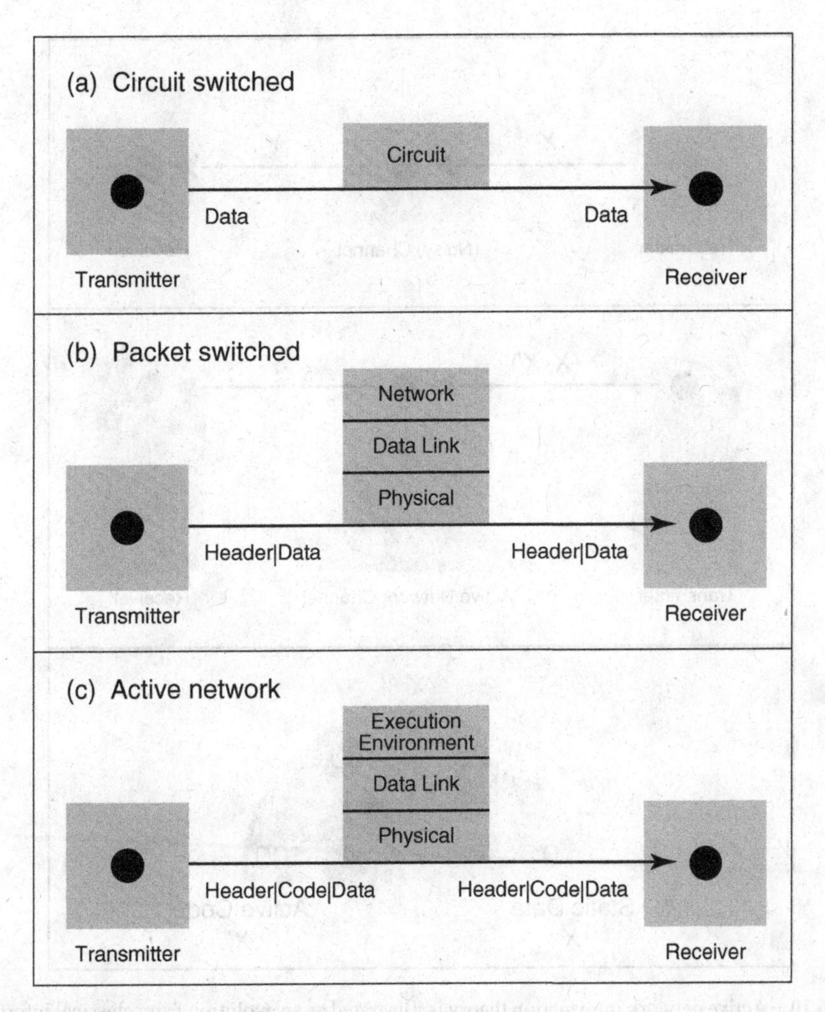

Figure 6.9 The active communication network architecture is shown as an evolution from static circuit- and packet-switching. The active network execution environment allows code within active packets to dynamically modify the operation of the channel. This allows the communication network to be extremely flexible and change its manner of operation on-the-fly. For example, active packets may carry their own protocol. Source: Bush, 2013a. Reproduced with permission of IEEE.

How much computation does a communication network require and how flexible should it be? What is the optimal trade-off between computation and communication? While much is being learned about active networks, these are still open questions that require more research (Bush, 2005). Partial solutions to these questions can be found in research that has roots in Kolmogorov complexity (Bush, 2002); namely, the bounds on the smallest program that computes a given outcome and communication complexity (Yao, 1979), which is the minimum information exchange required to solve a function in a distributed manner. One of the challenges in active networking is that communication and computation are bound together in a manner that is not always clearly separable; practitioners have instead developed rules-of-thumb for

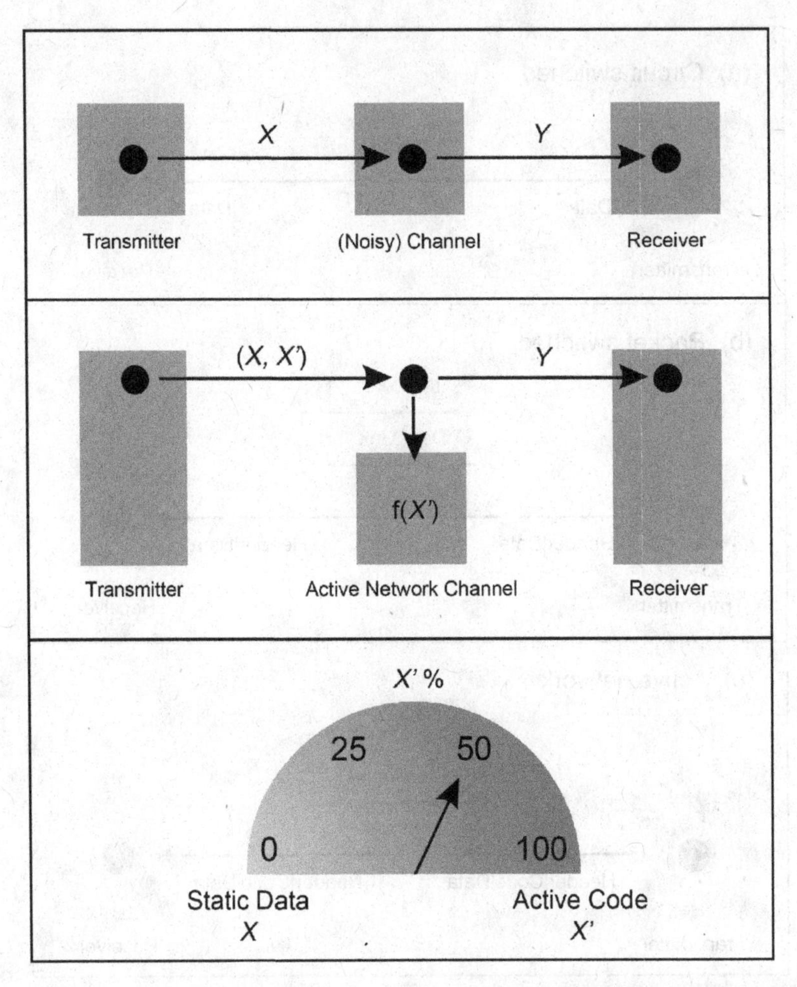

Figure 6.10 Active network information theory is illustrated as an evolution from classical information theory. A classical channel model is illustrated along the top of the figure. The middle of the figure illustrates an active channel. An active packet is comprised of (X, X'), where X' indicates executable code within the packet that may be deposited on a device in order to change the operation of the device's channel interface. The bottom of the figure illustrates the question regarding the proportion of an active packet payload that should be code versus static data. Source: Bush, 2013a. Reproduced with permission of IEEE.

determining when and where computation should be placed within a network, both for active and legacy networks (Bhattacharjee *et al.*, 1997).

A schematic representation for an active network channel is shown in Figure 6.11. The designation X represents information transmitted within packets and Y represents information received. The dashed line indicates the executable code portion of the packet that is dynamically installed and impacts the operation of the channel, which will be explained in more detail. Noise in the channel can be modified by an active packet's code while the packet is "in flight," resulting in a highly dynamic and self-healing system.

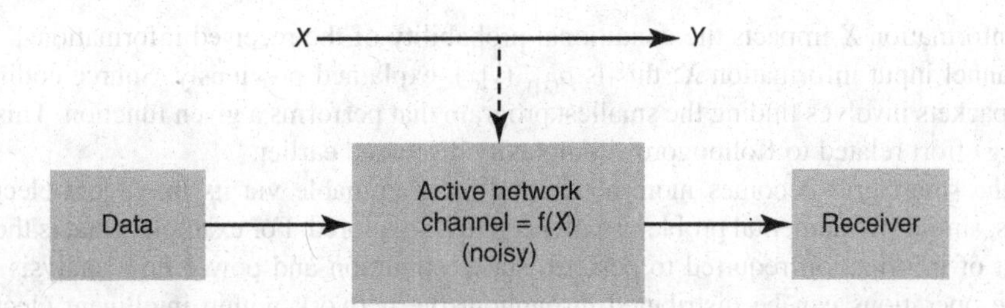

Figure 6.11 An active network channel can utilize the executable code in the input packet X to create or modify the operation of the communication channel. Instead of the classical case of transmitting X and receiving Y with some probability, active networks include a function $f(X)$ that impacts the probability of Y being received. Source: Bush, 2013a. Reproduced with permission of IEEE.

First, consider the input space X transmitted across the potentially noisy channel to the receiver, with received space Y. If X is a random variable representing the set of transmitted signals and Y is the set of received signals, then the conditional probability that Y is received given that X was sent is $p_{(y|x)}(y|x)$. Clearly, in order for communication to work, the information received should be equivalent to the information that was sent. From a theoretical standpoint, the signals to be transmitted X can be chosen, or engineered, to be the best possible set of signals for a particular channel. The joint distribution between the information sent and the information received is $p_{x,y}(x, y)$ and the probability of the choice of signal is

$$p_x(x) = \int_y p_{x,y}(x, y)\, dy. \tag{6.31}$$

The joint probability of the transmitted and received information is

$$p_{x,y}(x, y) = p_{y|x}(y|x)p_x(x). \tag{6.32}$$

The maximum mutual information, or the channel capacity, is (Shannon, 1948)

$$C = \sup_{p_x} I(X; Y), \tag{6.33}$$

assuming that the previous equations are satisfied for the particular channel under analysis.

So far, this is the well-known classical channel capacity, which applies to passive communication networks. However, in an active network, the input X may optionally encode information that modifies the channel itself, changing its operation. For example, parameter values may "tune" the operation of the channel, or executable code may be inserted into the operation of the channel. This may be intended to modify the robustness of error correction, fuse data while traveling within the network, or any number of other possible changes to the operation of the communication channel. The channel becomes a function $f(x)$ that operates upon the data flowing through it. The transmitted information X within an active network may be executable code that varies the trade-off in computational and communication resources in order to improve the communication system as a whole. Thus, the active portion of the input information X impacts the conditional probability of the received information Y given the channel input information X; this is $p_{(x|y)}(y|x)$, explained previously. Source coding for active packets involves finding the smallest program that performs a given function. This is an ongoing effort related to Kolmogorov complexity discussed earlier.

As the smart grid becomes more active and programmable via its intelligent electronic devices, similar fundamental problems will need to be explored. For example, what is the least amount of information required to perform state estimation and power flow analysis given that the operations can be distributed throughout the network within intelligent electronic devices? The same question can be asked for DR and all other smart-grid-related operations. Active network concepts could apply to the reconfiguration of the communication network throughout the smart grid – for example, when the channel changes dramatically, such as having to switch among many diverse paths in an emergency (for example, from wireless to power line carrier). The next section reviews theoretical background for wireless and wave guide communication, such as various forms of power line carrier. Keep in mind that the concepts of active networking can apply at the physical layer as well as the network layer.

6.6 Wireless Communication Introduction

This section dives into fundamental physical aspects of communication that apply to both wireless and wave-guided communication. These are a few key concepts useful for understanding specific communication technologies discussed later in this part of the book, namely electromagnetic radiation and the wave equation.

6.6.1 Electromagnetic Radiation

Electromagnetic radiation has an aspect that is only briefly mentioned here to foreshadow future applications; its quantum mechanical nature. The quantum mechanical nature of electromagnetic radiation has a role in both power systems and communication; however, these aspects are well beyond the current thinking for the smart grid and are discussed in detail in Chapter 15. Specifically, the quantum mechanical nature of electromagnetic radiation is the photon, which may be thought of as a wave packet. In PV DG, simply put, when a photon strikes an electron in a semiconductor, it provides the electron with enough energy to jump from its position within an atom of a semiconductor to form an electric current. From a communication perspective, the photon provides the quantum mechanical interaction necessary for entanglement used in quantum communication (Bush, 2010b). For the most part, we will remained focused primarily upon classical physics within the smart grid.

Let us begin with the difference between a reactive and radiative field. A reactive field stores energy for a circuit, while a radiative field is the energy lost by a circuit; radiative energy propagates into space. Consider a circuit comprised of an alternating power source and an inductor. The alternating power source supplies power to the inductor, which creates a field around the inductor. In an ideal inductor there is no power loss. Thus, the source supplies power to the inductor, which is stored in the field. The field then returns energy to the circuit. Owing to this cyclical behavior, the current and voltage will be 90° out of phase. This is known as reactive impedance, which will be discussed in more detail later. If another inductor is placed near the first inductor, it will absorb power from the electric field of the first inductor. The result is that power is lost from the first inductive field and supplied to the second inductor. We have created a transformer. The fact that one field reacts with another inductor suggests its name: "reactive" field. Using such a setup near a power line would be stealing power from the grid.

Now we can consider the impact upon communication. If the inductor is replaced with an antenna (for example, a loop or a dipole antenna), the field that is generated can propagate continuously into space. To the source of power within the circuit, this energy loss appears as resistance. A charge at rest or at a constant velocity will create only a reactive field. It takes acceleration of the charge to create a radiative field. An accelerating charge will create both a radiative and a reactive field. As mentioned previously, the wavelength λ is related to the frequency f by $\lambda = 1/f$. Now consider a conductor acting as an antenna. If the frequency is relatively slow, the wavelength will be relatively large; consider a wavelength that is equal to, or longer, than the antenna. In this case charge will flow back and forth across the antenna, creating electric field lines that are as long as the antenna. The power of the radiated field will be proportional to the current; that is, the amount of charge flowing, the length of the conductor acting as an antenna, since that holds the total charge, and the frequency, since the frequency controls the acceleration experienced by the charge: $P_{\mathrm{r}} \propto I/f$, where P_{r} is the radiated power, l is the length of the antenna, and f is the frequency. This is also equivalent to $P_r \propto I(l/\lambda)$.

As the frequency of oscillation increases, the charge flow through the antenna is no longer uniform in one direction. Instead, multiple waves exist along the length of the antenna. In other words, some portions of the antenna have current flowing in one direction, the positive part of a cycle, and some portions of the antenna have current flowing in the opposite direction, the negative part of a cycle. As the corresponding fields radiate, they either cancel or reinforce each other, just as multiple ripples colliding in a pond. A length of $\lambda/2$ is often chosen for the antenna because, given the combined waves just discussed, this is a good compromise for the diminishing yield in terms of power output. Also, at this length, the impedance is real, not complex; we will discuss impedance in more detail later. Finally, the antenna output pattern is a single broad lobe; that is, there are no side-lobes.

There is also terminology related to reactive and radiative fields: the near-field and the far-field. As the terms suggest, they are related to the characteristics of the field given the distance from the source of the field. The reactive field is near the source and is thus called the "near-field." The radiative field appears farther from the source and is thus termed the "far-field." The shape of the near-field is related to the geometric structure of the source. Farther away, the field becomes independent of the shape of the source, appearing as propagating outwards as a sphere. Farther yet, the field takes the shape of plane waves. The boundary between the near-field and the far-field is approximately $\lambda/2\pi$. Beyond 3π to 10π the field becomes nearly undetectable. Thus, for 60 Hz current found in North American power systems, the near-field extends to 833 km; almost all the field power is near-field or reactive. At 100 MHz it is 0.5 m, and thus mostly radiative, which is preferred for radio communication. Finally, at very high frequencies, such as light waves, almost all the field power is radiative and very little is near-field or reactive. Thus, we can see some of the interesting commonalities and differences in electromagnetic behavior between power systems and communication. Understanding these properties is very important for efficiently integrating power systems and communication.

6.6.2 *The Wave Equation*

Throughout this discussion, we have assumed the standard textbook explanation of electromagnetic propagation as waves. However, it may be instructive to question why this is so. How can we deduce that these really are waves? This is extremely important for the communication aspect of this book. It also provides insight into power-related aspects, such as understanding of frequency and phase. However, to explain the wave nature of electromagnetic fields generally requires fairly complex mathematics, which we will attempt to minimize as much as possible. The reader who does not feel comfortable with the mathematics of fields can safely skip this section.

Recall that electric fields may be quantified and visualized as lines on a map: lines pointing outward from a negative charge and inward toward a positive charge. These lines are vectors; they have a magnitude and a direction. There are many operators designed to analyze vectors and discover their properties; these have been used with great success in Maxwell's equations. Let us first discuss divergence. The intuitive notion of divergence is that it measures the degree to which a point in space acts as a source or sink of the vector field. Specifically, it turns the total outward flow described by a vector field at a given point into a single scalar value. A point from which a vector field originates will have lines that start within the point and lead outward. This is what divergence is measuring. Lines that start outside the point being measured do not contribute to the value of the divergence. Finally, lines that end at a point being measured contribute negatively to the divergence scalar value; that is, that point is a sink for the vector field. If the divergence is nonzero at some point, there must be a source or sink at that position.

More rigorously, the divergence of a vector field F, for example the electric field, at a point p is defined as the limit of the flow of F across the smooth boundary of a three-dimensional region V divided by the volume of V as V shrinks around the point p. This is what is described by

$$\mathrm{div}\mathbf{F}(p) = \lim_{V \to \{p\}} \iint_{S(V)} \frac{\mathbf{F} \cdot \mathbf{n}}{|V|} \, dS. \tag{6.34}$$

$|V|$ is the volume of V, $S(V)$ is the surface boundary of V, and the integral is a surface integral, where n is the outward unit normal to the surface. The result,div$F$, is, of course, a function of the location p within the field. Hopefully, it is clear to the reader that div$F$ can be seen as the source (or sink) density of the flux of the field F.

While the divergence measures whether a point acts as a source or a sink for a field, the curl measures the rotation of the field. While the divergence is a simple scalar value, the curl has a more complex, vector, result. This is because indicating direction requires a vector. Thus, the curl is essentially a clever little vector field operator that indicates the rotation of a vector field and returns a vector field as a result, because the direction of rotation of the measured field is taken at every point in the field.

The curl is precisely defined by

$$(\nabla \times \mathbf{F}) \cdot \hat{\mathbf{n}} \overset{\mathrm{def}}{=} \lim_{A \to 0} \frac{\oint_C \mathbf{F} \cdot \mathbf{dr}}{|A|}, \tag{6.35}$$

where $\hat{\mathbf{n}}$ can be any unit vector that will be defined to be the axis of rotation that we are measuring. $\oint_C \mathbf{F} \cdot \mathbf{dr}$ is a line integral defining the boundary of the area in question, and $|A|$ is the size of the area. If $\hat{v}$ is an outward-pointing in-plane normal, then $\hat{\mathbf{n}}$ is the unit vector perpendicular to the plane. We can again apply the right-hand rule: with the thumb pointing in the direction of $\hat{\mathbf{n}}$, the fingers will curl toward the direction of rotation being measured.

The simplest and most intuitive way to think about the curl of a field is to imagine placing a paddle wheel with its axis of rotation in the direction of $\hat{\mathbf{n}}$ into the field. The magnitude and direction of rotation of the paddle wheel will be the result of the curl at that point.

Now we have enough background to at least intuitively think about Maxwell's equations, which are at the heart of both power systems and communication. Although James C. Maxwell had originally derived many more equations, they have been boiled down to the following four equations. It is important at this point to recall the earlier intuitive explanation of fields and induction, as well as the vector operators. Equation 6.37, the first of the four equations, is also known as either Gauss's or Coulomb's law. Here we see the divergence operator being used, so we know this law is concerned with characterizing sources and sinks of field lines. Also, we see the **E** field, so it is referring to the electric field as opposed to the magnetic field. The equation is saying that the electric flux through any closed surface is proportional to the enclosed electric charge. ε_0 is known as the electric constant or permittivity of free space. For example, the force between two separated electric charges in a vacuum is given by Coulomb's law in Equation 6.36, where q_1 and q_2 are the charges, and r is the distance between them:

$$F_C = \frac{1}{4\pi\varepsilon_0}\frac{q_1 q_2}{r^2}. \tag{6.36}$$

ρ is the charge density or the amount of charge per unit length, area, or volume, depending upon the space being analyzed. Going back to the definition of divergence in Equation 6.34, it should be easy to see that Equation 6.37 shows that the source of an electric field is equivalent to the charge density creating the field divided by the permittivity of free space. Different materials have different values for permittivity. Notice that since permittivity appears in the denominator in Equation 6.37, it acts, in an intuitive sense, as a resistance to the formation of an electric field. Physically, permittivity characterizes the ability of a material's molecules to polarize within in an electric field, thus reducing the field strength; it relates to a material's ability to permit an electric field to form. As a side note, one may see the electric field represented by **D**. **D** is related to **E** by $\mathbf{D} = \varepsilon\mathbf{E}$ (in isotropic materials). **D** is known as the displacement field or the field characterization of a material's ability to orient, or migrate charges internally, in the presence of an electric field:

$$\nabla \cdot \mathbf{E} = \frac{\rho}{\epsilon_0}. \tag{6.37}$$

Since we have just discussed the electric constant ϵ, it is a good time to introduce the analogous constant for magnetic fields, the magnetic constant or μ. This relates directly to power systems in the form of two thin, straight, stationary, parallel wires, a distance r apart in free space. Suppose each carries a current I, then each wire will exert a force upon the other due to the magnetic field. Ampère's force law states that the force per unit length is given by Equation 6.38 because the ampere can be defined such that if two wires are 1 m apart and the current in each wire is 1 A, the force between the two wires is 2×10^{-7} N m^{-1}:

$$F_m = \frac{\mu_0 I^2}{2\pi r}. \tag{6.38}$$

Similar to the electric constant, the magnetic constant, or magnetic permeability, is the ability of a material to support a magnetic field. Similar to the electric constant and the electric and displacement fields, there is an auxiliary magnetic field **H** such that $\mathbf{B} = \mu\mathbf{H}$.

The next of Maxwell's equations is Gauss's law for magnetism:

$$\nabla \cdot \mathbf{B} = 0. \tag{6.39}$$

By inspection, we can see that this equation is very similar to Gauss's law in Equation 6.37. The divergence again plays a role, though for the magnetic field **B** rather than the electric field. Also, it is clear that this law states that the divergence is zero for the magnetic field. This means that there is no source or sink for the magnetic field; in other words, there is no equivalent to the point charge in an electric field. As discussed earlier, this simply means that there is no single magnetic charge, or pole, from which magnetic field lines begin or end; only dipoles exist. Magnetic fields are loops with no beginning and no end. The positive and negative magnetic charges, or poles, are inexorably bound together; they cannot be separated. One cannot cut a magnet in half and expect to have an individual pole, the two halves of the magnet will each have two poles, no matter how many times they are cut in half or how small the pieces of the magnet become. This is not to say, however, that there has been interesting research in attempts to find the elusive magnetic monopole (Giacomelli and Patrizii, 2003), which has so far not been found.

Next we come to the last half of Maxwell's equations; these are the equations that contain the curl operator. They describe relationships between the electric and magnetic fields; in particular, how one can be created from the other. Thus, these equations play a role in explicitly showing how electromagnetic waves are propagated, as will be explained shortly. However, for now, the goal is simply to explain these equations and attempt to understand them in an intuitive manner. Clearly, these equations are at the heart of power systems and electric power generation as well as electromagnetic communication, be it radio wave transmission, electrical, or optical communication.

Faraday's law of induction is the third of Maxwell's equations:

$$\nabla \times \mathbf{E} = -\frac{\partial \mathbf{B}}{\partial t}. \tag{6.40}$$

Recall the explanation of the curl operator given earlier in this section for Equation 6.35 and the intuitive notion of a paddle wheel being placed within the field and its direction of rotation. Also recall the explanation of the Lorentz force Equation 1.9. Equation 6.40 can be interpreted as saying that the time rate of change of the magnetic field **B** is the rotation of the induced electric field **E**.

Note how the result of the curl operator follows the right-hand rule, as discussed earlier. Thus, we can see that the faster **B** changes with respect to time and the more perpendicular it is with respect to the **E** (usually **E** is enabled as a wire moving relative to a magnetic field) the stronger the induced current is. Again, think of the electric field as the paddle wheel oriented within the changing magnetic field. If the magnetic field were able to apply more force to the paddle wheel, due to the magnetic field's flow and the orientation of the paddle wheel, then the induced current would also be stronger. Note that there are integral forms of each of

Maxwell's equations, but we focus only on the simpler differential forms in order to keep the explanation as intuitive and simple as possible.

The final equation of Maxwell's four equations is known as Ampère's law. As mentioned, the last two of Maxwell's equations describes the relationship among, and generation of, electric and magnetic fields. While the last law was stated in terms of a magnetic field creating an electric field, Ampère's law states that the electric field can create a magnetic field, either by electric current or by a changing electric field. This is shown by

$$\nabla \times \mathbf{B} = \mu_0 \mathbf{J} + \mu_0 \epsilon_0 \frac{\partial \mathbf{E}}{\partial t}, \tag{6.41}$$

where we see the rate of change of the electric field as the curl of the magnetic field. The fact that the curl of the magnetic field is equivalent to the two terms shown in the equation is immediately apparent. The first term is the charge density induced in a wire and the second term relates to the propagation of the electromagnetic field. $\mathbf{J}$ is the total current density. Note that this differs from ρ, which was the charge density. Also, as a side note, Ampère's original equation was simpler, ignoring the electric field term, but incorrect, until a correction was added by Maxwell.

Equation 6.41 allows us to see how a time-changing electric field creates a magnetic field and the curl of a magnetic field relates to both a current density and a time-changing electric field. Soon we will see how these last two equations of Maxwell explicitly define an electromagnetic wave.

An intuitive way to see why the equations tell us that electromagnetic fields propagate as waves is to first think about one of the simplest possible waves: a wave moving along a string or a rope. Imagine holding a string or rope that is attached at the other end and quickly giving the rope a shake. That shake imparts an acceleration to the matter comprising the rope. The interesting phenomenon that occurs is that the acceleration that displaces the rope travels along the length of the rope in a wavelike pattern. The wave may reflect back in the opposite direction after hitting the end of the rope; however, let us ignore this complication for the moment and consider only the original, forward motion of the wave.

Figure 6.12 shows a small portion of a wave traveling along a rope. While the wave moves along the rope, the rope itself does not move along the horizontal, or x, axis. The rope only

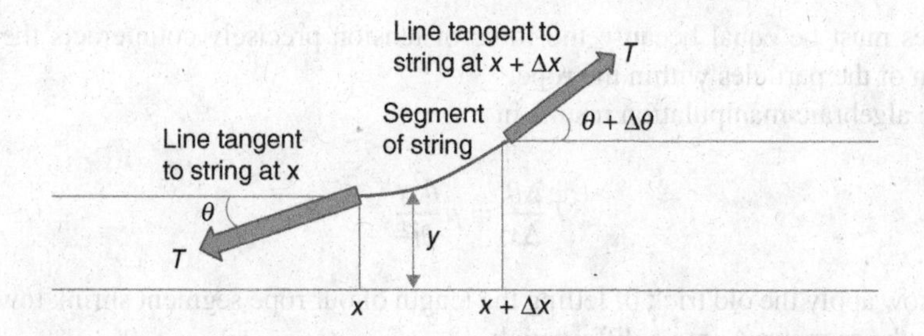

Figure 6.12 The displacement y is analyzed along a small segment of a rope that is held with tension at both ends. This illustrates the frequently occurring wave patterns that occur in both communication and power systems.

moves up and down, along the vertical, or y, axis. The portion of the rope shown in the figure extends from location x to $x + \Delta x$. As the wave travels along the rope, there is tension on the rope – tension caused by the disturbance of the wave and by the tendency of the rope to return to its original position because the rope is stretched. The direction of the forces of tension on a segment of rope are shown in Figure 6.12. There is an angle θ of the force of tension at the x end of the rope and an angle $\theta + \Delta\theta$ at the $x + \Delta x$ end of the rope.

Examining the right side of Figure 6.12, the upward force component caused by the wave is

$$F_y = T\sin(\theta + \Delta\theta) \quad \text{and} \quad F_x = T\cos(\theta + \Delta\theta), \tag{6.42}$$

where T is force of tension on the rope.

On the left side of the rope segment, the force of tension is acting in the opposite direction and has the following component:

$$F'_y = -T\sin(\theta) \quad \text{and} \quad F'_x = -T\cos(\theta) \tag{6.43}$$

The prime in F' is included to indicate the force on the left side of the rope as opposed to the right side, which has no prime.

We can assume that $\Delta\theta$ and θ are small. The value of the cosine near zero is one. Therefore, both F_x and F'_x are approximately equal and there is no net force along the x- axis. For the sine function, a small of value of θ results in a larger value of $sin(\theta)$. In this case, summing the y force components yields

$$F_y = T\Delta\theta. \tag{6.44}$$

Now that we have identified forces, we can apply Newton's famous second law, which relates the force to mass times acceleration. Suppose that we know the mass per unit length of the rope, μ, then $\Delta m = \mu\Delta x$. Also, we know that the rope is accelerating along the y-axis, $\partial^2 y/\partial t^2$. Thus, we can relate the force due to tension derived earlier with the force due to acceleration thus:

$$T\Delta\theta = \mu\Delta x\frac{\partial^2}{\partial t^2}. \tag{6.45}$$

These forces must be equal because the force of tension precisely counteracts the force of acceleration of the particles within the rope.

A simple algebraic manipulation results in

$$T\frac{\Delta\theta}{\Delta x} = \mu\frac{\partial^2 y}{\partial t^2}. \tag{6.46}$$

We can now apply the old trick of letting the length of our rope segment shrink toward zero, which turns the segment Δ into a differential:

$$T\frac{\partial\theta}{\partial x} = \mu\frac{\partial^2 y}{\partial t^2}. \tag{6.47}$$

Now is a good time to step back and consider where we are heading in the development of this equation. The goal is to gain intuitive understanding of a traveling wave so that we can understand how Maxwell's equations demonstrate that electromagnetic radiation takes the form of a wave propagating through space. We are attempting to gain intuition by looking at the propagation of a wave on a rope, since we can all easily see, feel, and more easily understand how such a wave propagates. Equation 6.47, which we have carefully derived so far, has the acceleration along the y-axis and the rate of change of θ along the x-axis. Ideally, we would like to see either the velocity or acceleration of the wave along the x-axis in order to see how the wave travels along the rope. Thus, we need to continue developing the equation to integrate such a term into the equation.

First, a very fundamental identity that we can use at the heart of calculus is the tangent, or slope, at a point along a curve:

$$\tan\theta \equiv \text{slope of curve} \equiv \frac{\partial y}{\partial x}. \tag{6.48}$$

Taking the derivative of the tangent with respect to x we have

$$\frac{\partial}{\partial x}\tan\theta = \frac{\partial}{\partial x}\left(\frac{\partial}{\partial x}\right). \tag{6.49}$$

Performing differentiation yields

$$\frac{1}{\cos^2\theta}\frac{\partial}{\partial x} = \frac{\partial^2 y}{\partial x^2}. \tag{6.50}$$

As mentioned earlier, $\cos\theta \approx 1$ for small values of θ; thus, we can make the following approximation:

$$\frac{\partial}{\partial x} \approx \frac{\partial^2 y}{\partial x^2}. \tag{6.51}$$

Inserting Equation 6.51 into Equation 6.47 with some simple algebraic manipulation yields

$$\frac{\partial^2 y}{\partial t^2} = \frac{T}{\mu}\frac{\partial^2 y}{\partial x^2}. \tag{6.52}$$

The equation in this form is extremely important. This is known as the wave equation and is used throughout physics. This equation relates the acceleration of the up and down movement of the rope with respect to time as a function of the acceleration of the wave along the rope. The speed of the wave along the rope v is

$$v = \sqrt{\frac{T}{\mu}}. \tag{6.53}$$

The denser the rope μ, the slower the speed. The more tightly the rope is pulled (that is, the more tension T), the faster the wave will travel.

To recap, we have gone from Newton's second law as shown in Equation 6.54 to the wave equation, shown in its general form in Equation 6.55:

$$F = ma = m\frac{\partial^2}{\partial t^2}, \tag{6.54}$$

$$\frac{\partial^2 \zeta}{\partial t^2} = v^2 \frac{\partial^2 \zeta}{\partial x^2}. \tag{6.55}$$

If it can be shown that Maxwell's equations yield a result in the form of Equation 6.55, then electromagnetic fields propagate as a wave. This turns out to be surprisingly simple. Let us begin by using the last two forms of Maxwell's equations that we have discussed – the equations that contained the curl operator, namely Equations 6.56 and 6.57:

$$\nabla \times \nabla \times \mathbf{E} = -\frac{\partial \nabla}{\partial t} \times \mathbf{B} = -\mu_0 \epsilon_0 \frac{\partial^2 \mathbf{E},}{\partial t^2} \tag{6.56}$$

$$\nabla \times \nabla \times \mathbf{B} = \mu_0 \epsilon_0 \frac{\partial^2 \nabla}{\partial t} \times \mathbf{E} = -\mu_0 \epsilon_0 . \frac{\partial^2 \mathbf{B}}{\partial t^2} \tag{6.57}$$

Next, we can use the following vector identity, where **V** is any vector:

$$\nabla \times (\nabla \times \mathbf{V}) = \nabla(\nabla \cdot \mathbf{V}) - \nabla^2 \mathbf{V}. \tag{6.58}$$

Applying the identity in Equation 6.58 to Equations 6.56 and 6.57 yields Equations 6.59 and 6.60:

$$\frac{\partial^2 \mathbf{E}}{\partial t^2} - c_0^2 \cdot \nabla^2 \mathbf{E} = 0, \tag{6.59}$$

$$\frac{\partial^2 \mathbf{B}}{\partial t^2} - c_0^2 \cdot \nabla^2 \mathbf{B} = 0, \tag{6.60}$$

where $c_0 = 1/\sqrt{\mu_0 \epsilon_0}$ is the speed of light in free space. Note that these equations are forms of the wave equation as shown in Equation 6.55.

Thus, both the electric and magnetic fields propagate as waves; this has long been known and utilized for communication, particularly power line carrier and wireless communication. Of course, this is also a form of wireless power transmission as well, so it applies directly to power systems.

6.7 Summary

The key takeaway from this chapter has been the relationship between communication and power systems and that this relationship goes beyond the obvious. There was not only an introduction to information theory, but also a suggestion as to what a power system information theory might look like. The chapter began with an almost trivial list of applications of communication and information theory to the power grid. Then it looked at an analogy between communication networks and the electric power grid. The chapter progressively became more detailed in the development of these analogies, establishing a mathematical relationship between energy and computation and between energy and communication. Finally, the chapter ended with an introduction to wireless power radiation and the wave equation, because these are fundamental concepts that appear not only in wireless communication, but also in the power grid itself in the form of noise and emission from switching elements.

Chapter 7 dives into the demand-side management aspect of smart grid; namely, DR and consideration of its communication requirements. Demand-side management can be envisioned as extending management of the power grid through the consumer to management of loads. There are myriad ways to achieve this goal; some require communication, such as demand-response, and others do not, such as dynamic demand. Metering infrastructure and power line carrier are old mechanisms, going back to the late 1800s. It is important to know the history or risk re-inventing it. More recent wireless AMI approaches, in particular, IEEE 802.15.4 and its variants, such as IEEE 802.15.4g are covered as well as the IEEE 802.11 variants. While demand-response has received considerable hyperbole in smart grid, there are other approaches toward achieving the same outcome. Dynamic demand for example, requires electronic devices to monitor their own alternating-current frequency to autonomously adjust their duty-cycle.

6.8 Exercises

Exercise 6.1 Information Theory

1. What is power flow entropy?
2. As the smart grid becomes more efficient, would you expect power flow entropy to increase or decrease? Why?

Exercise 6.2 Complexity Theory

1. Some would argue that the smart grid is making the power grid more complex. What would be an ideal measure of power grid complexity?
2. What does this complexity imply for smart grid communication? What does this complexity mean in terms of information flowing through smart grid communication channels?

Exercise 6.3 Compression

1. What roles could data compression play in the smart grid and for smart grid communication?

Exercise 6.4 WiMAX

1. Where is WiMAX best used within the power grid and why?

Exercise 6.5 Power Line Carrier

Power line carrier channels suffer from multipath effects similar to that of wireless communication. Differences in impedance will reflect a power line carrier signal back to the source, degrading the original signal.

From Philipps (1999), an approximate equation for the power line carrier channel transfer function is derived thus:

$$h(t) = \sum_{1}^{N} |\rho_v| e^{j\phi_v} \cdot \delta(t - \tau_v) H(f) = \sum_{i=1}^{N} |\rho_i| e^{-j2\pi f \tau_i}. \tag{6.61}$$

There are N Dirac pulses representing the superposition of reflections from N paths. ρ_i is a complex factor delayed by time τ_i. This represents the product of reflection and transmission along each echo path.

1. Based upon the above model, what happens as the frequency increases?
2. What happens as the number of reflected pulses increases?
3. What does this imply regarding potential impedance mismatches and the topology of a power line carrier network?

Exercise 6.6 Active Network

1. What is the main concept that defines active communication networks?
2. How is this similar to the operation of the smart grid?
3. What benefits does active networking provide for the power grid?

Exercise 6.7 Software Defined Network

1. What is a software defined network?
2. What does it have in common with active networks?
3. What unique benefits could SDN provide to the power grid?

Exercise 6.8 Energy and Communication

1. How is compression of an ideal gas similar to communication?
2. What ramifications could this have for smart grid communications?

Exercise 6.9 Wave Equation

1. Explain the main concepts behind the derivation of the wave equation.
2. Why is the wave equation important in power grid communications? List at least four different applications of the wave equation.

Exercise 6.10 Entropy in the Power Grid

1. What is power entropy as it relates to the power grid?
2. What impact does power entropy have on communications?
3. What desirable impact can communications have upon power entropy?

7

Demand-Response and the Advanced Metering Infrastructure

We believe that electricity exists, because the electric company keeps sending us bills for it, but we cannot figure out how it travels inside wires.

—Dave Barry

7.1 Introduction

This is the second chapter of Part Two, *Communication and Networking: The Enabler*, which delves more into communication and networking technology for the power grid. This chapter focuses specifically upon the topic of power consumption in the smart grid. This can be thought of as extending management of the power grid through the consumer to management of individual loads. The general concept is ultimately to schedule the use of loads in such a manner that the consumer is satisfied while at the same time the load operates in manner that is best for the power grid. There are a myriad ways to achieve this goal; some require communication, such as DR, and others do not, such as dynamic demand.

DR is designed to motivate the consumer to make the best local economic choices that are also the best choices for the power grid. Consumers would ideally purchase and consume power from the best sources at optimal times that would reduce peak demand, flow through the optimal transmission and distribution lines, use renewable sources, and perhaps even influence active or reactive power consumption appropriately. The concept is that this process should be automated and occur in near-real time, allowing market forces to equalize the supply and demand in an optimal manner. The thinking has been that this would require bidirectional communication between all actors involved; namely, consumer, power generators, and power carriers. Because communication is an integral part of this concept, privacy and security become significant issues.

Dynamic demand, on the other hand, requires electronic devices to monitor their own alternating current frequency and adjust when they turn on and off based upon frequency. Remember that frequency is related to supply and demand of power; when demand begins to

---

Smart Grid: Communication-Enabled Intelligence for the Electric Power Grid, First Edition. Stephen F. Bush.

exceed supply, the frequency will begin to drop. An electronics device simply has to monitor frequency and decide when best to schedule its activity to help keep the frequency at its nominal value. If all electronic devices did this, many, small, automated adjustments by loads themselves could significantly help reduce peak power consumption. Notice that no explicit digital communication is required in this case; communication is implicit, taking place through change in frequency.

A key to any of the load-management mechanisms previously described is proper monitoring, in which the so-called smart meter plays a role. Thus, this chapter next covers an introduction to the smart meter and particularly its communication via what is known as AMI. This chapter reviews some of the typical communication mechanisms used for AMI. There are two aspects to the smart meter: AMI communication carries data from the meter to the utility, while a HAN can integrate the power meter with a complete EMS within the home or consumer site. We begin with AMI and power line carrier, since this is one of the oldest mechanisms, going back to the late 1800s. Next we consider wireless AMI approaches, in particular, IEEE 802.15.4 and its variants, such as IEEE 802.15.4g. Then we also discuss the Internet Protocol over such a physical layer system. Then we cover advances related to the HAN. These are typically IEEE 802.11 protocols and their variants. In particular, we review IEEE 802.11n and IEEE 802.11ac. It should be noted that these technologies are advancing rapidly; the reader can expect a proliferation of double letter 802.11 standards to come. Because these protocols are primarily for direct consumer use, they can be replaced quickly by consumers as new technologies come along. It is much harder to replace the entire power grid infrastructure. Thus, we expect consumer communication to continue to change rapidly, whereas the smart grid communication infrastructure will not be as volatile. However, it should also be noted that the rapid changes are not fundamentally new ideas in communication but rather continued refinement of concepts developed decades ago. Thus, as noted throughout this book, it is more important to know the fundamentals of the technology rather than the specific protocols or product types. Management of the communication network for the smart grid will become increasingly challenging. We introduce that topic briefly here. It has been posited that electric vehicles will form an increasing segment of the automobile market. That means that electric vehicles become an additional consumer load that has to be handled by the power grid. An interesting aspect is that electric vehicles are the first significant consumer load to be mobile. Because electric vehicles are a consumer load, they are dealt with in this chapter. Finally, there are considerations for the use of cloud computing throughout the power grid. This chapter closes with a discussion of cloud computing applied to consumer-related activity. Exercises at the end of the chapter provoke further thought on the topics in this chapter.

7.2 Demand-Response

In the past, as we have seen, a careful equilibrium has been required between power demanded by consumers and power supplied through the power grid (Albadi and El-Saadany, 2007). Power utilities in the past had been vertically integrated, in full control of the power supply. The operating philosophy had generally been to use large, centralized generation to reduce cost and to supply all power demanded whenever it was demanded. The high cost of large, centralized generating plants made operating at less than full capacity uneconomical, and the notion originated of attempting to control demand rather than supply, since shifting demand could be a cheaper alternative. After electric power deregulation occurred, different organizations managed different components of power generation and transport. Each can offer their product at a market price. This has added further motivation to explore ways of influencing demand through price. The classic example is to reduce peak power by raising prices when demand is high in order to attempt to level demand.

While this book is not about market analysis of electric power, it is helpful to have a basic understanding of the topic, particularly with regard to the notion behind DR and ultimately the requirements for AMI communication. DR is used quite broadly to include any intentionally designed attempts to modify consumer use of power in terms of time of power use, instantaneous demand for power, or total power consumption. There are three general approaches to doing this. First, consumers can be motivated to use less power only during critical periods of stress on the power grid. Second, consumers can respond more continuously to market power price fluctuations. Finally, consumers that have onsite power generation or storage can be encouraged to utilize it, thus reducing demand from the power grid. Like many aspects of the smart grid, DR has received considerable hype as a solution to reduce cost and enable a clean environment. We need to step back and take a careful, rational analysis of what DR is and what it can and cannot do (Ruff, 2002).

The fundamentals of DR begin with the supply and demand curves from economics (Albadi and El-Saadany, 2007; Alizadeh *et al.*, 2010; Barbato *et al.*, 2011a,b; Belhomme *et al.*, 2008; Bu *et al.*, 2011; Caron and Kesidis, 2010; Choi *et al.*, 2011; Faria *et al.*, 2011; Goudarzi *et al.*, 2011; Hobby, 2010; James, 2008; Kallitsis *et al.*, 2011; Kefayati and Baldick, 2011; Koutitas, 2012; Langbein, 2009; Li and Qiu, 2010; Lisovich *et al.*, 2010; Lu *et al.*, 2011; Markushevich and Chan, 2009; Medina *et al.*, 2010; Mohsenian-Rad and Leon-Garcia, 2010; Nyeng and Ostergaard, 2011; O'Neill *et al.*, 2010; Ott, 2010; Palensky and Dietrich, 2011; Parvania and Fotuhi-Firuzabad, 2010; Peeters *et al.*, 2009; Pierce, 2012; Pourmousavi and Nehrir, 2011; Rahimi and Ipakchi, 2010; Reid *et al.*, 2009; Roossien *et al.*, 2008; Saad *et al.*, 2011; Samadi *et al.*, 2011; Sankar *et al.*, 2011; Shahidehpour and Wang, 2003; Spencer, 2008; Su and Kirschen, 2009; Vos, 2009; Wang *et al.*, 2011b; Webber *et al.*, 2011; Yue *et al.*, 2011; Zhang *et al.*, 2005;). Price-per-unit of product is on the ordinate (y-axis) and both the quantity of supply and the quantity of demand are on the abscissa (x-axis). The supply and demand curves are assumed to be independent of one another and are the price per given quantity of unit assuming the given quantity is available for sale on the supply curve and assuming the given quantity would be purchased by consumers on the demand curve. If one assumes perfect competition, then supply is determined by marginal cost; a supplier will continue to produce additional units as long as the cost of producing the unit is less than the price for which it can be sold in order to make a profit. The demand curve represents the amount of product an ideal consumer would purchase factoring in utility of the product to the consumer as well as all possible alternatives the consumer could make with regard to satisfying the utility via other means. This is known as opportunity cost; as long as the price goes down, the assumption is that as long as the marginal utility to the consumer of purchasing more of the product is greater than the opportunity cost, the consumer will purchase more of the product. The point where the two curves intersect is an equilibrium point; the point where the cost per unit satisfies both supply and demand is shown in Figure 7.1.

Now consider an example specific to electric power, as shown in Figure 7.2. This figure is a bit more complicated because there are two pairs of curves: two demand curves and two supply curves. Curve S_N is the "normal" supply curve. This is the supply curve for electricity

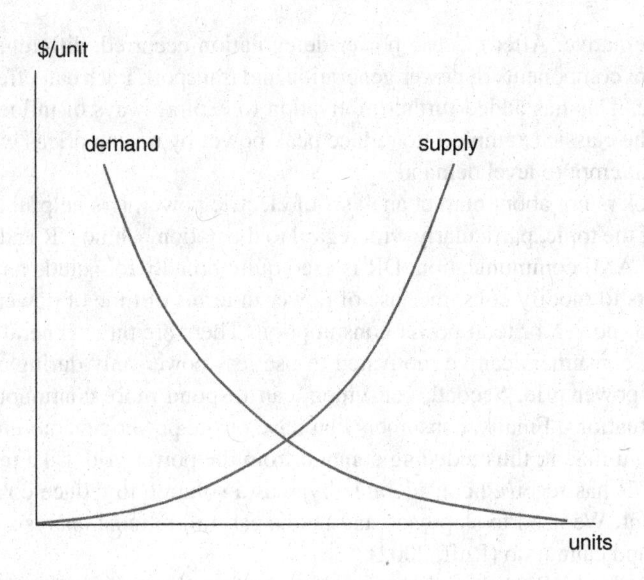

Figure 7.1 Generalized supply and demand curves are shown in terms of price per unit versus quantity of units. Supply tends to increase as the price per unit increases, while demand decreases as the price increases. The intersection of the curves is the theoretical equilibrium price per unit. Electric power is a product that ideally should follow the same behavior.

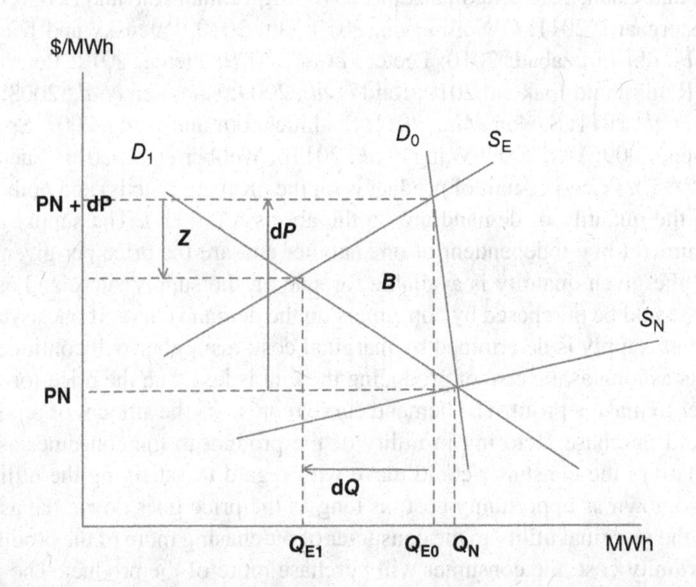

Figure 7.2 A DR plot is shown for an inelastic demand curve D_0 and an elastic demand curve D_1. Two electric power supply curves are shown: a "normal" supply curve S_N and supply under stress when there is excessive demand S_E. Source: Ruff, 2002.

production when the power grid is operating as expected; that is, there are no significant faults or stresses caused by excessive demand. But now consider S_E; this a different supply curve assuming that there is excessive demand or a significant event occurred in the power grid causing abnormal operation. Notice that if the change in supply is due to a change in demand, then the condition previously mentioned, that the supply and demand curves should be independent of one another, has been violated. However, continue with the analysis for now ignoring this fact. Notice that S_E results in an increased cost to produce the same amount of power; a prime example of this is the cost of starting new generators. Now consider the pair of demand curves, D_0 and D_1. These demand curves differ with respect to their elasticity, which is the consumer sensitivity to change in price. Electric power has generally been considered as relatively inelastic; in other words, consumers will continue to purchase the same amount regardless of fluctuations in price. Price elasticity in demand, because it measures sensitivity to price, is related to the rate of change or slope of the plot of quantity versus price. It measures the rate of change in quantity to change in price. D_0 is relatively inelastic; the quantity consumed changes little with respect to change in price. D_1 is more elastic; the quantity consumed changes more significantly with change in price. The next step is to consider what happens at the equilibrium points under various situations. One equilibrium point is the intersection of normal supply and inelastic demand S_N and D_0 and the price is P_N. If an abnormal event occurs to the power supply, the new equilibrium point jumps to the intersection of S_E and D_0 and the price is $P_N + \Delta P$. However, if the demand were somehow more elastic and curve D_1 were in effect, then during an abnormal event in the power grid the equilibrium point would be S_E and D_1, resulting in a price below $P_N + \Delta P$ by the amount Z. In terms of total cost savings to the public, the reduction in units of power is $Q_E^0 - Q_E^1$ and the total cost savings is the area of triangle B, assuming our simplified assumption of linear supply and demand curves holds true.

Many sophisticated market mechanisms have been proposed based upon variations of these fundamental concepts. These include the idea of a negawatt market; that is, a negative power market in which power that would have been consumed, but was not, is a commodity. Also, the idea of explicitly coupling emissions to cost has been suggested. However, this would take us deep into somewhat speculative economic theory and outside our scope. The importance of this analysis, from the perspective of smart grid communication, is the assumption that the demand curve will have a slope that is elastic enough to warrant the cost of the communication infrastructure necessary to enable the supply and demand activities to take place on an effective time scale. This has been the goal of the AMI discussed in Section 7.3.

7.3 Advanced Metering Infrastructure

The AMI has been the system within the smart grid that is to enable the implementation of DR (Sui *et al.*, 2009). From a higher level perspective, AMI transports information allowing the supply and demand curves to find their equilibrium. Ideally, this means that potentially massive amounts of demand information need to be collected from every consumer, and price information regarding supply needs to be sent to every consumer. While this may sound like it requires a sudden leap in technology, AMR has existed, at least in an ad hoc manner, for a long time. This includes partially manual techniques, such as local serial interfaces, infrared, or RFID, and fully automated techniques, such as via collecting readings utilizing power line carrier and early DMS programs, which used the plain old telephone system for bidirectional reading and control. This, like most aspects of the smart grid, is not actually a new idea.

Another aspect of the AMI system to keep in mind is that, similar to trends in most new technologies, what is currently thought of as the AMI will evolve to merge with other systems within the power grid, particularly with DA. This is sometimes referred to as the distribution management infrastructure (DMI) (Uluski, 2008). The general concept is that information from consumer smart meters can be utilized directly by the distribution system in order to optimize its operation; everything from estimating current and future load, to power quality, to detecting and measuring outages from smart meters would be useful to the distribution system. However, for purposes of clarity, AMI will initially be explained as a separate system with the goal of collecting consumer meter readings to be sent to the utility and supporting the transmission of price signals from the utility to the consumer.

The architecture of the AMI system is thus one in which a large number of spatially distributed consumer meters must periodically transmit relatively small amounts of information over a relatively long distance to a utility. Latency requirements for such information are more lenient than for other communication within the smart grid; for example, as compared with protection or control. The best type of communication network will be one that disperses its bandwidth resources to many spatially separated nodes efficiently and is also able to take optimum advantage of a large number of resource-constrained nodes. The marginal cost of adding a new node should be low, while meeting minimum communication requirements. Theoretically, this tends to favor broadcast media and mesh networks.

The typical solution is to assume a hierarchical communication network in which there are nodes distributed throughout a region in order to receive data from a number of consumers. These nodes have been called a variety of names, including "collectors," "data concentrators," and "data aggregation units (DAUs)". Thus, consumer meter information is aggregated before transmission on its path to its final destination. A larger bandwidth, backhaul communication channel serves to take the aggregated readings from the DAUs to the meter data management system (MDMS), where meter data are processed. Ideally, a mesh network – one that allows packets to hop from one node to another – has invariably been proposed for the lowest level of the AMI hierarchy; that is, from meters to the DAU. The idea is to enable a reliable, self-configuring system in which meters can forward information from other meters, if necessary, to ensure that they have a path to a DAU. This is examined further in the discussion on routing protocol for low-power and lossy networks (RPL) in Section 7.4.7.

Any number of communication technologies can and will be used for any given situation; 802.15.4, WiMAX, and power line carrier serve as only a few of many possible examples. Additionally, any number of communication protocols may reside upon whichever physical and media-access control technology happens to be used. A more lasting and useful discussion is to consider the problem from a theoretical point of view that asks how the infrastructure might best be used and what the communication requirements are (Niyato and Wang, 2012). Its method of operation will continue to change as more applications and means of integrating the AMI system into the other systems of smart grid operation are conceived.

One aspect of smart meters is that there is a large number of them. This fact will continue to spur innovation that seeks to leverage the benefits of scale with the limited communication and processing that the smart meter possesses. Ideas involving self-organization, emergence, and cooperative behavior have been explored for sensor networking. Since this is a chapter on DR and AMI, let us consider cooperative behavior from a cost and communication perspective. Consider the method of operation and case for cooperative communication transmission of meter data.

As we will note many times in this book, the terms HAN, NAN, MAN, and WAN are often used by people in the communication domain as though they mean something specific. Unfortunately, the opposite is true; there are no clear standard definitions for HAN, NAN, FAN, MAN, WAN, or any other (choose your favorite letter) AN. They do nothing other than specify a rough area over which a communication network should operate, and little else. However, given that there are no more-specific standard terms available, we will use those terms in describing general communication architectures. Following the general concept for an AMI system, meter data are transmitted from an HAN through an NAN to a DAU. The DAU transmits the aggregated data over a WAN to the MDMS where the data are processed to determine demand and electric power pricing.

The term "cooperative networking" indicates that messages can be overheard by all nodes within range of the original transmitter and any of these nodes may act as a relay by repeating the transmission. This technique provides reliability to ensure that messages reach their intended destination. Many mesh protocols utilize various forms of cooperative networking that allow intermediate nodes to help transfer information. We want to address the question of whether it makes economic sense for users to have nodes that are capable of cooperative networking. To do this, assume a simplified, mathematical model of our abstract AMI system and apply the Nash equilibrium. The Nash equilibrium assumes a noncooperative game in which players independently attempt to choose the best strategy to win. If each player has selected a strategy and no player can obtain a better result by changing their current strategy while everyone keeps their current strategy, then a Nash equilibrium exists. Let us see how this applies to communication for a mesh network metering system.

Assume there are I communities and the number of nodes in each community is N_i. Transmission of meter data from consumers takes place through an NAN (for example, Wireless Fidelity (WiFi)) to a DAU. The DAU transfers the aggregated meter data through a WAN (for example, Worldwide Interoperability for Microwave Access (WiMAX) IEEE 802.16j or 802.16m) to the MDMS. The MDMS estimates demand and allocates power. A decode-and-forward technique of cooperative communication is assumed. Any node designated as a relay station may help the DAU transmit aggregated meter data from the community on its path toward the MDMS. Let $r(C)$ be the communication rate given C active relay stations.

One of the keys to this process is that power supply and demand estimation takes place in two stages with different pricing at each stage. In the first stage, the utility reserves power given the best estimate of expected load. This first stage is known as the "unit commitment" stage. This takes place before power is actually generated. The price of power determined during unit commitment is known as the "forward price" p_{f}. Next, the second stage utilizes the observation of actual demand and makes corresponding adjustments to correct for any inaccuracies in the initial demand estimate. This second stage is known as "economic dispatch," and the "option price" p_{o} is the term for the price of power estimated during economic dispatch. The option price is generally higher than the forward price of power. Thus, there is an incentive for consumers to help ensure that demand is estimated accurately in order to obtain the lowest price. This is because a demand estimate that is too high gives the appearance of more demand than there really is and the price will be set higher than it should be. If the demand estimate is too low, than the option price required to correct for the actual higher demand will be more expensive than the forward price would have been. Let us continue with the mathematical model.

Let the actual power demand be $P_{i,n}$, which is the average power demand for node n in community i. Then the estimated total power demand is $D = \sum_{i=1}^{I} \sum_{i=1}^{N_i} P_{i,n}$. Now assume that there is some probabilistic communication packet loss L, which occurs uniformly and independently. Losing a packet means that whatever demand was indicated by that packet will not be counted. Then total power demand given packet loss is $D(L) = (1 - L)D$. Packet loss results in a lower estimated demand than the actual demand. Assume that the actual power demand probability distribution function is $f(z)$ for power demand z. Then in the first stage (that is, the unit commitment stage), power supply cost is $p_f D(L)$ because $D(L)$ packets have been received by the MDMS and the forward price is p_f. If no packets were lost, then this would be the actual cost to the consumer.

However, given that packets may be lost, the economic dispatch stage adds an additional cost to correct for the error estimate due to lost packets and the corresponding inaccurate estimation. The option price adds to the cost by the amount $\int_{D_{min}}^{D_{max}} p_o \max(0, z - D(L))f(z)\mathrm{d}z$. This value is simply the demand difference from the unit commitment stage integrated with the probability of that amount of demand multiplied by the option price, where D_{min} and D_{max} are the minimum and maximum power demands over which the probability distribution function is valid. Remember that the option price is generally higher than the forward price in this case.

Putting these results together yields

$$C^{pow}(L) = p_f D(L) + \int_{D_{min}}^{D_{max}} p_o \max(0, z - D(L))f(z)\,\mathrm{d}z, \tag{7.1}$$

which is the total cost of power including both the early unit commitment price and the later estimated economic dispatch price. The dispatch price is the difference in the amount of power actually demanded versus that which was reserved during unit commitment. $f(z)$ is a probability distribution function or probability of demand z; thus, integrating over max $(0, z - D(L))f(z)$ gives the average amount of power exceeding that which was initially allocated.

In order to obtain a lower total power price, consumers will want their nodes to help ensure successful transmission in order to obtain an accurate early unit commitment value. However, communication also has a cost, which includes the cost to purchase or build the communication infrastructure or to lease equipment. Thus, we return to the notion of the Nash equilibrium. Each "player" will want to minimize their total cost of both power and communication. Each player can choose a strategy, which is the probability with which to support the communication of a neighbor player's meter data. A low probability of cooperation means that communication costs will be saved, but the cost of power may be greater. A high probability of cooperative communication results in a larger communication cost but potentially lower power cost. However, the result also depends upon the choice that all other "players" decide to make as well. If no player can make a better choice given that all other players' strategies are constant, then a Nash equilibrium has been reached.

Consider this in more detail in our model. Let the strategy of consumer i be x_i. The strategy simply consists of choosing the probability of acting as a cooperative communication relay.

Thus, x_i is the probability of relaying a packet from the DAU to the MDMS. The goal, of course, is for each node to achieve the lowest total price per unit of power, including the cost of communication. It is helpful to introduce a vector of communication relay station transmission probabilities for each node of every community except that of community i as $\vec{x_{-i}}$. The subscript $-i$ indicates every node *except* for those in community i.

Let the cost of power for community i be $C_i^{\text{pow}}(L)$ considering only the nodes in community i. Let the cost of cooperative relay transmission be C_i^{rel}. This includes all costs related to communication. Then $C_i^{\text{rel}} x_i$ is the cost of communication relaying multiplied by the probability of acting as a relay. Now consider the power costs. Let $C_i^{\text{pow}}(L(x_i, \vec{x_{-i}}))$ be the cost of power allowing the strategy for nodes of community i to be variable and the nodes of all other communities to remain fixed. Then,

$$C_i^{\text{tot}}(x_i, \vec{x_{-i}}) = C_i^{\text{pow}}(L(x_i, \vec{x_{-i}})) + C_i^{\text{rel}} x_i \tag{7.2}$$

is the total cost of power and communication assuming nodes of community i choose a strategy and all other community nodes remain fixed.

Now consider optimal strategies. Let $x_i^* = \min_{x_i} C_i^{\text{tot}}(x_i, \vec{x_i})$ be the best-possible strategy for community i. It is the strategy yielding the least cost. Then $C_i^{\text{tot}}(x_i^*, \vec{x_i}) \le C_i^{\text{tot}}(x_i, \vec{x_i})$ is the Nash equilibrium. The packet loss in community i is

$$L^{\text{tot}}(x_i, \vec{x_{-i}}) = \frac{\sum_{i=1}^{I} N_i - R(x_1, \ldots, x_I)}{\sum_{i=1}^{I} N_i}. \tag{7.3}$$

The total number of packets of meter data over the estimation period is $\sum_{i=1}^{I} N_i$. The average transmission rate of the DAU is $R(x_1, \ldots, x_I)$. The set of all consumers is $S = \{1, \ldots I\}$. The transmission rate of the DAU when relaying is performed by all nodes in C is $r(C)$. The transmission rate of the DAU is

$$R(x_1, \ldots, x_I) = \sum_{\forall C \subseteq S} \left(\prod_{i \in C} x_i \right) \left(\prod_{j \in S-C} 1 - x_j \right) r(C). \tag{7.4}$$

The main point is that, under a broad set of conditions, it is beneficial for customers to help ensure their neighbors' communication reliability. Helping to ensure communication reliability leads to a lower price.

The general assumption toward implementation of AMI has so far been to communicate supply and demand information to a central location, the ISO, where a price is established and power is correspondingly generated, transported, and distributed to consumers. This is a relatively simple, centralized approach discussed in Section 5.2.1.

Additional insight into what an AMI communication system needs for support comes from the specification known as OpenADR. This is a communication data model for DR signals from a utility to consumers. This specification standardizes the interaction between price signals from the utility and consumers' preconfigured EMSs enabling the entire DR system to operate in an automated manner.

As an example, the utility could provide one or more of the following along with start and stop times:

PRICE_ABSOLUTE the price per kilowatt-hour;

PRICE_RELATIVE a change in the price per kilowatt-hour;

PRICE_MULTIPLE a multiple of a basic rate per kilowatt-hour;

LOAD_AMOUNT a fixed amount of load to shed or shift;

LOAD_PERCENTAGE the percentage of load to shed or shift.

Clearly, the first three notifications regard change in price and involve the consumer; the consumer participates by having a programmed response to these notifications. The last two notifications should typically be programmed to shed load as required. If one of the first three notifications is received indicating a rise in price, the consumer could have the system preprogrammed to reduce heating or cooling proportionally or reduce lighting. OpenADR also includes specifications for interaction with distributed energy resources.

7.4 IEEE 802.15.4, 6LoWPAN, ROLL, and RPL

For those unfamiliar with communication in the smart grid, one of biggest hurdles to overcome is understanding the "alphabet soup" of standards and their terminology. As so often happens in technical fields, jargon becomes an impenetrable shield to keep the uninitiated from understanding what in reality are often relatively simply, fundamental constructs. To make matters worse, communication standards and technical terminology are often assigned to, or inherited from, standards due to political, rather than technical, reasons. On top of all this confusion, there are many top-level standards entities working independently, sometimes creating standards that cover existing standards or ongoing development of standards in other organizations simply due to lack of awareness of developing standards across organizations. Unfortunately, it is difficult to avoid using the often illogically defined alphabet soup of standard names and terminologies. Attempting to explain the reason for the names and terminology in every instance would be a huge undertaking and take us too far outside the scope of our topic. However, some brief attempts have been made to explain the rationale, wherever possible, behind such communication standard names and terminology when they occur.

This section covers the IEEE 802.15.4 communication protocols, specifically noting IEEE 802.15.4g for the smart utility network. We begin with an overview of the IEEE 802 family of standards and work our way down to the IEEE 802.15.4g "Task Group." There tends to be some rationale in the standards naming convention, such that each dot (.) indicates a subfamily that becomes more specific in its specification and sometimes inherits the goals and standards from its superfamily, as we will see.

7.4.1 Relationship between Power Line Voltage and Communication

Roughly speaking, voltage in the power grid corresponds to the type of communication that is required. The correlation between voltage and communication occurs because higher voltage generally corresponds to longer transmission distance; that is, a reduction in I^2R loss. A longer transmission distance implies the need for a longer range communication system. On the other hand, a lower voltage power line implies shorter range transmission and likely proximity to large numbers of end-users. In this case, communication range is shorter while the number of end points is larger. In fact, one could go so far as to say that there is a natural design relationship; namely, voltage level in the power grid is inversely related to the communication network bits per square meter. A high-voltage power line will have relatively few bits per square meter (between widely spaced transmission towers), whereas a low-voltage power line will have a large number of bits per square meter (densely located customer meters).

7.4.2 Introduction to IEEE 802

The IEEE 802 family tree of standards originally intended to deal with both LANs and MANs. Furthermore, this section of the family tree was meant to focus on variable-size data packets, as opposed to standards dealing with fixed-size packets such as cell relay and asynchronous transfer mode (ATM). In isochronous networks, data is transmitted at regular, constant intervals. These types of systems are excluded from the 802 family. The number, "802," has no special significance other than being the next available assigned number to be assigned by the standards association. The 802 standards focus on the data link and physical layers of the infamous International Standards Organization ISO standard protocol framework. More specifically, the 802 standards divide the ISO data link layer into a logical-link control (LLC) sublayer and a media-access control (MAC) sublayer. The LLC sublayer, which resides above the MAC sublayer, provides a logical location for features such as flow control and automatic repeat-request (ARQ). It also provides multiplexing mechanisms that allow multiple network layers to coexist in the layer above. The MAC sublayer resides just below the LLC and provides addressing and channel access-control mechanisms that make it possible for multiple nodes to communicate over a shared physical-layer medium. Thus, the MAC sublayer acts as an interface between the LLC sublayer and the physical layer.

7.4.3 Introduction to IEEE 802.15

IEEE 802.15 Task Group covers wireless personal-area networks (WPANs). "X-area networks," where X can refer to a large number of possible geospatial terms, such as "local" (LAN), "body" (body-area network (BAN)), "campus" (controller-area network (CAN)), "wide" (WAN), "metropolitan" (MAN), and so on, is an attempt to classify networks by the geospatial "area" they were designed to be optimal at covering. The problem that occurs again, from a standards perspective, is that there are a large and growing number of terms used for such area networks. The terminology for area networks is not standardized and one must often guess at what precisely the terms mean. As a specific example, NAN can represent a "neighborhood-area network" or a "near-me-area network." A personal-area network (PAN) indicates interconnectivity of personal devices such as phones and personal digital assistants as well as providing an interface to a higher level network such as the Internet. Thus, the IEEE 802.15 Task Group would appear to be focused on short-distance wireless links.

7.4.4 IEEE 802.15.4

The IEEE 802.15.4 Task Group is focused on the physical layer and MAC for low rate-wireless personal-area network (LR-WPAN). In fact, the base goal that this task group intended to address assumed a data rate of only 250 kilobits per second over a distance of only 10 meters. The 802.15.4 standards were a change in philosophy from most standards at the time that had focused on supporting larger, faster data transport mechanisms. Instead, these standards focused on supporting more numerous, smaller, and lower data rate nodes, which, at the time, were seen primarily as elements of sensor networks. From a smart grid standpoint, the potential applications that come to mind are so-called "smart meters," as well as sensor networks for the power grid.

The IEEE 802.15.4 standard only specifies the physical layer and the MAC sublayer. It also does not include the entire set of IEEE 802 standards for this short-distance and low-data rate, since many of these features would not be desirable under these conditions. The primary goal for the standard was to specify a simple radio system that would not require radio technology skills to build or use. The standard would also specify a system that was robust with low power consumption. The network layer and above were left unspecified with the intention of allowing others to specify multiple approaches to the higher layers. Examples of higher layer protocols include IPv6 over low-power wireless personal area networks (6LoWPAN) and RPL, which will be discussed later in this chapter.

The low data rate for IEEE 802.15.4 comes primarily from enabling robust operation with a low duty cycle. This means that the node will "sleep" as much as possible in order to reduce energy consumption. This raises an important point regarding its use in the power grid; namely, the standard is antithetical to low-latency requirements. In other words, the standard achieves low energy consumption by increasing latency. Power grid applications that require low latency, such as power protection mechanisms, require that messages are transferred as quickly as possible. IEEE 802.15.4 standard networks are optimized for large numbers of users and low power consumption, but this comes at the expense of relatively slow message transfer rates. Thus, we can see that, unless significant changes are made, this would not be a good protocol for use in power systems protection. Since it is an IEEE 802 protocol, it uses the corresponding LLC mechanisms.

We will describe the IEEE 802.15.4 standard in some detail, as much of it is also assumed in the IEEE 802.15.4g Smart Utility Standards, which will be described shortly. There are two types of physical devices in IEEE 802.15.4: a full function device (FFD) and a reduced function device (RFD). The RFD is designed to be as simple as possible at the cost of functionality. Specifically, an RFD may only communicate with an FFD. There are three types of logical devices: PAN coordinator, router, and end device. The PAN coordinator, as evident by its name, is a key logical device for the network. It is an FFD that, at the very least, allocates addresses for all other nodes in the network. The router is an FFD device that, as its name implies, routes packets through the network. Finally, the end device may be either an FFD or an RFD and focuses simply on communicating its own information.

At the physical layer, the standard specifies 27 channels in three different frequency bands, which are shown in Table 7.1. Transmitter output power is specified to be 0.5 mW, or equivalently −3 dB referenced to 1 mW (dBm). This low power output is part of the standard's effort to achieve low power consumption as well as because, as mentioned, earlier, the design goal is to reach only approximately 10 meters. However, one note of great importance to smart

Table 7.1 IEEE 802.15.4 channel information.

Frequency bands (MHz)	868.3	902–928	2400–2483.5
Number of channels	1	10	16
Bandwidth (kHz)	600	2000	5000
Data rate (kbps)	20	40	40
Symbol rate (ksps)	20	40	62.5
Unlicensed usage	Europe	Americas	Worldwide
Frequency stability (ppm)	40	40	40

grid applications is that the standard allows for power output to be amplified to whatever legal limits are required. This has been done to extend the range for SCADA applications.

For communication to occur, both the transmit power and receiver sensitivity must be such that the signal can be detected. In IEEE 802.15.4 the receiver sensitivity is specified by the packet error rate (PER). The standard requires a 1% PER. This translates to −85 dBm for the 2400 MHz band and −92 dBm for the band below 1 GHz. These values are measured at the antenna terminals of the chip that implements the radio. The 1% PER requirement allows receiver sensitivity to be greatly reduced from current practice, allowing the receiver to draw less current, which again helps reduce power consumption.

Phase-shift keying (PSK) modulation is used in this standard. frequency-shift keying (FSK) modulation has been widely used in the market owing to its relatively inexpensive cost to implement. However, PSK performs better given a low signal-to-noise ratio. amplitude-shift keying (ASK), FSK, and PSK are all fundamental modulation techniques in which either the amplitude, frequency, or phase are modified, respectively, in order to convey information. More specifically, in IEEE 802.15.4, channels below one GHz use binary phase-shift keying (BPSK) and the 2400 MHz band uses offset-quadrature phase-shift keying (OQPSK). BPSK, also known as 2-PSK, is the simplest form of PSK modulation, encoding signals via two phases separated by 180°. The phases are spread as far as possible since a greater separation will require the most noise to cause the phases to overlap. While BPSK is robust, it can convey only two pieces of information, a 1 or a 0; thus it is not the fastest form of modulation. quadrature phase-shift keying (QPSK), also known as 4-PSK, simply extends the number of phases utilized to encode information to four, spread as far apart as possible. Thus, ideally, twice the amount of information can be transmitted than with 2-PSK or, alternatively, only half the bandwidth is required for the same amount of information.

In QPSK, the component that is in phase with the original carrier is referred to as the in-phase component or I. The other component, which is always 90° "out of phase," is referred to as the quadrature-component or Q. Quadrature simply means the state of being separated in phase by 90°. I and Q are orthogonal components of the signal. OQPSK is a technique to avoid large amplitude changes that can occur in a QPSK signal when both the I and Q components change at the same time. The concept behind the offset is to delay the Q component by half a symbol period, or one bit period. This ensures that the I and Q components will never change simultaneously and limits the amplitude change of the final signal, which is convenient for receiver hardware.

IEEE 802.15.4 utilizes direct-sequence spread-spectrum (DSSS) in addition to everything just described. A 16-chip sequence is used for sub-gigahertz frequencies and a 32-chip sequence for higher bandwidth transmission. In DSSS, the signal to be transmitted is multiplied by a "chip," a pseudorandom sequence (PN) of 1 and -1 values at a higher rate than the nominal bandwidth. The pseudorandomness of the sequence will create a signal that may look similar to white noise, like static. This noise-like signal can be used to exactly reconstruct the original data at the receiving end, by multiplying it by the same pseudorandom sequence (much like $1 \times 1 = 1$ and $-1 \times -1 = 1$). This process is known as "despreading" and is a correlation of the transmitted PN sequence with the PN sequence that the receiver believes the transmitter is using. The enhanced signal-to-noise ratio is known as processing gain. PN sequences are assigned in such a manner as to be orthogonal between radios sharing the medium that do not wish to communicate with one another. Orthogonality of the PN sequences results in no processing gain among interfering radios.

IEEE 802.15.4g, a variation of IEEE 802.15.4 for smart utility networks (SUNs), will be explained shortly. IEEE 802.15.4g uses much of the IEEE 802.15.4 standard, thus we need to complete the description of IEEE 802.15.4 and then compare with IEEE 802.15.4g.

Continuing with the IEEE 802.15.4 physical layer, the system uses carrier-sense multiple-access with collision avoidance (CSMA-CA) to help share the channel among transmitters. carrier-sense multiple-access (CSMA) is much like a group of people attempting to speak to one another on a conference call. People cannot see one another and they frequently begin talking at the same time. When they talk over one another, communication is garbled and no party receives a coherent message. Then there is the inevitable awkward silence as everyone waits for the other to speak, leaving dead air on the phone line. Then people begin speaking at the same time again, resulting in garbled messages and wasted phone time. A similar process can occur in communications, with nodes transmitting at the same time and interfering with one another resulting in garbled signals, a high bit error rate, and packet collision. Variations of CSMA systems can sense, at the physical layer, when multiple nodes are transmitting simultaneously and the signal is being corrupted. The transmitted nodes can then wait, or "back off," for a random amount of time until the line is clear to retry transmission. The collision-avoidance variation of CSMA enables the hardware to detect a transmission without actually transmitting and causing a collision. When transmission is sensed, a node backs off, or waits to transmit for a random amount of time, hoping to find the channel available. Of course, the situation is more complex in real life; for example, due to the hidden terminal problem. In this case, not all nodes can hear one another. Essentially, collisions will occur, but some of the offending nodes will not be able to sense that they are causing collisions. More elaborate techniques can be used to avoid this problem, but they are not in the IEEE 802.15.4 standard and so will not be discussed. Under good conditions, in which all nodes can hear one another, channel efficiency with CSMA is 36%, while under poor conditions, when nodes cannot all hear one another, channel efficiency goes down to 18%, which is similar to the ALOHA protocol. This was known when the standard was designed, and these choices were made intentionally to balance efficiency with low cost and small hardware size requirements.

The IEEE 802.15.4 physical layer defines four types of frames: data, acknowledgment (ACK), beacon, and MAC command. The names of the frames clearly indicate their function. The data frame, as its name implies, carries the data payload. The ACK frame is sent from the receiver back to the sender to indicate that a frame was successfully received. The beacon frame is a little less obvious. It is used by nodes that implement power-saving features and by nodes that are attempting to establish networks. More will be explained about network construction shortly. Finally, the MAC command frame is used to transport low-level commands.

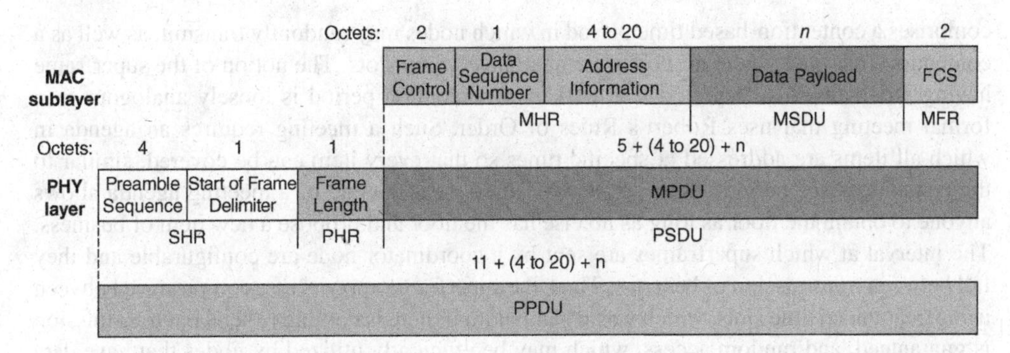

Figure 7.3 The structure of the IEEE 802.15.4 data frame. It is shown here because it is a fairly typical frame structure used by many protocols; generally, placement within the frame and size of fields tends to change from one protocol to another. Source: Adams, 2006. Reproduced with permission of IEEE.

All IEEE 802.15.4 network interfaces have 64-bit Ethernet-like addresses that uniquely identify the interface. However, the overhead of this long address can be reduced by allowing interfaces to exchange their 64-bit address for a unique 16-bit local address. The PAN coordinator is responsible for managing this exchange.

All frame types have a similar structure, shown in Figure 7.3. Consider the data frame as a representative example. The data frame is comprised of a synchronization header (SHR) that contains a 32-bit preamble sequence used to allow the receiver to acquire and synchronize with the arriving signal as well as a start-of-frame delimiter byte used to indicate the beginning of the frame. Next comes the physical-layer header (PHR), which begins with one byte indicating the length of the physical-layer service data unit (PSDU) in bytes. The SHR, PHR, and PSDU comprise the physical-layer protocol data unit (PPDU). The PPDU is comprised of the MAC header (MHR). The MHR has: (1) two frame-control octets, (2) a data-sequence number octet, and (3) 4 to 20 octets for the address. Recall that the address may be 16 or 64 bits. The MAC service data unit (MSDU) contains the data payload. The MSDU has a maximum capacity of 104 octets of data. The frame control octet through the end of the frame comprises the MAC protocol data unit (MPDU). The MPDU ends with the MAC footer (MFR) that contains a 16-bit frame check sequence (FCS).

The physical layer can be configured by the application as to whether or not to acknowledge successfully received frames. There are two basic types of nodes: low-power sensor nodes and higher power coordination and router nodes. The typical method of operation is that a sensor node will wake up due to an event, check the channel to make sure it is clear to send, transmit the message, wait for an acknowledge if configured to do so, and then may either receive any data intended for the node or go back to sleep.

The MAC layer contains over two dozen commands for implementing data transfer and managing the connection to the network. The MAC layer generates network beacons that allow other nodes to find and properly connect to the network. For example, the beacon may contain timing information for time-domain multiple access (TDMA) connections. The timing interval for the beacon is configurable, with one of the trade-offs being power consumption. Thus, the beacon is only sent from a coordinator or router assumed to be connected to the power grid. The beacon begins what is called a superframe interval. The superframe interval comprises a contention-based time period in which nodes may randomly transmit, as well as a contention-free time where all nodes are guaranteed time slots. The notion of the superframe having a contention-free period followed by a contention period is loosely analogous to a formal meeting that uses Robert's Rules of Order. Such a meeting requires an agenda in which all items are addressed at specific times so that every item can be covered, similar to the contention-free period. However, new business near the end of a meeting agenda allows anyone to obtain the floor as long as no else has the floor and propose a new item of business. The interval at which superframes are sent by a coordinator node are configurable and they fall between transmission of beacons. Thus, the superframe appears as a compromise between using guaranteed time slots, which waste bandwidth if there is nothing to send but transmission is guaranteed, and random access, which may be efficiently utilized by nodes that have data to send but there may be collisions.

As alluded to earlier, there are FFDs and RFDs. As always, communication is a trade-off among many competing dimensions. In this case, it is a trade-off between functionality and cost–power requirements. Simply put, an FFD has many features, but it draws more power, requires more costly hardware, and is a larger size. An RFD gives up features, but it requires less power and can be implemented in a smaller footprint; that is, it comes closer to the ideal sensor. However, it requires at least one FFD in order to form a network.

As mentioned, MAC commands establish and maintain the network. An IEEE 802.15.4 device starts up and searches for an existing network by scanning available channels for a beacon. If none is found and the device is an FFD, then it can establish a new network. If a beacon is found, the device will attempt to associate with the corresponding network. Every network has a PAN coordinator. Upon finding a network and being allowed to join the network, the MAC layer of the node joining the network will pass a command up to the network layer, causing an exchange of addresses between the new node and the network PAN coordinator. This can involve the exchange of a 64-bit address for a less burdensome 16-bit address. The PAN coordinator can also issue a MAC command that causes a dissociation of the device from the network.

IEEE 802.15.4 does not specifically identify the network architecture. All devices can connect directly with the PAN coordinator, creating a star network topology. Devices can also connect directly to each other forming a peer-to-peer topology. Finally, each device can attempt to connect to all other devices, forming a fully connected mesh network. Organizations such as the Zigbee Alliance, Internet Engineering Task Force (IETF), and the IEEE have worked on defining network architectures for IEEE 802.15.4. We briefly mention these efforts here and will go into more detail on the IETF-defined network shortly. Zigbee is inherently mesh based. The IETF has worked to define how IPv6 maps to IEEE 802.15.4 frames. The challenge has been to map IPv6 packets and addresses to the much smaller IEEE 802.15.4 frames that hold a smaller address. A technique known as header compression is used to map required information from the Internet Protocol (IP) header to the IEEE 802.15.4 frame header. IEEE 802.15.5 defines a mesh network architecture for IEEE 802.15.4 radios.

The term Zigbee refers to the waggle dance of honey bees after their return to the beehive. Recall that the function of the network layer is to utilize the lower layers of the protocol stack to enable multiple-hop paths through the network, namely routing. Since Zigbee is built on IEEE 802.15.4, its function is to control the MAC commands to route frames to their correct destination. Zigbee uses a simplified version of ad hoc on-demand distance-vector (AODV) routing (Ondrej *et al.*, 2006). The "on-demand" component of AODV means that routes are only determined when needed; all possible routes are not pre-established. This reduces the resources required to pre-establish routes at the cost of delay in establishing routes when they are needed. The "distance-vector" component of AODV means that it falls into the class of distance-vector protocols. This means that link changes are propagated to nearest neighbors rather than being immediately broadcast to all nodes. Again, this reduces overhead at the expense of potential delay in the actual network state, quickly reaching all nodes. In order to find a destination device, an AODV node broadcasts a route request to its immediate neighbors. Neighbor nodes repeat the process by broadcasting to their neighbors, and the process repeats until the destination is found. The destination node sends a response via unicast, rather than broadcast, along the lowest cost, or optimal, path back to the source. The source can then update its routing table with the next hop and cost to reach the given destination. Thus, AODV routing supports the IEEE 802.15.4 goals of reduced complexity and energy minimization at the expense of potential latency.

7.4.5 *Introduction to IEEE 802.15.4g Smart Utility Networks*

There are competing design requirements in the AMI versus DA networks. Specifically, AMI is designed for low energy consumption and a large number of users while tolerating high latency. The DA network has the exact opposite requirement: low latency is a primary requirement and there are fewer nodes to manage. AMI network routing (IEEE 802.15.4g) forms a tree with the collector as the root and "routers" with longer range transmission to forward data over longer distances. The AMI network allows a significant amount of time for the tree to reform when communication goes down; taking time to keep the tree optimal is more important in the AMI design philosophy than getting messages to their destination quickly. From a network standpoint, running DA applications directly over AMI will suffer high latency, unless DA messages override the AMI protocol.

IEEE 802.15.4g extends and modifies IEEE 802.15.4 in several different ways. The first is that IEEE 802.15.4g incorporates three different physical layers, rather than one, in an effort to increase flexibility and target more markets. There is a multirate-frequency-shift keying (MR-FSK) physical layer included for its transmission power efficiency because of the constant envelope of its signal. The multirate orthogonal frequency-division multiplexing (MR-OFDM) physical layer is included for higher data rate in channels that exhibit frequency-selective fading. Finally, the multirate-offset quadrature phase-shift keying (MR-OQPSK) physical layer is similar to the original IEEE 802.15.4 standard, which was designed for lower cost and easier implementation. Just as the number of types of physical layers has been extended in IEEE 802.15.4g, so too has the number of channels been extended to 12 frequency ranges. These ranges cover spectrum that is specific for use in Japan, China, and Korea, as well as in the USA.

The remainder of the IEEE 802.15.4g protocol deals with how to manage these extensions. In particular, the channel frequency ranges overlap with existing standards such as variations of IEEE 802.11 and other versions of IEEE 802.15, such as IEEE 802.15.1, IEEE 802.15.3, and IEEE 802.15.4c and d. The solution to mitigate the overlap and converge on the best common physical layer is to include the notion of a predefined common physical layer allowing IEEE 802.15.4g to detect other potential physical layers. The IEEE 802.15.4g devices can then use the common physical layer to agree to switch to a different physical layer and channel to avoid potential interference. The common predefined physical layer is called the common signaling mode (CSM). The CSM is a binary frequency-shift keying (BFSK) signaling scheme with a data rate of 50 kbps.

7.4.6 *Introduction to 6LoWPAN*

Wireless sensor networks, such as those implemented using IEEE 802.15.4 and its variants, were initially thought to be antithetical to supporting IP. This is because such sensor networks were targeted to comprise extremely large numbers of small, low-power nodes. IP was thought to be too heavyweight to operate over such constrained devices. It was assumed that proprietary protocols would have to be developed; Zigbee, mentioned earlier, is an example of one such proprietary protocol. 6LoWPAN ports IP over wireless sensor networks by adding an adaptation layer that bridges the heavyweight IP over the constrained sensor network devices. In a nutshell, 6LoWPAN accomplishes this by implementing so-called "header compression" to reduce overhead, by implementing fragmentation to support the IPv6 maximum transmission unit (MTU) requirement, and by implementing support for layer-two forwarding to transport the IPv6 datagram over multiple radio hops.

Of course, by porting IP over wireless sensor networks, it provides the ability for sensor networks to communicate directly with the Internet through routers, which simplifies the overall network and increases robustness as opposed to requiring an IPv6-to-Zigbee gateway that may need to maintain state. 6LoWPAN is specific to IP over IEEE 802.15.4. Recall that IPv6 extends the IPv4 address from 32 to 168 bits, fulfilling the need for many more addresses required to allocate unique addresses to the anticipated large number of sensor nodes. To meet a requirement that is contradictory to wireless sensor networks, IPv6 extends the MTU size in order to allow greater efficiency of anticipated higher bandwidths in the future. However, for the wireless sensor network the bandwidths are lower and require a smaller packet size.

Header compression is achieved, in general, by a concept the designers were probably not aware of, but comes from a concept derived from Kolmogorov complexity and active networking. Basically, this is the ability to replace the transmission of explicit information with additional processing to derive the information common to both sender and receiver. Simply put, information can be derived, via computation, from a common context or derived from information already transmitted in lower layers. Fragmentation is quite simple; IPv6 packets are fragmented into smaller chunks so that they can fit into the smaller IEEE 802.15.4 frames. Wireless sensor nodes are capable of forwarding data without requiring the overhead of implementing an IP router. Thus, IPv6 fragmented packets need to be routed properly through the wireless sensor network at the link level. 6LoWPAN does this by including lower layer address information within the adaptation layer packet. For header compression, redundant information in the 6LoWPAN adaptation layer, IP network, and IP transport-layer header information can be removed and the required information represented in a few bytes. While there is much more detail regarding other potentially useful protocols that can be discussed, the goal of this book is to provide fundamental and intuitive understanding of power systems and communications, not to be a handbook comprised of the details for every related protocol; that would be too tedious for the reader and take us too far from power systems. Next, let us move up the protocol stack and consider network and routing issues.

7.4.7 *Introduction to Ripple Routing Protocol and Routing over Low-Power and Lossy Networks*

Let us begin by getting the terminology straight. As usual in networking, there is an almost purposeful attempt to promote an alphabet soup of acronyms that can be redundant, sometimes pointless, and almost always confusing. While 6LoWPAN was the adaptation layer allowing Internet Protocol packets to reside on wireless sensor networks, RPL and routing over low-power and lossy networks (ROLL) focus on the routing of IP packets over low-power and lossy networks. ROLL is also the name of the working group that reached consensus on the protocol. On the other hand, RPL stands for the "ripple" routing protocol, an IP routing protocol designed to operate over any IP-supported physical layer such as IEEE 802.15.4 and its variants, as well as WiFi, power line carrier, or any other. Note that "ripple" here should not be confused with a much older ripple line carrier protocol developed in the early 1900s for power line communication.

The low-power and lossy network, such as that implemented via IEEE 802.15.4g, places severe constraints on the routing protocol. As previously mentioned, IEEE 802.15.4g smart utility networks are comprised of low-power devices communicating at a low data rate. The maximum transmission power for such devices is set to be 20 dBm (100 mW) and they can communicate at a maximum data rate of 250 kbps. Because the devices are attempting to save power, transmission range is limited and the duty cycle, the proportion of time the device is actively transmitting or receiving, is also limited. A typical goal is to be able to run for 20 years on a single battery. The duty cycle may be less than 1%; thus, the routing protocol needs to have as little overhead as possible and must converge to the correct route quickly. In addition, the focus is on wireless sensor networks with hundreds of thousands of nodes and requiring high reliability.

AODV has already been mentioned as a part of the proprietary Zigbee upper layer protocols. Instead of leveraging an existing routing protocol, as Zigbee did with AODV, RPL was developed. Like AODV, RPL is a distance-vector protocol that specifies how wireless sensor network nodes, such as those using 6LoWPAN over IEEE 802.15.4 or IEEE 802.15.4g will form routes. RPL is also considered a form of gradient routing.

As the number of nodes in a large-scale sensor network increases, the typically used ad hoc routing protocols, such as AODV, become burdensome, introducing significant overhead to the routing operation of the network. Sensor networks, by their nature, tend to involve large numbers of nodes reporting information to a central location. Thus, a tree structure, with sensors at the edges and a centralized collection of the information at the root, becomes a natural routing architecture. Instead of "tree," the fancier term "directed acyclic graph (DAG)" is used; this is simply a graph with directed edges (that is, edges that point in a specific direction), and the graph also has no loops or cycles (that is, there is no instance of a path that leaves a given a node and returns back to that same node). This implies that the graph must form a tree. A newly joining sensor must find its place in the DAG; that is, it must find out where to attach to the DAG so that the network is both efficient and no loops are formed. The network does this by assigning each node a "rank"; the rank must be higher than any of the parent nodes' ranks to which the newly joining node attaches. Thus, if the root node is rank zero, then the rank increases the further one moves from the root. The rank is roughly the distance of a node from the central, root node. Because a tree-based routing protocol is meant to improve upon existing ad hoc routing protocols and be applied to large-scale, low-power lossy networks, the typical goals of a routing protocol must be improved upon. These goals are to minimize the amount of state required so that constrained memory can hold all the information required as network size increases. The amount of state required should increase less than linearly with the size of the network. Additionally, overhead traffic must be minimized. For example, the creation or loss of a link should not require a broadcast to the entire network. Also, the routing operation itself should generate packets at a rate less than the rate at which data packets are generated. In other words, routing overhead should be bounded by the data traffic.

RPL attempts to meet the requirements just discussed utilizing two main components: (1) a destination-oriented directed acyclic graph (DODAG), basically a tree; (2) an objective function that defines how route costs are determined; that is, what is to be optimized when choosing a path. The root of the DODAG is called the LLN border router (LBR). Typically, all data flows from nodes in the tree toward this root node.

Construction of the DODAG begins when the LBR, the root of the tree, transmits a DODAG information object (DIO). This message contains: (1) the identifier for the LBR root node; (2) the rank of the node transmitting the message and a parameter called the "minimum-rank increase," both used to help the receiving node compute its own rank in the DODAG structure; (3) the objective function, which enables the receiving node to translate the metrics to be optimized into its rank.

The DODAG maintains an acyclic tree structure by requiring that nodes that join the network be of higher rank than their parent. A node's rank is thus related to its path distance from the root. Before joining the network a node listens to its neighbors' DIO messages and determines its rank by selecting which node will be its preferred parent. A "step" value is then added to the parent's rank. The step value is based upon the joining-node's path cost to reach its parent. A node must ensure that it has a higher rank than its parent's at all times. If a node's rank changes, then it must eliminate parents whose values are higher relative to the current node's rank. When a node has a packet to send, it then checks the actual path costs of its parents to find the optimal route to the LBR root.

As mentioned earlier, routing overhead traffic must be minimized. In order to keep the network reasonably stable given the possibility of an environment in which there are rapid link changes, a timer is used to slow the generation of update messages. This is known as the "trickle timer." The trickle timer dampens the rate at which DIO messages are generated and propagated. The trickle timer is controlled by a time interval I and redundancy counter C. I can range from I_{min} to I_{max}. The timer can fire randomly between $I/2$ and I. When the timer fires, C is reset to zero. However, until then, C increases by one each time it sees the same DIO message from a parent node. A DIO message is called "consistent" if it has the same rank and routing cost as the previous message. If C is below a threshold value and the timer fires, then I is doubled, thus taking longer before the timer will fire again. However, if a node receives a DIO message from a parent that is inconsistent (that is, with a different rank or route value than one previously received from the parent), then the node immediately takes action by resetting its trickle timer. The timer is reset by setting I back to its minimal value I_{min} and the timer is restarted. In this manner, messages are propagated quickly at the time a change occurs and at a rate that decreases as time progresses after the change.

After joining a DODAG, a node may change its rank. The key to changing rank is to avoid routing loops in the process. In the case of rank decrease, which is essentially moving closer to the root of the DODAG, a node simply has to ensure that it removes any parent nodes that are greater than its new, smaller rank. However, when increasing rank, a node needs to be careful if it also attempts to add new parent nodes. Increasing rank logically moves the node farther from the root and allows more possible nodes to become parents because more nodes will have a lower rank; that is, more nodes will be closer to the root of the DODAG. However, care needs to be taken when adding a new parent in this situation because the new parent may have been a child of the node whose rank is increasing. In other words, the new parent may also still be connected as a child. Any situation in which a node is both a parent and child implies there is a path from the node back to itself, which means that a loop exists. Loops must never exist in routing protocols, otherwise packets will travel forever around such loops; the loop becomes an absorbing state for packets. This illustrates the type of issues addressed by network and routing protocols in general. The next section looks at the well known 802.11 and how its characteristics impact applications in smart grid.

7.5 IEEE 802.11

The IEEE 802.11 protocol is an old and now nearly ubiquitous wireless standard used in developing LANs for devices ranging from laptops and wireless routers to cell phones and MP3 players. When protocols for wireless technology were first considered, it was assumed that the wireless physical layer would simply be an implementation of an existing protocol, such as Ethernet (IEEE 802.3). Ethernet is a wire-line protocol and uses carrier-sense multiple-access with collision detection (CSMA-CD). In IEEE 802.3, nodes transmit immediately when they have data to send and sense the line for possible collision with other nodes that may be transmitting at the same time. If a collision is detected, the node waits a random amount of time before attempting to retransmit the data to avoid sending at the same time as the other colliding node(s). The problem was that implementing this type of physical layer in a wireless environment would not work well due to the rapid attenuation, or reduction, in signal power, over relatively short distances. Collisions simply would not be detected. The IEEE 802.4 token-bus approach was also considered as a means to implement the wireless physical layer. This approach requires passing a single token from one node to another; only the node that holds the token may transmit. This avoids collision, but requires a complex implementation to handle the token; that is, to determine which node receives the token next, handling the case when the token disappears, or the case of multiple tokens. Thus, this approach was abandoned and a completely new approach had to be developed for the wireless physical layer.

IEEE 802.11 provides frequency-hopping spread-spectrum (FHSS) and DSSS in the unlicensed 2.4 GHz frequency band at 1 Mbps with an option for 2 Mbps. Similar to IEEE 802.3, IEEE 802.11 uses a listen-before-talk mechanism, known as a distribution coordination function (DCF). Unlike IEEE 802.3, it uses a collision-avoidance technique instead of collision detection. Since collisions cannot be detected in IEEE 802.11, a random interval of time occurs *before* transmission in the hope of avoiding collision instead of taking action after it occurs. The original IEEE 802.11 had a contention-free period with a point coordination function (PCF) in which a single node was appointed to poll nodes for transmission to guarantee transmission without collision. Hidden nodes caused such systems to fail and the contention-free period concept was not implemented.

IEEE 802.11 devices did not always interoperate due to different implementations of security mechanisms known as wired equivalent protection (WEP) and required a certification program that resulted in the WiFi Alliance in 2003 to enable full compatibility for the market. Since then, IEEE 802.11 has exploded in usage. Its success has led to a large number of amendments to the standard in order to extend the protocol in various ways to meet market needs. One of the obvious limitations of IEEE 802.11 for use in power systems is its relatively short range. Its range is limited to roughly 70 meters indoors and 250 meters outdoors (IEEE 802.11n). This is why the term "local" is in LAN. It becomes applicable to the power grid for local interconnectivity, such as connecting meters to a longer range network or perhaps to interconnect devices within a relatively small area. IEEE 802.11s addresses the range issue by allowing data to hop from LAN to LAN over longer distances, thus forming an interconnection of LANs. Another issue with IEEE 802.11 is that it was not designed to be low power or handle a high density of devices. IEEE 802.15.4 and IEEE 802.15.4g, as mentioned previously, address these issues. However, there has been effort to extend IEEE 802.11 over a longer range, making it feasible for use in DA; for example, IEEE 802.11y and Super WiFi.

7.6 Summary

This chapter covered a key element of smart grid DSM: DR. DR, as would any DSM system, requires a form of direct access to each and every consumer. If that is not challenging enough, it should also be bidirectional, enabling the consumer to receive prices and respond with choices. It could also allow the utility to control users' loads. Some of the many different communication technologies that have been proposed for this AMI were reviewed. The main take-away from this chapter should not be the details of any particular communication technology, but rather the more fundamental properties that they embody. Primary among these are the ability to leverage scale, to be able to efficiently aggregate data from large numbers of consumers. This includes the ability to utilize large size to increase reliability. Along these lines, the notion of game theory also played a role. Specifically, the motivation for each consumer to allow their communication node to forward information for themselves and on behalf of others was explored.

The next chapter covers DG and transmission. New power-electronic devices are changing the nature of DG and transmission, and also changing the corresponding requirements for communication. Although perhaps not obvious at first, distributed power generation and the power transmission system are related. Distributed generators should be capable of sharing power in a manner similar to the manner in which large, centralized generators use the transmission system to pool power resources. As distributed generation and microgrids become widely-deployed, the need for the classical, bulk power transmission system may be reduced. The technology evolution of both systems should be considered together. Control of distributed generation systems is crucial to understanding microgrid communication requirements. Also, new energy storage mechanisms and power electronics are being explored to help mitigate variability in renewable and distributed generation. Thus, control and communication requirements of power electronics and emerging energy storage systems are reviewed. From the transmission perspective, flexible alternating-current transmission systems (FACTS) and high-voltage direct-current (HVDC) transmission systems are explained. The ultimate evolution in power transmission technology is wireless—namely, high power, wide-area wireless power transmission. This fascinating topic and it implications for communication is explored in detail.

7.7 Exercises

Exercise 7.1 Advanced Metering Infrastructure

1. What are the main goals of the AMI?
2. What are the competing physical communication layers for AMI?
3. What are the typical interfaces that a power meter can have?
4. What are the key criteria for successful communication in the AMI?

Exercise 7.2 Early Power Line Carrier and Meter Reading

1. What is the difference between meter reading and the AMI?
2. What served as the first meter reading communication infrastructure?

Exercise 7.3 AMI and Protocols

1. What is cloud computing?
2. How is cloud computing being suggested for use in AMI?

Exercise 7.4 AMI and Demand-Response

1. What is DR? Define it in terms of supply, demand, and utility.
2. What role does scheduling play in DR? How will DR impact the diversity of duty cycles within the grid?
3. What are the similarities and differences between communication packet scheduling and energy usage scheduling?
4. Is there an interaction between communication packet scheduling and energy scheduling that has an impact on DR performance?

Exercise 7.5 ROLL

1. What is ROLL and what functionality does it provide?
2. What is the DAG algorithm and how is it used in ROLL?
3. What are the advantages of the DAG algorithm?
4. Where is ROLL/RPL best used within the smart grid and why?

Exercise 7.6 Electric Vehicles and Demand-Response

1. How could electric vehicles impact DR in a unique manner?
2. What is vehicle-to-grid communication? How do its requirements differ from smart grid communications requirements in general?

Exercise 7.7 Demand-Response Algorithms

1. What is the trade-off between electric power price elasticity and cost of the communication infrastructure? Illustrate using a DR curve and a geometrical argument.
2. Why could it be beneficial for your smart meter to operate in a mesh network supporting communication of other consumers' information packets (cooperative communication)?

Exercise 7.8 Meter Reading and Compression

1. What is the simple, main idea underlying compressive sensing?
2. How can compressive sensing be combined with MAC for meter reading?

Exercise 7.9 Demand-Response Effectiveness

1. How can the effectiveness of DR programs be measured?
2. How will the price elasticity of electricity play a role in DR effectiveness?

Exercise 7.10 Demand-Response and Control

1. Where should control of power in response to demand signals take place: at the generator, in the transmission system, in the distribution system, or among consumers, or all in of these locations? Should the DR control system be centralized or distributed? Why?
2. What are the implications of the answers to the previous question on the communication infrastructure?

Exercise 7.11 AMI and Cybersecurity

Privacy issues are of interest to some with regard to the use of smart meters and DR because the success of DR programs and customer energy-saving programs depend upon an accurate knowledge of energy profiles for devices and their usage.

1. How much information can be inferred from power usage information? What assumptions or ancillary data are required to make such inferences?
2. How can smart meter cybersecurity and privacy be enhanced?

8

Distributed Generation and Transmission

Electric power is everywhere present in unlimited quantities and can drive the world's machinery without the need of coal, oil, gas, or any other of the common fuels.

—Nikola Tesla

8.1 Introduction

This is the third chapter in Part Two, *Communication and Networking: The Enabler*, focusing upon distributed power generation and transmission. Note that these are, in a sense, anti-correlated components because as DG becomes ubiquitous the need for a classical power transmission system disappears; there is no need for long-distance bulk-power transfer if power sources are distributed within close proximity. However, there may always be a use for a highly connected power network in order to increase reliability and draw power from distant power sources in an emergency.

The chapter begins by discussing DG and the issues of integrating a growing number of highly variable, relatively small, renewable generators with the power grid. The question to consider throughout this section is what type of communication is required and how does it best integrate with DG as well as the transmission system. Note that this chapter considers DG to reside in the distribution system either with, or relatively near, the consumer. This means that the distribution system will need to become more like today's transmission system in terms of coupling power from generation facilities and handling bidirectional power flow. Thus, this chapter takes the unique position of covering both DG and transmission together. A natural solution to smoothen the variable rate of renewable energy generation is to include energy storage with renewable energy generators so that, during periods of relatively high energy output, excess energy can be stored to be released during periods of low energy output. Section 2.4 reviewed common energy-storage techniques. In Section 8.2 of this chapter, energy storage specifically related to distributed and renewable generation is reviewed. Section 8.3 covers the electric power transmission system in the smart grid. This includes an introduction

Smart Grid: Communication-Enabled Intelligence for the Electric Power Grid, First Edition. Stephen F. Bush.

to FACTSs, synchrophasors used in the transmission system, and wide-area monitoring and control. A FACTS enables a high degree of flexible control of power flows over transmission links, which enables greater optimization of power flows to increase efficiency. Synchrophasors are covered in detail in Section 13.2; however, they are introduced here as they pertain to the transmission system. Synchrophasors allow for a greater degree of observability into the state of the transmission system. All of these components point to the need for greater monitoring and control; because transmission systems tend to be very long, from tens to hundreds of kilometers, they require what is often vaguely classified as "wide-area" communication for monitoring and control. The transport of power does not require a physical line; wireless transmission of power takes place frequently. From low-power transmission for the purpose of communication to high-power microwave communication links. Section 8.4 covers this aspect of power transmission. Ideally, a fully tetherless power grid would allow for complete flexibility and mobility. One would no longer need to desperately search for a wall socket when running out of power or have to frantically find the nearest charging station for your electric vehicle. A longer term vision for wireless power transmission is discussed in Section 15.6. As mentioned previously, wide-area monitoring is a key component of smart grid communication for the transmission system, discussed in more detail in Section 8.5. Controlling the transmission system through a communication network is challenging owing to the variability of packet delay through a long-distance network. A classic control example is adjusting the temperature of the water in your shower given a long delay for the water to reach the current temperature setting. One may initially be too cold and turn the temperature far toward the warmer direction in order to increase the temperature. After a time, the temperature becomes too hot and one turns the temperature toward the cooler direction to compensate. However, over time, the temperature settles to a point that is too cold. Because of the delay in reaching the final temperature, one tends to continuously overcompensate. If the delay is constant, one can eventually learn how long to wait for a final temperature before readjusting the temperature. However, if the sensing or communication delay is variable, this makes the problem even more challenging. In a communication network, packets may take a variable amount of time to traverse the network; they may experience congestion or loss, in which case packets may need to be resent. Control over such a communication network, in which there is a variable packet delay, is known as networked control and is covered in Section 8.6. The chapter ends with a brief summary of key points and exercises for the reader.

8.2 Distributed Generation

The term "DG" and related terms such as "dispersed generation," "embedded generation," "decentralized generation," "distributed utility," "distributed capacity," and many others have been widely used, but unfortunately imprecisely defined. The notion is of a wide variety of relatively small, spatially dispersed generators working to supply power either as a supplement to current centralized generation or as a sole source of power. However, questions arise as to what capacity the generators must be to qualify as a component of a DG system versus a centralized generation system, where they should be located (for example, customer premises, distribution system, or on the generation side of a transmission system), their mode of operation (for example, when they operate and what rules they must follow), and their environmental impact (that is, whether they must be renewable or meet a minimum standard regarding environmental impact). A fundamental and lasting definition of DG has been lacking. As a specific example, consider whether a large wind farm generating thousands of megawatts whose power is transported hundreds of miles over the transmission system should be considered DG. The precise definition of a DG system can become challenging, but it should be addressed before continuing so that we know precisely what we are covering in this chapter.

Early work by (Ackermann *et al.*, 2001) provides a nice definition that will be used as a default throughout this book. Nine different dimensions of DG are explored in the search for a definition: purpose, location, power rating of the generators, power-delivery area, technology, environmental impact, mode of operation, ownership of the distributed generator, and penetration of DG. It is instructive to review each of these aspects. The purpose of DG is defined to provide active power. Recall that reactive power, often required by inductive loads in the power grid, has to be supplied as well. A distributed generator is not required to be a provider of reactive power. The definition of the location of the distributed generator is crucial. They may, in theory, be placed anywhere within the grid; however, the definition limits them to two locations: the customer premises and the distribution system. The reasoning is that one of the benefits of DG is that it can be located near the load it is supplying. However, this definition then requires a clear distinction between the distribution system and the transmission system, which is in reality not always clearly defined. Some distribution system voltages are higher than some transmission voltages, so voltage level is not a perfect metric. The definition falls back onto a legal/market definition of which portion of the grid is the distribution system and which is the transmission system. A wind farm that is directly connected to the transmission system is not considered a DG system.

Many people tend to consider a DG system based upon the power rating of the generator. The notion is that a distributed generator should be smaller than a centralized generator. However, the argument is made against any standard definition based upon generator power rating. The reason is that the rating of a distributed generator should be made relative to the power demand of the consumers or distribution system being supplied. The scale can vary significantly. In addition, the technology to build small-volume, but large-capacity generators is progressing rapidly. A suggestion based upon current technology is the following: a micro-distributed generator ranges from 1 W to 5 kW, a small distributed generator ranges from 5 kW to 5 MW, a medium distributed generator ranges from 5 MW to 50 MW, and a large distributed generator is in the range of 50 MW to 300 MW. But again, these values are only to provide a sense of capacity ranges given current technology and will change over time. The suggestion is to not base DG upon any specific capacity values.

Just as power rating is not part of the definition, the area of power delivery is excluded from the definition as well. The definition simply says that DG must occur on the customer premises or within the distribution system, as previously stated. Also, the technology for distributed generators varies widely, with dozens of different types of generators. The specific type of generator is thus not considered to be a part of the definition. Also, the impact on the environment is not considered part of the definition of DG. Any evaluation of the impact of the generator upon the environment, even for supposedly clean, renewable generators, results in a complex analysis that does yield a crisp, clear definition. The mode of operation of DG is also considered irrelevant to the definition. This is because there is significant variation in regulation governing the operation of DG so that any clear-cut distinction in operation, such as pricing or scheduling of power, would be lacking. A similar argument applies to ownership. Some would argue that the definition of DG should require ownership by either a consumer, small group of consumers, or a "small" business entity. However, the reality is that ownership varies significantly and would not yield a clear benefit toward clarifying the definition. The level of penetration required to meet the definition of DG is also irrelevant. Some might argue that, for generation to be considered truly distributed, the power grid should no longer have centralized generation or a transmission system, while others believe that it will never exceed a small fraction of total power production. Given the high degree of uncertainty, this metric also does not help support a solid definition. Thus, the final definition of DG is simply a source of active power directly connected to either the customer premises or the distribution system.

One of the problems with distributed generators in the distribution network is that the distribution system is typically unidirectional; current flows from the feeders in the substation to consumers (Dugan and McDermott, 2002). The protection system has been optimized at lowest cost to operate with current flow in one direction. Note that this is not the case for the transmission system. The transmission system typically interconnects generation plants and can take on a mesh topology. Thus, it is designed to handle bidirectional current flow with protection at both ends of the transmission line. On the other hand, the distribution system primarily has a radial topology. This consists of a single power flow from a feeder to consumers, where protection resides near the feeder and at various points along the radial line.

Protective relays are often distance relays; they protect a power line over a given distance from the relay. This is accomplished by assuming the power line has a constant impedance, where impedance is $Z = V/I$. When a fault occurs (that is, a short circuit), the current will rise and the voltage will drop significantly, causing the impedance to decrease. The angle of the current with respect to the voltage will lag significantly after a fault occurs. When the impedance falls below a given threshold, a fault is assumed to exist and the relay is activated in order to isolate the current flow. The magnitude of the impedance threshold corresponds to the distance along the line that a fault may be detected. The problem with DG is that a generator on a radial branch will inject current into the power line. The current from the distributed generator will feed a downstream fault and keep the voltage elevated from the perspective of the upstream relay. Thus, the upstream relay may not detect a downstream fault that would normally be within its reach, recall that reach was explained in Section 4.2.2, without the distributed generator. The result is that while adding DG is intended to increase reliability of the power grid, it can actually reduce the reliability if the system is not designed properly. A solution has been to require that all distributed generators must disconnect from the grid upon the occurrence of a fault, allowing the normal protection mechanisms to operate until the fault clears. Then the generator may rejoin the distribution system.

Another problem scenario occurs when the distribution system depends too much upon the distributed generator to maintain voltage. The distributed generator must disconnect after a fault in order to allow the "normal" unidirectional protection to clear the fault. However, if demand is high enough that the distributed generator is required in order to restore the normal voltage level, then the system will never return to normal. This is because the utility will be waiting for the voltage level to reach normal levels before allowing the distributed generator to rejoin the system; however, the normal voltage will not be reached without the power that the distributed generator had been supplying. This means that the amount of DG allowed in the distribution system is limited.

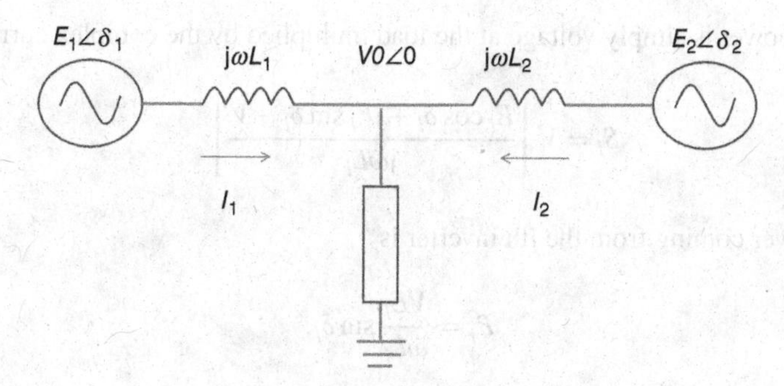

Figure 8.1 An inverter control scenario is illustrated in which two generators provide power to a single load represented by the grounded rectangle.

This is one reason why current limiters have been seen as a potential solution (Damsky *et al.*, 2003). A current limiter provides protection quite simply by limiting the current even in a fault condition. Thus, in the ideal case, there is no requirement for breakers, fuses, or reclosers and distributed generators would not be forced to disconnect during a fault.

DG control communication involves the control of the distributed generator inverters, which convert direct current into alternating current, that provide power to an electric power grid in order to drive the loads on the grid (Ci *et al.*, 2012; Marwali *et al.*, 2004). A simple case involves two distributed generators driving a single load, as shown in Figure 8.1. Recall from discussions regarding the power flow equation in Section 3.3.4 that active power is controlled by voltage angle and reactive power is controlled by the voltage magnitude of the generators, represented by δ_i and E_i respectively, where i is distributed generator 1 or 2 in this simple example. The impedance of the corresponding input lines are $j\omega L_i$, where ω is the frequency and L is the inductance of the line. Thus, by controlling E_i and δ_i, power to the load can be controlled as necessary to account for demand as well as any impedance in the power lines.

Typically, control can be implemented without explicit communication directly at the generators by having the generators sense the voltage magnitude and phase and adjusting to maintain a target voltage magnitude and phase. This is known as droop control, and it enables each distributed generator to provide a proportion of the demand. However, another approach is to explicitly allow distributed generators to share information with one another in order to control their inverters. First, let us consider control in more detail.

Equation 8.1 shows the complex power generated by each inverter for the load where $i = 1, 2$ and I_i^* is the complex conjugate of inverter i's current:

$$S_i = P_i + jQ_i = V \cdot I_i^*. \tag{8.1}$$

Taking the power line impedance into account, Equation 8.2 solves for the complex current from each distributed generator as seen by the load:

$$I_i^* = \left[\frac{E_i \cos\delta_i + jE_i \sin\delta_i - V}{j\omega L_i}\right]^*. \tag{8.2}$$

The complex power is simply voltage at the load multiplied by the complex current:

$$S_i = V\left[\frac{E_i\cos\delta_i + \mathrm{j}E_i\sin\delta_i - V}{\mathrm{j}\omega L_i}\right]^*. \tag{8.3}$$

The active power coming from the ith inverter is

$$P_i = \frac{VE_i}{\omega L_i}\sin\delta_i. \tag{8.4}$$

The reactive power coming from the ith inverter is

$$Q_i = \frac{VE_i\cos\delta_i - V^2}{\omega L_i}. \tag{8.5}$$

Figure 8.2 illustrates the complex, active, and reactive inverter power for each inverter.

Now, instead of droop control, consider explicit communication of each inverter's active and reactive power information to each other. Thus, each inverter knows its own voltage magnitude and angle along with that of all the other inverters from all other distributed generators feeding the same load. The inverter can plug this information into its own control algorithm and solve the equations for active and reactive power explicitly to determine its own optimal voltage magnitude and angle. This allows us to examine how communication latency impacts the system.

The Smith predictor provides an interesting and intuitive approach to handling delay in a control system. A typical closed-loop transfer function, whose derivation will be explained later in Section 8.6, is

$$H(z) = \frac{C(z)G(z)}{1 + C(z)G(z)}. \tag{8.6}$$

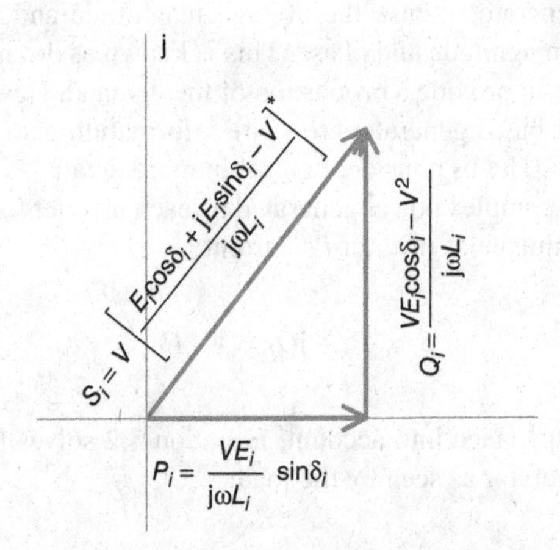

Figure 8.2 The complex power S_i for the inverter control illustrated in Figure 8.1 is shown as a function of the real power P_i and reactive power Q_i.

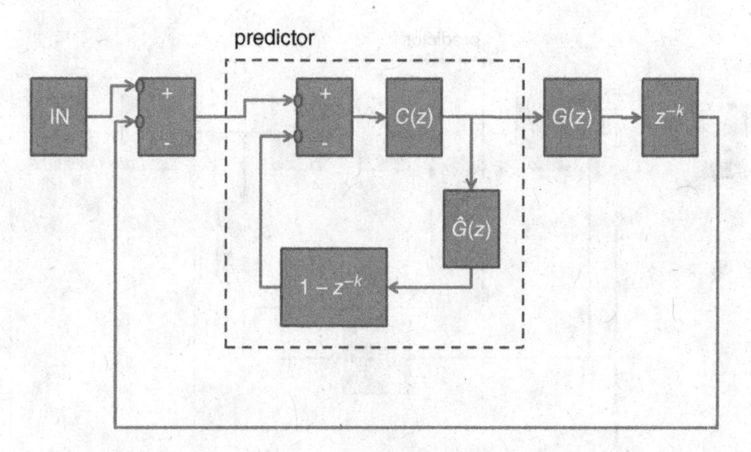

Figure 8.3 The Smith predictor handles delay in a control system and is modeled within the dashed box.

Now, suppose that there is a time delay in the feedback signal. This is represented by z^{-k} for a delay of length k. This means that the signal from the plant is $G(z)z^{-k}$; that is, its value is delayed by k time units. The goal is to design a controller with the plant delay $G(z)z^{-k}$ such that the transfer function is $H(z)z^{-k}$. This equality is shown by

$$\frac{\bar{C}Gz^{-k}}{1+\bar{C}Gz^{-k}} = z^{-k}\frac{CG}{1+CG}. \tag{8.7}$$

The solution for the corresponding controller is

$$\bar{C} = \frac{C}{1+CG(1-z^{-k})} \tag{8.8}$$

The controller is implemented in Figure 8.3. Notice that there are two control loops: an outer loop that includes delay, which is a typical feedback control loop, and an inner loop that corrects for the delay. $\hat{G}$ is used to indicate a model approximation of the plant. During the k time units for which there is no feedback due to delay in the outer loop, the inner is operationally providing simulated feedback. A simplified version of the implementation is shown in Figure 8.4. Here, it is clear that the outer and inner loops are equivalent and thus cancel one another, allowing the controller to take proper action even with feedback delay. Thus, we can see that distributed generation involves an understanding of power systems, control, and communication. The next section explains power system concepts that facilitate analysis.

8.2.1 Distributed Control

Various transformations commonly used in power systems are used to facilitate the design of control algorithms interfacing distributed generators with the power grid; for example, in order to maintain proper synchronization. One of these transforms is the *dq*0-transformation. These are essentially projections of phasors onto axes with different characteristics. These projections can then provide a simpler frame of reference from which to develop solutions. These transformations are important to know for communication in the power grid for several reasons. First, they provide a more efficient manner of either representing or compressing power

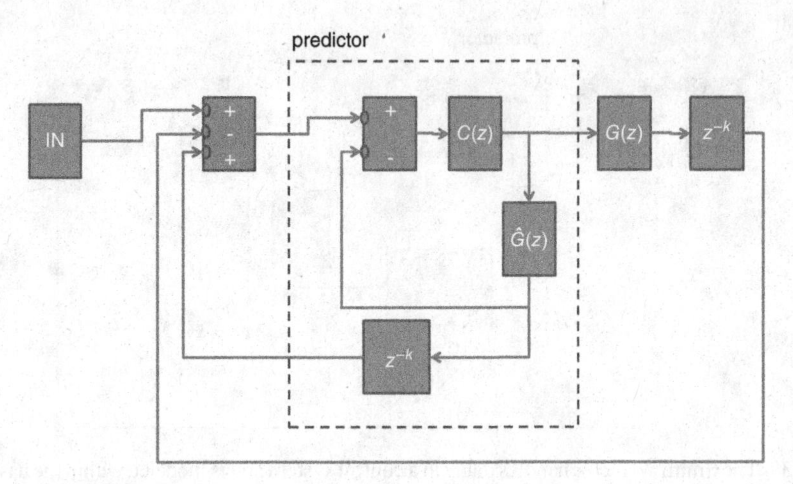

Figure 8.4 A simplification of the Smith predictor in which it becomes clear that the inner and outer loops are equivalent.

grid information to be transported. Second, just as they help the power engineer simplify design and development of control algorithms, they help the communication engineer become aware of how, when, and where communication is really needed and the corresponding communication performance requirements.

Consider the representation of periodic, sinusoidal voltage or current values as phasors in an *ABC*-phasor diagram. Recall that this shows the three-phase current as three phasors originating at zero and pointing 120° apart. A phasor assumes there is a constant angular velocity and voltage or current values are separated in phase from one another throughout their constant angular velocity. In this case, the *dq*0-transform values to be transformed. The *d*-axis is called the "direct" axis and the *q*-axis is called the quadrature axis, which, as its name implies, is 90° out of phase relative to the direct axis. The voltage or current phasor values are then projected onto these two axes, namely *d* and *q*. The 0 in *dq*0 is a constant value that adjusts for difference in magnitude; it has no phase component. An important point is that because the *dq* axes are rotating with the phasors and at the same angular velocity, the *dq*0-transform represents direct current, not alternating current. The mathematical transformation is

$$I_{dq0} = TI_{abc} = \sqrt{\frac{2}{3}} \begin{bmatrix} \cos(\theta) & \cos(\theta - \frac{2\pi}{3}) & \cos(\theta + \frac{2\pi}{3}) \\ -\sin(\theta) & -\sin(\theta - \frac{2\pi}{3}) & -\sin(\theta + \frac{2\pi}{3}) \\ \frac{\sqrt{2}}{2} & \frac{\sqrt{2}}{2} & \frac{\sqrt{2}}{2} \end{bmatrix} \begin{bmatrix} I_a \\ I_b \\ I_c \end{bmatrix}. \tag{8.9}$$

The inverse *dq*0-transform is

$$I_{abc} = T^{-1} I_{dq0} = \sqrt{\frac{2}{3}} \begin{bmatrix} \cos(\theta) & -\sin(\theta) & \frac{\sqrt{2}}{2} \\ \cos(\theta - \frac{2\pi}{3}) & -\sin(\theta - \frac{2\pi}{3}) & \frac{\sqrt{2}}{2} \\ \cos(\theta + \frac{2\pi}{3}) & -\sin(\theta + \frac{2\pi}{3}) & \frac{\sqrt{2}}{2} \end{bmatrix} \begin{bmatrix} I_d \\ I_q \\ I_0 \end{bmatrix}. \tag{8.10}$$

As previously mentioned, the $dq0$ transform is similar in concept. The $dq0$ transform is the projection of phase quantities onto a two-axis reference frame that is rotating at the same angular velocity; the $\alpha\beta\gamma$ transform is the projection of phase quantities onto a two-axis reference frame that is stationary, as shown by

$$I_{\alpha\beta\gamma} = TI_{abc} = \frac{2}{3}\begin{bmatrix} 1 & -\frac{1}{2} & -\frac{1}{2} \\ 0 & \frac{\sqrt{3}}{2} & -\frac{\sqrt{3}}{2} \\ \frac{1}{2} & \frac{1}{2} & \frac{1}{2} \end{bmatrix}\begin{bmatrix} I_a \\ I_b \\ I_c \end{bmatrix}. \tag{8.11}$$

The inverse $\alpha\beta\gamma$-transform is

$$I_{abc} = T^{-1}I_{\alpha\beta\gamma} = \begin{bmatrix} 1 & 0 & 1 \\ -\frac{1}{2} & \frac{\sqrt{3}}{2} & 1 \\ -\frac{1}{2} & -\frac{\sqrt{3}}{2} & 1 \end{bmatrix}\begin{bmatrix} I_\alpha \\ I_\beta \\ I_\gamma \end{bmatrix}. \tag{8.12}$$

What are some of the benefits of such transformations? Consider a balanced, three-phase alternating-current system. The $dq0$ transform changes this system into a simpler, direct-current system comprised of d and q components. The 0-component is zero if phases are balanced. This results in a representation that is simpler to work with and also requires less information to represent. Another advantage is that it turns out that the active and reactive power can be controlled independently by the d and q components.

Hopefully, knowledge of these fundamental techniques will help in understanding algorithms used in power systems, particularly for controlling distributed generators and their interface to the main power grid. This is because, generally speaking, there are two main control elements to a distributed generator: (1) the power input side of the generator, in which a controller is attempting to maximize the amount of power that is generated, and (2) the grid side of the distributed generator, which is attempting to convert the power into a from that is amenable to the power grid (Blaabjerg *et al.*, 2006). For example, the grid-side controller is handling control of active and reactive power transferred to the grid, power quality, and probably of most importance synchronization with the grid; that is, ensuring that the alternating current provided to the grid is in phase with the grid current. In addition, the grid-side controller can be even more grid-friendly by helping to maintain local voltage and frequency regulation. A significant concern is reclosing, or connecting to the grid, out of phase with the grid current, as illustrated in Figure 8.5. This can cause significant damage to all components involved, including electrical and mechanical components.

Inherently, direct-current distributed power generators, such as PV systems or fuel cells, require an inverter to convert direct current to alternating current. In these cases, grid-side control is primarily controlling the operation of the inverter (Timbus *et al.*, 2005). However, wind turbines and other types of generators with windings inherently generate alternating current and do not necessarily require an inverter to convert direct current to alternating current. However, it is often more efficient to use inverters with these types of distributed generators as well.

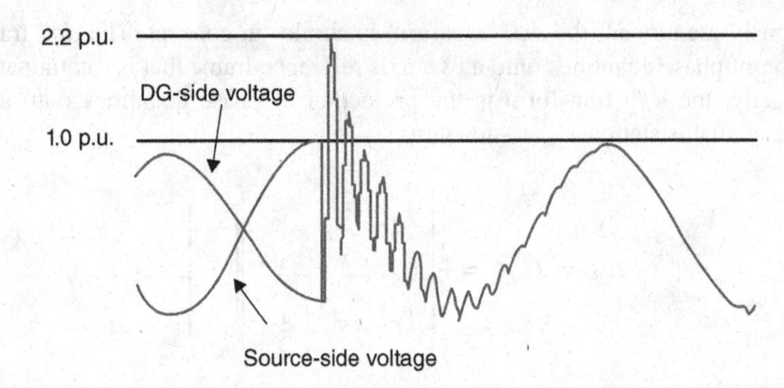

Figure 8.5 A distributed generator runs the risk of connecting or reconnecting to the power grid when it is out of phase with the power grid. The result can cause significant damage, as indicated in this plot of the voltage from both the distributed generator and the main power grid. Source: Walling, *et al.*, 2008. Reproduced with permission of IEEE.

It should be noted, that from a communication perspective, there is no significant, immediate requirement for either a large volume or long-distance communication in controlling the grid-side interface. Power system interface communication happens locally; that is, directly within the connection between the grid-side of the distributed generator and the power grid. Also, much of the communication is implicit or inferred; that is, physical values are sensed locally in order to make control decisions.

A typical grid-side control technique is to use two control loops: (1) a fast inner control loop and (2) a slower outer control loop. The fast, inner loop controls the grid current and the rapid changes related to it, such as power quality and protection. The slower, outer control loop controls voltage and manages the power flow, which has slower dynamics.

Three-phase alternating current from the power grid can be sensed and represented within the distributed generator control algorithm by the $dq0$ transform explained previously. It is the job of the distributed generator to track and maintain its output power to be in phase with the main power grid. This can be done via a proportional–integral controller acting upon the two direct currents from the $dq0$ transformation.

8.2.2 *Many Small Generators Working with the Grid*

While the goal of adding distributed generators to the power grid is to increase reliability by providing a backup source of power, improve efficiency by reducing the need to spin up expensive reserve generators, improve power quality, and to provide cleaner power generation by using renewable distributed power sources, there has been concern about DG causing the opposite effects if the system is not designed properly. In addition, there has been concern over grid stability as the number of distributed generators increases. As previously discussed, each distributed generator typically seeks to track the frequency of the main power grid, which is assumed to be well controlled by a large, centralized generator that has a massive amount of inertia in order to keep a steady phase. If power produced by DG were to become a significant portion of total power generated, the centralized generation facilities may not have the ability to set the beat, so to speak, and the system could become chaotic (Anwar *et al.*, 2011; Azmy and Erlich, 2005; Delille *et al.*, 2010; Huang, 2006; Marwali *et al.*, 2007; Reza *et al.*, 2004). In other words, each distributed generator would be tracking the main power grid phase, but there would no longer be a consistent phase to follow as each distributed generator began to add its own offset to the phase.

Simply setting a limit to the amount of DG to some proportion of overall power generation would not be a very efficient means of ensuring stability. Stability depends upon more subtle factors, such as the specific sizes of the distributed generators, the strength of the coupling between them, and the specific topology of power line interconnections. Communication used to aid stability can be thought of as additional, virtual power line connections between distributed generators. The problem regarding loss of grid synchronization due to a high degree of DG has been studied many different ways. We examine a very brief sketch of this technique, introduced in Section 6.4.4. The concept is an approximation, but it provides a tractable approach that also facilitates examining how communication fits into the solution in a fundamental way (Bush, 2013a). A simplified approach examines the problem as a set of coupled oscillators (Li and Han, 2011). The concept is to first create a graph with generators as nodes and connections between generators as edges. The swing equation describes the angle and acceleration of the generator rotor as a function of the mechanical force applied to, and the load experienced by, the generator; this characterization of the generator represents a node in the graph. The swing equation is discussed in more detail in Section 12.5.1.1. The swing equation is a second-order differential equation, so an adjustment is required to convert the problem to a first-order differential equation that can be represented by a graph Laplacian. Also, a Taylor series is used to approximate perturbations in rotor angles assuming they are relatively small; that is, generators operate near their stable point and do not deviate too far from it. Transmission lines can be modeled ideally by impedance and are represented by a graph edge. Generator interconnections with each other and with loads are described by a graph Laplacian matrix. Graph spectral techniques are applied to make general assumptions about whether disturbances in power flow will be amplified or dampened simply by the topology of the power network. The result can be reduced to examining the eigenvalues of the power grid network's graph Laplacian matrix. This approach is based upon Lyapunov stability; let us a take a minute to briefly provide some intuition behind the form of Lyapunov stability used in this approach.

As we can see, the analysis of stability of a complicated system can become intractable if all details are taken into account. The beauty of Lyapunov's approach is that stability can be proven without requiring detailed knowledge of the physical system, provided a proper Lyapunov function can be found. The notion is that energy from disturbances in a physical system, in our case distributed generators that may not be synchronized with another, may grow without bound and lead to an unstable system or dissipate until the system returns to an equilibrium. Lyapunov stability seeks to find the necessary and sufficient conditions for a system to be stable. The Lyapunov function is a key element in the approach and, in essence, represents the generalized energy of the system due to a disturbance; the goal is to determine the conditions under which it will dissipate and lead to stability. This occurs when the derivative of the Lyapunov function is negative.

Keeping the above in mind, consider a linear time-invariant system and assume the Lyapunov function is $V(x)$, where x is a state variable. We can choose the function to be positive definite, meaning that $V(x) > 0$, $x \neq 0$, and $V(0) = 0$, where we assume that 0 is the equilibrium point.

A stable system would require that the function be decreasing; therefore, its derivative is negative. Thus, we want the derivative of the function to be negative definite. We can represent these conditions in matrix form. A positive-definite matrix, represented by $P > 0$ has scalar values $x^{\mathrm{T}}Px > 0$ for all nonzero values of x. All its eigenvalues are positive and Equation 8.14 holds:

$$x^{\mathrm{T}}Px = x^{\mathrm{T}}(P^{1/2^{\mathrm{T}}}P^{1/2}x) = y^{\mathrm{T}}y, \tag{8.13}$$

$$y^{\mathrm{T}}y = ||y||^2 > 0. \tag{8.14}$$

Now consider the negative-definite matrix represented by $Q < 0$. All of its eigenvalues are negative. Also, if P, just discussed, is positive definite, then $-P$ is negative definite:

$$x^{\mathrm{T}}(-P)x = -x^{\mathrm{T}}(P^{1/2^{\mathrm{T}}}P^{1/2}x) = -y^{\mathrm{T}}y, \tag{8.15}$$

$$-y^{\mathrm{T}}y = -||y||^2 < 0. \tag{8.16}$$

Thus, for a negative-definite matrix, the eigenvalues are negative. As mentioned, this pertains to the derivative of the Lyapunov function. Thus, for a stable system, we want $\dot{V}(x(t)) = -x^{\mathrm{T}}Qx$ for Q positive definite. This tells us that given a positive-definite Lyapunov function $V(x)$, if the derivative is negative definite, then the system is asymptotically stable. Conversely, if the derivative is positive definite, then the system will be unstable.

Consider a linear time-invariant system of the form

$$\dot{x} = Ax. \tag{8.17}$$

The system is stable if and only if for any positive-definite matrix Q there exists a positive-definite symmetric solution P to the following Lyapunov equation:

$$A^{\mathrm{T}}P + PA = -Q. \tag{8.18}$$

This can be demonstrated simply by substitution from Equation 8.17 and matrix manipulation as follows in Equation 8.24:

$$V(x) = x^{\mathrm{T}}Px \tag{8.19}$$

$$\dot{V}(x) = \dot{x}^{\mathrm{T}}Px + x^{\mathrm{T}}P\dot{x} \tag{8.20}$$

$$= x^{\mathrm{T}}A^{\mathrm{T}}x + x^{\mathrm{T}}PAx \tag{8.21}$$

$$= x^{\mathrm{T}}[A^{\mathrm{T}}P + PA]x \tag{8.22}$$

$$= -x^{\mathrm{T}}Qx \tag{8.23}$$

$$A^{\mathrm{T}}P + PA = -Q. \tag{8.24}$$

Thus, we have a criterion for determining which DG power grid topologies will synchronize; that is, return to a stable state.

Adding communication to the above model requires the simplification of adding a communication term to the swing equation that represents a control command to reduce the difference in angle between the generators that are adjacent in the communication network. This is equivalent to adding additional entries in the graph Laplacian matrix for the communication links. The question can be asked, for a specific DG network, what are the minimum number of communication links, and where should they be placed, to ensure stability of the system? This can be answered easily following the same approach for the purely power-coupled DG network by finding the necessary and sufficient condition for synchronization based upon the largest eigenvalue.

8.2.3 Distributed Generation: Back to the Future

The history of power generation has been dominated by the synchronous machine; namely, the synchronous generator. The method by which the rotor was moved in the synchronous generator can vary greatly, ranging from coal, to hydro, to nuclear; a large, synchronous generator has always been the primary mechanism of power production. However, DG has been coming of age. These generators are fundamentally different. We discussed the induction generator in Section 2.2.4. There are also PV, microturbine, and fuel cell power generations sources, among many others to be discussed later.

Wind turbines have been in deployment since the 1980s. These inductive generators can only consume, not create, reactive power; adding capacitors can help to alleviate the problem. They also cannot control voltage or reactive power. They have difficulty controlling alternating current frequency and cannot start without power from the main grid. This means that wind turbines cannot be considered a replacement for traditional synchronous power. As long as the number of deployed wind turbines remains relatively small, they could be a viable option. Recent advances in power electronics, particularly power inverters, allow more control over reactive power, voltage, and frequency. Although the inversion process adds inefficiency, power loss is outweighed by the additional control it provides.

Microturbines are typically powered by natural gas and their rotors move at very high speed. They also use a two-step inversion process to control voltage, reactive power, and frequency. These units are commercially available in sizes that fit within residential basements. Adding combined heat and power, or co-generation, allows for greater efficiency. The concept is to utilize normally wasted heat from the generator in a productive manner.

PVs and fuel cells have no rotating parts; they produce direct current and thus require an inverter to connect to the grid. PV cells and fuel cells look to an electrical circuit like a battery. They are connected in series to increase the voltage generated. Sets of PV cells form PV modules and sets of fuel cells form fuel cell stacks with tens of volts. An inverter then converts the output to the required voltage and waveform.

With DG, there is a move away from the traditional approaches that had economy of scale. Purchasing one or a dozen PV cells or fuel cells, except for perhaps encouraging a small volume discount, does not reduce the marginal cost of power production unlike large wind farms or the concept of nanogrids, where massive amounts of nanogenerators, discussed in Section 15.3, are collectively producing larger amounts of power. The advantage of DG has to be found elsewhere. One advantage is that relatively small size and low cost allows generators to be placed when and where needed; long and costly power lines, with their associated power loss, are no longer required.

Another advantage that comes with their ability to be positioned where needed is that they can also serve to provide voltage support and reactive power. This could reduce the need for capacitors and voltage regulators, if the distributed generators are positioned and operated in such a manner as to accomplish this purpose for the utility. However, there is some evidence that quite the opposite could be true; DG could require the power grid to respond more quickly to changes due to the inherent variability of DG and thus add more wear and tear to existing tap changers for example.

DG, if implemented appropriately, could reduce the need for transmission and distribution, since power would be located where needed. This would reduce the use and expand the lifetime of transmission equipment as well as reduce the need for planned expansion of the transmission system. Of course, DG, as discussed earlier, significantly impacts protection mechanisms, particularly in the distribution system.

Another potential problem with DG, from the utilities' point of view, is that controlling and operating a large number of small, spatially dispersed generators with sporadic power output from a centralized location would require a tremendous amount of fast and reliable communication. Thus, utilities have treated DG more as a negative load, rather than real power generation. From the utility viewpoint, statistically varying distributed power output should help offset statistically varying loads. All the utility sees is (hopefully) a reduced load, ideally with fewer peaks as well.

Finally, DG should act to make the grid more resilient. Faults, whether natural or malicious, will be less likely to completely isolate customers from the grid. In the ideal case, DG would allow the grid to continue supplying power until main power is restored. However, as discussed earlier, this is a completely different way of operating the grid and opens up a host of issues related to islanding, including safety first and foremost, as well as liability, accounting, and control.

Another potential barrier to widespread DG from a cost standpoint is that, while it was mentioned earlier that DG has the advantage of reducing transmission and distribution usage because distributed generators can be positioned near the loads that need power, it is unclear, from a financial standpoint, how each of the organizations responsible for investment will view the potential benefits; for example, would the transmission and distribution entities have to make the investment while the generation entity reaps the reward? A similar problem exists for energy storage.

8.2.4 Photovoltaics

This section provides an introduction to the generation of electricity from light and considers some of the communication and computation require to make this conversion efficient. The mechanism is best explained by quantum mechanics. When packets of energy called photons are absorbed by a substance, the electrons in the substance may be excited enough to reach a higher energy level. This level may be high enough to allow them to reach what is known as the conduction band, allowing the electrons to form current in an electric circuit. The ability for photons to cause electrons to reach the energy level of the conduction band does not typically occur; instead, electrons, once excited by a photon will relax back to their initial state. Asymmetry is required in order to keep the electron in the conduction band. When enough electrons reach the conduction band, a potential difference is created, allowing current to flow through a circuit and potentially perform useful work. This is known as the PV effect.

The PV effect was first observed back in 1839 when Edmund Bequerel noticed that light striking a silver-coated platinum electrode immersed in an electrolyte generated an electric current. In 1876, William Adams and Richard Day discovered that selenium, when placed between platinum contacts, also generated electricity when light was present. Then, in 1894, Charles Fritts created the first large-scale solar cell using selenium between gold and another metal. Of course, Einstein's famous experiment in 1905 explained the PV effect.

Recall that asymmetry was mentioned as a key component of the electron excitation. All of the early mechanisms for creating the PV effect involved an asymmetric, electric, metal–metal junction now known as a Schottky barrier or diode. A translucent metal was layered above another metal; this allowed light to strike the electrons, and the barrier junction served to keep the electrons that were sufficiently excited from relaxing back to their original state.

The heart of a PV generation system is a single solar cell, in the form of a thin slice of semiconductor about 100 cm^2. This device operates like a diode when no light is present and generates electricity when exposed to photons. A typical cell can generate about 0.5–1 V and tens of milliamps of current per square centimeter. Since the voltage of a single cell is too small to be useful, many cells are connected in series to form a module, where a module typically has 28–36 cells in series. This results in a direct current of 12 V. Consideration has to be given to the fact that individual cells may be temporarily blocked from sunlight or simply fail. For this reason, cells must be isolated so that the module can continue to operate if individual cells fail. This is accomplished by placing a bypass circuit with a diode around each cell. Also, given that light intensity is variable throughout the day, a form of energy storage, such as a battery, is typically included with a PV system. Of course, the output of the PV system is direct current, and alternating current is typically assumed by consumers. Thus, an inverter is required to convert the direct-current power to alternating-current power.

Notice that use of a battery was mentioned as a form of energy storage. It is instructive to compare the PV cell with a battery to see how it differs. First, consider the current–voltage characteristic of a PV cell given what has been explained thus far. The PV cell can be characterized by its open-circuit voltage V_{oc}, the voltage across its open terminals, and its short-circuit current I_{sc}, the current that flows when a wire connects the terminals. If a load is added to the circuit with resistance R_{L}, then the cell will have a voltage that lies between zero and V_{oc} and current I such that $V = IR_{\mathrm{L}}$. The current is also a unique function of the voltage $I(V)$ of the cell. This is known as the current–voltage characteristic of the cell, and it varies with the intensity of light striking the cell. Thus, current and voltage are determined both by the resistance of the load and the amount of light the cell receives.

The PV cell differs from a battery in the fact that a battery provides a relatively constant voltage across its terminals regardless of environmental conditions and the resistance of the load. It is the battery current that tends to vary with the size of the load. The PV cell is quite the opposite; it depends upon a light source to excite the electrons, as described earlier, across the barrier and create a voltage. Thus, voltage is not constant but varies with the amount of light. This makes the PV cell a bit more complex, because its output depends upon both the resistance of the load and the amount of light. Thus, the battery can be considered a voltage generator and the PV cell a current generator. Equation 8.25 defines the photocurrent density J_{sc} as a function of the quantum efficiency $QE(E)$:

$$J_{\mathrm{sc}} = q \int b_{\mathrm{s}}(E) QE(E)\, \mathrm{d}E. \tag{8.25}$$

The quantum efficiency is the probability that a photon of energy E will excite an electron into the conduction band. The number of photons of energy in the range E to $E + \mathrm{d}E$ that are in a unit area at a unit time is the spectral photon flux density $b_s(E)$.

Both the quantum efficiency and the energy spectrum of light striking a cell can be given in terms of photon energy or wavelength λ:

$$E = \frac{hc}{\lambda} \tag{8.26}$$

were h is Planck's constant and c is the speed of light in a vacuum. It is ideal for the quantum efficiency of a PV cell to be highest at those wavelengths of light that carry the most energy.

When a load is placed across the terminal of a PV cell, a potential difference exists that creates a current flow in the opposite direction of the main PV current. This is known as "dark current" $I_{\mathrm{dark}}(V)$. This current is similar to current that would flow if the PV cell were in the dark and a voltage source were applied across its terminals. Recall the earlier discussion regarding the fact that asymmetry provided by a diode is needed for PV current generation. Dark current is in the direction of a forward bias across this diode. The dark current density is

$$J_{\mathrm{dark}}(V) = J_0[\mathrm{e}^{qV/(k_B T)} - 1], \tag{8.27}$$

where J_0 is a constant, k_B is Boltzmann's constant, and T is the absolute temperature. The total current, from superposition, is the sum of the short-circuit current and the dark current:

$$J(V) = J_{\mathrm{sc}} - J_{\mathrm{dark}}(V). \tag{8.28}$$

Equation 8.29 shows the total current with the dark current replaced by Equation 8.27:

$$J(V) = J_{\mathrm{sc}} - J_0[\mathrm{e}^{qV/(k_B T)} - 1]. \tag{8.29}$$

The potential difference across the PV cell is greatest at infinite resistance; that is, when there is no connection across the terminals. In this condition the PV and dark currents cancel. From Equation 8.29, one can derive Equation 8.30 by setting the currents equal to each other and solving for voltage:

$$V_{\mathrm{oc}} = \frac{kT}{q} \ln\left(\frac{J_{\mathrm{sc}}}{J_0} + 1\right). \tag{8.30}$$

As J_{sc} increases, the voltage increases logarithmically. The cell generates power when voltage is between zero and V_{sc}. When the voltage is less than zero – that is, current is flowing in the forward direction of the photocell diode – the device acts as a photodetector, allowing current to flow in proportion to the intensity of the light striking the cell. When voltage exceeds V_{sc}, the device can operate as a light-emitting diode (LED).

Now we can see that the internal PV cell barrier, or diode, plays a significant role in its operation. Recall that the amount of electric power generated depends upon both the intensity of the light and the resistance of the load. When load resistance is high, more current flows through the internal photocell barrier (or diode) and less through the load. Without the internal diode there would be no potential difference and no voltage across the PV cell.

The photocell power density is $P = JV$. Recall that there is a current–voltage curve for the photocell. Consider the maximum power that the PV cell can produce denoted by the subscript "m." Then $P_m = J_m V_m$, which is known as the "maximum power point" and depends upon the shape of the PV cell's current–voltage curve. This means that the optimum load would have a resistance given by V_m/J_m. A quantity known as the "fill factor" is given by

$$\mathrm{FF} = \frac{J_m V_m}{J_{sc} V_{oc}}. \tag{8.31}$$

Since J_{sc} is the short-circuit photocell current it is the maximum current, and V_{oc} is the maximum, or open-circuit, photocell voltage. Thus, these are the largest current and voltage values that can be attained; their product is the theoretical maximum power that can be generated. However, in reality, the product of current and voltage does not reach these maximum values and the ideal fill factor of one is not likely to be achieved in practice. One can also look at the efficiency as shown in Equation 8.32, where P_s is the light power density:

$$\eta = \frac{J_m V_m}{P_s}. \tag{8.32}$$

Equation 8.33 shows the efficiency as a function of the fill factor:

$$\eta = \frac{J_{sc} V_{oc} \mathrm{FF}}{P_s}. \tag{8.33}$$

One of the challenges in increasing the efficiency of PV cells is that substances with a high J_{sc} typically have a low V_{oc}.

The technique of adjusting a PV cell's voltage or current to remain at the maximum power point is known as maximum powerpoint tracking (MPPT). Since the power from a PV cell is direct current, it typically must pass through an inverter to convert the power to alternating current. Thus, it is common to incorporate MPPT with the inverter of the photocell power system. MPPT is an interesting problem with many different solutions. Simple solutions involve estimating the maximum power point a priori as a fraction of V_{oc} and dynamically adjusting the voltage to achieve the given fraction of V_{oc}. More sophisticated methods include gradient search techniques, in which the voltage is perturbed and adjusted in the direction of increasing total power. Clearly, techniques such as particle swarm optimization (PSO) could apply here. Other techniques involve periodically sampling the current–voltage curve for a system and adjusting the voltage to maximize power along the curve. The problem is difficult because the current–voltage curve changes over time. There are many factors influencing the curve beyond the primary ones already mentioned. For example, subtle differences in manufacturing, partial shading from a tree branch blowing in the wind, changes in temperature, the appearance of multiple maxima on the current–voltage curve, as well as age and damage to solar panels all impact the curve. Communication is required to transmit information from monitors residing at the PV cells to the MPPT algorithm located at the inverter. In addition, control information may need to be sent to the inverter from the utility or the consumer. As discussed later, widespread use of inverters in the power grid allows them to be used for many functions beyond simply converting from direct to alternating current. They also help with increasing power quality, including increasing the power factor. This subject will be revisited in more detail when power electronics are discussed in Chapter 14. Finally, note that the PV cell is receiving energy in the form of photons; the notion of the photocell as comprised of many nanoscale antennas has been proposed (Simovski and Morits, 2013) and is discussed under the topic of nanogeneration in Chapter 15. The PV cell becomes a platform for nanoscale communication, and the concept of a fundamental coupling of communication with PV generation is a potential area of research. Currently, there are numerous competing technologies and standards being applied to all aspects of the power grid, ideas and technologies that involve a close, fundamental integration of communication and power systems that hold more promising potential.

However, it was not until the 1950s that semiconductor technology had advanced enough to make PV power generation a feasible consideration, although the price per watt was still too expensive to be commercially viable. Currently, new materials and thin-film techniques are helping to reduce the cost.

8.3 The Smart Power Transmission System

The remainder of this chapter covers the power transmission system and communications. This includes an introduction to FACTS for controlling power flow over transmission lines, monitoring and control of the transmission system, and the types of communication traditionally used. This section also includes an introduction to the emerging technology of wireless power transmission. Wide-area monitoring and control is required in order for communication over the vast expanse typically covered by the transmission system. The section ends with an introductory discussion of networked control systems. This is an important concept because the communication network will experience delay when attempting to control a system over a wide area.

It may seem odd at first to discuss DG with transmission when we have defined DG to reside within the distribution system. The reason these topics are combined is that, even though we consider DG to reside within what is now considered the distribution system, the distribution system will have to act more like today's transmission system. In other words, today's transmission system couples large generation plants so that excess power can be shared among power plants and among different synchronous zones. In a similar manner, today's distribution system would ideally allow distributed generators to cooperate and share power in a bidirectional manner as well. In this sense, there may be techniques from today's transmission system that could benefit the design of future distribution systems that incorporate DG.

8.3.1 The Flexible Alternating Current Transmission System

This section introduces FACTS (Edris *et al.*, 1997). This is essentially a power flow control mechanism and, in essence, a form of power routing. In the beginning, power flowed from generators to consumers. There was little, if any, transmission system and there was little reason to control the route that power took along its path from the generator to the consumer. As generators became larger and more centralized they were also located farther from population centers. Transmission became a separate entity from distribution at this point, and the need to control bulk power flow and the route that it took became more acute. With deregulation, the simplified concept was that power would flow in the direction of those that pay the highest price. The phase-shifting transformer (PST), FACTS, and HVDC have allowed for more flexibility in controlling power flow. Let us see how this is done.

The active power transported over a transmission line is given by

$$P = \frac{|U_s| \cdot |U_r|}{X_L} \sin(\delta), \tag{8.34}$$

where U_s and U_r are the voltage at the opposite ends of a transmission line. δ is the angle between the phase of the voltages. Finally, X_L is the impedance of the line.

Equation 8.34 shows that there are several ways to control the active power flow over the line. This can be done by changing one of the voltages at either end of the transmission line, changing the phase angle between the voltages at either end of the transmission line, or by changing the impedance of the transmission line.

Similarly, the reactive power flow over a transmission line is

$$Q_s = \frac{|U_s|^2}{X_L} - \frac{|U_s| \cdot |U_r|}{X_L} \cos(\delta) \tag{8.35}$$

and can be controlled by adjusting the same parameters.

The techniques used for power control can be broadly categorized into mechanical switching, thyristor-controlled switching, and fast switching via a power converter. Each of these techniques has implications for communication and control. Mechanical switching involves such activities as changing taps on a transformer. This is relatively slow, on the order of several seconds at best. Thyristors, being solid-state devices, can switch faster, on the order of a few cycles of the power frequency. Faster switching, such as that enabled by an insulated-gate bipolar transistor (IGBT), allows for nearly instantaneous power control or on the order of less than one power cycle. A variety of other solid-state power-electronic devices are emerging as well and are discussed in detail in Chapter 14.

Voltage phase-angle changes can be accomplished quite simply by inserting a series voltage in quadrature, that is, 90° out of phase, with the voltage at one end of the transmission line. The impedance can be controlled most simply by adding a capacitor in series with the transmission line. By changing capacitance, the impedance through the line changes. Thyristor-switched and thyristor-controlled series compensation can be used. The switched version simply turns capacitors on or off in series with the transmission line. This means that only discrete values of capacitance can be used. The controlled version allows for continuous control of the amount of capacitance, and thus finer control of the impedance of the transmission line and, in turn, finer control over power flow across the line. Finally, it should be noted that significantly varying the amplitude of the voltage at either end of the line is generally not an option since voltage is designed and regulated to remain at a fixed value. Voltage can vary due to stress on the system, and voltage support can be provided, which in effect causes more power to flow where it may be needed. However, this should be neither a typical nor desirable part of normal power flow control.

These types of power flow control mechanisms are analog in nature. HVDC transmission lines and power converters/inverters can provide a higher resolution of control and result in a packetized form of power control that may more closely resemble communication network packets. This was introduced in Section 3.4.4.

8.4 Wireless Power Transmission

This section introduces the notion of wireless power transmission. This is the transport of electric power from a power source to a load without wires or any form of human-constructed conductors (Ahmed *et al.*, 2003; Benford *et al.*, 2007; Brown and Eves, 1992; Brown, 1984; Budimir and Marincic, 2006; Carpenter, 2004; Dessanti *et al.*, 2012; Dionigi and Mongiardo, 2011; Drozdovski and Caverly, 2002; D'Souza *et al.*, 2007; Grover and Sahai, 2010; Gundogdu and Afacan, 2011; Ishiba *et al.*, 2011; Komerath and Chowdhary, 2010; Komerath *et al.*, 2012; Komerath and Komerath, 2011; Komerath *et al.*, 2011, 2009; Kubo *et al.*, 2012; Kumar *et al.*, 2012; Lee and Lorenz, 2011; Li, 2011; Lin, 2002; Mohagheghi *et al.*, 2012; Mohammed *et al.*, 2010; Neves *et al.*, 2011; Pignolet *et al.*, 1996; Popović, 2006; Sample *et al.*, 2010; Shinohara, 2011, 2012; Shinohara and Ishikawa, 2011; Siddique *et al.*, 2012; Smakhtin and Rybakov, 2002; Vaessen, 2009; Waffenschmidt and Staring, 2009; Wang *et al.*, 2011a; Yadav *et al.*, 2011; Yamanaka and Sugiura, 2011; Yu and Cioffi, 2001; Zhang *et al.*, 2012; Zhong *et al.*, 2011; Zou *et al.*, 2010). There are several interesting aspects about wireless, high-power transmission over relatively long distances. First, it is a beautiful and direct demonstration of the synergy between wireless communication and power systems. Wireless communication is the transmission of power from a transmitter to a receiver. The only difference is that wireless communication is concerned with minimizing power and maximizing information transmission, while power systems are concerned with maximizing power and minimizing information transmission, where information in this case means entropy of the signal. Wireless communication systems already utilized wireless power transmission for RFID. Passive RFID relies upon the transmission of wireless power to the RFID label or tag in order to power its operation. Inductive power transmission at relatively close range has been used commercially for a long time.

Wireless power transmission of high power over longer distances would dramatically increase convenience and flexibility of consumer devices, the power distribution system, the power transmission system, and ultimately the generation of power, allowing satellites to beam renewable solar power from orbit. Power cords could be eliminated, electric vehicles could be recharged while in use, power lines would be unnecessary, and prosumers and distributed generators could beam power directly to where it was needed most or the price was highest. Most people today, upon hearing of wireless power transmission for the first time, view it as either science fiction or express concern over the perceived dangers of invisible power accidentally beamed through living organisms or people. These concerns are often expressed by people who would not think twice about holding a cell phone to their head or operating and opening a microwave oven. Like many aspects of power systems that may seem new today, such as a direct-current power grid or microgrids and DG, wireless power generation has a long history, and developing the technology would in a sense be going back to an idea conceived by a genius at the very beginning of power systems, Nikola Tesla.

A brief history of wireless power transmission begins in 1862 with the formulation of Maxwell's equations that form the basis for wireless power transmission (Shinohara, 2011). Soon after, the Poynting vector was formulated, which characterizes electromagnetic energy flux density; in other words, the rate of flow of energy per unit area. This is measured in watts per square meter. This flux exerts physical pressure, although very weak from our perspective. The next major player was Nikola Tesla, who reasoned a century ago that all electric power could be transmitted wirelessly. He carried out many famous experiments toward developing the concept, one of which was transmitting 300 kW of power via 150 kHz radio waves. It is speculated that Nikola Tesla experimented with his eponymous Tesla coil as a transmitter of power conducted via the Schuman resonance, a natural, efficient wave guide between the Earth and the ionosphere in which electrical energy from lightning continuously bounces around the earth, as a means to transmit commercial electrical power, thus forming a free, natural power grid. However, after Tesla's experiment, the focus generally turned toward application of wireless transmission to communications, a much easier application. It was not until the 1960s that William Brown began to reinvestigate the topic of wireless power transmission using microwaves at 2.45 GHz with magnetrons and klystrons as power sources. He developed the rectifying antenna, also known as a rectenna, which was able to receive and rectify microwaves. William Brown's applications were primarily for aerospace; for example, to transmit power to helicopters and for beaming power through outer space. In 1975, 30 kW of microwave power was received over a distance of 1.6 km at 2.388 GHz. However, the cost of the system at that time precluded its use as a widely used commercial product. Since then, it has been recognized that there is enough radio energy continuously being transmitted for communication purposes (for example, broadcast radio and other communication) that this energy can be harvested without the need for explicit power transmission. Also, shorter range transmission, on the order of a few meters, using kilowatts of power can be implemented wirelessly and relatively efficiently using resonance, as will be explained shortly.

There is a difference between near-field and far-field power transmission, and these differences follow directly from Maxwell's equations and the equations involving electric fields and those involving magnetic fields. More specifically, electric fields produced by electric charge differ from electric fields generated by a changing magnetic field. Also, magnetic fields produced by electric currents differ from those generated by changing electric fields. The near-field is the region relatively close to a power radiation source where the electromagnetic field is dominated by fields generated by current and electric charge. The far-field is the region relatively far from the radiation source where electric and magnetic fields create and reinforce one another; that is, the electromagnetic wave is dominant and the previous fields generated by current and charge no longer reach.

In the near-field, techniques such as inductive and resonant coupling can be used to wirelessly transmit power. Inductive coupling is the well-known technique of using a transformer coil to create a changing magnetic field that sweeps past a secondary coil and thus induces an electric charge in the secondary coil. This requires that the primary and secondary coils reside in close proximity and also typically requires a magnetic core. As the coils move farther apart, most of the power is wasted in the impedance of the primary coil.

Resonant coupling improves efficiency by adding capacitance to each coil. Each coil becomes an inductor–capacitor (LC) circuit; as is well-known, such a circuit will oscillate with energy stored periodically as a magnetic field in the inductive element and then transferred to the electric field of the capacitor and back again to the inductive element. If both circuits oscillate at the same frequency, then one will cause the other to resonate. This resonance significantly improves wireless power transmission efficiency of the inductive coupling over longer distances. Remember that this is still a near-field effect.

Eventually, like a swinging pendulum, oscillations will disappear due to both resistance and radiation of energy. A parameter known as the Q factor characterizes the oscillation. A specific resonator will oscillate with a given amount of energy over a given range of frequencies. The Q factor is a dimensionless value that characterizes the resonator's bandwidth relative to its center frequency. A high Q factor indicates a lower rate of energy loss relative its stored energy, and thus oscillations dissipate more slowly. In other words, a high Q factor resonator has less damping and will "ring" for a longer time than the same resonator with a lower Q factor. More specifically, the Q factor is the ratio of the energy stored in the resonator to the energy supplied, per cycle, to keep the signal amplitude constant, at the resonant frequency f_r and the stored energy is constant with time. This is shown by

$$Q = 2\pi \times \frac{\text{Energy stored}}{\text{Energy dissipated per cycle}} = 2\pi f_r \times \frac{\text{Energy stored}}{\text{Power loss}}. \quad (8.36)$$

We have a primary coil that is oscillating at a high frequency and inducing current that oscillates in the secondary coil that can now be located farther away from the primary coil. Only a relatively small portion of the field from the primary coil has to reach the secondary coil to achieve high-efficiency energy transfer. In this case, the Q factor for the electrical RLC circuit of the coils is

$$Q = \frac{1}{R}\sqrt{\frac{L}{C}}. \quad (8.37)$$

While the Q factor deals with the energy of the oscillation in the resonant wireless transfer system, the coupling coefficient deals with how much of the electromagnetic field reaches from the primary to the secondary coil. The fraction of the flux of the primary coil that cuts through the secondary coil is known as the coupling coefficient k. This value can obviously range from zero to one. Qualitatively, the coupling can range from tight coupling, in which the coupling coefficient is near one, to even closer coupling, known as overcoupling, in which the secondary coil is close enough to cause collapse of the primary coil's field. For wireless power transmission, we typically would want longer distances, in which there is loose coupling and the coupling coefficient is approximately 0.2 or even as low as 0.01. Thus, a figure of merit for the wireless power transmission system is kQ. With purely inductive coupling, there is no resonance and thus there is no Q. Thus, with resonant coupling, even if k is small, due to a large distance or small size for example, increasing Q can still increase the efficiency of power transmission.

Now we consider further distances of wireless power transmission in which we move into the far-field. As previously mentioned, the far-field is created by the mutual-reinforcing activity of the electrical and magnetic waves as they propagate through space. It should be noted that the actual distance at which far-field behavior exhibits itself depends upon the wavelength. As an example, visible light exhibits far-field behavior at distances greater than a micrometer from the source because its wavelength is on the order of 0.4–0.7 μm. Also, for the sake of completeness, it should be noted that the transition between near- and far-field is not sharp; rather, there is a zone in which both near- and far-field effects are present.

As mentioned earlier in the brief history of wireless power transmission, which we will consider as a far-field process using microwave power transmission for now, the rectifying antenna, or rectenna, is used to receive the radiated power. The rectenna is comprised of an antenna, a low-pass filter to block higher harmonics, and a diode or multiple diodes with an output filter to rectify the received microwave power to direct current. Rectennas with over 90% efficiency have been developed at the 2.45 GHz band for microwave power transmission.

While we had previously discussed resonant coupling as a means of improving transmission efficiency, the microwave power transmission technique is a far-field phenomenon and the transmitter and receiver are not directly coupled. Instead, beam efficiency is governed by the Friis transmission equation:

$$\frac{P_r}{P_t} = G_t G_r \left(\frac{\lambda}{4\pi R}\right)^2 = \frac{A_t A_r}{(\lambda R)^2}. \tag{8.38}$$

Here, P_r and P_t are the received and transmitted power, G_t and G_r are the gain of the transmitting and receiving antennas, λ is the wavelength, and R is the distance between the transmitting and receiving antennas. A_t and A_r are the aperture areas of the transmitting and receiving antennas. Notice that Equation 8.38 returns the ratio of received-to-transmitted power, which is the efficiency of the wireless transmission system.

The antenna aperture characterizes how effectively an antenna receives radio waves. Also, recall from the Poynting vector that electromagnetic radiation flows through space with a given flux density. The antenna aperture is a measure of the ability to capture this flow of radiation. Imagine a two-dimensional panel positioned at a right angle to flow of radiation. The antenna aperture is the area of the imaginary panel that cuts through, or blocks, the same amount of incoming radiated power as that actually received by the antenna. In effect, it allows us to transform the antenna into an area that would receive the equivalent amount of power. If PFD is the power flux density – that is, the amount of power passing through a unit area – and P_o is the output power of the antenna, then

$$A_{eff} = \frac{P_o}{\mathrm{PFD}} \tag{8.39}$$

returns the effective aperture, in meters-squared, of the antenna.

An antenna's gain is a measure of its directivity (or ability to focus its power) and its efficiency (that is, its ability to transmit or receive all the power directed to it). Gain is a unitless quantity and is simply the product of efficiency and directivity.

Clearly, phased-array antennas are well known in communication technology and a feasible approach for directing power radiated toward a target. The difference is that wireless power transmission is focused upon maximizing the direct-current RF efficiency, something that for communications is not a primary design goal.

8.5 Wide-Area Monitoring

The power systems transmission system, because it interconnects capacity in different regions, generally tends to extend over long distances. Thus, management of the transmission system generally requires what is rather imprecisely denoted "wide-area" communication architectures and WANs. The exact definition of local area, metropolitan area, wide area or any of a number of other "area" network technologies is somewhat loosely defined; however, the idea is to convey some sense of the distances of the communication involved. From a power systems standpoint, the term "wide-area monitoring (WAM)," "wide-area monitoring system (WAMS)," or "wide-area monitoring, protection, and control (WAMPAC)," or any number of other terms that begin with "wide-area" are all used to denote large-scale, global management of power systems, including the transmission system. The widespread introduction of the PMU and the measurement and transmission of synchrophasors spurred the growth of wide-area management in power systems. Chapter 13 is dedicated to synchrophasor applications, thus this section will only touch lightly on that topic, saving the details for the aforementioned chapter on that topic. However, even before the widespread introduction of PMUs in the transmission system, communication was required, typically implemented in an ad hoc manner, to provide wide-area networking capability. As one example, protection across transmission lines was required. We review the need for wide-area communication and then discuss some of the historically more commonly used wide-area networking approaches for the transmission system.

Before discussing transmission system communication technologies, the notion of bandwidth–delay product will be introduced because this communication network characteristic becomes significant in WAN communication systems. The bandwidth–delay product is simply the product of the bandwidth of a communication channel and the propagation time of a bit through the channel. The units are thus in bits. Essentially, this is the maximum number of bits that may be held, instantaneously, within the communication channel. A communication channel that spans a large physical area will tend to have a large delay. Such channels also typically are high bandwidth. This means that significantly large amounts of data may be in the channel at any time. From a communication control standpoint, this means that any control information flowing through the channel may be followed by a significant amount of information already in the channel flowing immediately behind it. Thus, communication control information such as ACK packets or changes in priority to streams flowing in the channel, and so on, will not be able to impact bits already in the channel, which may be a significant amount of information. Thus, communication control mechanisms need to be designed carefully for such large bandwidth-delay channels.

Communication technologies for WANs, including those used in the electric power transmission system, tend to have a large bandwidth–delay product. Satellite, microwave, and fiber optic communication systems can span long distances, but are typically used for backhaul systems; that is, carrying a large volume of data over long distances. The challenge with both WAN communication and wide-area power system monitoring and control had been the communication time delay and the lack of time synchronization. From a communication perspective, more power is required and propagation time begins to dominate in wide-area networking. Efficiency over such large bandwidth–delay product networks suffers unless large volumes of data are transported. From a power systems perspective, this meant that achieving precise time synchronization across such WANs was a significant challenge. Of course, with GPS, time synchronization is no longer an issue, as long as a GPS signal can be received.

Lower bandwidth–delay product communication approaches, such as power line carrier, were used for point-to-point protection schemes in the transmission system. Since the communication signal flows along the transmission line, a lower bandwidth signal, but with less delay, can be used to convey small amounts of important information, such as the presence of an electrical fault for protection mechanisms.

Four types of wide-area communication that have a long history in transmission system operation are discussed: power line carrier, satellite communication, fiber optic communication, and microwave communication. It is no accident that many of these communication approaches have also been used to transmit power as well as information. In particular, research into space-based power generation and transmission to Earth is an active research area, microwave systems have long been considered a means to transmit power, and, although it is relatively small scale, transmission of power via fiber-optic lines has been considered

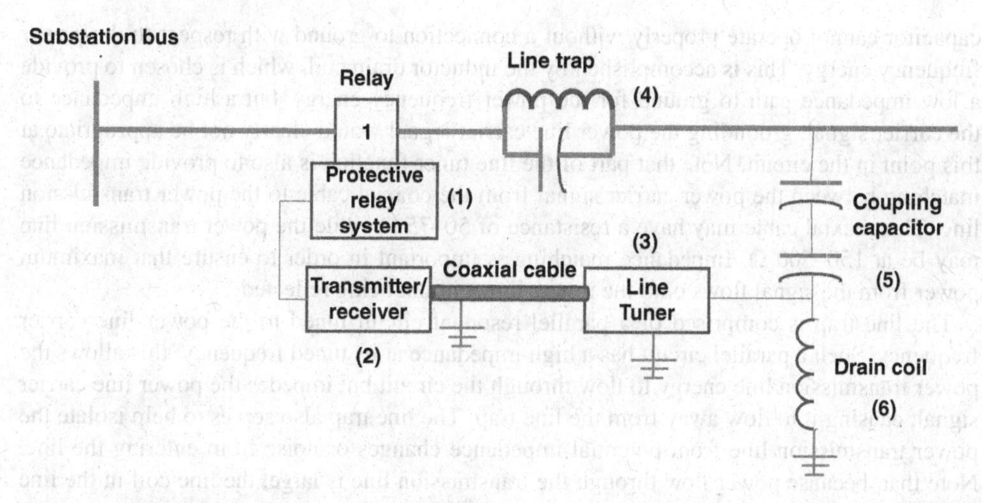

Figure 8.6 The main components of a power line carrier system include (1) the protective relay system, (2) the transmitter/receiver, (3) the line tuner, (4) the line trap, (5) the coupling capacitor, and (6) the drain coil.

as well. Power line carrier is obviously utilizing the power system as a waveguide. Let us consider each of these wide-area technologies in more detail.

Power line carrier has long been a solution for communication in the power transmission system because, as previously mentioned, it utilizes the same conductor as the power system as a waveguide to form the communication channel. Section 8.5 discusses power line carrier in more detail. There is likely to be less interference on a relatively simple power transmission line than along a more-topologically complex distribution system network that also has many devices injecting noise into the distribution power lines.

The main components of a power line carrier interface are shown in Figure 8.6 (Sanders and Ray, 1996). In this figure, there is a protective relay system on the far left that feeds a signal into a transceiver, labeled (2). A coaxial cable carries the signal to the line tuner. The line tuner adjusts the signal for optimal transmission over the power line and feeds the signal through the coupling capacitor onto the power line. The line trap is used to block the signal from going in the wrong direction; it directs the energy of the signal along the proper path. Now that we have seen the system as a whole, we can discuss each of the components in a little more detail.

First, the line tuner typically resides in the switch yard and can be a considerable distance away from the transceiver, which is typically in a cabinet in the control center. Thus, the coaxial cable can be long and needs shielding from noise. Additionally, it is possible for there to exist multiple transmitters sharing the same coaxial cable; in this case, isolation circuits are used to keep the signals from overlapping.

The line tuner is one of the unique key elements of the power line carrier system. The goal is to safely provide a low-impedance path for the power line carrier signal energy to reach the transmission line while providing a high-impedance path for the main power frequency energy. This is done in conjunction with the coupling capacitor and the drain coil. The line tuner and coupling capacitor form a resonant circuit that is tuned to the carrier frequency. The coupling capacitor cannot operate properly without a connection to ground with respect to the power frequency energy. This is accomplished by the inductor drain coil, which is chosen to provide a low-impedance path to ground for the power frequency energy, but a high-impedance to the carrier signal; grounding the power line carrier signal would clearly not be appropriate at this point in the circuit. Note that part of the line tuner function is also to provide impedance matching between the power carrier signal from the coaxial cable to the power transmission line. The coaxial cable may have a resistance of 50–75 Ω while the power transmission line may be at 150–500 Ω. Impedance matching is important in order to ensure that maximum power from the signal flows onto the power line without being reflected.

The line trap is comprised of a parallel resonant circuit tuned to the power line carrier frequency. Such a parallel circuit has a high impedance at its tuned frequency; this allows the power transmission line energy to flow through the circuit but impedes the power line carrier signal, causing it to flow away from the line trap. The line trap also serves to help isolate the power transmission line from potential impedance changes or noise from entering the line. Note that, because power flow through the transmission line is large, the line coil in the line trap must be physically large.

Notice that, other than line traps, once the power line carrier signal is placed on the line, the system operates like a communication network bus system. In other words, the signal will propagate wherever the transmission line leads with little control over the direction it takes. There is no inherent notion of routing except for physically switching the connection between power lines.

When waves traveling in opposite directions pass through each other, they will, in general, create a standing wave. An impedance mismatch, perhaps caused by an improperly terminated line, will cause a wave to reflect back to the source passing through the originally generated waves upon its reflection back to the source. This will create problems because it can attenuate the desired waves or create noise source that has to be overcome. Thus, issues of impedance and wavelength become important in power line carrier systems. A transmission line may be 100 km long or more. We know that the power frequencies are either 50 or 60 Hz. Power line carrier frequencies are typically 500 times greater than the power frequency. To put it another way, power transmission lines are actually electrically short at power frequencies and long at carrier frequencies. The relationship is

$$\lambda = \frac{0.98c}{f}, \tag{8.40}$$

where c is the speed of light.

For a 250 kHz signal, a 100 km transmission line is 85 wavelengths long, while at 60 Hz the line is only 0.02 wavelengths long.

While bandwidth increases as the carrier frequency increases, attenuation due to line loss also increases. This is largely because of shunt capacitances along the transmission line whose impedance decreases as the frequency increases. The line losses are also subject to weather conditions; frost is particularly significant because the signal will tend to propagate along the ice instead of the conductor, which significantly increases attenuation of the signal. Another source of attenuation occurs when an earth-return path is used. In this case, soil conductivity can change over time due to change in moisture, as one example. It should also be noted that attenuation in underground cables tends to be larger than in overhead lines.

As mentioned previously, line impedance plays a significant role in the operation of power line carrier. The characteristic impedance of a transmission line is the ratio of voltage to current of a traveling wave assuming the transmission line is infinitely long. The characteristic impedance, shown in Equation 8.41, is important because the power line carrier interface equipment must match the characteristic impedance for optimum transfer of the signal:

$$Z_0 = \frac{V}{I} = \sqrt{\frac{R + j\omega L}{G + j\omega C}}. \tag{8.41}$$

Note that, if a transmission line is properly terminated with the correct impedance, there will be no reflected energy and the transmission line will appear as though it were infinitely long.

Of course, noise along the transmission line is one of the greatest challenges that a power line carrier has to overcome. There can be both continuous noise and impulse noise. As its name implies, continuous noise is present for long periods of time, but generally has a low amplitude and varies slowly. Impulse noise exists for a brief duration, but can have a relatively large amplitude. While the power distribution system tends to have many connections directly to consumer equipment, such as motors and other devices that inject noise, transmission lines are relatively isolated from such noise. However, transmission lines are not noise free. For example, corona discharge occurs every half-cycle of the power frequency waveform and is a source of continuous noise. In addition, switch operation can inject impulse noise into the transmission system. These are relatively sudden, large injections of wave energy that cause significant degradation to the power line carrier equipment. Finally, there is noise from a growing technology; namely, HVDC transmission. HVDC lines require converters to transform power from alternating current to direct current at the source of the line and from direct current back to alternating current at the end of the line. This conversion process creates harmonic noise that is typically just below the power line carrier band, but still capable of causing problems for the power line carrier signal. For example, converter harmonics can be resonant with the coupling capacitor and drain coil (refer to Figure 8.6). This can saturate the drain coil and cause the power line carrier signal to become grounded, which, as we noted at the beginning of this section, should not occur.

Finally, there is the issue of power faults that occur along the transmission line. We should not forget that, like the SNMP in communication network management, the whole point of power line carrier is to manage the system when faults occur. If power line carrier cannot operate through a fault, then its purpose becomes pointless from a power systems perspective, particularly since one of its biggest roles in the transmission system is to implement protection. So, how well does it work when a fault occurs? An electric arc is created by the breakdown of the gas in the surrounding atmosphere allowing a conductive path where none was intended. This is a concern for high-voltage transmission lines, while perhaps less so for the distribution system, which operates at a lower voltage but has many other sources of electrical faults. When an electric arc begins, the noise can be severe for the first 4 ms. However, once an electric arc is established, a conduction path is created and the noise becomes minimal. In fact, the final magnitude of noise after a fault can be less than normal operation before the fault since the voltage may be reduced by the fault. However, it is interesting to keep in mind that, for the first 4 ms, the signal may not be able to pass through the communication system due to the impulse noise from the fault. In order to improve reliability of the power line carrier communication, the carrier signal can be coupled to more than one power line, utilizing multiple phases of a three-phase system for example.

Impedance mismatch results in signal loss known as mismatch loss

$$ML = 20\log \frac{Z_0 + Z_1}{2\sqrt{Z_0 Z_1}}, \tag{8.42}$$

where Z_0 is the characteristic impedance of the transmission line and Z_1 is the impedance of the power line carrier circuit.

Frequencies that have been used in power line carrier on transmission lines range from 30 to 500 kHz. However, the best frequency to use will depend upon a variety of design considerations. These include the bandwidth requirements of the applications, possible interference from other sources, the coupling method used, and attenuation and distance requirements. The Utilities Telecom Council (http://www.utc.org/plc-database) maintains a database of power line carrier frequencies in an attempt to minimize interference by allowing users to coordinate installation, configuration, and frequency selection.

Noise on the power line decreases as frequency increases; however, as the frequency increases, as previously mentioned, attenuation also increases. Finally, it is good to review the sources of attenuation for a power line carrier system. This helps in estimating the performance of the system by considering a link budget analysis. First, when multiple transmitters share the same coaxial cable, isolation of the signals results in signal power loss. Next, there is a coupling loss, which includes loss through the coaxial cable, line tuner, coupling capacitor, and drain coil (see Figure 8.6). Then there is the transmission line attenuation of the signal, which may be weather dependent. Transposition – that is, changing the order of alignment of the power lines in a multiline system – is done to reduce cross-talk and improve transmission of power. However, transposition involves crossing the power lines, and this reduces the power line carrier signal strength. Thus, the number of transpositions should be included in any link budget analysis as well.

Satellite communication appears as a potential solution to power transmission system communications because satellite communication can efficiently span the wide areas that transmission systems typically cover (Holbert *et al.*, 2005; Madani *et al.*, 2007; Marihart, 2001). Satellites and power systems have had a long history. Weather satellites have long provided information regarding demand for electric power; in a similar manner, they also can provide information related to the supply of power from renewable generation sources such as PV and wind energy (Krauter and Depping, 2003). Thus, satellites, with their lofty view, are ideal for predicting and optimally controlling renewable power supply and demand, as well as the large power flows that take place across transmission lines to compensate for mismatches in supply and demand. GPS satellites have long been used as a source of time synchronization, allowing power system events that are widely disseminated through space to be accurately compared with one another. One of the disadvantages of satellite systems is the engineering contradiction related to their benefit of wide-area coverage; namely, their great distance from the Earth induces a relatively large communication latency. This is particularly true for geosynchronous satellites, which must reside at a height of 35 000 km above the Earth in order to remain stationary relative to the rotation of the Earth. This results in a communication latency of 400 ms simply due to time of flight at the speed of light between the Earth and the geosynchronous satellite. For this reason, low Earth orbit (LEO) satellites are considered for time-critical communication. The LEO satellites are typically located at 500–1500 km above the Earth and have a delay on the order of 25 ms, which is on the order of the latency through a fiber-optic link. As the space industry becomes privatized and costs are reduced, it becomes more likely that low-cost, small-satellite payloads will become commonplace and drive the market for more satellite-based power system communication applications.

Numerous applications for satellite communication in power systems are possible. Some examples are old ones. For example, wide-area protection mechanisms and power restoration are possible. In particular, the PMU exists only because GPS enables the wide-area synchronization required to implement timing. Many more applications may be possible by considering the communication of synchrophasor information via satellite as well. Also, as mentioned, earlier, weather prediction and the ability to detect activity over a wide area give the satellite power to provide advance warning of impending problems, including impending disruption due to space weather, known as geomagnetically induced current.

The use of satellite communication to manage remote power distribution systems has also been utilized. Remote PV systems are managed via satellite (Krauter and Depping, 2003). The Inmarsat-C satellite communication system is used for the remote management of distribution systems (Beardow *et al.*, 1993). The use of hybrid satellite–terrestrial communication for use by power utilities is described in Holbert *et al.* (2005). When considering security, reliability, and vulnerability, satellite communication is considered an approach that increases the security and reliability of the entire combined power and communication infrastructure (Hui Wan *et al.*, 2005). Lightning detectors on satellites can locate range, bearing, and intensity of lightning strikes by direct observation. The National Lightning Detection Network (NLDN) uses satellites to relay information regarding electric and magnetic field information to the NLDN control center. A satellite-based GPS has been used to measure transmission line sag and accurately infer line overload, allowing transmission lines to carry loads closer to their maximum capacity. With regard to GPS, the standard positioning service (SPS), was purposely degraded before May, 2000. After this date, by presidential order, the degradation was ended and full, accurate positioning information was open to the public with the provision that the US military could institute regional denial-of-capability if needed. The GPS can provide synchronization within $\pm 25\,\mu s$, which is within 0.5° degrees at 60 Hz.

LEO satellites have also been used to improve stability within the power grid by implementing a space-based power system stabilizer (Holbert, *et al.*, 2005). A power system stabilizer is a control system designed to mitigate potentially dangerous power oscillations in the transmission system. It is interesting to note the causes for such oscillation. Typically, power system stabilizers can be implemented locally within the winding of generators, known as a damper winding. However, with long transmission lines and greater interconnectivity, the damper winding approach has become ineffective in damping oscillation. Also, another interesting cause that has been identified is the use of more automatic controls throughout the power grid. These automatic control systems can interact in unexpected ways to actually create negative damping and amplify oscillation. Finally, there is the concern that DR could, as an untended side-effect, cause more frequent adjustments in power and more bulk exchanges across transmission lines. A proposed solution is a supervisory power system stabilizer (SPSS), which is a wide-area control system for global control of power system stabilizers.

However, along with potentially high cost and large latency, there are other challenges with satellite communication for use in power systems. Large antennas and satellite dishes may be required. Also, on a related note, the regions closer to the Earth's poles require their antennas to be pointed at low angles, which can be obstructed by interfering objects within the line of sight between the antenna and satellite. Finally, there is the issue of geomagnetic storms, or "space weather" that can occasionally interfere with satellite communication, or in extreme cases cause permanent damage to a satellite. Geomagnetic storms are covered in detail in Section 15.2. However, satellites can also provide a benefit to the power grid with respect to space weather. This is because geomagnetic storms can also cause disruption and damage to the power grids on Earth. Strong magnetic fields from space weather can induce current within cables and power lines of the electric power grid and damage equipment, such as transformers. Satellites that can provide advance warning of impending geomagnetic storms will allow utilities to take appropriate action to protect their systems.

The advantages of LEO satellites relative to higher orbit satellites, are low communication latency, multi-satellite hand-off, smaller antennas are required, lower power consumption is required for communication equipment, and lower cost. Lower latency means that the bandwidth–delay product is smaller and the system becomes more amenable to IP operation. There are several LEO satellite constellations that exist today, including Iridium, Emsat, Globalstar, and Thuraya among others.

We briefly mention another active research area that relates satellite communication with the power grid; namely, space-based power generation (Komerath *et al.*, 2012). Space-based solar power is an interesting example of transitioning the satellite from its use as a sensing and communication system into a power generation system. The concept is to collect solar power from space and transmit the power to Earth. Solar power collection from a satellite has the advantage of being approximately 144% better in collection of solar energy than from Earth because Earth-based systems are filtered by the atmosphere. Also, a satellite can capture solar radiation constantly, whereas Earth-based systems are of course blocked at night. Clouds, weather, and other obstructions are not an issue for satellite-based systems. From a power transmission standpoint, the power collected by a satellite can be transmitted to Earth via a narrow beam directly to where power is required. In other words, power could be supplied on demand precisely where it is needed; for example, in order to reduce peak load. This concept has raised new and fascinating challenges, particularly related to cost and wireless power transmission, discussed further in Section 15.6.

Microwave communication systems have long been used as backhaul communication for telecommunication networks. A backhaul serves as the link between the high-capacity core switching system and the potentially widely scattered, smaller communication nodes in the edge networks. Microwave links have had the bandwidth capacity to handle large amounts of aggregate traffic between core and the edge networks.

One of the first challenges in understanding microwave communication involves simply defining what it is. The prefix "micro" in "microwave" understandably leads one to expect that microwave communication involves electromagnetic frequencies whose waves have a length on the order of a micrometer. Contrary to this notion, the prefix "micro" does *not* refer to electromagnetic waves on the order of a micrometer. Instead, the prefix is used much more loosely to refer to wavelengths that are relatively small compared with those used in standard radio broadcasting. The term microwave in the communication field actually refers to wavelengths that are on the order of centimeters in length. In fact, there tends to be no hard-and-fast rule regarding the boundary between infrared, terahertz, real microwaves, and ultra-high frequency radio waves. Many of the defined ranges overlap with one another. Microwaves are considered to be in the 300 MHz to 300 GHz range; but for microwave communication, 3–30 GHz is typically used.

Because microwaves are higher frequency and shorter, they are highly directional and have the capability of supporting a high-bandwidth channel. This also means that they tend to require line of sight and have been commonly used, as mentioned, for large-data transmission in backhaul networks and for satellite-to-ground communication. Microwaves also have a long history and relationship with power. Consider the heating element of a common microwave oven where in which heat is generated by the power. This device contains a series of distinct cavities (that is, holes) that can be thought of as acting like waveguides whose ends are short-circuited. The result is that only electromagnetic waves of a precise length will resonate within the cavities. The wavelength will be one-half the length of the cavity. A hot cathode driven by a high direct current emits electrons that are driven in a circular motion by an applied magnetic field such that the field passes through the cavities. A waveguide couples the waves from the cavities to the load, which may be the cooking volume within a microwave oven. The history of the development of microwaves in the microwave oven are far from mundane and illustrate the relationship among power, communication, and microwave transmission (Osepchuk, 2010). Other sources of microwaves besides the cavity magnetron include the klystron, traveling-wave tube, and gyrotron. The klystron and traveling-wave tube are vacuum-tube-based approaches to generating microwaves, while the gyrotron uses microwave amplification by stimulated emission of radiation, known as a maser.

Microwave communication has been used to form communication links along high-altitude and relatively unobstructed lengths of power transmission lines and has been increasingly used to transport carrier Ethernet (MEF, 2011; Wells, 2009). Microwave links can provide gigabit capacity using packet microwave in IP-based radio access networks (RANs). Hybrid microwave links provide both packet data and time-division multiplexed voice traffic. Microwave links are cognizant of fiber-optic synchronous optical networking (SONET) interfaces, discussed in the next section, and can provide and maintain SONET clock synchronization, which may be useful for power grid applications.

Communication via fiber optics is another technique that has long been utilized for the power grid transmission system. Although fiber–optic communication has already been covered in detail, because it has a long history with the power transmission system it is also discussed here. Optical fiber has many advantages for use in the power grid, including high bandwidth, the potential for relatively long-distance communication without the need for relays or amplification, and is inherently immune to electromagnetic interference. Additional advantages, as we will see later in this section and in Chapter 14 on power electronics, are that optical devices can also be used as current and voltage sensors. Disadvantages are primarily due to both cost and the need to lay and manage tremendous lengths of yet more cabling. A technique to help mitigate the latter problem has been to try to find ways to more-directly integrate fiber optics into the process of installing power line cables (Ostendorp *et al.*, 1997). Note that this is a typical progression of technology in general known as "integration with the supersystem"; namely, technologies evolve to find synergies with their environment or with one another.

Some examples of the integration of fiber optics into power line cabling include encasing the optical fiber within the ground wire for power transmission lines. This is known as an optical ground wire (OGW). It combines the functions of grounding and communication by integration of the optical fiber with the ground wire. The optical fibers form the central portion of the integrated cable and they are surrounded by steel and aluminum wire layers. This integrated communication and ground wire is run along the top of transmission towers. This serves to connect the towers to earth ground and also serves to protect the actual power transmission lines from lightning strikes. The high-bandwidth communication that the optical fiber provides can be used for managing the transmission system, for transmission of other utility information, including voice, or for lease to third-parties for telecommunication usage. Optical fiber acts as an insulator and by its very nature can serve useful functions as an insulator to transmission line arcing, the induction of current from lighting discharges, as well as potential cross-talk.

Typically, actual power transmission line cables are known as aluminum conductor steel-reinforced cable (ACSR). The outer strands are aluminum because of their high conductivity, low density, and low cost, while the center strands are made of steel for their strength, required to support the aluminum strands. Much of cabling involves mechanical issues related to solving the engineering contradictions of low cost, high strength, and low power loss. Our focus remains on the communication aspects, although it should be noted that wireless communication and wireless power transmission could solve many these problems. Another type of integrated power and communication cabling involves wrapping the optical fiber around the phase conductor or ground wire. This utilizes an existing cable for support. Careful consideration has to be given to bending of the optical fibers and the mechanical changes that will occur to the wrapped cable system. For example, galloping or whipping of the cable due to wind can occur, placing mechanical stress upon the entire cabling system.

Finally, the all-dielectric self-supporting cable is another type of fiber optic cabling system that is widely deployed. In this case, the terms "all-dielectric" and "self-supporting" both indicate that the cable is composed entirely of dielectric material. No extraneous metallic cabling or wiring is added in order to provide additional strength to the cable. This provides an alternative to OGW and has a lower installation cost.

The driving force behind fiber optic communication has been the significant amount of bandwidth that it provides (Massa, 2000). So it is worth considering the underlying operation of fiber optics in more detail. The invention of fiber optics for communication took place in the early 1970s and has grown tremendously since then, with the widespread deployment of fiber in the 1980s. To provide a sense of the development of the technology, consider that early fiber-optic systems operated at 90 Mb/s or 1300 simultaneous separate voice channels. By the 2000s, it was operating at 10 Gb/s and higher with the equivalent capacity of 130 000 simultaneous voice channels. Additional advances, such as dense wavelength-division multiplexing and erbium-doped fiber-optic amplifiers (EDFAs) have increased the transmission rate to over a terabit per second, equivalent to 13 million simultaneous voice channels, over distances longer than 100 km. There is no question that the channel bandwidth is very large with this technology.

The advantages of fiber over other forms of electromagnetic communication, including copper, are the fact that the signal has relatively low attenuation over long distances through optical fiber. The energy per bit of information communicated is lower. A voice-grade copper wire requires signal amplification every few kilometers, while a fiber-optic line can extend for over 100 km without processing. Fiber has a much smaller diameter than copper, has larger bandwidth, and is lightweight, which was previously mentioned as very important in installation with power transmission lines. Because optical fiber has no metallic component it is also ideal in power electronics applications; there is little chance of electromagnetic interference.

How does fiber optics achieve these benefits for communication? Fundamentally, a fiber optic system converts an electrical signal into a light signal that travels over an optical fiber cable. At the other end of the cable, a receiver detects the signal and converts it back into an electric signal. This sounds simple enough; however, the details of extracting the most channel capacity for least cost from such a system are a bit more complex.

As we will see, the underlying contradiction to be resolved in engineering design revolves around cost, capacity, and distance. For example, increasing capacity increases cost and can reduce distance. Typical sources for the signal are either LEDs or lasers. The wavelengths used are in the near-infrared spectrum, just beyond visible light and invisible to the human eye. The wavelengths are determined simply by which wavelengths travel best through the optical fiber, where the term "best" will be clarified as we proceed. Frequency ranges are centered around 850, 1310, and 1550 nm. The ranges of operation are typically 50 nm above and below these wavelengths.

The first problem with optical fiber is attenuation, expressed in decibels, as shown by

$$\mathrm{Loss}_{\mathrm{dB}} = \log \frac{P_{\mathrm{out}}}{P_{\mathrm{in}}}, \tag{8.43}$$

where P is the power of either output or input to the fiber. Since attenuation increases with distance it is often expressed as decibels per kilometer. A simplification is to assume that input power is 1 mW. In this case, the units are expressed as dBm, as shown explicitly in

$$\mathrm{Loss}_{\mathrm{dBm}} = \log \frac{P_{\mathrm{out}}}{1\ \mathrm{mW}}. \tag{8.44}$$

This makes computing link budgets – that is, accounting for all the losses along the channel – an easy process. Each individual component loss can be subtracted. It is also handy to keep in mind that for every 3 dB of loss the power is reduced by half.

Now let us consider the signal in the fiber in a bit more detail. There are three types of fibers: step-index multimode, step-index single mode, and graded-index. These types of fibers are related to the index of refraction, which impacts how a light wave passes through the fiber. The term "step-index" in step-index multimode fiber indicates that there is a sharp change in the refractive index between the core and the cladding, or outer layer surrounding the core. This causes light waves to be reflected in a straight line from one wall of the optical core to another down the length of the fiber. The term "multimode" means that there are many different reflective paths that the light beam can take through the fiber. In this type of fiber, the core is relatively large and easy to work with and to interconnect different transmitters and receivers. But keep in mind that the core is still on the order of the size of a human hair. Also, this type of fiber can be driven by both LEDs and lasers. The disadvantage of this type of fiber is that, because light can take many different paths, there is modal dispersion. This is analogous to multipath in wireless systems, in which a signal takes multiple paths to arrive at the same destination; this causes the signal to become distorted and results in a reduction in bandwidth.

A single-mode step-index fiber is similar to the previous type of fiber, but its core is narrow enough that it has only a single mode or path through the fiber. This eliminates modal dispersion and allows for greater bandwidth over longer distances. However, another signal distorting phenomenon can occur; namely, chromatic dispersion. In chromatic dispersion, individual wavelengths of light propagate at different velocities and take different amounts of time to reach the receiver. Solutions to mitigate this problem include transmitting at a wavelength at which the fiber has a constant index of refraction, namely 1300 nm, using a source that emits at only one wavelength, or, finally, employing a compensation mechanism within the fiber. Because single-mode step-index fiber has such a small diameter, it can be difficult to work with and is driven solely by lasers; LEDs are not used because it is difficult to couple the signal to the fiber without high signal loss.

The final type of fiber is known as graded-index fiber. Here, the index of refraction between the fiber core and the outer cladding changes gradually rather than having a sudden, sharp change. The result is that, instead of the light wave reflecting directly from one wall of the fiber to the other, the light waves bend in a curved, parabolic manner such that they are periodically refocused as they travel through the fiber. The result is a reduction in modal dispersion. This reduces the noise and increases the bandwidth of the fiber.

From the discussion so far, we can see that advances in extracting more bandwidth over longer distances through fiber have involved clever approaches to reduce dispersion. There are two sources of dispersion within an optical fiber: material dispersion and waveguide dispersion. Material dispersion is the frequency-dependent response of the material to light waves. The other type of dispersion, waveguide dispersion, is related to the geometry of the waveguide, the optical fiber in this case. It turns out that material dispersion and waveguide dispersion can have opposite signs, which means that there is an optimal wavelength at which they precisely cancel. This occurs at 1310 nm in an optical fiber. However, the other factor, attenuation, reaches a minimum at a wavelength of 1550 nm. EDFAs operate at 1550 nm and the ideal case would thus be to design a fiber system such that minimum dispersion coincides with minimum attenuation. In fact, this occurs in a type of fiber known as zero-dispersion-shifted fiber. The waveguide dispersion is modified so that the zero-dispersion wavelength is shifted to 1550 nm. The EDFA is an optical amplifier that uses a laser to directly boost the optical power of a signal without having to convert the signal to electrical form.

Dispersion is the spreading of a pulse through the optical fiber. The greater the spreading, the greater the chance that pulses will overlap with another, causing information to be indistinguishable and thus reducing the amount of bandwidth through the fiber. Dispersion is measured as shown in Equation 8.45 in units of time, typically in nanoseconds or picoseconds:

$$\Delta t = (\Delta t_{\mathrm{out}} - \Delta t_{\mathrm{in}})^{1/2}. \tag{8.45}$$

Dispersion increases with distance through the fiber, so the measure should account for total dispersion as a function of distance as shown by

$$\Delta t_{\mathrm{total}} = L \times \frac{\mathrm{Dispersion}}{\mathrm{km}}, \tag{8.46}$$

where L is the length of the fiber. So, we can see immediately that a shorter fiber can potentially provide more bandwidth; the challenge is to increase bandwidth over longer distances. If there are different types of dispersion, they aggregate as shown by

$$\Delta_{\mathrm{tot}} = [(\Delta t_1)^2 + (\Delta t_2)^2 + \cdots]^{1/2}, \tag{8.47}$$

where t_n is each type of dispersion.

Pulse-code modulation is used to convert analog to digital code for fiber optic communication. Digital values can be transmitted in any number of standard encoding formats from non-return-to-zero to Manchester encoding. Recall that pulse-width modulation in power systems is a technique to digitally control power to electrical devices that need careful control by changing the width of the duration of the time that power is sampled from its alternating-current cycle. This is related to pulse-code modulation in communication, in which analog signals are sampled and converted to digital form.

Once information is encoded for transmission over the fiber-optic cable, it becomes apparent that, given the tremendous available bandwidth, many different channels can share the fiber simultaneously. This is accomplished through multiplexing, and there are two types of multiplexing: time-division multiplexing and wavelength division multiplexing (WDM). Of course, time-division multiplexing divides the channel into time-slots and assigns channels based upon time-slot. Fiber-optic systems utilize the SONET to do this. WDM divides the channels by placing them on different wavelengths. Dense wavelength-division multiplexing is the technique of transmitting many closely spaced wavelengths within the same fiber. The ITU defines the standard frequency spacing as 100 GHz. This frequency spacing results in a 0.8 nm wavelength spacing, where $\Delta\lambda = \lambda\Delta f/f$. Typical systems could multiplex up to 128 different wavelengths yielding 2.5 Gb/s and 32 at 10 Gb/s. A wavelength-division multiplexer separates or combines signals that are being transported at different wavelengths. A fiber Bragg grating is used to separate wavelengths using diffraction. A small portion of the optical fiber can be constructed such that it reflects a specific wavelength and passes all other wavelengths. This allows channels to be added or dropped from the fiber as needed.

As previously mentioned, either an LED or a laser provides the light source for the fiber-optic system. The signal to be transmitted is applied to the light source by a modulator. A waveguide Mach-Zehnder interferometer is one type of modulator. It can be fabricated on a substrate of lithium niobate ($LiNbO_3$). An RF signal can be impressed upon the light wave by splitting the light wave into two paths. One path passes through unmodulated. The other path passes through electrodes. When a voltage is applied to the electrodes, the index of refraction is changed in proportion to the applied voltage. This causes the light wave to experience a phase delay relative to the unmodulated light wave. When the waves combine near the output of the device, they may combine constructively, indicating an "on" signal, or destructively, indicating an "off" signal. Note that we can turn this device on its head and say that it is using light to measure voltage. Thus, this device could conceptually be used to measure voltage in the power grid. In fact, optical devices exist to measure both voltage and current and thus to construct optical synchrophasor measurement units (Nuqui *et al.*, 2010).

The fiber-optic receiver detects light striking it; absorbed photons excite electrons with enough energy to rise from the valance band to the conduction band (recall discussion of photovoltaics in Section 8.2.4). This enables a small current to flow when a bias voltage is applied that must then be amplified. Of course, there is much more detail in the development of a sensitive detector in order to be a high-performing fiber-optic receiver; this detail is related to the classic ability of a receiver to be sensitive enough to detect a signal but without giving a false output in the presence of noise. Quantum efficiency is a ratio of the electrons generated by the detector to the number of photons striking the detector. Dark current is current generated by the detector with no light applied. Response time is the amount of time it takes for the detector to respond to the optical input. This depends upon the rise time of the signal.

Finally, a typical link budget analysis can be undertaken to determine power loss through the system and the margin required to ensure a clean signal. Typically, this involves the power transmitted from the source, the source to fiber loss, the fiber loss per kilometer, the connector or splice losses, and the fiber-to-detector loss. Power margin is the difference between the received power and the receiver sensitivity. Now that we have covered common wide-area communication techniques, let us consider the potential for control over such long distance networks.

8.6 Networked Control

This section begins the introduction to networked control. More detail related to networked control and stability is found in Section 12.2. Networked control is the implementation of a control system over a network such that portions of the control system can be located remotely from one another. In an extreme case, sensors, actuators, and the controller may each be separated from one another by hundreds of miles and communicate with one another over a communication network. Typically, just as there has been a separation of communication and network engineers and power engineers there has been a separation of communication and network engineers and control engineers. Thus, it is important to have at least a very basic understanding of control theory because much of the power grid is concerned with optimizing efficiency and reliability through communication and control. The emphasis upon communication in the smart grid is often precisely so that control systems can be distributed through the grid; sensors (for example, current transformers, voltage transform, PMUs, and any number of other types of sensors), actuators (for example, inductors, capacitors, switches, relays, and FACTS devices), and controllers, whether located in control centers, substations, or embedded within intelligent electronic devices, all require communication in order to work together.

A simplified illustration of a networked control system is shown in Figure 8.7. A networked control system is sensitive to the latency of messages transmitted through the network; even though communication engineers work hard to minimize this latency, it still persists, particularly over the long distances in a wide-area network. In classical control theory, the control system is typically analyzed via transfer functions in the frequency domain, where frequency is

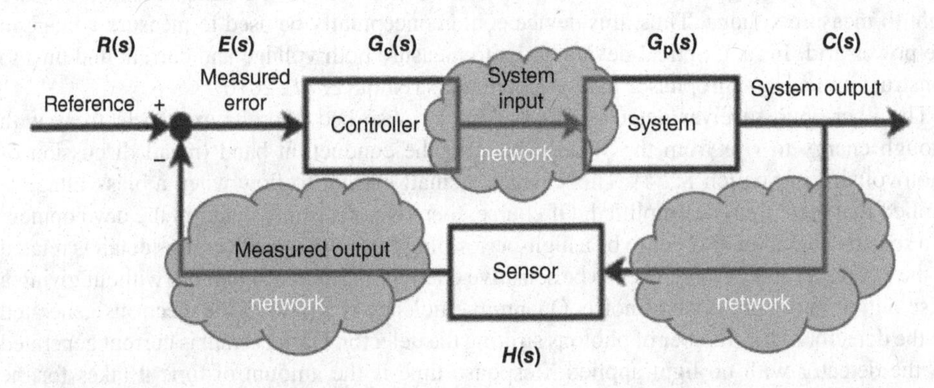

Figure 8.7 A networked control system has components that are remotely located and operate over a communication network.

a complex value (Engelberg, 2008). Referring to Figure 8.7, let $H(s)$ be the Laplace transform of the feedback through the "Sensor" box, $G_c(s)$ be the "Controller" box, and $G_p(s)$ be the plant or "System" box. The goal is to understand the relationship between the communication network, primarily represented by the delay through $H(s)$, and the controller $G_c(s)$.

First, assume temporarily that the communication network is perfect; it delivers packets without error and instantaneously. This allows us to ignore the communication network and focus on the basics of the control system. Since the goal of a control system is to control the relationship between the input and the output of the system, it should make intuitive sense that it would be desirable to look at the ratio of the output to the input. To derive this, begin by looking at the output of the system $C(s)$ and work backwards to see the following derivation:

$$C(s) = G_p(s)G_c(s)[R(s) - H(s)C(s)]. \tag{8.48}$$

Next, solve for $C(s)$ by expanding Equation 8.48 and bringing the $C(s)$ on the right around to the left, as shown in

$$C(s) = \frac{G_p(s)G_c(s)}{1 + H(s)G_c(s)G_p(s)}R(s). \tag{8.49}$$

Now, to obtain the ratio of output to input, simply divide the equation above by $R(s)$:

$$T(s) = \frac{\text{output}}{\text{input}} = C_{(S)}/R_{(S)} = \frac{G_p(s)G_c(s)}{1 + H(s)G_c(s)G_p(s)}. \tag{8.50}$$

Note that $H(s)$ represents the feedback loop and can be assumed to simply be one. In this case there is a direct comparison of the output with the input and there is no delay in communicating the feedback information. Now, continuing to keep things simple, assume the controller is defined as $G_c(s) = K$; that is, the controller purely adds gain. Making these assumptions yields

$$T(s) = \frac{KG_p(s)G_c(s)}{1 + KG_p(s)}. \tag{8.51}$$

If the goal in this simple example is to ensure that the output follows the input as closely as possible, then $T(s) = 1$ and $KG_p(s) = 1 + KG_p(s)$. This can only be true as K approaches infinity. Thus, a high-gain controller, at least in this specific example without noise or delay, seems to result in the best control system.

Consider Figure 8.8, in which noise $N(s)$ enters the system immediately before the plant $G_p(s)$. The transfer function can be determined just as before by working backwards, as shown by

$$C(s) = G_p(s)\{N(s) + G_c(s)[R(s) - H(s)C(s)]\}. \tag{8.52}$$

Notice that the noise term $N(s)$ is now included in the equation. Expanding Equation 8.52 results in

$$C(s) = G_p(s)N(s) + G_p(s)G_c(s)R(s) - G_p(s)G_c(s)H(s)C(s). \tag{8.53}$$

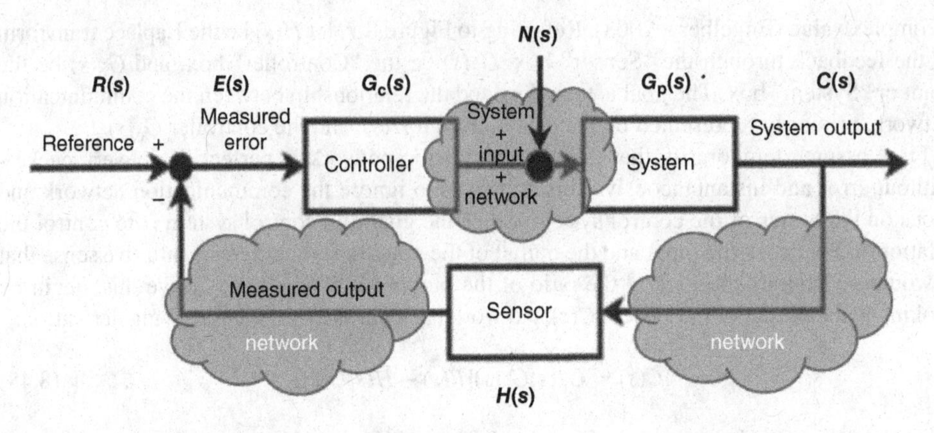

Figure 8.8 The communication network introduces noise into the operation of the control system. Information may be delayed, dropped, or reordered due to conditions in the network.

The next step is to rearrange Equation 8.53 to get the output of the system clearly in terms of the inputs to the system $R(s)$ and $N(s)$:

$$C(s) = \frac{G_p(s)G_c(s)}{1 + G_p(s)G_c(s)H(s)}R(s) + \frac{G_p(s)}{1 + G_p(s)G_c(s)H(s)}N(s). \tag{8.54}$$

Assuming as before that the controller is pure gain K and $H(s) = 1$ yields the simplification

$$C(s) = \frac{G_p(s)K}{1 + G_p(s)K}R(s) + \frac{G_p(s)}{1 + G_p(s)KH(s)}N(s). \tag{8.55}$$

Now we can apply the earlier assumption that K is very large. If this is true, then the factor multiplying $R(s)$ goes to one and the factor multiplying $N(s)$ goes to zero, as shown in

$$C(s) = R(s) + 0 \cdot N(s). \tag{8.56}$$

Hopefully, this little exercise introduced, in a simplified manner, the notion of control systems and how classical control systems can be analyzed. The next question, from a communication point of view, is what happens when the communication network is no longer perfect? The network may experience noise on the line and drop packets, it may experience congestion and packets may queue up along the way packets maybe reordered en route; any number of common events can cause the network to deliver packets after a relatively long delay or not at all.

Consider a time shift due to communication latency $y(t - T)$, where T is the latency. The Laplace transform is $\mathcal{L}(y(t - T)) = e^{-Ts}Y(s)$. The feedback $H(s)$ then takes the general form of $\alpha y(t) + Ay(t - T)$, where α is the information that arrives at the correct time and A is the information that is delayed due to network latency. As long as $\alpha > A > 0$, it can be shown that the system will be stable. However, if A becomes large relative to α, the system may become unstable.

Characterization of stability comes from examining the transfer function as a ratio of polynomials:

$$\frac{\alpha s + \beta}{as^2 + bs + c}. \tag{8.57}$$

The inverse Laplace transform is of the following form:

$$\frac{e^{(-b/2a)t}}{a}\left[\alpha \cos\left(\sqrt{\frac{4ac-b^2}{4a^2}}t\right) - \frac{\alpha b - 2a\beta}{\sqrt{4ac-b^2}} \sin\left(\sqrt{\frac{4ac-b^2}{4a^2}}t\right)\right] u(t). \tag{8.58}$$

The important point to notice about function 8.58 is that its rate of growth is determined by the first term, the exponent to Euler's number; namely

$$\frac{-b}{2a}. \tag{8.59}$$

The frequency of oscillation, from the argument to the trigonometric functions in Equation 8.58, is

$$\sqrt{\frac{4ac-b^2}{4a^2}}. \tag{8.60}$$

Therefore, it is clear that any pole of Equation 8.57 in the positive real portion of the complex plane, namely the right half-plane, will cancel the negative value in Equation 8.59 and create unbounded growth as t increases, and thus result in an unstable function. The system becomes unstable when the region of operation in the Nyquist plot encircles a pole or zero.

Networked control performance can be measured by how well the control system output follows its reference. This can be achieved via the following measures: integral squared error (ISE) Equation 8.61, integral absolute error (IAE) Equation 8.62, and integral time-weighted absolute error (ITAE) Equation 8.63:

$$\int \epsilon^2 \,\mathrm{d}t, \tag{8.61}$$

$$\int |\epsilon| \,\mathrm{d}t, \tag{8.62}$$

$$\int t|\epsilon| \,\mathrm{d}t. \tag{8.63}$$

The ISE will penalize large errors more than small ones because of the squaring operation. Minimizing error in this manner will lead to quick response but with a potentially large amount of low-amplitude oscillation. The IAE may not reduce error quickly, but it will reduce low-amplitude oscillation. Finally, the ITAE weights error that exists after a long time period as opposed to error that has a short duration. This encourages the system to converge more rapidly. Other measures include: steady-state offset, percentage overshoot, maximum absolute overshoot, rise time, period of oscillation, settling time, and decay ratio. Each of these measures has precise definitions and provides a practical indication of the performance of the control system and, most importantly from our perspective, will be influenced by the underlying communication network.

8.7 Summary

This chapter on DG and transmission began by looking at communication for inverter control. Communication for control systems has been covered in progressively more detail as the book proceeds. In the areas of DG and transmission, communication for fast but stable control are key requirements if network communication is going to be able to play a leading role in the power grid. While there are many forms of DG, this chapter looked at PVs as a representative example of DG. Again, we saw that inverter control was a key aspect of smart grid communication. On the transmission side, this chapter looked at FACTSs as well as HVDC transmission. Again, we saw that inverter control and power-electronic communication interfaces were paramount to the smart grid. Since transmission occurs over a large area, wide-area communication was reviewed. The chapter ended by advancing the discussion on networked control.

Chapter 9 covers distribution automation (DA). As has been alluded to in previous chapters, the distribution system is where much of the action in the smart grid is taking place. This is because many of the smart grid systems meet within the distribution system. Because the distribution system is greatly exposed to the environment and consumers, protection, reliability, and self-healing are key concepts. The classical distribution system is a large, complex system that must safely transform and distribute power to individual consumers. The distribution system forms the "last mile" of power transport. In addition, the distribution system has been a natural interface for many different "smart grid" applications. The distribution system is where "the rubber meets the road" with regard to smart grid and communication. This opens up many opportunities for distribution automation, such as many new opportunities to combine smart grid applications in new ways. Protection coordination is a significant component of the distribution system and new ways of automating protection and incorporating self-healing are discussed. Communication that has the most suitability and flexibility for the rapid technology changes expected in the distribution system are discussed.

8.8 Exercises

Exercise 8.1 Fiber Dispersion

A 2 km length multimode fiber has a model dispersion of 1 ns/km and chromatic dispersion of 100 ps/(km nm). If it is used with an LED of line width 40 nm:

1. What is the total dispersion?
2. Calculate the bandwidth of the fiber.

Exercise 8.2 Power Line Carrier

Consider an overhead line that is 250 Ω and that is connected to a cable at 25 Ω.

1. What is the signal loss due to impedance mismatch in dB?

Exercise 8.3 Resonant Coupling

1. What is the Q factor for a resonant coil system in which the resistance is 10 Ω, the capacitance is 1 μF, and the inductance is 10 mH?

Exercise 8.4 Microgrids

1. Explain the ramifications of attempting to utilize communication for generator control versus droop control. What are the dangers and reliability issues involved in using communication for explicit generator control?

Exercise 8.5 Nanogrids

1. Determine to what extent the concepts in this chapter apply to small-scale power control (see Section 15.3.2 for more on nanogeneration).

Exercise 8.6 Synchronization

1. Determine the stability of the linear time-invariant system with the state matrix in Equation 8.64. *Hint:* use a Lyapunov function.

$$A = \begin{bmatrix} 0 & 1 \\ -6 & -5 \end{bmatrix}. \tag{8.64}$$

Exercise 8.7 Impact on Distribution

1. Anticipate how the microgrid concept could help solve problems with distributed generation. Note that the impact of microgrids on distribution is explained in Chapter 9.

Exercise 8.8 Islanding

1. Why has islanding been considered potentially dangerous? Explain how communication may be used in identifying islands.

Exercise 8.9 FACTS

1. Explain the main methods of flexibly controlling power over a transmission line.

Exercise 8.10 FACTS

1. How are advances in high-power solid-state electronics impacting FACTS?

Exercise 8.11 WiMAX Mesh

1. How does the WiMAX Mesh protocol differ from WiMAX?
2. What are the advantages and disadvantages of WiMAX Mesh compared and contrasted with WiMAX?

Exercise 8.12 IEEE 802.11

1. What does an 802.11 frame structure look like?

Exercise 8.13 IEEE 802.11s

1. What is the primary difference between 802.11 and 802.11s?
2. Where and what applications would 802.11s be best suited for in the smart grid?

Exercise 8.14 IEEE 802.15.4

1. What type of medium-access control does IEEE 802.15.4 use and how does the medium-access control operate?

Exercise 8.15 6LoWPAN

1. What physical/media-access control layer was 6LoWPAN designed to reside upon?
2. What functionality does 6LoWPAN provide?
3. What applications would 6LoWPAN be best suited to in the smart grid?

9

Distribution Automation

This 'telephone' has too many shortcomings to be seriously considered as a means of communication. The device is inherently of no value to us.

—Western Union internal memo, 1876

We can't solve problems by using the same kind of thinking we used when we created them.

—Albert Einstein

9.1 Introduction

This is the fourth chapter in Part Two, *Communication and Networking: The Enabler*. The focus in this chapter is on the power distribution system and smart grid. In the past, much of the research and analysis had gone into the meatier systems of the power grid; namely, large, centralized power generators and the transmission system, which generated and carried bulk power. The distribution system, in the mean time, received relatively little attention, an appendage required to make the final transition to the end customer. However, the concept of the smart grid is breathing new life into the distribution system. Distributed and renewable generation at the customer premises, self-healing protection mechanisms, and DA are among the many ideas impacting the distribution system. From a communication perspective, the distribution system can cover the area of a small city. Thus, what are loosely termed as MANs are most applicable here. Several key parts related to the evolving distribution system are covered in other chapters. The basics of pre-smart-grid distribution were discussed in Chapter 4, while DG was discussed in Section 8.2. Also, AMI, which is communication for smart meters within the distribution system, was discussed in Section 7.3. All of these systems point to distribution as a key player in the smart grid. The distribution system is where "the rubber meets the road" so to speak. It is where consumers meet the power grid and where the results of generation and transmission impact the consumer.

The first topic covered is DG within the distribution system. An overview of the type of equipment, configuration, and challenges is given. Section 9.2 covers protection within the distribution system. The distribution system is one of the most exposed portions of the power

Smart Grid: Communication-Enabled Intelligence for the Electric Power Grid, First Edition. Stephen F. Bush.

grid; power lines are within easy reach of buildings, trees, animals, and customers themselves, requiring significant consideration for balancing protection of objects and living organisms that could come into contact with dangerous amounts of power if a fault occurs versus keeping as many customers as possible energized for as long as possible; that is, minimizing the impact of an electrical fault. A smart grid should be able to detect faults as well as heal itself after faults occur. Protection coordination has been configured manually, using expert knowledge, rules of thumb, and trial-and-error approaches. Section 9.3 considers automated optimization approaches for protection coordination. Distribution substations have been one of the first components of the power grid to be updated with advanced communication in the form of Ethernet switches. The next section reviews substation automation and communication. Wireless communication is also an option within the distribution system, and this is discussed in Section 9.3.6. Wireless communication capable of covering the area within a distribution system is reviewed. The final section of this chapter covers communication-network-specific aspects of the distribution system. This includes issues focused upon packet routing and the impact of variable latency in the distribution system. Note that the AMI was covered in Section 7.3. Since the distribution system is the "last mile" from a power standpoint to reach the end customer, the AMI and the distribution system naturally work together to also reach the customer meter. This section also closes the loop on that topic. The chapter ends with a brief summary and exercises for the reader.

In its simplest form, DA refers simply to greater automation of processes within the distribution system. However, automation of power distribution processes has been evolving since the power grid began operation. By this definition, DA is certainly not a recent smart grid activity. Sometimes other terms, such as "advanced distribution automation (ADA)," are used to attempt to emphasize more recent activity that involves greater integration of communication, computing, and networked control. A relatively short-term vision for DA is a distribution system that is highly automated with a more flexible electrical system architecture and supported by open-architecture communication networks. DA should result in a system that is multifunctional and takes advantage of new capabilities in power electronics, IT, and system simulation. Real-time state-estimation tools should be used to perform predictive simulations and to continuously optimize performance, including demand-side management, efficiency, reliability, and power quality in real time and help bridge the communication–power architectures.

Beyond automating existing distribution functions as explained in Chapter 4, DA should also be integrating other developing smart grid systems; for example, DG discussed in Section 8.2 and AMI covered in Section 7.3. Because several such systems uniquely overlap in the distribution system, this is an indication that the distribution system is uniquely poised to evolve quickly and to benefit from smart grid advances. One can expect that the integration of systems within the distribution system will yield new terminology that merges existing system names such as advanced distribution infrastructure (ADI), indicating an integration of DA and AMI, or distribution energy storage system (DESS) for energy storage managed by the distribution system. Finally, DA will evolve to develop entirely new and yet-to-be-defined automated systems and applications by building upon the integrated smart grid systems and applications.

Generally speaking, automation started at the substation and has extended outward toward the consumer and smart meters through AMI systems; it will likely continue to move from substation to feeder to consumer systems. Communication and control will eventually integrate with DR and real-time pricing systems. Integration will be a longer term process in which information and equipment will be consolidated, and current distinct hardware and software systems will merge to reduce cost and eliminate redundancy. Lower communication costs could be a driver toward increasing smart grid communication integration. This is because the cost of communication must be offset by a resulting improvement in system reliability and/or improved operation and management efficiencies. Three typical communication approaches have been distribution line carrier, land lines, and wireless communication systems. While distribution line carrier has been cost-effective in meter reading and load control, as well as some protection mechanisms, it suffers from the fact that communication is lost when a line is severed. This is known as the open-circuit problem for distribution line carrier. Unfortunately, when a line is severed, communication is lost precisely when it is most critical. Land lines include telephone lines and fiber-optic connections, where leased-lines are often used to reach substations. However, cost becomes an issue and they are less often used for DA. Fiber-optic lines tend to be used because of their high bandwidth capability and noise immunity; but of course, they are costly as well. Wireless communication has the advantage of enabling communication almost anywhere at relatively low cost. Private wireless systems can be expensive but, in return, allow a utility full control over the wireless system. Public wireless systems can be less expensive, but they are not under the utilities' full control.

As we can see, distribution systems are evolving to become complex systems that will include everything from DG to the AMI as well as almost everything in between. A concern is that the distribution system could become a hodgepodge, or ad hoc collection, of crudely interconnected systems with communication networks thrown on top for good measure. There needs to be a clear, theoretical understanding and model of what a distribution is and how it is evolving to prevent the formation of such an ad hoc system.

There are three general architectures, or frameworks, for considering how the distribution system could evolve: (1) active grids, (2) microgrids, and (3) virtual utilities. An "active grid" is also known in the literature as an "active network" and should not be confused with an active communication network, which is also called an active network and is discussed in Section 6.5.2. The term "active grid" will be used in this chapter instead of "active network" to be clearly distinguishable from an active communication network. However, the concept is the same in both power and communication: an active system is one that is more controllable with more flexibility and degrees of freedom. Active networks in both power and communications comprise open, programmable interfaces enabling a highly dynamic system. The IP is a nonactive network in which messages flow from source to destination without programmable control once the network is configured; so, too, pre-smart-grid distribution systems assumed passive power flow once the distribution system was constructed and configured. Instead, an active grid allows bidirectional power flows to be adjusted while the system is in normal operation. An evolution toward an active grid could begin with simply controlling generator input to the active grid. Then, full control over all DG sources along with a coordinated dispatch system and voltage profile optimization could be implemented. Finally, the active grid could be divided into a highly interconnected set of local areas that are responsible for their own management and that negotiates exchanges with other local areas within the active grid for power.

Another model, the microgrid model, was introduced in Section 5.5.1. The microgrid is an independent energy generation and distribution system, able to operate with or without a connection to the main power grid. It is quite similar to a local-area in the above active grid model.

Finally, the virtual utility, similar to the virtual power plant concept discussed in Section 5.5, is the management and control of a set of distributed generators, including renewable energy sources, such that the set of distributed generators appears as a single utility to the consumer. The virtual utility transmits and responds to price signals as a single coordinated system.

One way to understand DA is to examine a typical DMS. Management of a distribution system involves many different applications that monitor and control the entire distribution system; a DMS typically incorporates all these applications into a single, coherent platform with a consistent look and feel. The goal, of course, is to improve the reliability and quality of service by reducing outages, minimizing outage time, and maintaining acceptable power quality, including frequency and voltage levels, throughout the distribution system. DMS applications include FDIR, IVVC, a topology processor, distribution power flow, load modeling/load estimation (LM/LE), optimal network reconfiguration, CA, switch order management, short-circuit analysis, relay protection coordination, optimal capacitor placement/optimal voltage regulation, and even various simulators. These applications began independently as extensions of the SCADA systems that extended from the transmission system into the distribution system. Because each of these applications were separate systems, operators had to contend with many different, and potentially confusing, interfaces and controls. As the distribution system grew in importance within the overall power grid, integrating these disparate applications into a common package made sense. The trend continues for monitoring, data acquisition, and control to extend further into the electrical network to the distribution pole-top transformer toward individual customers by means of AMI and DR and home EMSs on HANs. This provides higher resolution, more "granular" data. The assumption is that higher resolution data will enable improved optimization. Integration will continue with other systems, including GISs, OMSs, and MDMSs, among others. FDIR will require more optimization, and distribution network configurations will become more complex, moving from radial to mesh distribution network topologies as more DG and consumer–producers come online. IVVC and voltage optimization will progressively improve. LM/LE will change as consumer behavior becomes less predictable. As consumers respond to DR price signals, old assumptions about consumption profiles may no longer be valid. All DMS applications are likely to be used more frequently and reach down into individual customer premises. DG and electric vehicles will add complexity to all of these applications. In support of all of these changes there will be more visualization and dashboard metrics. There will be more interaction among DMS applications, and new applications are likely to emerge by leveraging combinations of existing applications. The smart grid is not only about doing what has been currently done better, but about discovering new possibilities. This will come as applications are developed that share the communication infrastructure, filling in product gaps and leveraging existing technologies to a greater extent in order to achieve greater synergy. Given the broad spectrum of applications in a DMS, standardization has been crucial. IEC 61968 defines a standard for information exchange among distribution system applications. The interface reference model of IEC 61968 defines standard interfaces for each class of application. It extends the common information model (CIM). More information on standards applied within the distribution system are discussed in Chapter 10. Next, we consider performance metrics that impact the distribution system.

9.1.1 Performance Metrics

Performance metrics have been designed to quantify the impact of faults on the customer base (Cheney *et al.*, 2009). The system average interruption duration index SAIDI is the sum of all customer interruption durations divided by the number of customers:

$$\mathrm{SAIDI} \equiv \frac{\sum r_i N_i}{N}. \tag{9.1}$$

The index i is a load point and r_i is the average restoration time at load point i.

The momentary average interruption event frequency index (MAIFI) is similar to the SAIDI except that it defines the customer impact in terms of the number of "momentary" outages, where the length of a moment can be arbitrarily defined. Thus, MAIFI is the number of interruptions greater than a specified duration divided by the number of customers:

$$\mathrm{MAIFI} \equiv \frac{\sum U_i N_i}{N}, \tag{9.2}$$

where U_i is the number of interruptions exceeding a given time at load point i.

The impact of smart grid communication on reliability within the distribution system needs to be better understood; new metrics are being developed that take communication into account. The operation of the distribution system is already quite reliable, given its exposure to the environment and heavy use by consumers. DG and DR will certainly add stress to the distribution system. A key question is whether communication will improve the reliability. While we know that SAIDI and system average interruption frequency index (SAIFI) metrics, discussed in Section 12.5.1, attempt to measure the reliability of the distribution system, they do not explicitly take communication into account. One proposed measure is the average communication failure frequency index (ACFFI) (Aravinthan *et al.*, 2011), which extends the commonly used distribution reliability metrics. In communications, the signal-to-noise ratio defines when the signal falls below the noise level and valid information fails to be properly communicated. This can result in dropped bits and bit error. If redundant information provided by channel coding cannot restore the corrupt bits, then the entire message may be in error. Retransmission techniques can be used to retransmit the message; however, this results in increased latency. The ACFFI metric attempts to capture this behavior in the form of a traditional distribution reliability metric applied all the way to devices on the consumer premises, assuming AMI enables this capability. ACFFI is defined as

$$\mathrm{ACFFI} = \frac{\sum w_i n_i}{\sum w_i N_i}, \tag{9.3}$$

where w_i is a predefined weight for appliance of type i, N_i is the number of appliances in a household of type i, and n_i is the number of times a communication message is missed or detected below threshold for an appliance of type i. Thus, ACFFI is the ratio of the total weighted number of missed events to the weighted number of appliances. Notice that information from different types of appliances may be weighted differently; a device with heavy energy use, or that is more important to manage, may be configured with a higher rating. This brings up the notion of consumers being able to purposely interfere with communication in order to minimize their electric bill. Such a metric could call for consumers to be responsible for reliable communication because they would be penalized for an abnormally high rate of unreliable communication of the smart meter.

Another metric that follows a similar to approach to distribution reliability metrics is the average communication interruption duration index (ACIDI) defined as

$$\mathrm{ACIDI}_k = \frac{\sum_{\forall i \in k} \text{Communication failure duration for each node due to defect } i \text{ in cluster } k}{\text{Total number of nodes in a cluster}}. \tag{9.4}$$

While ACFFI measures the rate at which communication interruptions occur, ACIDI measures the duration of interruptions where k is a sensor cluster. The idea is that spatially local sensors would form communication clusters for purposes of communication efficiency. The average communication interruption duration is simply the sum of the ACIDIs:

$$\text{ACIDI} = \sum \text{ACIDI}_k. \tag{9.5}$$

Finally, since the main goal of the distribution system is to reliably provide power, a corresponding metric that explicitly incorporates communication is the energy not served due to communication failure (ENS-C):

$$\text{ENS-C} = \frac{\sum_i \text{Energy not served while communication failure } i}{\text{Total energy not served}}. \tag{9.6}$$

This is the amount of energy lost due to a communication failure. This measure can be used as a design trade-off with alternative, cheaper communication technologies. Lower cost communication technologies may have a corresponding increase in ENS-C that offsets the initial lower cost.

These metrics are useful because they provide a direct comparison among communication technologies as they apply to the distribution system and they can be applied to analyses and simulations related to DA. These metrics inherently take into account the impact of cybersecurity, latency, and scalability of the communication system.

9.2 Protection Coordination Utilizing Distribution Automation

Protection coordination has been discussed in Section 4.2.7. Since this chapter is about smart grid DA, protection coordination is explored from the point of view of communication. It begins with a detailed review of time–current curves and then FDIR.

9.2.1 Time–Current Curves

Protective relays have a long history; their early 20th century terminology continues to carry over to today's technology even when the implementation has changed significantly. The older, underlying models for protective relaying provide an underpinning for our understanding and relay settings today and in microprocessor-based relays. Thus, it is important to understand the terminology and where it came from. The setting for protective devices is often characterized by an inverse time–current curve whose terminology comes from the induction disk relay (Benmouyal *et al.*, 1999). The induction disk relay, pictured in Figure 9.1, is a very old device. The concept of operation for this device is used in the terminology of TCCs today. First, notice that there is a metallic disk capable of rotation either clockwise or counterclockwise, depending upon the applied forces. Next, notice that there are two metal contacts, one fixed and one that travels with the disk. In Figure 9.1, clockwise rotation would bring the contacts toward one another. A spring provides tension in the opposite, counterclockwise direction. Note also that there is a stop position controlled by a time dial in the upper portion of the

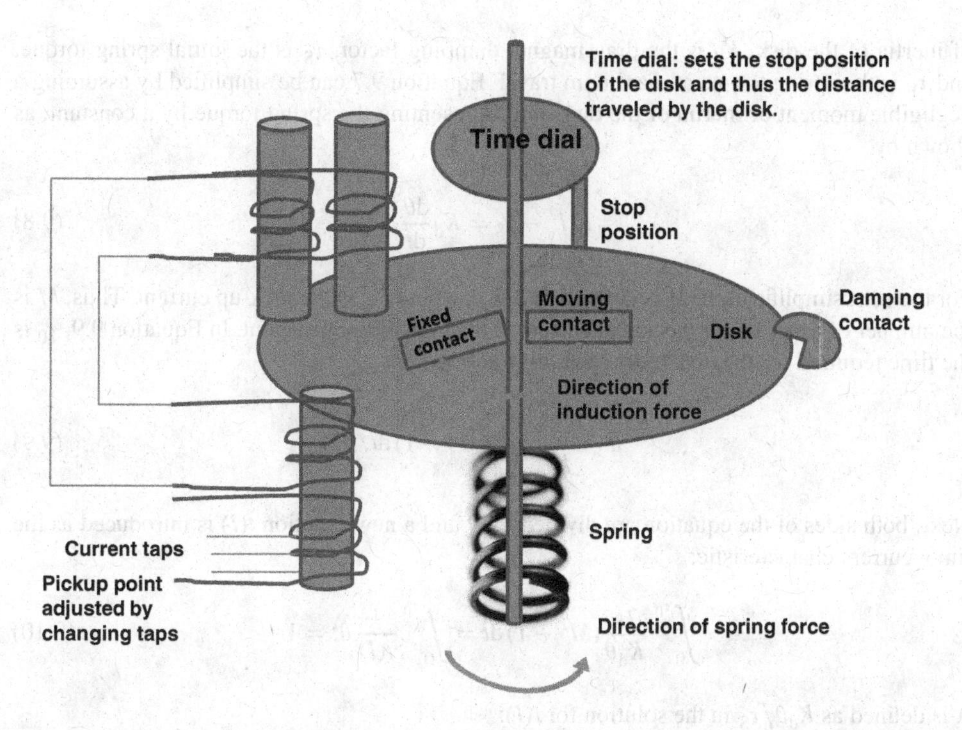

Figure 9.1 The induction disk relay is comprised of a disk that spins with a force proportional to the current induced on its surface. Settings can be adjusted to control when contacts open and close. Concepts and terminology from this classic device carry over into microprocessor-based relay systems.

figure. The spring forces the moving contact against the stop position. Next, note the primary and secondary coils on the left side of the induction disk relay. These induce eddy currents in the metallic disk, causing it to rotate in the clockwise direction. There is a choice of current taps into the primary coil that control the strength of the induced current rotation. The current at which the disk first begins to rotate is known as the "pick-up current." The operation of the device should now become clearer; when the current entering the tap becomes large enough, it will overcome the force of the spring and cause the disk to rotate and the moving contact to approach the fixed contact. The amount of time it takes for contact to occur depends upon the distance the disk must rotate or "travel." This is controlled by the time dial setting. When the moving contact reaches the fixed contact, the protection circuit opens in order to allow the fault to clear. The damping contact is a magnet that serves to slow the motion of disk rotation. Once the fault is cleared, the spring will rotate the disk such that the moving contact will open and return to its initial stop position.

The underlying model for this protection system is described by

$$K_{\mathrm{I}} I^2 = m \frac{\mathrm{d}^2}{\mathrm{d}t^2} + K_{\mathrm{d}} \frac{\mathrm{d}\theta}{\mathrm{d}t} + \frac{\tau_{\mathrm{F}} - \tau_{\mathrm{S}}}{\theta_{\max}} + \tau_{\mathrm{S}}, \tag{9.7}$$

where θ is the angle indicating the amount of disk travel, $\theta_{\max}$ is the disk travel to contact-closure, K_I is a constant that relates the torque on the disk to the current, m is the moment of inertia of the disk, K_d is the drag magnet damping factor, τ_S is the initial spring torque, and τ_F is the spring torque at maximum travel. Equation 9.7 can be simplified by assuming a negligible moment of inertia of the disk and representing the spring torque by a constant, as shown by

$$K_I I^2 - \tau_S = K_d \frac{d\theta}{dt}. \tag{9.8}$$

For the next simplification, M is defined as I/I_{pu}, where I_{pu} is the pick-up current. Thus, M is the number of times larger the actual current is than the pick-up current. In Equation 9.9, τ_0 is the time required for the disk to travel a full rotation:

$$\theta = \int_0^{\tau_0} \frac{\tau_S}{K_d}(M^2 - 1)\, dt. \tag{9.9}$$

Next, both sides of the equation are divided by θ and a new function $t(I)$ is introduced as the time–current characteristic:

$$\int_0^{\tau_0} \frac{\tau_S}{K_d \theta}(M^2 - 1)\, dt = \int_0^{\tau_0} \frac{1}{t(I)}\, dt = 1. \tag{9.10}$$

A is defined as $K_d\theta/\tau_S$ in the solution for $t(I)$:

$$t(I) = \frac{\left(\frac{K_d \theta}{\tau_S}\right)}{M^2 - 1} = \frac{A}{M^2 - 1}. \tag{9.11}$$

As mentioned earlier, once the relay has been activated it will proceed to reset. The time that the disk takes to return to the original reset position is given by

$$|t_r| = \frac{K_d \theta}{\tau_S}. \tag{9.12}$$

The full TCCs are shown in Equations 9.13 and 9.14, where t is the trip time in seconds, M is the number of multiples of the pick-up current, TD is the time dial setting, and p is a constant exponent of M replacing the square to emulate specific curve shapes. Equation 9.13 holds when $0 < M < 1$, as derived above. Equation 9.13

$$t(I) = \mathrm{TD}\frac{A}{M^2 - 1}. \tag{9.13}$$

The difference in Equation 9.14 is that $M > 1$, indicating that the current has exceeded the pick-up current and caused the coils to become saturated. Thus, a constant B is added to account for this effect.

$$t(I) = \mathrm{TD}\left(\frac{A}{M^p - 1} + B\right). \tag{9.14}$$

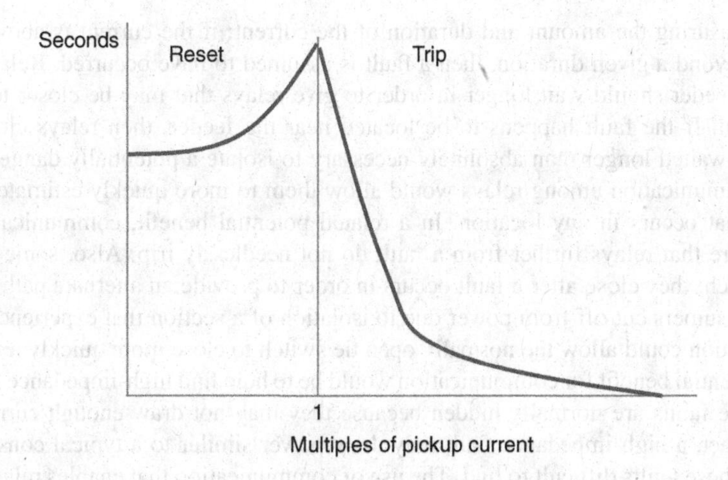

Figure 9.2 The TCC defines when a protective relay trips and resets as a function of time and current. The pick-up current is a threshold beyond which the relay will trip.

An illustration of a TCC is shown in Figure 9.2. Multiple relays may operate in a cooperative manner to isolate faults. It is important to isolate only the fault segment and not disrupt power to more customers than necessary. Thus, precise coordination of time–current curves is necessary in order to achieve sufficient protection and coordination. Adjustment of the curves takes place using Equations 9.13 and 9.14 that were just derived.

9.3 Self-healing, Communication, and Distribution Automation

As we have seen, one of the most important aspects of DA is protection and switching; that is the focus of this section. Protection systems must be able to properly detect and isolate a fault within the grid topology and determine whether distribution components have enough connectivity and capacity to feed customers disconnected by fault isolation. The process of detecting, isolating, and finally attempting to restore power is the goal of FDIR. Many distribution systems are comprised of a main trunk fed by three-phase current with single-phase laterals. Laterals branch from main feeders and are also known as "taps" or "branches." The laterals may be three phase, two phase (two of the three phases from the feeder with a neutral), or single phase (one phase from the single-phase feeder and a neutral). Laterals are usually protected with fuses; power faults in laterals should not cause interruption at the feeder level. Single-phase laterals are used to connect the main trunk to customer locations. A smart grid should self-heal; that is, it should quickly and flexibly reconfigure when necessary to minimize damage caused by a fault. This has typically been accomplished by local sensing and control. As described earlier in this chapter, time–current curves are adjusted in such a manner as to trigger a recloser to open when necessary. One of the challenges, explored in more detail in this section is to determine whether, when, and where techniques that incorporate efficient coupling of communication and power protection control provide added benefit. There are benefits of communication in the distribution system that can potentially lead to improved performance. For example, recall that local sensing of fault conditions typically requires measuring the amount and duration of the current; if the current is above a preset threshold beyond a given duration, then a fault is assumed to have occurred. Relays located closer to a feeder should wait longer in order to give relays that may be closer to the fault time to open. If the fault happens to be located near the feeder, then relays closer to the feeder have waited longer than absolutely necessary to isolate a potentially dangerous fault. Ideally, communication among relays would allow them to more quickly estimate and react to a fault that occurs in any location. In a related potential benefit, communication could help to insure that relays further from a fault do not needlessly trip. Also, some relays are normally open; they close after a fault occurs in order to provide an alternate path for power to reach consumers cut off from power due to isolation of a section that experienced a fault. Communication could allow the normally open tie switch to close more quickly and reliably. Another potential benefit for communication would be to help find high-impedance faults (Ko, 2009). These faults are normally hidden because they may not draw enough current to trip a relay. In fact, a high-impedance fault may draw power similar to a typical consumer, and that makes these faults difficult to find. The use of communication that enables relays to share more precise real-time current information could allow high-impedance faults to be detected. Communication could also enable a relay that may have failed to operate correctly to either become self-aware of its faulty operation or the failure may be detected by the collective effort of other relays. Finally, asset monitoring of all protection equipment will improve reliability. For example, this includes monitoring such properties as recloser temperature, oil, pressure, and insulation (Santillan *et al.*, 2006).

Most protection devices, such as fuses and reclosers, utilize time–current-activated control. A lock-open type of recloser is assumed (Peirson *et al.*, 1955). This type of recloser has an operating cycle consisting of two fast open operations followed by two time-delayed open operations, with the intent of giving the fault time to clear. If at any point in its operation the fault clears, the recloser immediately closes and resumes normal operation. A key element in the configuration of a system of reclosers is the careful design of time–current curves in order to set the optimal time to begin a recloser's open and reclosing sequence relative to its position in the configuration. An overcurrent condition must be measured over a given time period in order to determine the likelihood of a fault. The time–current curve is shifted further away from the origin as reclosers are located closer to the substation. This decreases the likelihood that a recloser nearer to the substation will open when a recloser closer to a fault should open. One question is how well a set of time–current curves can be optimized to perform alone versus with the aid of communication. In essence, a TCC prescribes a form of communication. Overcurrent sensing acts as a communication channel; it can "sense through" the adjacent line segment. However, a fast pick-up may increase the likelihood of a false opening (General Electric, 1977; Russell, 1956). Thus, receiver operating TCC concepts need to be considered such as false positives and false negatives. A false positive would needlessly cause an outage to consumers, which would reflect poorly on the SAIDI reliability metrics. A false negative would allow a fault to cause significant damage that could take a long time to repair. This would also reflect poorly on reliability metrics. How does the use of communication improve time–current curves? In the remainder of this section we explore how the dynamic coupling between the power distribution network and the communication network plays a fundamental role that can serve to illustrate smart grid communication in many other applications. In the next section, typical distribution configurations are reviewed. Then a specific algorithm that detects, isolates, and mitigates a fault is explained. An analysis of the communication load for this algorithm is developed and a simple mesh communication model is implemented using IEEE 802.11. Then, an overview of WiMAX is presented, which includes a discussion of WiMAX mesh mode and a comparison with IEEE 802.11. Next, the DA performance metrics are explained and related to the communication model.

Section 9.3.1 reviews typical small-scale power distribution networks. Section 9.3.2 begins to consider the interaction with a communication network and reviews a simple algorithm for FDIR assuming the communication network is a ring. The goal is to advance beyond simple ring communication topologies and examine the impact of different power distribution topologies and communication topologies. To this end, Section 9.3.3 discusses a simple analytical model of combined power and communication networks. Section 9.1.1 reviewed common FDIR metrics which naturally capture both the duration of a power outage and the number of customers affected. One of the main contributions of this section is to derive the matrix form of those previously defined scalar metrics. The matrix form allows both the power distribution network and the communication network structures to be directly integrated into the definition of the metrics. This is a key idea which allows the dynamics of both network structures to be analyzed. This result is extended to include stochastic fault analysis and can include the notion of both false alarms and missed faults. From a communication point of view, the matrix analysis derived later in this chapter facilitates analysis of the impact of the power distribution network upon load and routing in the communication network. This analysis lays the foundation for other network science concepts to be applied to the smart grid, such as graph spectra for larger network structures.

9.3.1 Distribution Topologies

The topology of electric distribution systems, introduced in Chapter 4, at the substation level is briefly reviewed. It is both the activity and topology of these networks that impact the requirements for the communication network. Activity refers to the likelihood of the occurrence of a fault; the topology has an impact on the impact of a fault and the potential use of tie switches to restore power to isolated segments. One of the simplest topologies is a radial distribution system (Taylor *et al.*, 2001), as shown in Figure 9.3. There are three main components in this simplified model: the substation, labeled SS; the recloser, labeled R*n* (where *n* is a recloser identifier); and sectionalizers, which are not explicitly shown. Power from the electrical substation is transmitted to customers through feeder lines, labeled F*n*. The recloser can be set to follow a TCC that controls the condition under which the recloser initiates its open and reclosing sequence (Witte *et al.*, 1992). As explained earlier, in Chapter 4, the sectionalizer does not interrupt current flow, but instead works with the recloser to isolate both sides of a faulted power line segment.

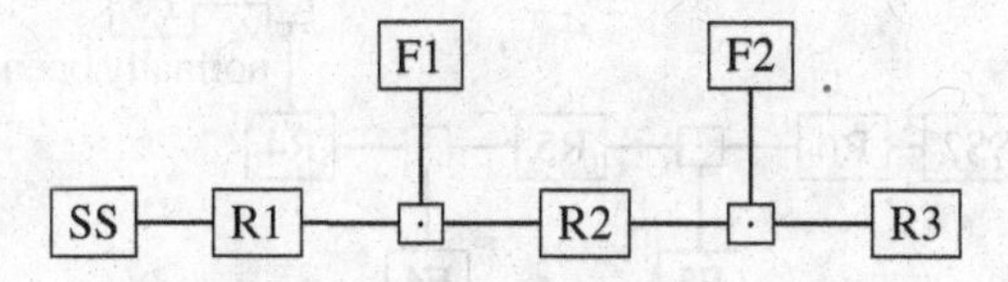

Figure 9.3 A radial distribution network. SS is the substation, R identifies reclosers, and F identifies feeders.

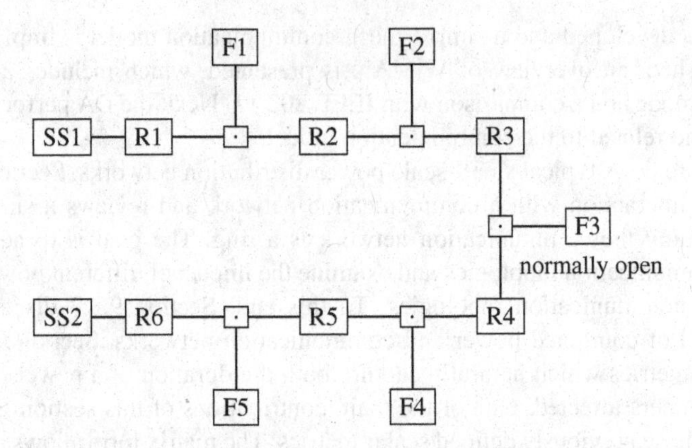

Figure 9.4 A loop distribution network. There are two substations, SS1 and SS2. Recloser R4 is normally open, but can close, if necessary, in order to share power between the upper and lower segments of the loop.

A loop configuration is shown in Figure 9.4. In this configuration, there are two substations, labeled SS1 and SS2. Note that recloser R4 is normally in the open position. This serves as a tie switch; if necessary, the tie switch can be closed, allowing power to flow across the upper or lower section of the radial system. The next section reviews a simple communication algorithm for FDIR, which will then be analyzed, leading to a progressively more complex matrix analysis in Section 9.1.1.

9.3.2 *An Example Algorithm*

Now consider an algorithm that implements FDIR (Hataway *et al.*, 2006). To motivate and illustrate this, suppose a fault occurs in the location shown in Figure 9.5. Faults may occur anywhere between R1 and R6 in which an overcurrent is detected long enough to cause one of the reclosers to lock out. When the recloser locks out, it remains open. At that point, a bit of information is transmitted to two adjacent reclosers. Overcurrent sensing starts for the upstream device. The normally open tie switch is closed to restore power to the line segments without faults.

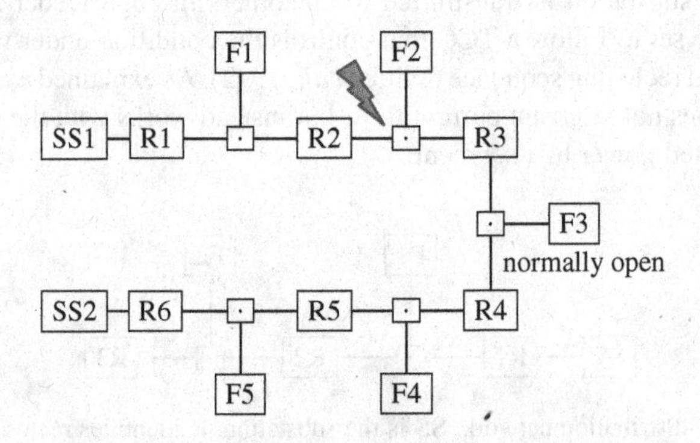

Figure 9.5 A loop distribution network with a fault. The fault is indicated on the R2–R3 segment.

Let us examine this process in more detail. First, the fault draws a significant amount of current. Recloser R2 senses that the current is abnormally high for long enough to be classified as a power fault. This causes recloser R2 to initiate its reclosing sequence. It attempts to open and then close automatically three times in the hope that the fault will clear and current will drop within a normal range. However, in this case, after the third attempt, a fault current level is still detected. This causes the recloser to permanently lock out and remain open. A sectionalizer on the opposite end of the R2–R3 segment will also open, ensuring that the segment remains free of power. Note that segments R2–R3 and R3–R4 are deenergized due to the lockout of R2. R2 then transmits a message to R1 and R3 indicating that it has opened. By this time, the overcurrent sensors in R1 and R2 sense the fault. The message transmitted by R2 to R1 is actually ignored since the overcurrent sensing in R1 takes precedence in this case. When R3 receives the transmitted message, it trips since it has not sensed an overcurrent; when R3 opens, the faulted segment is fully isolated. Next, R3 transmits a message downstream to relay R4. Note that R4 is normally open, allowing the lines supplied by the two feeders to operate independently. When R4 receives the message from R3, indicating a fault, R4 closes and restores power to the R3–R4 segment. Note that if R4 were closed, it would have relayed the message downstream until a normally open tie switch was found and closed.

The distribution system reliability metrics should help in determining whether the approach improves the system. Specifically, the MAIFI, explained in Equation 9.2, should be reduced as little as possible for the distribution system when the fault occurs. This is true if the outage is below the user-defined threshold for an outage duration as defined in the MAIFI metric. The total restoration time is the time for R2 to lock out in addition to the time for the normally open tie switch R4 to close. The opening of R3 is done in parallel with the previous steps. If all of these steps can be performed rapidly enough, then only the customers for which there is no alternative route to power will notice the fault.

If the fault were further downstream – that is, it occurred between R3 and R4 – the process would be similar, except that tie switch R4 would remain open in order to isolate the faulted segment. The algorithm is summarized in Algorithm 1. The REQUIRE statement indicates a

Algorithm 1 Recloser fault mitigation for a loop distribution topology.

Require: overcurrent-sensed:
  send bit to immediately adjacent neighbors
Require: bit-received:
  if overcurrent-sensed **then**
    ignore bit
  end if
  if nearest-fault-neighbor AND overcurrent-not-sensed AND not-a-tie-switch **then**
    open and pass bit around ring
  end if
  if not-a-tie-switch **then**
    pass bit around ring
  end if
  if tie-switch AND not-nearest-fault-neighbor **then**
    close tie-switch
  end if

state that must exist, such as an overcurrent condition or reception of a bit from another node; *nearest-fault-neighbor* indicates the current node is an immediate neighbor of a faulted segment; *not-a-tie-switch* indicates the current node is not a tie switch, while *tie-switch* indicates the current node is a tie switch. Note that the algorithm assumes a ring network architecture for the communication network. A star or mesh network would require either a logical ring structure or a modification of the algorithm to incorporate the ability to address specific nodes. The next section introduces an analysis of the impact of communication upon performance.

9.3.3 A Simplified Distribution System Protection Communication Model

This section considers a simple, scalar analysis of the combined electric power and communication network system, which will be extended to matrix form in the following section. First, consider the problem in general. The impact of a fault depends upon the distribution of customers connected to each feeder and the segment upon which a fault occurs. For this analysis, assume customers are uniformly distributed along feeders. For a completely automated restoration system, the restoration time is dependent on the fault detection time and the number of message transmissions required to restore power to as many customers as possible. Let the number of approximately 1-km segments between reclosers be s. There will be N/s customers on each segment, where N is the total number of customers. A simple assumption is that the likelihood of a fault is proportional to the length of a power line segment. A longer segment will be more likely to suffer a fault given its greater exposure to the environment. However, for this simple analysis, assume that all lengths are equal. Also assume that the communication network is utilized to independently verify a fault sensed by a time–current curve. Given the uniform likelihood of a fault on any segment, there are $s/2$ reclosers to check on average; that is, $s/2$ transmissions are required to verify that the closest recloser to the fault has been identified. A binary search would yield $\lceil \log_2 s \rceil$ exchanges. If a tie switch is required in the case of a loop distribution system, then an additional transmission is required to close the tie switch. All customers downstream of a fault in a radial configuration will lose power. Thus, the mean number of customers without power after a fault, given these simplified assumptions, is $s/2 \times N/s$, or $N/2$. Assuming a constant number of users N, a small number of feeders and segments s implies quick communication and coordination, while a large s implies the ability to isolate the line closer to the actual fault and keep more customers online. Thus, there is an optimal value of s given these trade-offs when designing the distribution system.

Using the process given in Algorithm 1, consider the number of communication transmissions, load, fault current, and time to isolate a fault given the size of the distribution system. Assume a ring network in which messages flow sequentially through distribution system devices along the ring. In other words, distribution system devices forward messages around the ring until the messages reach their destination. This is in contrast to a fully connected mesh network, in which any node may connect to any other node within its transmission range. The number of transmissions required is the number of nodes from the locked-out recloser R_f to tie switch R_t, as shown in

$$n_{\mathrm{t}} = R_{\mathrm{t}} - R_{\mathrm{f}} + 1. \tag{9.15}$$

The communication load is simply the number of transmissions n_t of IEC 61850 generic object-oriented substation events (GOOSE) messages of length 123 bytes with an additional 176-byte RSA encryption:

$$l = n_t(123 + 176)8. \tag{9.16}$$

The total latency d is simply the communication load divided by the total bandwidth bw for each message plus the time it takes for the reclosers to process each message r:

$$d = \left(\frac{l}{bw\eta} + n_t\right) r. \tag{9.17}$$

The protocol requires reliable communication; η is the efficiency of a communication reliability technique known as Go-Back-*N* ARQ. An ARQ protocol requires an ACK that a message has been received in order to determine if the message needs to be resent. A Go-Back-*N* protocol does not require the transmitter to stop and wait for acknowledgment, but rather allows communication transmission to continue in order to more effectively utilize the communication channel. Messages that are transmitted ahead of the last acknowledged message form a sliding window. The communication efficiency is governed by the size of the Go-Back-*N* ARQ window, which should equal the delay–bandwidth product of the channel and the probability of a PDU error, which is based upon both wireless physical communication transmission errors and congestion within the network. The communication efficiency of Go-Back-*N* is approximated by $1/\{1 + N[P/(1 - P)]\}$, where N is the window size and P is the probability of PDU error. In this simplified analysis, it is assumed that the wireless physical channel is perfect and congestion is proportional to the communication traffic load.

The effective bandwidth is the actual bandwidth available for message transmission; that is, after the bandwidth has been reduced by all protocol overhead and physical error. There is a subtle trade-off: if protection device operation time is long, there is less load on the network, resulting in better network performance. As the protection device operation time becomes shorter – that is, the device operates faster – it also increases load on the communication network, which can lead to congestion and retransmission, and thus increase the communication latency.

9.3.4 The Communication Model

Message sizes can be small; simply receiving a message indicates a fault, while typical header information provides the source of the message or the direction from which the fault occurred. In the example above, there was a communication transmission from R2 to R1 and R3. Then there was a communication transmission from R3 to R4. Thus, in this small, example distribution system, there were only three message transmissions. As another example, one could consider the communication network to be IEEE 802.11 with GOOSE messages (Yang *et al.*, 2009) specified by IEC 61850 (Mackiewicz, 2006) over a distance of 1 km with relays located at intervals of 100 m between reclosers. If AODV routing is used, the first packet transmitted will require time to establish the routing path and experience a longer latency; the following messages will have experienced less latency due to routing.

Table 9.1 Summary of symbols used in the analysis.

Symbol	Dimension	Description
$\overline{\mathbf{PoA}}$	$r \times r$ matrix	Distribution adjacency matrix
$\overrightarrow{\mathbf{SS}}$	$1 \times r$ vector	Substation vector
$\overline{\mathbf{CoA}}$	$r \times r$ matrix	Communication adjacency matrix
$\overline{\mathbf{R}}$	$r \times r$ matrix	Tie-switch indicator matrix
$\overline{\mathbf{F}}$	$r \times r$ matrix	Fault indicator matrix
$\vec{\Psi}_{F}$	$1 \times r$ vector	Isolated segments due to true faults
$\vec{\Psi}_{FP}$	$1 \times r$ vector	Isolated segments due to false positives
$\vec{\Psi}_{FN}$	$1 \times r$ vector	Isolated segments due false negatives
$\vec{N}$	$1 \times r$ vector	Customers on each segment
$\vec{1}$	$1 \times r$ vector	All 1s vector
$\vec{R}$	$1 \times r$ vector	Feeders restored after tie switch(es) closed
$\overline{\mathbf{C}}$	$r \times r$ matrix	All pairs of shortest paths
$\overline{\Theta}_{\mathbf{F}}$	$r \times r$ matrix	Probability of a true fault
$\overline{\mathbf{\Theta}}_{\mathbf{FP}}$	$r \times r$ matrix	Probability of false alarms
$\overline{\Theta}_{\mathrm{FN}}$	$r \times r$ matrix	Probability of missed faults
η	scalar	Efficiency of ARQ
r_{m}	scalar (seconds)	Manual restoration time
r_{t}	scalar (seconds)	Automatic recloser time
f	scalar (recloser)	Faulted recloser (single fault)
t	scalar (switch)	Tie switch (single switch)

To incorporate more detail, let us develop the operational analysis in matrix form with symbols defined in Table 9.1. Matrices are notationally distinguished from vectors; matrices are accented with an overbar and vectors with an arrow. Let there be r reclosers protecting r feeders. Let $\vec{N}$ be of dimension $(1 \times r)$, where each element of the vector is a feeder connection as shown in Figure 9.5. Let $\overline{\mathbf{F}}$ be a matrix of dimension $(r \times r)$, where each element indicates the presence or absence of a fault on the r feeder segments, represented by a 0 or 1. The index of the tie switch will be denoted by matrix $\overline{\mathbf{R}}$ of dimension $r \times r$, where a 1 indicates a tie-switch connection and all other connections are 0. The interconnectivity of the electric power distribution network can be represented by a graph described by an adjacency matrix $\overline{\mathbf{PoA}}$ of dimension $(r \times r)$. The graph is assumed to be undirected. A representation of the communication network can be found in the connectivity matrix $\overline{\mathbf{CoA}}$, which is also of dimension $(r \times r)$ and indicates the interconnectivity and latency of recloser communication.

The connectivity matrix $\overline{\mathbf{C}}$ indicates the total transmission latency from the node in the row index to the node in the column index. In other words, each element of the connectivity matrix represents the total communication latency in transmission between the row and column indices. $\overline{\mathbf{C}}$ can be derived from the adjacency matrix for the communication network, $\overline{\mathbf{CoA}}$. If there are n nodes, then $\overline{\mathbf{C}}$ is derived from $\overline{\mathbf{CoA}}^{n}$, but with a difference in the manner of matrix multiplication. Specifically, the element-wise multiplication operation in the matrix multiplication is replaced with addition and the addition operation normally used in matrix multiplication is replaced with minimization; that is, returning the minimum-value element. There is a latency of duration $\overline{\mathbf{C}}(f, t)$ in transmitting from the faulted segment f to the tie switch t. The vector of feeders served during normal operation is derived thus

$$\overrightarrow{\mathbf{SS}} \cdot \sum_{i=1}^{r} \overline{\mathbf{PoA}}^{i}. \tag{9.18}$$

Vector and matrix dot products are indicated by a middot. Because $\overline{\mathbf{PoA}}$ is an adjacency matrix, Equation 9.18 yields the nodes traversed by power released from the corresponding nodes in $\overrightarrow{\mathbf{SS}}$ after r hops; that is, the reachability from the substation.

The vector of feeders receiving power after a fault is determined by

$$\vec{\Psi} = \overrightarrow{\mathbf{SS}} \cdot \sum_{i=1}^{r} \left(\overline{\mathbf{PoA}} - \overline{\mathbf{F}}\right)^{i} \tag{9.19}$$

This is similar to Equation 9.18 except that faulted segments have been subtracted from the adjacency matrix. Note that $\vec{\Psi}$ is the vector defined in Equation 9.19 where nonzero values are replaced with zero, since they continue to receive power, and zero values are replaced with ones, to indicate that they are isolated from the distribution system. It important to note that $\overline{\mathbf{F}}$ here indicates *actual* faults, as opposed to probabilistic faults, which makes the analysis simpler. A more complex version of this analysis could consider the details of fault sensing; for example, via a TCC. In this case, more time is required in order to detect smaller current faults; at any instant in time, there is the possibility for a false-positive or false-negative fault. This is considered later. The operation time r_t includes the time required to sense the fault.

The vector of feeders restored by closing tie switches is

$$\overline{\mathbf{R}} = \overrightarrow{\mathbf{SS}} \cdot \sum_{i=1}^{r} \left(\overline{\mathbf{PoA}} - \overline{\mathbf{F}} + \overline{\mathbf{T}}\right)^{i}. \tag{9.20}$$

$\overline{\mathbf{R}}$ is an adjacency matrix and similar to $\overline{\mathbf{F}}$ except that it adds new links rather than subtracts them. Similar to $\vec{\Psi}$, $\vec{R}$ is an indicator vector in which nonzero values are replaced with zeros and zero-values replaced with ones.

The total number of customers is

$$\vec{1} \cdot \vec{N}. \tag{9.21}$$

The inner product of $\vec{N}$ with the all-1s vector results in the summation of all feeder customers.

Putting together the terms from Equations 9.19, 9.20, and 9.21, the SAIDI is

$$\mathrm{SAIDI} = \frac{r_t \overline{\mathbf{C}}[f, t] \eta \vec{\Psi}_{\mathbf{F}} \cdot \vec{N} + r_m \vec{R} \cdot \vec{N}}{\vec{1} \cdot \vec{N}}, \tag{9.22}$$

where $\vec{\Psi}$ and $\vec{R}$ represent the power distribution network and $\overline{\mathbf{C}}$ represents the communication network architecture. Note that $\overline{\mathbf{C}}$ is derived from $\overline{\mathbf{CoA}}$, the communication adjacency matrix, and $\vec{\Psi}$ and $\vec{R}$ are derived from $\overline{\mathbf{PoA}}$, the power distribution network adjacency matrix.

A slight rearrangement of Equation 9.22 results in

$$\text{SAIDI} = \frac{(r_t\overline{\mathbf{C}}[f,t]\eta\vec{\Psi}_{\mathbf{F}} + r_m\vec{R}) \cdot \vec{N}}{\vec{1} \cdot \vec{N}}. \tag{9.23}$$

The result in Equation 9.23 can be utilized for any size and any topology power distribution and communication network with multiple faults. For a complex distribution network, it may be useful to find the optimal set of tie switches to close among a set of possible choices. In this case, the goal is to solve for the $\overline{\mathbf{R}}$ that meets a given a criterion, such as minimizing SAIDI. Rearrangement of Equation 9.20 gives

$$\overrightarrow{\mathbf{SS}} \cdot \sum_{i=1}^{r} \left(\overline{\mathbf{PoA}} - \overline{\mathbf{F}}\right)^{i} + \overrightarrow{\mathbf{SS}} \cdot \sum_{i=1}^{r} \overline{\mathbf{R}}^{i} \tag{9.24}$$

in which $\overline{\mathbf{R}}$ is separated into its own summation.

Section 9.3.5 considers extending this analysis to probabilistic cases; faults are no longer deterministically defined in $\overline{\mathbf{F}}$, but rather probabilistically defined in $\overline{\mathbf{\Theta}}$.

9.3.5 *Probabilistic Interpretation of the Distribution Protection Communication System*

Equation 9.23 may be used to examine the impact of setting TCCs in a probabilistic sense. The fault matrix $\overline{\mathbf{F}}$ may represent a probability of fault, based upon the magnitude and duration of overcurrent.

$\overline{\mathbf{\Theta}}_{\mathbf{FP}}$, $\overline{\mathbf{\Theta}}_{\mathbf{FN}}$, and $\overline{\Theta}_{\mathbf{F}}$ are $(r \times r)$ matrices used to capture the impact of TCC settings. They represent the probability of a false positive, false negative, and true positive, respectively, for each protected segment. $\overline{\mathbf{\Theta}}_{\mathbf{FP}}$ indicates the probability of the relay opening when it should not, thus needlessly isolating customers. $\overline{\mathbf{\Theta}}_{\mathbf{FN}}$ indicates the probability of the relay failing to open when a fault is present, thus damaging equipment and failing to properly open any tie switches that could have mitigated the isolated segments. $\overline{\Theta}_{\mathbf{F}}$ indicates the probability of a true fault.

The false-positive matrix $\overline{\mathbf{\Theta}}_{\mathbf{FP}}$ is treated in a similar fashion to $\overline{\mathbf{F}}$, as it represents the probability of a fault, which may differ along each segment. Consider transforming these adjacency matrices to a probability vector $\vec{\Psi}$ of isolated links due to faults. The approach is to compute the probability of successful power transmission $1 - \overline{\Theta}_{\mathbf{FP/FN/F}}$.

Let the elements of $\overline{\Theta}_{\mathbf{FP/FN/F}}$ be the probability of the feeder segment opening for any reason. Then $1 - \overline{\Theta}_{\mathbf{FP/FN/F}}$ is the probability of not opening, which is only a true no-fault condition. All other conditions (true positive: opening due to a real fault; false positive: opening for a false alarm; false-negative: missing a real fault) cause open sequencing or an open condition to occur. Equation 9.25 represents the probability of a false positive over any path through the grid:

$$1 - \sum_{i=1}^{r} \overline{\mathbf{\Theta}}_{\mathbf{FP/FN/F}}^{\,i}. \tag{9.25}$$

Specifically, the first row shows the probability of a false positive from the substation to any point through the grid. This applies similarly to false negatives and the probability of true faults. This provides an indication of the probability of each segment being isolated due to true faults, missed faults and thus equipment damage, or false alarms and thus needless isolation of a segment. Segments further from the substation will always have a greater chance of becoming isolated. Thus, we can now speak of probabilistically mitigating isolation. For example, assuming the first recloser is connected to a substation, the top row of the resulting matrix from Equation 9.25 yields $\vec{\Psi}_{FP/FN/F}$. It is also possible to consider a stochastic matrix P for probability of transition for which the columns sum to one and the initial state is a vector p, which must sum to one. $p' = pP$. Then p is the probability of not being isolated and P is the probability of a fault. The problem is that the rows and/or columns in a stochastic matrix must sum to one. In the scenario described here, the $\overline{\mathbf{\Theta}}$ values are independent values assigned as an adjacency matrix, and thus it is not necessarily stochastic.

Equation 9.26 shows SAIDI given the probability of false positive and false negative:

$$\text{SAIDI} = \frac{(r_\text{t}\overline{\mathbf{C}}[f,t]\eta\vec{\Psi}_{\mathbf{FP/FN/F}} + r_\text{m}\vec{R}) \cdot \vec{N}}{\vec{1} \cdot \vec{N}}. \tag{9.26}$$

In the probabilistic case, it is useful to consider a value-at-risk approach to SAIDI-like metrics. For a given probability and time horizon, value-at-risk is defined as a threshold value such that the probability that the restoration time over the given time horizon exceeds this value is the given probability level.

The complete equation for SAIDI is

$$\text{SAIDI} = \frac{\{r_t d(\overline{\mathbf{CoA}})[f,t]\eta z[\overline{\mathbf{SS}} \cdot \sum_{i=1}^{r} (\overline{\mathbf{PoA}} - \overline{\mathbf{F}})^i]\} \cdot \vec{N}}{\vec{1} \cdot \vec{N}} + \frac{\{r_m z[\overline{\mathbf{SS}} \cdot \sum_{i=1}^{r} (\overline{\mathbf{PoA}} - \overline{\mathbf{F}} + \overline{\mathbf{T}})^i]\} \cdot \vec{N}}{\vec{1} \cdot \vec{N}}, \tag{9.27}$$

where $d()$ is a function that takes the adjacency matrix and returns the connectivity-, or distance-matrix, and $z()$ takes a vector and replaces zero values with ones and all other values with zeros. There are a few items to note in Equation 9.27. The structure of the communication adjacency matrix $\overline{\mathbf{CoA}}$ has a direct impact on the connectivity matrix $d(\overline{\mathbf{CoA}})$ and the latency of the route taken by FDIR messages $d(\overline{\mathbf{CoA}})[f,t]$. The structure of the power grid adjacency matrix $\overline{\mathbf{PoA}}$ in conjunction with the fault matrix $\overline{\mathbf{F}}$ and tie-switch matrix $\overline{\mathbf{T}}$ impacts the number of isolated segments.

There are competing goals indicated by $\overline{\mathbf{\Theta}}_{\mathbf{FP}}$ and $\overline{\mathbf{\Theta}}_{\mathbf{FN}}$: detect all faults as quickly as possible without initiating false alarms. An increase in both false positives and false negatives increases SAIDI. Segments that we can now quantify with high a probability of fault or isolation may receive more attention from collaborative communications.

Determining the probability of fault and isolation also allows a determination of the probability of events for reconfiguration within the FDIR process. Returning to the main theme of a coupled-network system, namely (1) the electric power distribution network and (2) the communication network, the probability of a fault on the power distribution network determines the probability of a corresponding message being transmitted on the communication network. In the absence of a priori fault information, all power grid segments are equally likely to experience a fault; that is, the system is at its highest entropy and the communication network must correspondingly be designed to handle messages from any segment. As more a priori information about the likelihood of faults becomes available, the entropy decreases and the communication network can expect to dedicate more resources to the segments with higher probability of fault, which includes impact on the physical communication topology and routing. The concept of predictability and network behavior is discussed in Bush and Smith (2005).

This matrix analysis allows us to examine the self-healing performance based upon the topologies of the communication and power networks. The $\overline{\mathbf{PoA}}$ and $\overline{\mathbf{CoA}}$ matrices represent the topologies of the power and communication networks, respectively. Thus, as illustrated in Figure 9.6, using the simple example distribution network, the impact of different communication network topologies can be quantified in relation to an arbitrary power network topology.

There are numerous types of communication technologies that could support smart grid communication in the distribution system, ranging from optical fiber, discussed in Section 8.5, to power line carrier, covered in Section 8.5, to wireless technologies. In this section, the topic of wireless communication will be discussed. Wireless communication is a broad topic ranging from WiFi IEEE 802.11 to IEEE 802.15.4, covered in Section 7.4, to WiMAX IEEE 802.16 and so-called "Long Term Evolution." A key element in the latest incarnation of wireless techniques is orthogonal frequency-division multiplexing (OFDM). OFDM is widely applied, not only to wireless technologies, but also over wired media, including power line carrier, digital television and audio broadcasting, digital subscriber line (DSL) broadband Internet access, wireless networks, and mobile communication. Knowledge of OFDM provides a key insight into many of these technologies being applied to the distribution system.

Beginning with a high-level, intuitive perspective, we can see by the acronym that OFDM involves frequency-division multiplexing (FDM). In fact, OFDM involves a combination of modulation and multiplexing. Modulation is the mapping of "information" to perturbations in the carrier signal, including phase, frequency, and amplitude changes. Multiplexing is simply the act of sharing available bandwidth among independent channels. In OFDM the signal to be transmitted is split into multiple, independent channels. Thus, OFDM can be considered a special case of FDM. A simple analogy is to compare FDM with a faucet – a single, large stream of water – and OFDM to a shower – comprised of many, smaller, independent streams of water. Carrying this analogy further, one can block the faucet with your thumb, but the shower head is more difficult to block completely. Rather than putting all our eggs in one basket, so to speak, OFDM splits the signal into multiple channels with the hope that, in the

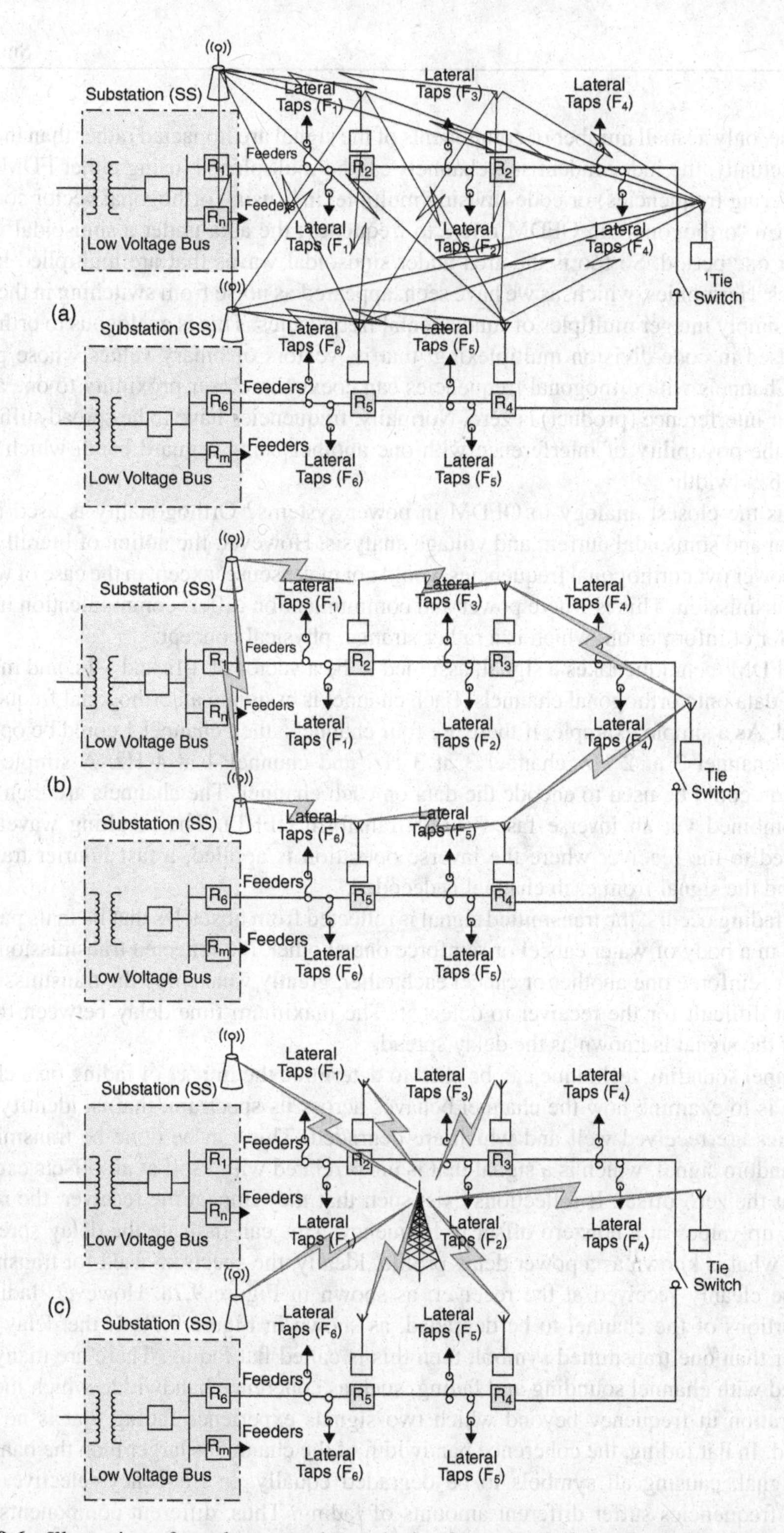

Figure 9.6 Illustration of regular network topologies for a simple loop distribution network: (a) a mesh communication network topology, (b) a ring, and (c) a star. There are other, more interesting communication topologies; for example, topologies with random or scale-free node-degree distributions. Our goal is to understand how the interaction between communication and power distribution topologies impacts FDIR. The communication topology is represented by $\overline{\mathbf{CoA}}$ in Table 9.1.

worst case, only a small number of components of the signal are impacted rather than the entire signal. Actually, the independent sub-channels can be multiplexed, using either FDM (using noninterfering frequencies) or code-division multiplexing (using orthogonal vector codes).

The term "orthogonal" in OFDM refers to frequency; the area under a sinusoidal wave is zero over one period. So too is the area under sinusoidal waves that are multiplied by their harmonics. Harmonics, which, as we have seen, appeared as noise from switching in the power grid, are simply integer multiples of fundamental frequencies. This is analogous to orthogonal vectors used in code-division multiplexing; that is, vectors of binary values whose product is zero. Channels with orthogonal frequencies can coexist in closer proximity to one another since their interference (product) is zero. Normally, frequencies have to be spread sufficiently to avoid the possibility of interference with one another using a guard band, which wastes valuable bandwidth.

What is the closest analogy to OFDM in power systems? Orthogonality is used in state estimation and sinusoidal current and voltage analysis. However, the notion of literally transmitting power over orthogonal frequencies would not make sense, except in the case of wireless power transmission. This is where power and communication differ: communication involves the transfer of information, which is a rather strange, physical concept.

The OFDM technique takes a signal, assumed to be a vector of +1s and −1s, and maps the vector of data onto orthogonal channels. Each channel is assigned an orthogonal frequency as discussed. As a simple example, if there are four channels, then channel 1 could be operating at 1 Hz, channel 2 at 2 Hz, channel 3 at 3 Hz, and channel 4 at 4 Hz. A simple BPSK modulation could be used to encode the data on each channel. The channels are then added, being combined via an inverse fast Fourier transform (IFFT). The resulting waveform is transmitted to the receiver where the inverse operation is applied, a fast Fourier transform (FFT), and the signal from each channel is decoded.

When fading occurs, the transmitted signal is reflected from obstacles that lie in its path. Just as waves in a body of water cancel or reinforce one another, the reflected transmission waves can either reinforce one another or cancel each other, greatly weakening the transmission and making it difficult for the receiver to detect it. The maximum time delay between received copies of the signal is known as the delay spread.

A channel sounding technique can be used to determine the impact of fading on a channel. The goal is to examine how the channel behaves across its spectrum; that is, identify which frequencies are received well and which are degraded. This can be done be transmitting a pseudorandom signal, which is a signal that is uncorrelated with itself at all offsets except, of course, at the zero offset. If reflections exist such that they impact the receiver, the receiver will pick up values at a nonzero offset in frequency. This can indicate the delay spread and provides what is known as a power delay profile. Ideally, the spectrum used for transmission should be cleanly received at the receiver, as shown in Figure 9.7a. However, fading can cause portions of the channel to be degraded, as shown in Figure 9.7b. If the delay spread is smaller than one transmitted symbol, then this is called flat fading. There are many terms associated with channel sounding and fading, such as coherence bandwidth, which measures the separation in frequency beyond which two signals experience fading that is no longer correlated. In flat fading, the coherence bandwidth of the channel is larger than the bandwidth of the signal, causing all symbols to be degraded equally. In frequency-selective fading, different frequencies suffer different amounts of fading. Thus, different components of the signal are impacted differently. One of the significant advantages of OFDM is its ability to

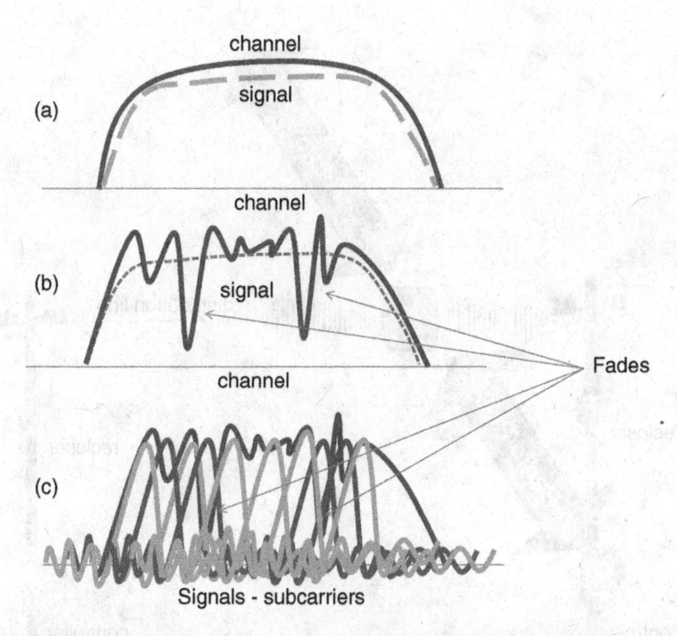

Figure 9.7 The OFDM technique maps data onto different sub-channels, where each sub-channel is assigned an orthogonal frequency. This provides a form of frequency diversity that helps to ensure that all sub-channels are not blocked by various forms of interference. This technique is used in many different types of physical media, including wireless communication and power line carrier.

handle frequency-selective fading. Going back to our analogy of the thumb trying to stop the water in the shower, only specific OFDM sub-channels will be adversely impacted by frequency-selective fading and data in the other channels can be recovered, for example, by higher level coding. In other words, instead of an entire wider bandwidth channel being adversely impacted, only smaller bandwidth sub-channels are impacted; the remainder of the sub-channels are successfully received, as shown in Figure 9.7c.

The increasing trend toward automation, flexibility, and self-healing can be found in wireless communications and mobile power line radios. A communication architecture for the electric power distribution system that has been proposed many times involves the use of multihop radios mounted at strategic locations on pole tops within the distribution system, sometimes integrated with distribution equipment, such as protection equipment. The challenge with such an architecture is that the radios reside at relatively low heights and non-optimal locations compared with cellular-communication radio towers. Multihop radios at pole-top height are subject to significant amounts of fading, which is attenuation of the signal strength due to both reflection (from objects along the radio-wave propagation path) and shadowing (caused by objects obstructing radio-wave propagation). Fading for nonmobile radios can be mitigated by using different types of diversity, either in time, frequency, or space (for example, adding channel coding, which requires additional bandwidth overhead, transmitting repeatedly until the information is received, transmitting along multiple frequencies, or transmitting along multiple paths). All of these approaches are problematic in that they reduce the potential available bandwidth, and require radios that are more expensive, with more complex protocols.

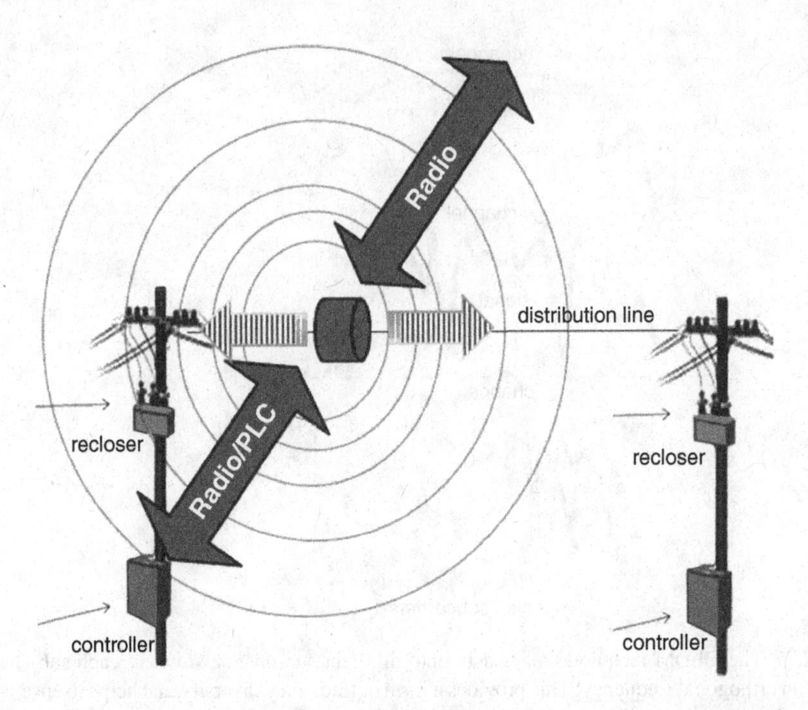

Figure 9.8 A mechanical means of mitigating communication interference is for the antenna and/or radio of a power line robot to move slightly along the distribution power line in order to improve multihop radio reception throughout a power distribution system.

A simpler, potentially cheaper approach to mitigating fading is to allow one degree of freedom of motion, as illustrated in Figure 9.8, which illustrates a power line robot. This motion can allow radios (or, more specifically, their antennas) to move into optimal positions to avoid fading. Thus, the advantages are the ability to achieve and maintain excellent communications connectivity throughout the wireless communications system within the challenging distribution side of the power grid. The same technique can be applied to the transmission system; however, the height of power transmission towers is such that they are less susceptible to fading.

Automated power line inspection and repair equipment, including power line crawling robots, have been developed for both the electric power transmission and distribution systems (Elizondo *et al.*, 2010; Montambault and Pouliot, 2010; Roncolatto *et al.*, 2010; Wang *et al.*, 2009). However, in past work, communication has been considered an ancillary service to the function of the robots' main task. Here, the mobile radio has the primary function of mitigating fading for smart grid power distribution communication. However, depending upon cost and benefit, nothing precludes the mobile power line radio from incorporating sensing and control elements along with its fading mitigation function. The concept is comprised of wireless radios or antennas mounted upon simple low-cost power line crawler devices capable of autonomous movement along the power line in order to optimize radio communication. The robotic crawler device is an inexpensive, hollow cylinder placed around the power line such that it cannot be easily dislodged from the power line, and utilizes a variety of mechanisms to propel itself along the power line. These may include, but are not limited to, an on-board motor with a wheel that grabs the power line and rotates to move the entire unit, and magnetic field levitation and propulsion using the power line's electric and magnetic field properties. If the entire radio is placed on-board the robot, then transmission of information from the radio to stationary devices on the grid may take place through the power line itself, either utilizing the power line as a waveguide or power line carrier. Alternatively, only the radio's antenna needs to be placed on the robotic mobile unit. In this case, the radio unit remains fixed (that is, integrated within, or near, a recloser, for example), and the antenna adjusts its position on the power line in order to improve communication. As the distribution radio system is comprised of many mobile power line radios, there will be a dynamic communication network topology generated by the underlying power line topology. This is one of many possible examples of increasing smart grid controllability and complexity with communication taking a more active role within the power grid. Next we consider a wireless distribution communication solution known as WiMAX.

9.3.6 Introducing WiMAX

We began by considering the well-known IEEE 802.11 WiFi protocol (introduced in Section 7.5) in the distribution system. An alternative communication architecture uses the IEEE 802.16 standard family, also known as WiMAX (Kas *et al.*, 2010), which was designed to serve as the last-mile connection for wireless broadband access. Radio coverage extends 5 miles with a bandwidth of 70 Mbps. The IEEE 802.16 family of standards includes support for mobility; a mesh support mode allows users to route through one another to reach a base station (BS). IEEE 802.16d is the fixed-location standard and IEEE 802.16e is the mobile WiMAX standard. The IEEE 802.16j standard defines a relay mode for IEEE 802.16e (Chang *et al.*, 2009).

Consider the differences between IEEE 802.11 and IEEE 802.16. IEEE 802.16 networks have a range in kilometers, while IEEE 802.11 radios have a range in hundreds of meters. One of the biggest differences is that IEEE 802.16 uses TDMA, while IEEE 802.11 allows frame collisions and uses carrier sensing. Thus, IEEE 802.11 has the potential for hidden and exposed terminals, which are addressed using an request to send/clear to send mechanism, while IEEE 802.16 uses a three-way handshake to establish a dedicated channel. Also, in IEEE 802.16, data and control channels are separate, so control traffic does not contend with data traffic. IEEE 802.11 is a best-effort (BE) type of communication, while IEEE 802.16 has four QoS classes. The issue of interoperability between IEEE 802.11 and IEEE 802.16 has yet to be addressed in a standard manner (Li *et al.*, 2007).

From the above comparison, it is clear that WiMAX is connection oriented, while IEEE 802.11 is not. The WiMAX MAC connections are identified by unique connection identifiers (CIDs), and IP addresses are mapped to CIDs. In WiMAX point-to-multipoint mode (PMP) mode there is a BS and multiple SSs. In PMP mode, communication exists only between subscriber stations and the base station; no communication is allowed directly between SSs.

The mesh mode option allows multihop capability through SSs; there is support for this in both IEEE 802.16d and IEEE 802.16e. For the mobile standard, IEEE 802.16j is a draft standard that supports multihop capability. This draft specifies that the network forms a tree with the BS as the root; relay stations (RSs) form the branches between the mobile stations (MSs) and the BSs. In plain IEEE 802.16 mesh mode, communication from SSs to the BSs can take place through multiple SSs. Each SS becomes a router, forwarding other traffic for other SSs to the BS. Using this we can analytically determine performance and compare with IEEE 802.11 in Hayajneh and Gadallah (2008) and Vu *et al.* (2010).

There are four service classes: unsolicited grant service (USG), real-time polling service (rtPS), extended-real-time-polling service (ertPS), non-real-time polling service (nrtPS), and BE. USG provides a real-time, constant-bit rate (CBR) service. rtPS is for real-time applications that generate variable-size packets, such as MPEG video. ertPS allocates a dedicated bandwidth similar to USG; however, the allocation can be dynamically increased or decreased similar to rtPS. nrtPS is similar to rtPS but with longer periods between slots. It is ideal for delay-tolerant applications. The BE service offers no minimum service requirement; slots are used as they become available. For stationary smart grid devices, such as reclosers, a fixed-network configuration among recloser radios can be established and the USG service can provide the CBR service required. Another alternative is to use BE service and construct smart-grid-specific reliability mechanisms.

The physical layer data rate is given by

$$\mathrm{PHY}_{\mathrm{rate}} = \frac{M n_{\mathrm{data}} r}{T_{\mathrm{S}} C_{\mathrm{sector}}}, \tag{9.28}$$

where

$$T_{\mathrm{S}} = T_{\mathrm{b}} + T_{\mathrm{g}}, \tag{9.29}$$

$$T_{\mathrm{b}} = \frac{S_{\mathrm{fft}}}{B C_{\mathrm{sampling}}}, \tag{9.30}$$

$$T_{\mathrm{g}} = \mathrm{cp} T_{\mathrm{b}}, \tag{9.31}$$

and where B (MHz) is the bandwidth, C_{sampling} is the sampling coefficient, T_{b} (μs) is the useful symbol time, T_{g} is the guard time, T_{S} (μs) is the overall OFDM symbol time, n_{data} is the number of carriers in an OFDM symbol, M is the M-ary modulation, r is the coding rate, and C_{sector} is the cell-sectorization (the proportion of the spectrum used in each cell) coefficient (Xhafa *et al.*, 2005). S_{fft} is the size of the FFT and cp is the cyclic prefix. For the sampling factor, see So-In *et al.* (2010, Tables 1–3). For FFT sizes, see Barber *et al.* (2004).

The MAC layer rate is given by

$$\mathrm{MAC}_{\mathrm{rate}} = \frac{B_{\mathrm{total}} - B_{\mathrm{overhead}}}{T_{\mathrm{OFDMA}}}, \tag{9.32}$$

where B_{total} is the total number of bits transmitted in one time-division duplex (TDD) time frame, B_{overhead} is the number of bits that conveys the control information, and T_{OFDMA} is the orthogonal frequency-division multiple access (OFDMA) frame time. The frame time comes from Jain and Al Tamimi (2008).

It should be noted that there are more security vulnerabilities in WiMAX mesh mode than in PMP mode Maccari *et al.*, 2007, Rengaraju *et al.*, Zhou and Fang, 2006). Given the large number of node interconnections in mesh mode, there are more opportunities for man-in-the-middle attacks and node spoofing. Another simple attack is bandwidth spoofing, in which a node simply requests more bandwidth than it needs. This can occur in both PMP and mesh mode.

9.3.7 WiMAX Mesh Mode

The term "mesh network" and "ad hoc network" appear to be used interchangeably. Each node in such a network is capable of acting as an independent router, enabling a highly interconnected network topology. PDUs can hop from one node to another through a mesh or ad hoc network in order to reach their final destination. This increases performance and reliability by allowing the choice of many potential routes, including routes that can be taken around broken links. Wireless mesh networks enable node mobility and utilize IEEE standards such as IEEE 802.11, IEEE 802.16, and cellular technologies or combinations of more than one type. Today, IEEE 802.11 is the most widely deployed mesh routing system; WiMAX mesh mode is an alternative solution.

Distributed scheduling allows each node to make its scheduling decisions locally. The WiMAX mesh mode standard is designed to ensure that distributed communication takes place without collision within a neighborhood of nodes. Spatial reuse is enabled for nodes that are two or more hops away. As typical for distributed approaches, distributed scheduling tends to be more efficient and adaptable than a centralized approach.

The IEEE 802.16 WiMAX mesh mode operates in a different manner from the standard WiMAX PMP mode. In PMP mode, SSs must communicate to a BS, which serves as their access point into the network. In mesh mode, SSs can communicate directly with, and through, one another. In mesh mode, the BS communicates as another SS node. The only exception is that, in mesh mode, the BS provides additional services, such as advertising the network configuration and authenticating a new SS that attempts to join the network. For the purpose of understanding mesh communication, the BS and SS are indistinguishable.

In WiMAX mesh mode, two scheduling mechanisms have been defined: centralized and distributed. In centralized scheduling, the BS forms the root of a logical tree along with all other nodes and performs the centralized scheduling of bandwidth for all nodes. As is typical for centralized mechanisms, the coordination is easier to control since the BS forms a single, central controller for the entire network; however, overhead required to keep the BS apprised of complete information for the entire network is significant. In addition, the BS becomes a single point of failure.

Mesh operation takes place in the WiMAX MAC layer using the same underlying physical layer as the more typically used PMP mode of operation. MAC frames use time-division duplexing with fixed-duration slots on up to 16 noninterfering channels. There is a distinction between data and control messages.

A key message in mesh mode operation is the mesh-distributed scheduling message (MSH-DSCH). These messages are exchanged in a collision-free manner in order to perform a distributed election procedure to determine access to the data frames. Two parameters that control the rate of exchange of these messages are the number of neighboring nodes and a system parameter named XmtHoldoffExponent (Cao *et al.*, 2007).

While the MSH-DSCH message enables collision-free message exchanges among nodes within the WiMAX network, other messages are used in the configuration and maintenance of the network. Nodes may enter or leave the network. In this case, mesh network entry request messages (MSH-NENTs) are transmitted by nodes that wish to enter the network. mesh network configurations (MSH-NCFGs) are used to send updated network parameters to nodes and are transmitted in a collision-free manner, similar to MSH-DSCH messages. Using these messages, each node eventually gains complete information about its two-hop neighborhood.

Nodes may attempt to enter the network at any time; thus, MSH-NENT messages may arrive at any time and may collide; a collision-resolution mechanism is necessary and specified by the standard. The importance of control messages is recognized by the standard and they receive additional protection by means of a more robust coding rate and a longer preamble and guard time.

The operation of MSH-DSCH message exchanges occurs as follows. There are fields within an MSH-DSCH message known as information elements (IEs). Because these messages are broadcast, all nodes within a neighborhood receive the message and have knowledge of the information content. There are four types of IEs: *request*, *grant*, *confirmation*, and *availability*. A three-way handshake is used to request bandwidth as follows. The *request* IE is used to request bandwidth; that is, slots from a neighbor. The neighbor may respond with a *grant* IE that includes a range of slots offered. The original requester then responds with a *confirmation* IE that it will use those slots. When the original, requesting node confirms the grant, it will use the channel specified by the neighbor to transmit, using the confirmed slots. PDUs being carried by WiMAX, such as IP datagrams, can be fragmented or aggregated into IEEE 802.16 frames as-needed in order to improve performance. Fields exist within the MPDUs to implement frame dropping in order to react to congestion as well as implement frame priority.

The *availability* IE identifies more detail regarding the availability of a slot. IEEE 802.16 defines four states to identify availability of a slot in a data frame: 11, availability for transmission/reception; 01, availability for transmission only; 10, availability for reception only; 00, not available (Ge *et al.*, 2008).

Queuing theory analysis has been employed to examine scheduling in a WiMAX mesh network (Bastani *et al.*, 2009). The assumption is that SS nodes are sending messages to the BS. The analysis assumes two queues at each node: one for packets originating from the node and another for packets flowing through the node from the network. This is known as the relay queue. There are also two levels of packet scheduling assumed: a network scheduler that allocates time slots to each SS and a node scheduler that decides when to transmit from each queue within a node. Operation of the node scheduler is assumed and the analysis focuses on the network scheduler. The relay queue is assumed to be $M/M/1/K$, which indicates a single-server queue with a buffer size of K. Delay and throughput trade-off is analyzed under these assumptions.

A stochastic model of WiMAX mesh mode performance is developed in Cao *et al.* (2007). Channel contention is a function of the number of nodes, an exponent value that determines a hold-off time allowing other nodes to access slots, and the network topology. Using this knowledge, a probabilistic framework can be developed. It is also assumed that the transmit time sequences of all the nodes in the control subframe form statistically independent renewal processes. The WiMAX Forum ns-2 extension is described in Guo *et al.* (2009). It does not implement mesh mode; however, ns-2 WiMAX mesh mode extensions exist in Cicconetti *et al.* (2006, 2009).

9.4 Summary

Key aspects of DA were covered in this chapter, with an emphasis on power system protection and self-healing. Even though WiMAX was discussed in this chapter, it is only a representative communication mechanism. The main point is not to be overly concerned with the details of WiMAX, but rather to learn what might be common fundamental features of a DA communication network. For example, WiMAX covers a metropolitan area; thus, it is designed for sufficient coverage of a distribution system. Mesh mode options consider leveraging a potentially large number of nodes to improve overall operation. It also utilizes OFDM efficiently. Finally, it considers the need for QoS. Many other communication technologies could handle these requirements as well.

Chapter 10 reviews standards related to smart grid communications. Standards can be a somewhat boring topic; however, the key aspects are quickly reviewed; only the most significant standards are discussed in detail. This allows one to see what industry feels is most important and requires a common interface for interoperability. The goal of standards is to promote innovation by providing a common framework for ideas as well as the obvious goal of achieving interoperable components. Thus, areas left unstandardized are sometimes more informative than those that are standardized. Standards are a practical requirement for a successful implementation of the "smart grid." Many smart grid-related standards are still in development and will continue to undergo considerable change. Each major standard organization developing smart grid-related standards is reviewed including NIST, International Electrotechnical Commission, International Council on Large Electric Systems, Institute of Electrical and Electronics Engineers, American National Standards Institute, International Telecommunication Union, Electric Power Research Institute, and the International Telecommunication Union. Many ad hoc industry groups have also created lasting standards that impact power systems and communication. Modbus is introduced as an early example. The Microsoft Power and Utilities Smart Energy Reference Architecture also provides an interesting holistic information technology perspective on the smart grid.

9.5 Exercises

Exercise 9.1 Reactive Power and Voltage Regulation

1. How are capacitance and voltage related? Provide an equation.
2. How does the answer to the above question relate to controlling reactive power?
3. What implication do the above answers have on controlling both reactive power and voltage simultaneously?
4. What impact(s) do the answers to the above questions have on requirements for communication?

Exercise 9.2 AMI and DA

1. Why have AMI and DA been considered separate systems when they both exist in the electric power distribution system?

Exercise 9.3 Communication Reliability

1. Will communication make the power system more or less reliable? Why?
2. Under what conditions will communication make the power system more reliable?

Exercise 9.4

1. Use the induction disk relay to explain the curve in Figure 9.2. What is happening in the induction disk relay during a fault? What is happening during a reset?

Exercise 9.5 IEEE 802.1Q tagging

1. What is a virtual LAN?
2. What two features does IEEE 802.1Q provide?
3. Explain how these features may be useful for communication in DA?

Exercise 9.6 Routing Protocols

1. What are the general characteristics of a reactive, proactive, and geographical routing protocol?
2. Which of the above routing protocols would likely be best for use in an electric vehicle network? Why?
3. Which of the above routing protocols would be best on a pole-mounted system? Why?

Exercise 9.7 Adaptive Relay

1. Show the relationship among the inverse TCC curve, power load entropy, and communication required for adaptation of an adaptive relay. Hint: consider using the Markov inequality.

Exercise 9.8 Inverse Time–Current Curve

1. How should the time dial settings be related to one another to achieve coordinated protection in multiple protective relays along a long radial distribution system extending from a single feeder?
2. How would communication among protective relays improve operation of the scenario above?

Exercise 9.9 WiMAX in DA

1. Of all the parts of the power grid, the distribution system is relatively close to the ground and exposed to the public and clutter in the environment. What are some of the channel problems this causes and how does OFDM help mitigate these problems?
2. Suppose that a protection relay emits a large amount of instantaneous electromagnetic interference when triggered to open or close. What techniques could a WiMAX system use to ensure operation when it is most needed during a fault?
3. WiMAX utilizes TDMA while IEEE 802.11 uses carrier sensing and allows for packet collisions. What are some of the advantages and disadvantages of TDMA over CSMA for protection in DA?
4. What would be the advantages of the WiMAX mesh mode of operation?

Exercise 9.10 Matrix Reliability Analysis

In Equation 9.23, the value of SAIDI in a DA system is represented by vectors and matrices that represent the topology of the distribution system and the communication network and their performance – for example, speed of operation of relays (automatic recloser operation time) – as shown in Table 9.1.

1. What does $\overline{\mathbf{C}}[f, t]$ represent in Equation 9.23? How is $\overline{\mathbf{C}}$ derived from the corresponding adjacency matrix for the communication network?
2. How many islands exist in the distribution system if the number of zero eigenvalues of the graph Laplacian of the power distribution topology is four (recall the graph Laplacian from Section 6.4.4)?

10

Standards Overview

The nicest thing about standards is that there are so many of them to choose from.
—Andres S. Tannenbaum

10.1 Introduction

This is the last chapter in Part Two on communication and networking for the power grid, and it closes the topic by looking at standards. This book is emphatically not about standards, but rather introducing fundamentals. However, standards are a practical requirement for a successful implementation of the "smart grid." At the time of this writing, most standards are still under development and will continue to undergo considerable change as they evolve. The fundamentals introduced in this book help in understanding standards as they develop. Standard documents can be boring and tedious; we will only overview key points of standards that appear to have future impact on the fundamentals of power grid operation and we will expand on important points within standards that provide general, fundamental background for communication in smart grid. These are standards that have a direct bearing upon smart grid communication and are expected to be widely deployed. Standards can be viewed from an organizational perspective (that is, the standards organization under which the standard was developed), as well as from a functional standpoint (that is, the functionality provided by the standard). There are advantages to both approaches. Standards often have a history of development under a single standards organization, and so it can be helpful to know the provenance and relationship among the organization's standards. On the other hand, one also wants to know what the standards actually do. In this chapter, standards are classified by organization with clear functional explanations.

Since NIST was given the leading role in coordinating smart grid standards, it is a logical place to begin the discussion of standardization (Section 10.2). A large number of international standards fall under the IEC. Section 10.3 discusses standards from this organization. While CIGRE does not create standards, it does influence and participate in standards development. Thus, a brief introduction is included here as well (Section 10.4). In Section 10.5, key IEEE standards are discussed. Then there is a discussion of standards from ANSI (Section 10.6), ITU,

Smart Grid: Communication-Enabled Intelligence for the Electric Power Grid, First Edition. Stephen F. Bush.

and EPRI. Section 10.7 briefly introduces standards activity under development by the ITU that is related to the smart grid. EPRI is funded by the utilities industry and, like the International Council on Large Electric Systems, does not create standards, but rather has influence on their development. This is covered in Section 10.8. There are many ad hoc industry groups that have created lasting standards that impact power systems and communications. In Section 10.9.1, Modbus is introduced as an example. And finally, Section 10.9.3 reviews the Microsoft Power and Utilities SERA), which provides an interesting, holistic information-technology perspective on the smart grid. The chapter ends with a brief summary and exercises for the reader.

10.2 National Institute of Standards and Technology

NIST was given the role of coordinating the development of an interoperability framework for the smart grid in the 2007 Energy Independence and Security Act. This was a large undertaking and required NIST to work with industry and government stakeholders to identify existing and developing standards as well as gaps in standards and technology. Release 1.0 of the roadmap occurred in 2010. This reference model partitioned the smart grid into seven domains: bulk generation, transmission, distribution, markets, operations, service provider, and customer. It also identified main actors and applications within each domain. The reference model also attempted to identify interfaces among the domains and information to be exchanged.

Another job that NIST attempted was determining the priority of the standards required. Ultimately, hundreds of standards are required, but some were more urgent than others to ensure success. The priority areas were: DR and energy efficiency, wider area situational awareness, electric storage, electric transportation, AMI, distribution grid management, cybersecurity, and network communications (FitzPatrick and Wollman, 2010; Nelson and FitzPatrick, 2010; NIST, 2009; Wollman *et al.*, 2010). It is instructive to see what some of the priority action plans were and judge their success. Listed here are only a few of the more relevant ones to communication: standards related to the ability to upgrade smart meters, standard for DR signals, standards for energy-use information, DNP3 mapping to IEC 61850, harmonization of IEEE C37.118 (the synchrophasor standard) with IEC 61850 and with precise time synchronization, guidelines for use of IP in the smart grid, guidelines for use of wireless communication in the smart grid, energy storage interconnection guidelines, interoperability standards for plug-in electric vehicles, and harmonization of power line carrier standards. Clearly, there is a vast and growing number of smart grid standards; it would be impossible to cover them all in a single, readable book. This book covers only the most widely used smart grid protocols relevant to communications.

Operation takes place through the NIST Smart Grid Interoperability Panel (SGIP), which is an open process for stakeholders to interact with NIST in developing the smart grid framework. The SGIP initiates a priority action plan (PAP) to address short-term gaps identified in the standards or framework. Release 2.0 of the NIST framework and roadmap for smart grid interoperability standards appeared in February 2012. The document continues work defined in the first release. An interesting item of note regarding the second release is its two next steps; it identifies electromagnetic disturbances and interference as critical, long-term challenges to be addressed, as well as reliability, implementability, and safety. Regarding electromagnetic disturbances, it is recognized that smart grid devices and communication solutions will place an unprecedented amount of electromagnetic sources into operation. This will result in exposure to potential electromagnetic disturbances and interference. Such sources can include everything from geomagnetic storms to malicious sources such as high-power generation directed through antennas or high-altitude electromagnetic pulse (HEMP) weapon systems. Many other, more frequently occurring sources include switching and fast transient currents, electrostatic discharge, lightning, and RF interference. Vehicles will soon have standard computer communication with the infrastructure and other vehicles. The massive introduction of electromagnetic sources throughout the power grid and vehicles will require careful attention to electromagnetic compatibility. The second item of note, namely reliability, implementability, and safety, addresses the concern that the vast number of standards and the corresponding complexity could have unforeseen and unintended effects.

10.3 International Electrotechnical Commission

The IEC is a nonprofit, nongovernmental, international standards organization that covers electrical, electronic, and related technologies, which they term "electrotechnology." More specifically, it covers technologies such as power generation, transmission, and distribution to home electronics, semiconductors, solar energy, nanotechnology, and many others. It also manages organizations that certify equipment conforms to its standards. There are over 100 IEC standards relevant to the smart grid. However, there are a few key IEC standards that appear to be core standards from the IEC perspective. We summarize them here beginning with IEC 62351.

IEC 62351 is a series of standards documents comprised of 11 parts that covers cybersecurity for the IEC Technical Committee 57 standards series, which encompasses a large part of the IEC smart grid communication standardization effort (Cleveland, 2012). Specifically, this includes IEC 60870-6 or TASE.2 ICCP discussed in Section 10.9.3, IEC 61840 over MMS, IEC 61850 GOOSE and sampled values discussed in Section 1.2.1.9, IEC 60870-5-104 and DNP3 discussed in Section 10.5, and IEC 60870-5-101 and Serial DNP3.

IEC 62351 makes extensive use of transport layer security (TLS) Version 1.2 (Dierks, 2008). As the name suggests, TLS provides security at the transport layer to client–server applications. It is designed to prevent eavesdropping, tampering, and message forgery. It utilizes a block cipher and keyed-hash message authentication code (HMAC). HMAC is also used for security in DNP3 (see Section 10.5).

A block cipher code can be understood as follows. First, consider the original plain text that needs to be securely transmitted. Then partition the plain text into blocks: $P = [P_1, P_2, \ldots, P_L]$, where P_i are plain text partitions. Next, encrypt each block independently: $C_i = E_K(P_i)$, where $E_K(\cdot)$ is an encryption function. Then $C = [C_1, C_2, C_3]$, where C_i is an individual, encrypted block. Cipher block chaining adds more entropy by including the previously encrypted block with the current, plain text block being encrypted: $C_j = E_K(P_j \oplus (C_j - 1))$, where $\oplus$ is the exclusive-or operator. Note that there must be a C_0, which is predefined. HMAC is described in Section 10.5.

TLS begins with a handshake protocol to enable consensus between the client and server on protocol version and cryptographic and compression algorithms, as well as optionally authenticate one another. TLS utilizes public key encryption to generate shared keys. Once the handshake protocol completes, a record protocol takes over to handle data transmission, securely utilizing the shared keys established during the handshake protocol. More specifically, the record protocol fragments the data into blocks, compresses the data, applies a message authentication code, described in Section 10.5, encrypts the data using the CBC block cipher, just described, and transmits the final result.

As mentioned, IEC 62351 is comprised of 11 parts. Part 1 is an introduction and Part 2 is a glossary. Parts 3–6 divide the problem of security into underlying protocol components; namely, standards that use TCP/IP, standards that use MMS, and so on. Part 3 covers security profiles for standards using TCP/IP – namely, IEC 60870-6 (TASE.2 or ICCP) – IEC 60870-5 Part 104, DNP3 over TCP/IP, and IEC 61850 over TCP/IP. Part 3 specifies the use of TLS as the cybersecurity solution. Part 4 covers standards utilizing MMS – namely, IEC 60870-6 (TASE.2 or ICCP) – IEC 61850 using MMS and again specifies the use of TLS. Part 5 covers DNP3-specific security, where TLS is again a recommended solution for DNP3 over TCP/IP, but is too heavyweight for serial connection. A lighter-weight mechanism that includes authentication is specified for serial connections. Part 6 covers IEC 61850 aspects not based upon TCP/IP; namely, GOOSE and sampled values. Since GOOSE messages have a stringent time requirement of 4 ms for critical power protection operation, full cybersecurity protection could slow down operation such that it would not meet the time-critical, 4 ms requirement. Instead, authentication is the only cybersecurity requirement.

Part 7 of IEC 62351 is interesting in that it addresses network management, a significant aspect of smart grid communication. The reason that network management is included within the cybersecurity standards set of documents is that a well-managed network is a necessary component of cybersecurity. A network that is unmanaged is also likely to be unmonitored; anomalies, whether malicious or accidental, will go unnoticed until they become serious issues. SNMP is widely utilized for Internet network management. SNMP is designed to be simple and low overhead, residing on UDP/IP. SNMP clients can query SNMP agents to obtain a wide variety of management information, such as the number of packets flowing through an interface, as one specific example. The SNMP agent resides on a managed device such as a router, or a smart grid IED, for example. Objects that can be queried on a device are defined in an SNMP management information base (MIB) that contains an ASN.1 description of each object. Thus, an MIB provides a clear description of the management information that device provides. Some SNMP MIBs have been standardized for use in common Internet devices, while manufacturers are also free to create their own MIBs specific to the management information related to their devices. MIB objects are related to one another in a single tree structure with a common root; thus, standard MIBs can be easily extended with vendor-specific MIB information. There is currently no standard MIB for the smart grid. Each vendor's device will have its own communication interface with its own unique set of managed objects; the fear is that there will be many different objects representing the same fundamental power system concept in different ways, leading to inefficiency and potential confusion. Finally, a well-designed smart grid MIB will serve as a common means of assessing whether a power system problem is an electrical problem or an application or network problem.

Part 8 of the series specifies role-based access control (RBAC) for power system management. RBAC is an old concept that ensures a user has no more permission than that required to do a person's job function. This is known as the "principle of least permission." This part of the standard defines a comprehensive set of mechanisms to implement RBAC, including credentials for users, types of roles and permissions, transport of role information, extension of data models required to implement RBAC, and verification of credentials.

Part 10, security architecture, provides overall cybersecurity guidelines and maps the interrelationship of all the cybersecurity components and their interactions.

Finally, Part 11 in the series covers cybersecurity for extensible markup language (XML) files. XML files are generated in a CIM, where data in CIM objects can be transmitted via

XML, and IEC 61850 substation configuration language (SCL), where substation devices and their configuration and interconnection are represented. Thus, both CIM and SCL XML files are stored and transmitted through the network and require cybersecurity protection.

IEC 62056 covers data exchange for meter reading, tariff, and load control. There are many electric meter protocol standards; listing them all would be tedious and of little benefit. It is important to keep in mind that we are only discussing a selected, small sample of protocols based upon their expected impact, novel concept, or relevance to the smart grid. IEC 62056 is a meter communication protocol that supersedes an earlier standard, IEC 61107. IEC 61107 is an older, simpler, meter protocol widely used in Europe. It transmits ASCII over a serial port in half-duplex mode, where the serial interface communication can be optical, utilizing an LED and photo-diode reader, or wire based, using EIA-485.

IEC 62056 is a series addressing international standardization for meter reading. In reality, it is the international version of device language message specification (DLMS)/companion specification for energy metering (COSEM), and these specifications are maintained by the DLMS User Association. COSEM contains information regarding the application and transport layers of the DLMS protocol. The DLMS User Association organized their information into colored books: the Blue Book describes the meter object model and identification system, the Yellow Book describes conformance testing, the Green Book describes the overall architecture and protocols, and finally the White Book contains the glossary.

The IEC organizes the information as follows.

- IEC 62056-21: direct local data exchange describes how to use COSEM over a local port (optical or current loop);
- IEC 62056-42: physical layer services and procedures for connection-oriented asynchronous data exchange;
- IEC 62056-46: data link layer using the high-level data link control protocol;
- IEC 62056-47: COSEM transport layers for IPv4 networks;
- IEC 62056-53: COSEM application layer (withdrawn by IEC);
- IEC 62056-61: object identification system (known as OBIS);
- IEC 62056-62: interface classes.

IEC 61850 was designed to run over wired Ethernet within the LAN installed in a substation. For IEC 61850 to operate over a wider area – for example, throughout the distribution or transmission system – IEC 61850 packets would have to be translated to a different WAN protocol or the Ethernet network would have to somehow be greatly extended over the wider area over which transport is required. Thus, a straightforward means of extending IEC 61850, without having to change the protocol, is to simply extend Ethernet itself; that is, create a wide-area Ethernet LAN. This is, in fact, an approach that has been recommended and implemented. Another term for such wide-area Ethernet is "Carrier Ethernet," although this is used more to indicate the transport of Ethernet frames by common carriers; that is, large telecommunication companies. The basic concept is rather simple, to tunnel Ethernet frames over wide-area communication networks. One example of this approach is to utilize the virtual private LAN service (VPLS), which provides Ethernet-based multipoint-to-multipoint communication over IP multiprotocol label switching (MPLS) networks.

First, let us consider the nature of MPLS. MPLS routes data based upon relatively short path labels rather than longer network addresses and, by avoiding complex table lookup, implements a more efficient routing process through a network. The labels represent virtual paths rather than endpoints. MPLS also encapsulates PDUs from numerous other network protocols and allows operation with transmission system 1 (T1), ATM, frame relay, DSL, SONET, and Ethernet. The goal is to create a universal data link layer that is independent of specific technologies such as those previously mentioned; it is universal in the sense that it also handles both circuit-switched and packet-switched traffic. Because MPLS attempts to provide both data link and routing capabilities, it is sometimes called a layer 2.5 protocol; that is, between the data link at layer 2 and the network at layer 3. For readers familiar with ATM, MPLS was designed to supplant ATM, having learned from its weaknesses. MPLS no longer required the small, fixed cell-sizes of ATM, but still allowed for traffic control and out-of-band signaling that have been useful for both ATM and frame relay.

As mentioned, one of the keys to MPLS is its label routing. Packets are prefixed with an MPLS header that contains one or more labels. The labels in the header are known as a "label stack"; each entry in the stack is comprised of four fields: (1) a label value of 20 bits, (2) a traffic class of 3 bits, (3) a bottom-of-stack flag at 1 bit, and (4) a time-to-live field of 8 bits. As an MPLS packet flows through the network, the packet is switched based upon a label lookup. Because the packet's label can control the path a packet takes through the switching fabric, no time-consuming CPU lookup is required, as would be required for an IP address. Thus, routers that look only at the label are known as label-switch routers (LSRs). A label edge router (LER) pushes an MPLS label onto an incoming MPLS packet and removes it from outgoing packets. In essence, label-switched paths are quite similar to virtual circuits in ATM and frame relay. At each hop, the top label in a packet's label stack is examined by a router; it may be swapped with a new label, a new label pushed on top, or the label may simply be popped; that is, removed from the stack. Thus, the topmost label controls the next-hop operation. When the last label is popped from the stack, the packet has left the MPLS tunnel. Note that the last MPLS router, after the last remaining label has been popped from the stack, must then route the packet using the packet's native routing mechanism, typically IP. Thus, MPLS effectively tunnels an arbitrary PDU, or packet, through the MPLS network. The PDU may be an Ethernet frame; MPLS is an effective way to implement wide-area Ethernet. However, management of label-switching/routing for an MPLS is not magic; it requires a standardized protocol to manage the labels. This can be either the label distribution protocol (LDP) or the resource reservation protocol for traffic engineering (RSVP-TE). Finally, recall that IEC 61850 GOOSE messages require multicast; thus, it will be useful to utilize MPLS multicast.

MPLS is only one example of a tunneling protocol that IEC 61850 can utilize in order to extend the reach of substation automation information. Since Ethernet has its origins as a wired bus, the idea that VPLS uses is the concept of connecting geographically dispersed sites via a virtual bus, or pseudo-wires. The pseudo-wires can simply be MPLS as previously described or layer 2 tunneling protocol version 3 (L2TPv3), described shortly. However, Ethernet is, by nature, a broadcast technology: every receiver on the bus can potentially receive a message. Some pseudo-wire technologies, such as L2TPv3 are point-to-point tunnels through layer 2 of the protocol stack. VPLS enables any-to-any connectivity by extending LANs from each connected site to create a single, bridged, LAN. Thus, Ethernet broadcast and multicast are supported.

The idea behind layer 2 tunneling protocol (L2TP) is to allow the layer 2 protocol (namely, data link layer packets) to reside on layer 3, the network layer. This allows layer 2 data link operation to occur transparently across a network. As an example, this could allow a serial link protocol to be extended over a network. In fact, L2TP was initially designed for transporting point-to-point protocol (PPP) over layer 3, where PPP implements IP connectivity over a variety of different types of serial connections.

L2TP was designed to be simple. It provides only minimum-required services since the protocol being tunneled through it will provide whatever additional services are required. For example, encryption and authentication are not implemented since the layer 2 protocol passing through the tunnel is assumed to handle whatever services are necessary. In L2TP terminology, an L2TP access concentrator (LAC) connects over L2TP to a remote host known as the L2TP network server (LNS). The LAC can accept serial connections from users and establish an L2TP tunnel for the serial connections to the LNS; the LAC is the initiator of the tunnel, or tunnel client, and the LNS is the tunnel server. The LNS performs security checks, locally, upon incoming connections. L2TP transmission occurs over UDP/IP, and reliability must be implemented by the protocol being tunneled through L2TP. Thus, L2TP is another candidate for tunneling GOOSE within Ethernet frames over a wide area.

10.4 International Council on Large Electric Systems

CIGRE is known by its French acronym, which comes from Conseil International des Grands Reseaux Electriques. It was founded in Paris in 1921 and is an international organization dedicated to high-voltage electricity. CIGRE's main objective can be summarized as designing and deploying the power system for the future, including optimizing today's equipment while respecting the environment and facilitating access to information.

10.5 Institute of Electrical and Electronics Engineers

The IEEE is a nonprofit professional association comprised of 38 technical societies dedicated to the advancement of technology and innovation. The IEEE was officially formed in 1963 with the merger of the Institute of Radio Engineers (IRE), founded 1912, and the American Institute of Electrical Engineers (AIEE), founded in 1884. While the IEEE is well known for its publications, journals, and conferences, it also plays a significant role in standards development in its numerous electrical, electronics, and related fields through the IEEE Standards Association.

The two societies of most relevance to smart grid are the IEEE Power and Energy Society (PES) and the IEEE Communications Society (ComSoc). Interestingly, the PES can claim to be the oldest society of the IEEE and was formerly known as the IEEE Power Engineering Society; it changed its name in 2008. The IEEE Communications Society, or "ComSoc" as it is affectionately known, had its official start in 1952 with the formation of IRE, mentioned earlier.

The IEEE, along with other standards bodies, is working with NIST to develop the smart grid standards roadmap, as well as testing and certification standards. The IEEE has more than 100 published standards and standards under development relevant to the smart grid. Over 20 IEEE standards were specified in the NIST Framework and Roadmap for Smart Grid Interoperability Standards Release 1.0. It would be too much to cover every standard in detail; the goal of this book is to focus on underlying fundamentals from which one can understand the relevance, usefulness, and efficiency of various standards. Here, we briefly review only a select set of the most popular IEEE standards.

The IEEE 1815-2010 Standard for Electric Power Systems Communications – DNP3 is an updated version of the distributed network protocol (DNP) standard; it was updated in light of recent smart grid activity and cybersecurity concerns. DNP3 has been in use primarily in North America since 1993, when it was first released as an open standard. IEC 60870-5, a SCADA protocol to accomplish the same function, was under development at the time. However, industry could not wait for the IEC 60870 standards process to reach completion; DNP3 was released earlier and placed into operation sooner. DNP3 utilized portions of IEC 60870 that had been completed (IEEE, 2010); thus IEC 60870 and DNP3 are quite similar. Also, around the same time, EPRI's Utility Communications Architecture (UCA) 1.0 had been recently released. However, UCA 1.0 was never widely used by utilities; it also led to concerns regarding its bandwidth efficiency. Interestingly, this eventually led to UCA 2.0 that then evolved to become IEC 61850.

DNP3, a SCADA protocol, is used to implement communication between a master station and RTUs and IEDs. Another protocol, the ICCP, discussed in Section 10.9.3, also part of IEC 60870-6, interconnects master stations or control centers. DNP3 runs over simple, serial physical media – for example, RS-232 or RS-485 residing on physical layers such as copper, fiber, radio, or satellite – as well as over Ethernet. DNP3 provides layers that are equivalent to the OSI data link layer (layer 2), transport layer (layer 4), and application layer (layer 7).

The DNP3 data link layer inserts a 16-bit CRC every 16 bytes within the data link frame. The data link frame has an address as well as control information, such as ACK, negative acknowledgment (NACK), link reset, link is reset, request data link confirmation ACK, request link status, and reply to a link status request.

The transport layer handles fragmentation and reassembly of a message into multiple data link frames. Each frame has an indicator that signifies the beginning of a message, end of a message, or both, for a message that fits within a single frame. A sequence number is also included within each frame to allow the receiver to detect dropped frames. Thus, the data link layer is heavily protected by CRC sequences to detect errors and dropped frames.

The application layer of DNP3 has many features that address issues related to the smart grid, including cybersecurity, the ability to assign priority levels for more critical messages, bandwidth savings by allowing IEDs to utilize unsolicited event reporting, inherent functions that implement time synchronization, and the use of objects in a manner that is compatible with IEC 61850. These features make DNP3 more sophisticated and complex than Modbus, while perhaps simpler and less restrictive than IEC 61850, in which time-critical GOOSE messages are restricted to Ethernet frames.

While early SCADA protocols such as Modbus are primarily poll-request systems; DNP3 can save bandwidth by configuring RTUs or IEDs to send unsolicited messages when events of interest occur. Thus, for example, data values only need to be sent when they change value or exceed a specified bound. The overhead of a periodic-poll message from the master is eliminated as well as the overhead from potentially uninformative response messages; namely, resending a value to the master station that has not changed. When a poll-request paradigm is used, the application layer typically makes the decision whether to retry if a response is not received. Lower layers do not typically automatically retry since the application may decide to move on to its next poll and will not be expecting a response from a lower layer. However, it is optional to enable lower layer retries. On the other hand, an unsolicited message is initiated by an RTU or an IED. The master station will not be expecting such a message; it is up to the sending unit to ensure the transmission was successful. In this case, lower layer retries are encouraged.

DNP3 priority levels utilize the notion of classes; namely, class 0, 1, 2, or 3. Class 0 is for static data; that is, data that do not change during normal operation. The master station can poll for event data from class 1, 2, or 3. The meaning behind the different classes of event data remains open for the user to define; however, a typical approach is to assign class 1 to high-priority data, class 2 to medium-priority data, and class 3 to low-priority data. Each class can then be polled at a rate set to correspond with its priority. It is important to note that the standard does not indicate that a priority mechanism applies at lower levels within the network. In addition to the classes, there is also an integrity poll. The integrity poll requests information from all the classes, including class 0. The integrity poll is intended to be used infrequently; typically, for example, upon startup to check the complete state of a device.

As previously mentioned, DNP3 includes functions that support and implement time synchronization. Propagation delay over the physical media is measured, as well as the processing time at the RTU or IED. A distinction is made between non-LAN and LAN time synchronization. The LAN time synchronization is simpler than the non-LAN method since the LAN is assumed to use high-speed switched-Ethernet; physical media propagation delay is assumed to be negligible, but with a relatively large variance given changing traffic load and possible frame collisions.

DNP3 implements cybersecurity at the application layer at the level of individual messages. In other words, it is possible to authenticate some messages and not others. This allows cybersecurity overhead to apply only where it is deemed necessary. The concept is straightforward; if a message marked critical is received, it must be challenged by the receiver. The receiver will then issue a challenge message to the transmitter of the original message. The transmitter of the original message must then send an authentication response in answer to the challenge message. If the challenger determines authentication is successful, the message will be acted upon normally and a normal response returned to the transmitter of the original message. If authentication fails, the challenger will notify the sender of the failed authentication, allowing the transmitter a chance to retry. However, only a finite number of retries should be allowed from the same transmitter.

The specific authentication mechanism used is known as an HMAC. Roughly and intuitively speaking, it can be described as a hash function applied to a concatenation of a key and the message to be transmitted, where the key is assumed to be known only by valid senders and receivers. The actual encoding process is a bit more complex, and is described in detail in NIST FIPS (2002). The result is that the challenger can repeat the process with the original message and its key to determine if the same result is obtained. If the result is the same, then the message is valid; if the result differs, then something has changed: either one of the parties has an invalid key or the data being transmitted has been modified. This technique is not perfect; it is possible for an attacker to repeatedly attempt the authentication process with different keys until a valid key is found. That is why the challenger should place a finite limit on the number of authentication attempts from the same transmitter. DNP3 also handles encryption, distribution, and management of keys. The authentication key is known as a session key, and a separate key known as the update key is used to handle session key distribution.

Similar to IEC 61850, DNP3 utilizes object-oriented data and device profiles. Device profiles are described using XML and aid in system configuration. A vendor can describe a device using DNP3 objects in XML and share it with a utility so that the utility can determine whether the device meets their requirements. Alternatively, a utility can create an XML document that describes the type of device required for a vendor. The XML document allows devices to be easily configured and interconnected for operation. Finally, the DNP3 standard requires a mapping between DNP3 and the IEC 61850 object model using the XML path language (XPath).

DNP3, as previously described, consists of three layers: data link, transport, and application. These three layers can be placed on top of any other layered protocol, such as the IP transport layer; namely, UDP/IP or TCP/IP. DNP3 is often placed on top of TCP with the assumption that this will provide extra reliability, because UDP is only a BE transport protocol. However, one should carefully consider whether the additional reliability of TCP justifies the overhead. DNP3 link-layer frame confirmation is not used when running over IP; the application layer handles retries as it would if running over a wired connection.

At this point, a few words about error correction are in order. Almost all SCADA protocols incorporate error detection, but few, if any, use forward error correction; that is, techniques that provide redundant bits in order to correct for error at the receiver. This is perhaps partly historical, but also because it is felt that forward error correction would be too costly in terms of added delay. Almost all SCADA protocols use CRCs to detect error. A CRC can be intuitively thought of as adding parity bits: additional bits appended to a protocol data unit that allow the receiver to determine whether any of the bits have been corrupted in transit. The operation of a CRC is intuitively straightforward once the idea of representing a binary value as a polynomial is understood. The polynomial coefficients are either 0 or 1. The data to be protected is treated as such a polynomial and divided by a generator polynomial, also known as the CRC polynomial. Choosing a generator polynomial is the key part of the error protection design. The remainder of the polynomial division is transmitted along with the data it is protecting, much like a simple parity code. The receiver knows the generator polynomial that was used and performs the same division of the data by the generator polynomial; if the remainder computed by the receiver differs from the one that was transmitted, then an error has been detected. Unfortunately, there is no divisor that can detect all possible combinations of bit errors. However, the CRC mechanism has worked reasonably well and is particularly important for critical information that may be impacted by electromagnetic interference. However, it is important to emphasize that no error detection or correction mechanism is perfect; it is always possible for bits to be corrupted in such a manner as to *appear* to be a valid transmission; that is, to appear as valid code accepted by the receiver even though the message is corrupt, holding invalid data that could cause unexpected behavior of the system. Understanding this fact and designing a good CRC code is often overlooked. This is perhaps because it is a difficult problem, but mainly because standards tend to specify precisely which CRC must be used. However, the standards working group has to make a compromise when specifying a CRC. There is a trade-off between the size and complexity of the CRC code relative to the data being protected and the degree of error protection. A measure of the degree of protection is the Hamming distance. The Hamming distance is the minimum number of bit changes required to change a valid code into another valid code. Ideally, the Hamming distance should be large, and preferably infinite, although this is impossible. The larger the Hamming distance, the larger the number of bits that can be changed in error and still be detected. A complicating factor is the distribution of the bits that are corrupted in a PDU. The bits in error may be uniformly distributed throughout the PDU or they may all be in consecutive order, known as a burst error. A common CRC is the 16-bit checksum known as CCITT-16.

How do we know the amount of protection the CRC is providing? One way to view the error protection provided by the CRC is to consider the Hamming weight. The Hamming weight is the number of errors, out of all possible corruptions, that will be undetected; that is, that will appear as a valid code word. As an example, consider the commonly used CCITT-16 generator polynomial used to protect a 48-bit data sequence. The Hamming distance is 4; it is possible for a minimum of four bit-errors to be undetectable. However, the Hamming weight provides more intuition about the strength of the CRC. First, we know that since the Hamming distance is 4, the Hamming weight for a single, double, or triple bit error must be zero. However, for four bit errors, the Hamming weight is 84, which means that there are 84 possible combinations of four-bit errors that will be undetectable.

We can make some general observations regarding CRC protection (Jiang, 2010). These involve: (1) error pattern coverage, (2) burst-error detection, and (3) probability of undetected error. The error pattern coverage is the ratio of non-codewords to the total number of codewords and is

$$\frac{2^n - 2^k}{2^n}, \tag{10.1}$$

where n is the total length of a PDU including the checksum, and k is the length of the data not including the checksum. Thus, n must be at least as large as k; and the larger the value of k, the more useful data can be transmitted. The error pattern coverage provides the probability that data, when corrupted, is not included in the set of valid codewords causing it to be rendered undetectable. Therefore, for good error coverage, we want the ratio to approach 1 and k to be small relative to n. Herein lies the trade-off between protection and bandwidth overhead.

As mentioned, bit errors may not be uniform, but rather appear in consecutive strings of errors known as bursts. Consider a burst error of length b. Clearly, b must be less than n, the total length of the PDU. A CRC code where k is the length of the data being protected (not including the checksum) and n is the total length of the PDU including the CRC code will be able to detect all bursts such that $b \leq n - k$. Note that $n - k$ is simply the length of the checksum. It will also be able to detect the fraction $1 - 2^{-(n-k-1)}$ of all bursts of length $b = n - k + 1$; that is, one bit more in error than the length of the CRC code. Finally, if $b > n - k + 1$, then the CRC code will be able to detect the fraction $1 - 2^{-(n-k)}$ of such burst errors.

Finally, we can consider the probability of an undetected error. For a binary symmetric channel, a simple channel model in which a binary bit may be flipped, with a small probability of channel error and relatively large code length, the probability of undetected error approaches $2^{-(n-k)}$ regardless of the channel quality. Given the CRC code analysis above, we can see that, while standardizing on a single CRC code is often done for expediency, a more efficient system might use an adaptable CRC code, one that adapts to the expected channel model to provide the level of protection required while minimizing overhead.

10.6 American National Standards Institute

The ANSI is a private, nonprofit organization that oversees the development of standards in the USA as well as serving to coordinate international standards for use in the USA. Other standards organizations also receive accreditation from ANSI to ensure that their results meet international standards requirements.

Thus, ANSI does not directly develop standards, but rather oversees the processes of standards developing organizations through a process of accreditation to ensure that all processes are open, balanced, reach true consensus, and, in general, use proper procedure. The label "ANSI" can be used when it determines that a standard was developed meeting the previous requirements. As an example, IEEE standards meet ANSI requirements and are thus also "ANSI standards."

At the time of this writing, there are nearly 10 000 ANSI standards; attempting to cover each one that relates to the power grid or smart grid would be impossible. Many of the smart grid standards covered in this chapter are also ANSI standards. However, we focus on one particular smart grid standard often prefixed with the acronym "ANSI"; namely, the ANSI C12 standard series for smart meters.

ANSI C12.18 specifies an optical interface to meters along with a protocol specification for electronic metering (PSEM). C12.19 defines the meter data structure, commonly referred to as "tables." C12.21 specifies the modem and PSEM for metering. Clearly, techniques that can leverage HANs and avoid laying additional wiring will make a compelling argument in their favor. Finally, C12.22 specifies a network protocol for bidirectional communication with the meter.

AMR and AMI have long used a variety of communication media, including RF, both public and private carrier, power line carrier, and two-way paging among others. The result was that utilities ended up stuck with a proprietary single-vendor solution. C12.22, the network meter standard specifies an application-level protocol independent of the underlying communication system along with a physical and data link-level protocol communication system. The goal of the system is to transport data in C12.19 tables via PSEM messages. PSEM is a session-based protocol allowing two parties to transmit requests and responses. Extended PSEM or EPSEM enables the PSEM to operate over a shared channel with multiple nodes.

ANSI C12.19, *Utility Industry End Device Data Tables*, defines data structures that are transported by C12.22. C12.22 specifies encryption, authentication, credential management, intrusion detection, and logging and auditing of all changes to the configuration. Physical security, namely antitamper information, is also incorporated, such as inversion (running the meter backward), removal (opening or removing the meter), and blink counts (detection of whether a meter has been deenergized more often than its neighboring meters).

C12.19 and C12.22 are amenable to two-way communication and enable update of meter configuration and firmware modification over the network. A C12.22 "network" has nodes with C12.22 addresses and there is an accompanying multicast communication protocol. Of course, meter operation depends upon accurate time information, so time synchronization as well as remote programming of the meter are included. C12.19 is also concerned with handling data storage. DR requires programmed-load control based upon pricing parameters, and this is included as well. Network management, involving the ability to gather statistics and issue automated alarms, is also included. The C12.19 and C12.22 specifications also enable measurement not only of power, but also of more interesting power quality metrics, such as total harmonic distortion, sags, swells, interruptions, harmonics, phase, voltage RMS values, and the ability for near real-time monitoring. The standard also covers reporting outages and restoration times, as well as the ability for a meter to infer its location in latitude and longitude. C12.19 and C12.22 are cognizant of the need for keeping the cost of the meter low and operating in a constrained environment. Finally, the specifications assume the standard ASN.1 encoding and association control service element (ACSE) OSI presentation layer call-establishment mechanism.

The IETF request for comments (RFC) 6142 defines the mapping of the C12.22 application layer messaging protocol onto the IP. More specifically, it covers the mapping and encoding of ANSI C12.19 device table elements onto IP networks.

IEEE 2030-2011, *IEEE Guide for Smart Grid Interoperability of Energy Technology and Information Technology Operation with the Electric Power System (EPS), and End-Use Applications and Loads*, is a comprehensive standards guide that covers smart grid interoperability. Another standard, IEEE 1547, focuses on distributed resources and their interconnection with the grid. Both of these guides focus on a comprehensive, system-level view.

The IEEE 2030-2011 guide can be thought of as taking the high-level conceptual reference model developed by NIST and adding another level of detail by dividing it into different perspectives – namely, communications architecture, power systems architecture, and IT architecture – through what is referred to as a smart grid interoperability reference model.

In addition to the main IEEE 2030-2011 document there are three additional documents of interest: IEEE P2030.1 *Guide for Electric-Sourced Transportation Infrastructure*, IEEE 2030.2 *Guide for the Interoperability of Energy Storage Systems Integrated with the Electric Power Infrastructure*, and IEEE P2030.3 *Standard for Test Procedures for Electric energy Storage equipment and Systems for Electric Power Systems Applications*.

IEEE C37.118-2005 originally specified aspects related to both the measurement and real-time transfer of synchrophasor data. The 2010 version of the standard has split these aspects into two standard documents, C37.118.1-2011 *IEEE Standard for Synchrophasor Measurements for Power Systems* and C37.118.2-2011 *IEEE Standard for Synchrophasor Data Transfer for Power Systems*, in order to decouple measurement and quality standards from real-time transmission standards. The idea behind the decoupling was to allow for the possibility of alternative communication mechanisms to be developed while continuing to use the same measurement and quality standards. In this chapter, we are focused upon the communications aspects, and thus the second standard. However, we should review the basics from the first standard. Synchrophasors, introduced in Section 13.2, are a vector representation of a waveform. The vector length is the cosine wave amplitude and the vector angle is the cosine wave phase angle. The goal is simply to specify a standard message format for transmission of these values. Thus, this standard specifies only an application-layer message format and does not directly deal with underlying network issues, other than to allow mapping the message format over many different possible protocols from serial lines to IP transport protocols. Four different message types are defined: data, configuration, header, and command. Data messages convey the actual phasor measurements, configuration messages contain meta information about the PMU, such as calibration factors and a description of data types, header messages carry human-readable descriptions related to the PMU, and command messages convey machine-readable commands to the PMU that control or reconfigure the PMU. Messages are of variable size and contain their length in a standard field after the initial beginning-of-message frame indicator (SYNC) field. The message has a relatively simple format: SYNC (2 bytes), FRAMESIZE (2 bytes), IDCODE (2 bytes), SOC (4 bytes), FRACSEC (4 bytes), a variable number of DATA fields, ending in a message frame checksum (2 bytes). Note that we do not consider the precise formatting of all messages, as this is only intended to serve as an overview of the standard. The message is described in more detail later and is shown in Figure 13.7. The frame size is thus variable; however, a typical frame size is considered to be from 40 to 70 bytes from a single PMU or up to 1000 bytes from a phasor data concentrator (PDC). Communication over a serial line adds 25% overhead, whereas, at the other extreme, TCP/IP adds 44 bytes per frame, or 50% overhead.

The synchrophasor message frame, as defined above, can be mapped onto TCP/IP, UDP/IP, or any other protocol capable of transporting the frame in a timely manner. Because this standard is focused upon real-time operation, when operating over UDP/IP, dropped packets and message frames are simply ignored unless the user chooses to implement a user-defined recovery mechanism. It is possible to use a hybrid TCP/UDP mechanism in which TCP is used for control (that is, for commands, header, and configuration messages) and UDP is used strictly for data messages. The idea is to gain the reliability and security of TCP for the control information and low overhead for data streaming.

IEEE C37.112-1996, *IEEE Standard Inverse-Time Characteristic Equations for Overcurrent Relays*, defines the characteristics of measuring and detecting fault current in a relay or circuit breaker (Benmouyal *et al.*, 1999). This standard is not a communication standard per se; however, it specifies and provides insight into commonly communicated power system information; namely, the likelihood of a fault or notification that an actual fault has occurred. The inverse-time overcurrent relay produces an inverse TCC; it integrates a function of current with respect to time. A value known as the pick-up current (see Section 9.2 for more detail) determines the level of current at which the integrated value is positive. In other words, the pick-up current is a threshold above which a potential fault exists. A value known as the time-dial determines the value of the integral at which the relay or breaker is tripped. A plot can be drawn showing time as a function of multiples of the pick-up current at which the relay would trip. Similar to a trip, there is also a reset; there is a time–current curve controlling when the device should reset after a trip.

IEEE C37.90.2-2004, *IEEE Standard for Withstand Capability of Relay Systems to Radiated Electromagnetic Interference from Transceivers*, is concerned with the issue of wireless communication equipment inadvertently causing a relay to trip or otherwise malfunction due to radiated energy (IEEE, 1995). The goal is to ensure that all protection equipment is immune from RF energy from any form of communication in the vicinity. IEEE C37.90.2-2004 takes into account field strength, test frequencies, modulation, sweep rate, equipment setup and connection, test procedures, criteria for acceptance, and documentation for test results.

The IEC 61588 Ed.2 (2009-02) (IEEE Std 1588-2008), *IEEE Standard for a Precision Clock Synchronization Protocol for Networked Measurement and Control Systems*, is a standard for time synchronization. This is an application of extreme importance for the smart grid; many applications require precise time synchronization. Time synchronization accuracy can range from that required to distinguish one phase from another in a three-phase system, to timing required in integration of the inverse TCC curve for protection, to comparing synchrophasor measurements, to simply recording accurate time for programmed activities and billing purposes; these are only a few among many of the applications requiring precise timing.

In this clock synchronization standard, it is assumed that clocks are connected via a single network and that there is a "grandmaster" clock to which all clocks must synchronize. Since this protocol is known as the precision time protocol (PTP), PTP messages are exchanged to accomplish the synchronization. PTP assumes all messages are multicast using UDP/IP by default, allowing all clocks to receive each other's messages. UDP/IP provides the necessary low-overhead, time-sensitive messaging required at the cost of potentially dropping packets. This protocol allows a large number of clocks to be divided into domains in which a best master clock (BMC) is chosen within each domain or network segment. This provides a hierarchical approach to dividing up the problem of maintaining synchronization. The rest of the algorithm is straightforward: communication delays are estimated between each clock and its master and a corresponding correction is made. A common simplifying assumption, that may not always be true is that delay is symmetric; that is, that the delay from a clock to its master is equal to the delay between the master and the clock.

IEEE C37.238-2011, *IEEE Standard Profile for Use of IEEE 1588 Precision Time Protocol in Power System Applications*, is another use of PTP designed to provide time synchronization among substations across wide geographic areas via Ethernet communication.

IEEE P1909.1, *Recommended Practice for Smart Grid Communication Equipment – Test Methods and Installation Requirements*, documents testing and installation procedures. Safety, electromagnetic compatibility (EMC), and environmental and mechanical tests are covered in this standard.

The IEEE P1906.1, *Recommended Practice for Nanoscale and Molecular Communication Framework* standard, is currently developing the foundation for nanoscale communication networks. This is clearly one of the more visionary technologies under development, but it could have a significant impact on smart grid communication. A common framework will greatly aid in developing useful simulators for nanoscale communication and nanogenerators discussed in Chapter 15. The standard includes interconnecting systems of multiple types of nanoscale simulators. Because nanoscale networking involves a large number of diverse technical fields, a common abstract model is required in order to enable theoretical progress to proceed from different disciplines with a common language. More specifically, the IEEE P1906.1 recommended practice is on target to create: (1) a consistent definition of nanoscale networking; (2) a conceptual model for ad hoc nanoscale networking; (3) common terminology for nanoscale networking, including (a) a definition of a nanoscale channel highlighting the fundamental differences from a macroscale channel, (b) abstract nanoscale channel interfaces with nanoscale systems, (c) performance metrics common to ad hoc nanoscale communication networks, (d) a mapping between nanoscale and traditional communication networks, including necessary high-level components such as a map of major components: coding and packets, addressing, routing, localization, layering, reliability. The vision, discussed in more detail later, will be the generation and transport of power at very small scale in a more flexible, ad hoc manner.

10.7 International Telecommunication Union

The ITU, or Union Internationale des Télécommunications in French, is an agency of the United Nations responsible for communication technologies with membership comprised of 193 member states. Among other things, the ITU manages the use of the international radio spectrum, works toward international agreement in satellite orbits, and helps to improve communications in developing countries. For this chapter, we focus on the ITU's additional role in developing international communication standards. The ITU is comprised of three sectors: (1) radio communication known as ITU-R, focused on the radio spectrum and satellite orbits, (2) standardization, known as ITU-T, the oldest sector previously known as the Comité Consultatif International Téléphonique et Télégraphique (International Telegraph and Telephone Consultative Committee) (CCITT) from its French name, used before 1992, and (3) development, known as ITU-D, which works toward the development of communication infrastructure in developing countries.

The establishment of the ITU-T Focus Group on Smart Grid, or FG Smart, took place in Geneva in February 2010 with the objective of collecting and documenting information and concepts that would be helpful for developing recommendations to support smart grid from a telecommunication and information and communications technology (ICT) perspective.

There were three working groups and five deliverables, namely: (1) a smart grid overview, (2) terminology, (3) use-cases, (4) requirements, and (5) architecture (Martigne, 2011).

The smart grid overview describes three functional layers: (1) an energy layer comprised of devices, sensors, and controllers, as well as advanced metering and intelligent grid control; (2) a control/connectivity layer comprised of the communication network and information access, comprised of data with its associated syntax and semantic descriptions; and (3) an application/service layer comprised of applications and programs that manage the previously described energy layer. The key areas identified for standardization are automated energy management, including DG, intelligent grid management, smart meters and the AMI, the information and communication infrastructure, applications and services, and finally security for control of the smart grid.

The smart grid overview provides a smart grid model in which there are five domains: (1) customer, (2) smart metering, (3) grid, (4) communication network, and (5) service provider. Between these domains there are four reference points: (1) grid – communications enables exchange of information between the grid domain and service provider domain; (2) smart metering – communication network enables the exchange of information and control with the customer domain; (3) customer – communication network enables interaction between operators and the service provider; (4) communication network – service provider enables communications between services and applications to all other domains. Finally, there is an optional fifth reference point between smart metering and customer through an energy service gateway. The smart grid architecture document extends the smart grid overview three-layer model document by essentially filling in more detail to form an architecture.

The ITU-T Focus Group on Smart Grid has relationships with other standards bodies, including the IEC, ISO, and the European Telecommunications Standards Institute (ETSI), among others. The ETSI M2M Technical Committee, for example, will be examining the impact of applications and use-cases as well as cybersecurity for the smart grid.

ITU-T has several ongoing (at the time of this writing) activities related to the smart grid (Brown, 2011); namely, study groups on (1) M2M comprised of activities related to ubiquitous sensor network (USN), IP home networks, and USN applications and services, (2) smart metering, (3) vehicle communications, including the networked-vehicle and vehicle-gateway platform for telecommunication services, (4) home networking, including power line carrier, and finally (5) future energy-saving networks. ITU-T standards G.9955 and G.9956 describe physical and data link layer specifications for narrowband OFDM power line carrier for alternating- and direct-current power lines for frequencies below 500 kHz. The importance of these standards is that they support communication over low- and medium-voltage lines through transformers in both urban and longer distance rural communication.

10.8 Electric Power Research Institute

EPRI is not a standards-setting organization; however, it is mentioned in this chapter for completeness as it plays a role in power systems research and development, as well as in helping to shape public policy. EPRI was founded by the electric utility industry and is a nonprofit organization that conducts research on issues related to the electric power industry. While based in the USA, it has international participation. EPRI was established in 1973 following US Senate hearings in the wake of the Northeast blackout of 1965 with the realization that more fundamental power systems research was needed to avoid such large-scale blackouts in the future. More specifically, the United States Senate Commerce Committee, in essence, threatened to create a government power research agency if the industry could not develop its own research and development program within one year. Government research, without pressure to provide operational solutions, tends to be come lazy and open ended with a large cost to the public without corresponding value to industry and the public (Starr, 1986). To help mitigate this, the result was the formation of EPRI by the utility industry. Because EPRI is funded by a pool of electric utilities, the cost for the research comes directly out of the cost of electricity and results need to be demonstrated. As one example, EPRI provided the initial research and development in the 1970s for FACTS, now usurped as one of the components of "smart grid." As an example of its public policy role, EPRI has the largest program in the world studying the impact of electromagnetic fields on human health.

EPRI realizes the importance of standards in the smart grid and works with standards-setting bodies to help develop standards that achieve efficiency and interoperability, but without creating barriers to innovation or requiring long periods of time to develop or implement. EPRI has made high-level, common-sense suggestions regarding standardization that are worth noting (EPRI, 2010). First, layered approaches, well known in communication, have proven to be an effective method for rapid integration and reuse of innovative approaches. By standardizing only necessary, key aspects, standards are able to be applied more widely and efficiently. Standards must take scalability into account for smart grid; the infrastructure will continue to grow rapidly. Finally, of course, cyberphysical security must be addressed and any new threats or vulnerabilities that develop, either due to increasing attacker sophistication or to the complexity of the power grid, must be anticipated. The EPRI IntelliGrid program has been a primary EPRI project, focused on supporting standards development for the smart grid and has interacted with NIST's efforts. From the EPRI perspective (EPRI, 2010), generator scheduling and transmission are already automated and structured as organized markets. The distribution side is only partially instrumented and has more room for automation. EPRI has many smart grid demonstration projects, including projects in distribution automation demonstrating integration of distributed energy resources.

EPRI also has demonstration projects in smart grid transmission using synchrophasors to help maintain reliability as more DG sources are added to the power grid. EPRI is also cognizant of the fact that utilities will need to continue utilizing older portions of the grid for a long time to come. Therefore, monitoring with new types of sensors will allow aging asset health and impending failure to be quickly determined. The bottom line is that EPRI research impacts a broad range of communication and information technologies for the power grid and related standards development.

10.9 Other Standardization-Related Activities

There has been a lot of other standards activity related to smart grid. This section introduces (1) Modbus, an example of an old, simple, widely used, de facto industrial process standard, and (2) SERA, an attempt to design a computing-oriented framework for the smart grid, recent power line carrier standards activity, the CIM, and finally, the inter-control center communications protocol.

10.9.1 Modbus

The Modbus protocol was developed in 1979 and has its roots in the first programmable logic controller, produced by Modicon, which stood for MOdular DIgital CONtroller. Given its long history and initial use in power line carrier applications, Modbus data types may seem strange today, having been derived from use in driving relays. For example, a single-bit physical output is called a coil, and a single-bit physical input is called a contact. Modbus is a relatively simple application-layer protocol originally designed to operate over serial communication lines, but has since been used over many modern communication protocols, including Ethernet and IP.

Modbus standardizes message format between master and slave devices in a SCADA network. Messages contain a device address, command, and data information along with a checksum to ensure messages have been properly received. The specific message format changes depending upon the network Modbus resides upon. For example, if used with TCP/IP, a checksum is not required and is not included. Modbus also allows broadcast messages; slaves respond when individually addressed, but they do not respond to broadcast messages. When slaves are addressed individually, they respond with a message indicating their status or an exception code, thus implementing a request-response type of protocol. Standard serial Modbus includes two modes: ASCII and RTU. In ASCII mode, each character byte of a message is sent as two ASCII bytes. Timing is critical, including the time allowed to receive the characters of a message. Modbus ASCII mode allows up to a second between characters for a message transmission.

A Modbus message frame marks the beginning and ending of a message. Each word of a message is placed in a data frame that also includes a start bit, stop bit, and parity bit. The address is contained in eight bits and allows for valid addresses from 1 through 247. The function-code byte allows commands numbered from 1 to 255. As we have seen, Modbus uses a simple parity check for each data frame; however, it also uses a CRC-16 code for the complete message frame in RTU mode and a longitudinal redundancy check (LRC) in ASCII mode. An LRC code detects error by operating upon a sequence of parity bits, one for each character in parallel. Finally, Modbus messages over serial lines are limited to 256 bytes and over TCP/IP to 260 bytes (Modbus, 2006).

10.9.2 Power Line Carrier

The IEEE 1901-2010, *IEEE Standard for Broadband over Power Line Networks*, specifies the use of power line carrier, also known as broadband over power line (BPL). The standard is designed to be broadly applicable to all types of BPL devices, including last-mile connections, in-building, LANs, within vehicles, and a broad range of smart energy applications, among others. A goal of the standard is not only power line carrier interoperability, but also a balanced and efficient utilization of the communication channel among the power line carrier devices. A further goal is to incorporate necessary cybersecurity mechanisms.

10.9.3 Microsoft Power and Utilities Smart Energy Reference Architecture

The Microsoft Power and Utilities SERA is a reference architecture that attempts to bridge the standards development work at NIST with Microsoft technologies. It is more than a communications or interoperability architecture; rather, it was a holistic view of the state of smart grid standardization at the time and how Microsoft components fit within that view. Thus, it encompasses communication, computational platforms, visualization, information modeling, databases, and many other aspects outside the specific realm of communications. It does not claim to attempt to provide a complete, detailed architecture, but rather a loosely defined architecture that is flexible enough to adapt to diverse and changing standards-development situations. The holistic components of SERA include a performance-oriented infrastructure, an holistic, life-user experience, energy network optimization, and a partner-enabling, rich application platform and interoperability. One of the goals is for Microsoft is to provide a common computational platform and enabling their partners freedom to focus on application solutions.

Interoperability can be considered more general than communications; it involves not just the transfer of information, but the ability for the sender and receiver to agree upon the meaning of the information. An ontology is a formal representation of knowledge as a set of concepts within a given field of expertise or domain and the relationships among the concepts. These concepts and relationships not only provide a description of the domain, but also allow reasoning to take place within that domain. As a description, it provides a shared vocabulary, much like a standard. Ontologies are used throughout artificial intelligence, including systems engineering and the development of system architectures. They typically include descriptions of instances or specific individual entities, classes or concepts, attributes, and relations.

The CIM is a standard under the umbrella of the IEC. This standard is related to ontologies because its goal is to allow applications to exchange information about the configuration and status of an electrical system. This means that applications need to have a common understanding of the meaning of information. In fact, it is defined by a common vocabulary and an ontology representing fundamental knowledge in the domain of electric power.

CIM is the power systems version of an ontology. It began in the 1990s when it became clear that proprietary energy management systems were inflexible and incompatible with other grid systems, and it was becoming too expensive to add new applications to energy management systems. EPRI began the control center application program interface with the goal of creating a standardized interface for energy management systems and other applications within the control center so that applications could easily plug into the EMS. One benefit of plug-in applications is that system software components and applications can be easily changed or replaced, allowing the investment in working systems and applications to remain in place. One of the results of this project was an information model for power grid systems known as the CIM. The goal of CIM is to provide a common language so that information can be represented in XML and shared among applications. The IEC has incorporated the CIM into its standards and it is known as IEC 61970.

Included within the CIM IEC 61970 standard is a resource description framework (RDF) schema of the CIM. RDF provides a machine-readable version of an information model and is used extensively in development of the semantic web. The goal of the semantic web is to allow machines on the web to attempt to "understand" information and manipulate it more efficiently on our behalf, to create a web that links knowledge rather than simply static files. Thus, CIM is natural point of interface between the power grid and the semantic web and has been extended beyond its initial use for EMSs.

In order to provide a deeper understanding of CIM, consider its classes in more detail. Classes reside in packages and there are eight main packages. The `Core` package contains classes shared by all applications. The `Topology` package extends the `Core` package and models connectivity; that is, it describes a physical definition of how equipment is interconnected. The `Wires` package extends the `Core` and `Topology` packages and models transmission and distribution networks. This is useful for state estimation, load flow, and optimal power flow applications. The `Outage` package extends the `Core` and `Wires` packages and models current and planned network configuration. The `Protection` package extends `Core` and `Wires` and, as its name suggests, models protection equipment, such as relays. The `Measurement` package contains classes that describe dynamic measurements exchanged between applications. The `Load Model` package models energy consumers and system load. This is useful for load forecasting and load management. The `Generation package` contains classes in two sub-packages: `Production` and `GenerationDynamics`. `Production` models generators and `GenerationDynamics` models prime movers, such as turbines and boilers. The `Domain` package provides a data dictionary for units used in attributes throughout classes in other packages. The `Financial` package models energy transaction settlement and billing. The `Energy Scheduling` package models electric power exchange between companies. This is primarily used with accounting and billing. The `SCADA` package is specifically relevant to communications and defines a logical view of SCADA for CIM. This class models information related to remote terminal units and substation control systems.

There are three general types of relationships among classes: generalization, association, and aggregation. Generalization can be seen as a relationship between a specific and more general class. A nice example of a chain of generalizations is between a `Breaker`, which is a specific type of `Switch`. A `Switch` is a more specific form of `ConductingEquipment`, and `ConductingEquipment` is a more specific type of `PowerSystemResource`. `PowerSystemResource` is a primitive in the `Core` package. Each of the more specific classes can inherit attributes and relationships from the more general classes. An association is simply a relationship between two classes with a named role for the association. For example, a `Measurement` class `HasA MeasurementUnit`, where `HasA` is the name of the association. Aggregation is a whole–part relationship; for example, multiple `TopologicalNode` classes comprise a `TopologicalIsland`.

As an example, consider CIM used for load flow computation. The CIM `Topology` package has classes containing data that represent the power system topology. Specifically, nodes have a `ConnectedTo` association that may be utilized to find which nodes are connected directly together and the proper topology to use for the analysis can be determined. The CIM data model also includes all relevant power line and transformer parameters. Other required information can be found from the `Generation` and `Scheduling` classes. An important point is that the application will know exactly what data are available, where to obtain it (that is, which package, class, and attribute defines it), and precisely what the data mean and how it relates to other data.

IEC 60870, also known as ICCP or tele-control application service element.2 (TASE.2), specifies a standard solution to the problem of data exchange among electric utility control centers. ICCP was developed by North American utilities along with EPRI and many SCADA and EMS vendors to allow for real-time information transfer over WANs. In 1992, ICCP was submitted to the IEC and become known as TASE.2. TASE.1 was developed initially to meet European Common Market requirements in 1992. TASE.2 incorporated the use of MMS and is the current version in use at the time of this writing.

Since deregulation of the power industry and now with smart grid and the expectations for tighter and finer resolution control among many different entities and organizations, the need for inter-control center communications becomes more critical. ICCP enables the exchange of real-time and logged power system information such as measured values, status and control information, energy accounting data, scheduling data, and operator messages.

ICCP is also used to enable information exchange for the import and export of power between regions that may have different organizational groups, such as generators, transmission utilities, and distribution utilities. This is particularly important for transmission across major power inter-ties.

ICCP uses a client–server paradigm. The control center that requests information is a client and the control center providing information is the server. A control center may act as both a client and server. ICCP resides in the application layer of the OSI stack; as mentioned, TASE.2 uses MMS to provide message services. Since MMS uses TCP/IP, TCP/IP must be supported; however, lower layers are left unspecified. A typical implementation is TCP/IP over Ethernet (IEEE 802.3).

ICCP uses its own object definitions to describe information being transferred. It is partitioned into nine blocks, where blocks, other than Block 1, are optional. The blocks are comprised of:

1. Periodic system data – contains status points, analog points, quality flags, a time stamp, change of value counter, and protection events. Association objects are also included to control ICCP sessions.
2. Extended data set condition monitoring – provides report-by-exception for the data types in Block 1.
3. Block data transfer – provides block transfers of Block 1 and Block 2 data types instead of point by point, which can help reduce bandwidth.
4. Information messages – provides for simple text and binary file exchanges.
5. Device control – describes objects for on/off, trip/close, raise/lower, and other similar control operations, as well as setting digital setpoints.
6. Program control – allows an ICCP client to remote-control programs executing on an ICCP server.
7. Event reporting – contains objects for extended reporting of error conditions to a client, as well as device state changes at a server.
8. User objects – objects related to scheduling, accounting, outage, and plant information.
9. Time series – objects that enable a client to request a report from a server of historical time-series data between a start and end date.

The objects in these blocks are different from the IEC 61970 CIM, and harmonization of the objects in IEC 61850, IEC 60870 ICCP, and IEC 61970 CIM has yet to occur.

10.10 Summary

The art of standardization is knowing precisely what to standardize in order to reduce risk, while simultaneously encouraging innovation. We can see different, new frameworks attempting to take shape to define limited aspects of the evolving grid, as well as many old, well-established standards trying to update to anticipate merging with the evolving power grid, as well as many developing, detailed, communication protocol specifications. The wise reader will first understand the fundamental physics of operation and apply the standards when and where they make the most sense. The IEEE P2030 *Draft Guide for Smart Grid Interoperability of Energy Technology and Information Technology Operation with the Electric Power System*

(EPS), and End-Use Applications and Loads is a work-in-progress to develop an overall smart grid framework, including, of course, communication.

Chapter 11 focuses on the word "smart" in smart grid by reviewing machine intelligence in the power grid. It will examine what is meant by intelligence and how it relates intimately with communication and information theory and how it might relate to power systems. The appellation, "smart," in "smart grid" conjures up the notion of a power grid with "intelligence," in the form of artificial or machine intelligence. If that is indeed the goal, then power grid communication and networking must be designed to support such intelligence in the long run. Techniques involving artificial intelligence and machine learning for the power grid are explored while highlighting the role of communication and networking. Communication itself will benefit from advances in machine intelligence. Machine-to-machine communication, the semantic web, cognitive radio, and cognitive networking serve as attempts to incorporate intelligence into communication and are extensions of the active networking concept. The fundamental nature of complexity and communication in the form of active networking and communication complexity plays a role in power system information theory as the power grid itself becomes more active.

10.11 Exercises

Exercise 10.1 Synchrophasor Standard – Base Frequency

1. Can one directly compare a synchrophasor value from a 50 Hz system with a synchrophasor from a 60 Hz system? If not, why not?

Exercise 10.2 Synchrophasor Standard – CRC Protection

1. Recall that C37.118-2010 message frames can hold a variable number of data fields and that a constant 16-bit CRC-CCCITT code is used for error detection. Assume that each data field is 14 bytes. How does the error detection performance change as a function of the number of data fields transmitted? Derive a plot showing the error detection performance as a function of data fields.

Exercise 10.3 DNP3 – Cybersecurity

1. Consider a critical DNP3 message using an HMAC. This mechanism not only provides authentication, but also detects whether the message value has changed, either through malicious intent or by noise in the channel. Recall that CRC codes also detect message corruption. Is it redundant to use the CRC code as specified in the DNP3 standard when the keyed-hash message authentication mechanism is also being used? If so, explain the nature of any redundancy.

Exercise 10.4 IEC 61850 – Origin

DNP3 and IEC 61850 are existing standards for distribution system communications. Both have a historical lineage that involves the IEC, EPRI, and the IEEE.

1. Briefly outline the historical evolution of DNP3 and IEC 61850, including their relationship to IEC 60870 and UCA.

Exercise 10.5 IEC 61850 – Publish–Subscribe Mechanism

1. IEC 61850 defines a publish–subscribe mechanism for GOOSE messages. Explain how the IEC 61850 publish–subscribe mechanism works.

Exercise 10.6 IEC 61850 – Multicast

1. IEC 61850 does not prescribe a multicast mechanism. What techniques can be used to efficiently send a message to many subscribers using IEC 61850?

Exercise 10.7 DNP3 – Message Priorities

1. Does DNP3 implement message priority? If so how? Compare and contrast your answer with how IEC 61850 implements message priorities. [Explain in text.]

Exercise 10.8 DNP3 – Error Detection

1. Explain two mechanisms that DNP3 uses to implement error detection.

Exercise 10.9 DNP3 – CRC

1. Explain how the DNP3 CRC error detection mechanism is designed. What trade-offs must be made in choosing a CRC code? What is the error pattern coverage, probability of undetected error, and maximum length burst error that it can detect?

Exercise 10.10 Modbus – CRC

1. What is the probability of an undetected error in a maximum-size RTU Modbus message over a serial line? Which protocol provides better error protection: Modbus or DNP3?

Part Three

Embedded and Distributed Intelligence for a Smarter Grid: The Ultimate Goal

11

Machine Intelligence in the Grid

> The utility model of computing – computing resources delivered over the network in much the same way that electricity or telephone service reaches our homes and offices today – makes more sense than ever.
>
> —Scott McNealy

11.1 Introduction

You have now entered Part Three of the book, delving into the notion of intelligence in the smart grid. Communication is still the central focus; however, communication exists to support applications where the applications are promising to provide "intelligence." We have already discussed communication with respect to classical control commonly used in the power grid. Now we need to consider what the evolution toward machine intelligence, particularly distributed forms of machine intelligence, will mean both for the power grid and for communication in particular. The appellation "smart" in the term smart grid, conjures up the notion of a power grid with "intelligence," in the form of artificial or machine intelligence. If that is indeed the goal, then power grid communication and networking must be designed to support such intelligence. Communication itself will benefit from advances in machine intelligence and learning as well. It is necessary to understand and anticipate machine-learning aspects likely to be implemented in the smart grid and the role that communications will be required to provide. This chapter explores techniques involving artificial intelligence and machine learning for the power grid while highlighting the role of communication and networking.

We begin by discussing computational models for the power grid. This involves looking at where computational complexity lies within the current power grid and how it is likely to evolve as machine intelligence advances. Then we discuss a bit about the nature of complexity and communication in the form of active networking and also introduce communication complexity. Communication complexity provides an indication of the minimal amount of communication required to perform a distributed function and thus provides an indication of communication requirements for algorithms. We discuss a few specific machine intelligence algorithms and how they might be utilized in the power grid. This includes algorithms such as

Smart Grid: Communication-Enabled Intelligence for the Electric Power Grid, First Edition. Stephen F. Bush.

PSO and neural networks, as well as neural networks and expert systems along with several other well-known algorithms that had their origins in the study of artificial intelligence. The next topic switches to machine intelligence applied to communication within the smart grid. This includes the notion of M2M communication, the semantic web, and both cognitive radio and cognitive networks. Since this is a chapter on machine intelligence, the topic of semantic reasoning for the power grid is introduced here. The chapter ends with a concise summary of the main points and a set of exercises for the reader.

11.2 Machine Intelligence and Communication

The term "smart grid" engenders the notion of intelligence embedded within the power grid. However, as we will see, even the definition of a basic term such as "intelligence" has never been clearly defined. This section reviews various definitions of machine intelligence as well as its relationship to both communication and the power grid. For example, how much communication is required for an entity to be intelligent? Or conversely, if an entity is "intelligent," does it require less communication than if it were not, since it would be smart enough to infer information without explicitly requiring communication? These are fascinating, high-level questions that take us well outside the scope of smart grid and deep into the realm of early artificial intelligence. However, they are worth briefly reviewing because they are necessary if one takes the term "smart" in "smart grid" seriously.

11.2.1 What is Machine Intelligence?

One way to begin the discussion of machine intelligence is to consider that if the smart grid is to make the power grid more intelligent, then how is machine intelligence defined and measured? In a general sense, the notion of quantifying machine intelligence began with Alan Turing in 1950, with the well-known test to determine whether a computer could be sophisticated enough to fool a human into thinking that the device was another human. Since then, other tests, including the reverse Turing Test were devised. The reverse Turing Test measures whether a computer can distinguish whether it is conversing with a human or another computer. The power behind the Turing test is its simplicity, its ability to provide a practical, measurable result. A criticism of the Turing test is that it is anthropomorphic; it appears to assume human-like intelligence is the only kind of intelligence there is or that is worth measuring. Most artificial intelligence research has abandoned the idea of creating general intelligence and instead focuses upon practical applications that can be tested with more specific, and much less ambitious, techniques than the Turing test.

However, this still leaves open the notion of what machine intelligence is, how it can be measured, and to what degree it can be utilized in the power grid. An effort that has extended information theory, known as algorithmic information theory, is related to communication and has been used to quantify machine intelligence in new and interesting ways (Hernández-Orallo and Dowe, 2010). Algorithmic information theory takes its inspiration from Kolmogorov complexity, which relates the notion of complexity of information by the size of the smallest program that computes the information. The formal definition of Kolmogorov complexity is very simple and is shown in Equation 11.1, where p is a program that runs on machine U and generates x, $l(p)$ is the length of the program p, and $U(p)$ is the result of executing program p on U:

$$K_U(x) := \min_{\substack{p \text{ such that} \\ U(p) = x}} l(p). \tag{11.1}$$

Originally, U was specified to be a universal Turing machine; that is, a Turing machine that can simulate any other computational machine by simulating any specific Turing machine along with its program. A universal machine is one that can emulate any other machine, and this provides the generality for the definition. The key behind this definition is that a more complex sequence x will require a larger smallest program to implement and it is the size of the program that best reflects the degree of complexity of x. This is a beautiful notion and one that researchers return to time and again. Unfortunately, the problem with this definition is that computing the smallest program has been intractable. Basically, the ideal result is an uncomputable function; thus, the basic idea has been modified and approximated in many different ways in order to allow for computable implementations.

A more tractable variation that is sometimes used is known as Levin's *Kt* complexity. In this version, time is used to help make up for the inability to determine the absolute smallest program. The term time(U, p, x) is added to the definition of complexity and it represents the time that U takes to generate x, as shown in

$$Kt_U(x) := \min_{\substack{p \text{ such that} \\ U(p) = x}} \{l(p) + \log \text{time}(U, p, x)\}. \tag{11.2}$$

However, the notion of the existence of a machine U can be assigned an a priori probability based upon the universal distribution and the value of the sequence x produced, as shown in

$$p_U(x) := 2^{-K_U(x)}, \tag{11.3}$$

where $p_U(x)$ is the probability of the existence of the universal machine U given sequence x. The idea follows from Occam's razor, in which the simplest (that is, least complex) hypothesis that accounts for all observations is the most likely explanation. This was another, beautiful contribution developed by Ray Solomonoff (Solomonoff, 1964). Maximizing the universal distribution will determine all regularity in the sequence x, and this has led to many variations, including the notion of minimum message length and data compression as means of measuring intelligence. Simply put, minimum message length estimates the complexity of a sequence utilizing two parts: (a) the size of the model or hypothesis that describes how the sequence was generated and (b) the size of the error between (a) and the actual sequence. The size of the smallest combination of those parts is the best estimate of the complexity of the sequence.

This leads to the notion that the more a machine knows about a sequence of data, the more efficiently it will be able to compress the sequence. More generally, the greater the machine intelligence, the greater its ability to compress information.

While the topics discussed so far have been rather abstract and deep, there are literally thousands of machine-learning algorithms and programs available for use, with hundreds more published yearly. This certainly gives researchers and developers job security in terms of applying new algorithms and programs to the power grid. One way to make sense of all these machine-learning algorithms and programs is to find a relatively simple way to classify them (Domingos, 2012). It is a bit ironic that we have to use classification of machine-learning algorithms to make sense of them, because that is how machine-learning algorithms typically operate, as we will see.

There are three general features of machine-learning algorithms: (1) representation, (2) evaluation, and (3) optimization. Representation is how a classifier is represented in a machine-readable form. Examples include simple instances, hyperplanes, decision trees, neural networks, and graphical models. The choice of representation places limits on what can and cannot be learned simply by whether or not it can be represented. Evaluation relates to the choice of an objective function that is used to distinguish the performance of classifiers. Examples include accuracy/error rate, precision and recall, squared error, information gain, and Kullback–Leibler divergence. Optimization is the algorithm used to search among the classifiers for the best-performing ones. Examples, include combinatorial optimization, such as greedy search or branch-and-bound, and continuous optimization.

Typically, machine intelligence involves learning from examples with the goal of generalization. This means that the solution should learn to recognize the overall characteristics of what is "thought" to be important and not specific details of the training set. It is not enough to learn a training set perfectly, because the goal is to learn to correctly classify information that has never been "seen" by the classifier before.

The famous "no free lunch" theorem relates to generalization and recognizes that all algorithms have identically distributed performance when objective functions are selected randomly; thus, all algorithms have identical mean performance in general. From a communications perspective, the "no free lunch theorem" has been applied to routing algorithms when the initial explosion of new, ad hoc routing algorithms occurred and a deeper understanding of why this was happening was required (Bush, 2005).

11.2.2 Relationship between Intelligence and Communication

Recent definitions of machine intelligence have focused upon the degree to which a machine can learn to compress information. Clearly, this has direct implications for communication. An obvious implication is that communication among smart machines will likely be highly compressed; that is, the information will likely have high entropy or more "complexity." Another potential implication is that communication among machines with greater machine intelligence will be less frequent and require less bandwidth. The reason is that a characteristic of machine intelligence is the ability to form better predictions with less training data. The need for communication could decrease as machines are able to infer all information they need. Each bit of information obtained from a communication link becomes less "surprising" and more predictable to an intelligent machine for a given task. Many of these futuristic impacts upon communication are extremely important to plan for, but are often overlooked; more detail can be found in Bush (2000).

11.2.3 Intelligence in Communications

As we have seen, intelligence and communication-related concepts are intimately linked. Communication provides data required by machine intelligence in order to learn, while machine intelligence can be used to improve communication. Cognitive radios and cognitive networks are probably the most explicit examples of machine intelligence applied to communication.

However, it can be worth looking back at trends involving intelligence, complexity, and communications that have guided the development of communication networks because there may be applicable concepts in the power grid.

The end-to-end principle suggests that application-specific functions should not be implemented within a communication network, but rather should be placed outside the network; for example, within edge nodes of the network (Saltzer *et al.*, 1984). The intuition is that adding additional functionality within the network increases the complexity of the entire network and requires those not using the additional functionality to suffer the cost of the increased computation and complexity. The end-to-end principle has been taken to extremes with the notion of maintaining a "dumb network," the idea, for example, that all Internet protocols be as simple as possible. Unfortunately, as we have seen earlier in this section, the ability to measure complexity has proven intractable; thus, keeping protocols within a network simple is a subjective, often anthropomorphic, process. The question with the smart grid will naturally arise at to whether: (1) smart grid communication should follow this principle and (2) whether the smart grid itself has anything to learn from this principle.

It is interesting to compare and contrast the end-to-end principle with the active network communication framework (Bhattacharjee *et al.*, 1997; Bush and Kulkarni, 2001). An active network allows applications to inject packets into a communication network that contain program code that changes, or tailors, the operation of the communication network to better support the application. This would appear, prima facie, to be contrary to the end-to-end principle because it allows the potential for the network to be embedded with intelligence when and where it is needed. The communication network itself has the potential to become intelligent and to rapidly adapt and modify its intelligence. There are many variations of active networking, including software-defined networking. In fact, active networking supports, rather than violates, the end-to-end principle. This is because active networks place intelligence only when and where it is required and to leverage information only efficiently accessible within the network and benefits a wide variety of applications.

Finally, as another potential end-to-end principle conflict, cognitive radios are attempting to place machine intelligence close to the physical layer (Clancy *et al.*, 2007). These are intelligent radios intended to learn from their environment and adapt to it. The term "cognitive radio" has become a buzzword and applies to many different radio concepts since it was first coined in Joseph (2000). Many simple examples have been standardized, such as the ability to change modulation schemes from 16-quadrature amplitude modulation (QAM) to QPSK to BPSK as the signal-to-noise ratio decreases in the IEEE 802.11 standard. How much intelligence this actually embodies is open to debate. As we have mentioned, the ability to generalize is critical to machine intelligence. A cognitive radio extends a software radio and should include a cognitive engine comprised of a knowledge base, reasoning engine, and a learning engine. The knowledge base is the radio's long-term memory. As an example, if the reasoning engine is implemented as an expert system, then the knowledge base is the set of rules and the learning system is capable of managing the rules in the knowledge base; for example, updating or adding new rules to the knowledge base. The cognitive radio typically focuses upon aspects of benefit to the radio system, which are maximization of channel capacity and optimization of access to the available spectrum. Maximization of channel capacity typically involves, as already noted, selecting an optimal modulation type and coding rate. Spectrum access involves locating center frequencies, bandwidths, and transmission times while both maximizing capacity and minimizing interference with neighboring radios. In both the maximization of channel capacity and optimization of spectrum access, the radio's knowledge base can be updated with rules such that the radio learns over time by how channel conditions change and when other users transmit.

11.2.4 Intelligence in the Power Grid

Recall the discussion in Section 6.3 regarding the relationship between energy and computation. A general, fundamental, quantifiable relationship between the amount of computation required to achieve energy efficiency was derived. However, this was only a theoretical bound on the best that could be achieved. Actual implementation, as always, holds considerable challenge and cannot hope to achieve perfect theoretical efficiency. In this light, consider the discussion in Section 11.2.3 regarding machine intelligence and communications. Would a power grid operated similarly to a dumb communication network, an active network, or a cognitive radio be most desirable? Engineers need to consider how machine intelligence may be utilized in the power grid, how it relates to communication, and what might be learned from machine intelligence and communication networks. Many machine-learning algorithms will be presented in detail in the remainder of this chapter; our goal in this section is to consider these topics from a high-level perspective.

Certainly the idea of applying machine intelligence and machine learning to the power grid is not new. In fact, almost every area of the power grid has received attention from research in machine intelligence (Saxena *et al.*, 2010). This includes power system operation, planning, control, markets, automation, distribution system applications including DG, and power and weather forecasting applications. An entire book could be written on machine intelligence in the grid alone. However, these have all been more or less applications of machine intelligence to specific portions of grid operation. What may be interesting from a higher level perspective are measurements that attempt to represent the overall intelligence of the power grid; that is, how "smart" the smart grid is, as a whole, as it continues to evolve (Dupont *et al.*, 2010). Such measures, if they are specific, measurable, attainable, relevant, and time-bounded, will drive the direction for development of the smart grid.

11.3 Computing Models for Smart Grid

The various subfields of artificial intelligence, including machine learning and computational intelligence, have been widely applied to power systems long before the term "smart grid" was coined. Artificial intelligence has been studied and applied to power systems since at least the mid 1980s, when it was realized that the power grid was growing too complex for human operators to manage effectively, particularly in crisis scenarios (Wollenberg and Sakaguchi, 1987). Since then, machine learning and computational intelligence have been applied to nearly all aspects of the power grid, including optimal power flow, state estimation, stability, protection, generation, transmission, and, of course, load prediction and energy management systems, to name a few examples. An exhaustive list of artificial intelligence techniques applied to power systems would require a separate book in itself; thus, the goal in this chapter is to briefly review some of the most useful computational intelligence and machine-learning techniques and keep an eye toward how they relate to communication in the power grid.

The simplest, high-level consideration for computational intelligence in the smart grid era is the notion that every power systems device, or IED, will have sensing, communication, and control and that these devices will be ubiquitous throughout the power grid. The result will be a massive increase in the amount of information and control, requiring not just faster processors, but the widespread use of more-efficient, intelligent computational algorithms.

Beyond the sheer increase in information and control, a concern is whether classical power grid computational algorithms can handle the stochastic nature of the evolving power grid. For example, control of intermittent distributed power generation and changing user patterns due to DR mechanisms as simple examples. However, added to the more stochastic nature of the smart grid, there is the stochastic nature of communication; packets are potentially delayed, dropped, duplicated, or misordered en route to their destination. This problem alone would cause havoc upon a classical control system. Thus, the smart grid is seeking computational intelligence that can handle all aspects of power grid variability and randomness, including self-healing capabilities. Computational intelligence techniques tend to be less brittle (Bush *et al.*, 1999) than classical optimal algorithms, with the ability to adapt and learn in a changing environment. Other problems that smart grid computational tools will have to face include low-fidelity models of the environment, complexity, the large size of problems, and operator actions that cannot be reduced to concise mathematical form, but rather rules of thumb.

One of the earliest and perhaps most straightforward techniques is decision analysis. Decision analysis is a rather broad term for a large set of procedures and tools to help identify, represent, and formally assess important aspects of a decision. The goal is to maximize the utility of an "action axiom"; that is, an axiom that represents the criteria for taking an action. Decision trees are a common technique that the reader is most likely to have encountered from decision analysis; however, the field considers itself to be much broader. Decision trees, in their simplest form, list all possible decisions and their outcomes, tending to form large trees for complex decisions. Nodes represent decisions that have to be made and branches represent making a particular decision. One of the problems in decision making is that there is uncertainty in the available information at any point in the tree. One can evaluate the impact of obtaining more precise or accurate information on the outcome using the decision tree; that is, adding more sensors or increasing bandwidth, for example. Other techniques that fall under the umbrella of decision analysis are multicriteria decision analysis and analytical hierarchical programming. Multicriteria decision analysis addresses the issue of problems in which there are multiple criteria with which to judge solutions. In this case, there may not be a unique optimal solution.

Subjective logic is another good tool for dealing with uncertainty in decision making. Subjective logic explicitly separates uncertainty from belief and disbelief when determining the likelihood of an outcome. Subjective logic provides a consistent mathematical framework that explicitly handles uncertainty and belief. Thus, subjective logic is suitable for modeling and analyzing situations involving uncertainty and incomplete knowledge. Operations in subjective logic involve subjective opinions about propositions. A binomial opinion applies to a single proposition, and is represented as a beta distribution, as will be explained shortly. Subjective logic operates with four values: (1) an a priori probability a, (2) a belief b, (3) a disbelief d, and (4) uncertainty u (Jøsang, 2012). The a priori probability a is the probability of the truth of a proposition in the absence of any belief or disbelief. In other words, a priori knowledge is independent of any subjective experience or evidence. Belief is the degree to which one subjectively feels a proposition is based upon experience and weight of evidence. Similarly, disbelief is the degree to which one feels a proposition is false. Uncertainty is an explicit measure of the ignorance one has with regard to a proposition being either true or false. These values are assumed to apply to a frame of reference in which propositions are exhaustive and mutually disjoint.

The simplest way to use subjective logic is to consider a frame of reference in which there are two possibilities. This can be seen as the result of a beta distribution, which takes two shape parameters, α and β. These parameters can be mapped to the a priori probability, belief, disbelief, and uncertainty. The result is a beta probability density function that provides a useful way to quantify and analyze subjective logic. The beta distribution can be easily viewed as the posterior distribution of the parameter p of a binomial distribution after observing $\alpha - 1$ independent events with probability p and $\beta - 1$ with probability $1 - p$. Thus, a relatively larger α parameter has a tendency toward a greater probability of one (the proposition being true), while a relatively higher value of β places a higher probability on values closer to zero (the proposition is false). This assumes that the prior distribution of p is uniform. The simplest form is

$$\mathrm{Beta}(\alpha, \beta) = \frac{1}{\mathrm{B}(\alpha, \beta)} x^{\alpha-1}(1-x)^{\beta-1}. \tag{11.4}$$

In the following discussion, β is one of the parameters of the beta distribution, B() is the beta function, and Beta(α, β) is the beta distribution. The beta function B appears as a normalization constant to ensure that the total probability integrates to unity and $\mathrm{B}(x, y) = \frac{\Gamma(x)\Gamma(y)}{\Gamma(x+y)}$, where $\Gamma(n) = (n-1)!$.

When more than two mutually exclusive events are involved in a frame of reference, the beta distributions form the marginals of a Dirichlet distribution. The Dirichlet distribution of order $K \geq 2$ with parameters $\alpha_1, \ldots, \alpha_K > 0$ has a probability density function $f(x_1, \ldots, x_{K-1}; \alpha_1, \ldots, \alpha_K) = \frac{1}{\mathrm{B}(\alpha)} \prod_{i=1}^{K} x_i^{\alpha_i - 1}$ for all $x_1 \ldots x_{K-1} > 0$ satisfying $x_1 + \cdots + x_{K-1} < 1$, where x_K is an abbreviation for $1 - x_1 - \cdots x_{K-1}$. Note that the normalizing constant is the multinomial beta function:

$$\mathrm{B}(\alpha) = \frac{\prod_{i=1}^{K} \Gamma(\alpha_i)}{\Gamma\left(\sum_{i=1}^{K} \alpha_i\right)}, \qquad \alpha = (\alpha_1, \ldots, \alpha_K). \tag{11.5}$$

11.3.1 Analytical Hierarchical Programming

Classical optimization techniques such as linear programming, nonlinear programming, dynamic programming, and Lagrangian methods are currently used in power systems. However, as previously discussed, the concern is whether they will be flexible and powerful enough to handle the evolving, smart grid, requirements. One of the simplest and most widely used optimization techniques is linear programming. The objective function is shown in Equation 11.6 and the constraints in Equation 11.7

$$\max c^{\mathrm{T}} x, \tag{11.6}$$

$$\text{such that } Ax \leq b. \tag{11.7}$$

Because the problem is assumed to be linear, the constraints form a convex polytope; this forms the feasible space in which to search for the optimum value specified by the objective function. There are several techniques for solving a linear programming problem, in which the simplex technique is one of the most popular. Linear programming is restricted to static problems; it is unsuitable to the types of nonlinear, stochastic problems faced by the smart grid.

Nonlinear programming sets up an optimization similar to linear programming, as shown in Equations 11.8 and 11.9:

$$\max_{x \in X} f(x), \tag{11.8}$$

$$f : R^n \rightarrow R, X \subseteq R^n. \tag{11.9}$$

The Karush–Kuhn–Tucker conditions supply the necessary and sufficient conditions to determine optimality and are typically employed to find the optimal solution. Nonlinear programming is more computationally burdensome than linear programming and suffers from similar problems; namely, variables must be static values. Also, solutions are often approximate for complex optimizations, with the potential for inaccurate results.

Integer programming is another variation of linear programming whose setup is shown in Equations 11.10 and 11.11:

$$\text{Maximize}\, P(x) = \sum_{j=1}^{n} c_j x_i, \tag{11.10}$$

$$\text{Subject to} \sum_{j=1}^{m} \left(\sum_{i=1}^{n} a_{ij} x_j \leq b_i \right). \tag{11.11}$$

The key characteristic of this optimization is that all x_j are integer values in pure integer programming or mixed-integer programming if only some are integer values. Solution techniques typically involve forms of exhaustive search and a corresponding large combinatorial explosion. Again, integer programming suffers from the same problems as other forms of optimization previously discussed.

11.3.2 Dynamic Programming

Dynamic programming is another useful computational intelligence optimization technique, and one that is well known in communication networking for finding the shortest path to route a packet through a multihop communication network via the Bellman–Ford algorithm.

A necessary condition for optimality utilized by dynamic programming, also known as the Bellman equation, allows for an optimization problem to be solved recursively, by effectively breaking down a complex problem into a set of smaller, easier, optimization problems that build up to form the final optimal solution. The equation specifically partitions the value of an optimization problem into the value at the current time and the value of the optimization problem to be solved for the remaining time. The term "Bellman equation" refers specifically to discrete-time optimization problems; there is also a corresponding equation for continuous-time optimization problems known as the Hamilton–Jacobi–Bellman equation. We briefly introduce the discrete Bellman equation in order to give the reader insight into how communications may have to conceptually support such computational techniques.

Begin by considering a dynamic programming optimization problem. In the representation for such a problem, the state at discrete time t is x_t. The decision process will start at time 0; the initial state is thus x_0. As mentioned above, dynamic programming is a recursive process; therefore, at any time, the set of remaining possible actions depends on the current state, which is $a_t \in \Gamma(x_t)$. The action a_t represents the value of a control variable. The state changes from x to a new state $T(x, a)$ when action a is taken. The "payoff" from taking action a in state x is $F(x, a)$. There is also a factor $0 < \beta < 1$ that represents the need to find a solution quickly. Intuitively, the value of this weight controls whether a solution is chosen quickly rather than waiting for a better one later. The infinite-horizon decision problem is shown in Equation 11.12.

$$\begin{aligned} V(x_0) &= \max_{\{a_t\}_{t=0}^{\infty}} \sum_{t=0}^{\infty} \beta^t F(x_t, a_t), \\ &\text{subject to} \\ a_t &\in \Gamma(x_t)\ \forall t = 0, 1, 2, \ldots, \\ x_{t+1} &= T(x_t, a_t)\ \forall t = 0, 1, 2, \ldots \end{aligned} \tag{11.12}$$

$V(x_0)$ is the optimal value of the objective function; it is a function of the initial state variable at time 0. The dynamic programming method breaks this decision problem into smaller subproblems using the "principle of optimality," in which states (Bellman, 1952):

> An optimal policy has the property that whatever the initial state and initial decision are, the remaining decisions must constitute an optimal policy with regard to the state resulting from the first decision.

We can now view Equation 11.12 in light of the principle of optimality. The equation has two parts: (1) the first decision and (2) all remaining decisions. Having determined the first decision, there is one less remaining decision to be made. The remaining decisions can also be partitioned into two parts: (1) the next decision to be made and (2) all remaining decisions after that decision. This recursive approach continues until no further decisions are remaining and an optimum is found. Therefore, Equation 11.12 is rearranged such that the current and future decision terms are partitioned:

$$\begin{aligned} V(x_0) &= \max_{a_0} \left\{ F(x_0, a_0) + \beta \left[\max_{\{a_t\}_{t=1}^{\infty}} \sum_{t=1}^{\infty} \beta^{t-1} F(x_t, a_t) \right] \right\} \\ &\text{such that } a_t \in \Gamma(x_t), x_{t+1} = T(x_t, a_t)\ \forall t = 1, 2, \ldots \\ &\text{subject to } a_0 \in \Gamma(x_0) \quad \text{and} \quad x_1 = T(x_0, a_0). \end{aligned} \tag{11.13}$$

Equation 11.13 is showing the first step in the recursive process where the optimal value for a_0 is determined and enables the computation of $x_1 = T(x_0, a_0)$. The entire process may be repeated as time increments. The equation can be simplified to form the Bellman equation by noticing that what is inside the square brackets on the right side is the value of the time one decision problem, starting from state $x_1 = T(x_0, a_0)$. Therefore, we can rewrite the problem as a recursive definition of the value function thus:.

$$V(x_0) = \max_{a_0}\{F(x_0, a_0) + \beta V(x_1)\} \quad \text{subject to } a_0 \in \Gamma(x_0) \text{ and } x_1 = T(x_0, a_0). \tag{11.14}$$

A further simplification occurs in Equation 11.15, where the time-index subscripts are dropped and the value of the next state is replaced by the value function $T(x, a)$:

$$V(x) = \max_{a\in\Gamma(x)} \{F(x, a) + \beta V(T(x, a))\}. \tag{11.15}$$

By solving for the value function $V(x)$, function $a(x)$, which describes the optimal action as a function of the state, is also determined.

11.3.3 Stochastic Programming

A technique that extends linear programming into the space of probabilistic problems is stochastic programming. In other words, stochastic programming is a framework for optimization that enables the incorporation of uncertainty. The goal is to incorporate parameters or variables that are probabilistic, known only to be within given bounds, and then to determine an optimal solution that is feasible within the probabilistic bounds of the given parameters. The optimal solution is typically one that maximizes an expected value of the objective function based upon the random-variable parameter values.

As one can easily imagine, many of aspects of the smart grid are probabilistic. Renewable power generation depends upon the amount of sunlight or wind that happens to be available over a period of time; as we all know, predicting the weather is a highly probabilistic activity. The amount of load is also probabilistic, particularly predictions related to DR and the corresponding allocation of power generation and energy storage (Kristoffersen, 2007; Xiao *et al.*, 2000).

A two-stage, linear program incorporates the above requirements for stochastic optimization. In a two-stage linear program, optimization is determined in the first stage, then a random event occurs that impacts the first-stage result. A second-stage optimization is then performed that attempts to compensate for the impact of the random event on the first-stage optimal result. A key assumption in two-stage stochastic programming is that optimization should be done using existing data; it should not require waiting for future observation and measurements.

More specifically, the concept is to choose x to determine optimal control for what happens now and then allow for random events to occur; next, after the outcome of the random event is known, take a recourse action y to correct for the random event. Thus, there are two periods or stages. In the first stage, the data for the first period are known with certainty while the data for the second period are stochastic.

The structure of the first-stage problem is

$$\begin{aligned} &\min c^{\mathrm{T}}x + E_{\omega}Q(x,\omega) \\ \text{subject to } &Ax = b \\ &x \geq 0. \end{aligned} \tag{11.16}$$

Here, the stochastic program is designed to minimize known costs of the first period $c^{\mathrm{T}}x$ and the expected value of the cost of the second, future, uncertain period $E_{\omega}Q(x, \omega)$ as well as the second-period recourse decision, defined in Equation 11.17. The first-period constraints are defined by $Ax = b$. In Equation 11.17, the correction or recourse cost $Q(x, \omega)$ depends upon x the optimum value from the first-stage optimization and ω, the random event that may change the result of the optimization:

$$\begin{aligned} &Q(x,\omega) = \min d(\omega)^{\mathrm{T}}y \\ \text{subject to } &T(\omega)x + W(\omega)y = h(\omega) \\ &y \geq 0. \end{aligned} \tag{11.17}$$

Notice the outcome of the optimization in the second-stage optimization is $y(\omega)$. Here we can see that cost $d(\omega)^{\mathrm{T}}y$ is being minimized, but it is subject to a recourse function constraint that $T(\omega)x + W(\omega)y = h(\omega)$. This constraint serves to account for the cost of any action required to correct for having made a potentially nonoptimal choice in the first stage. For example, if we were deciding how much energy to store given load and generation predictions, this constraint would account for the cost of having to purchase the power to recharge the energy storage system after having made what turned out to be the nonoptimal choice of expending the stored energy without replenishing it at an earlier lower price. This form of stochastic programming utilizes the property of "non-anticipativity," meaning that x in the first-stage scenario is independent of y in the second-stage scenario. In other words, the future is uncertain and no information known in the future can be used to optimize the initial result.

It is important to note that we are not limited to two stages; there can be multistage stochastic optimization problems. To see this more explicitly, the problem can be represented in a deterministic form by introducing a unique second-period y variable for each scenario:

$$\begin{aligned} &\min c^{\mathrm{T}}x + \sum_{i=1}^{N} p_i d_i^{\mathrm{T}} y_i \\ \text{subject to } &Ax = b \\ &T_i x + W_i y_i, \ i = 1, \ldots, N \\ &x \geq 0 \\ &y_i \geq 0, \ i = 1, \ldots, N. \end{aligned} \tag{11.18}$$

In this equation there are N different stages and p_i is the probability of the occurrence of stage i. The non-anticipativity principle still applies; the first-stage optimization cannot utilize information with certainty from any succeeding, future stage. The optimization takes place over all feasible x and y_i simultaneously and, thus, in the first stage, x must be optimal over all possible future stages.

11.3.4 Lagrangian Relaxation

A technique used to help solve complex linear programs is known as Lagrangian relaxation. It is an approximate solution that, in some sense, trades off accuracy for computational tractability. The concept is fairly simple; begin with a standard linear programming problem $x \in \mathbb{R}^n$ and $A \in \mathbb{R}^{m,n}$ as shown in

$$\begin{aligned} &\max c^{\mathrm{T}} x \\ &\text{subject to } Ax \leq b. \end{aligned} \tag{11.19}$$

Next, the relaxation aspect comes in by allowing some flexibility in the constraints. The constraints in A can be partitioned in such a manner that $A_1 \in \mathbb{R}^{m_1,n}$, $A_2 \in \mathbb{R}^{m_2,n}$, and $m_1 + m_2 = m$, as shown in

$$\max c^{\mathrm{T}} x \tag{11.20}$$

$$\text{subject to } A_1 x \leq b_1 \tag{11.21}$$

$$A_2 x \leq b_2. \tag{11.22}$$

The constraint in Equation 11.22 can be moved into the objective function, as in

$$\max c^{\mathrm{T}} x + \lambda^{\mathrm{T}} (b_2 - A_2 x) \tag{11.23}$$

$$\text{subject to } A_1 x \leq b_1. \tag{11.24}$$

Notice that $\lambda = (\lambda_1, \dots, \lambda_{m_2})$ is a vector of nonnegative weights. Now that the constraint is in the objective function, the constraint can be violated; however, there is a penalty for such a violation and reward for meeting the constraint. The challenge is then to select the best set of weights λ. The weights can be incorporated into the optimization problem such that we choose weights that seek to minimize Equation 11.24. In other words, the weights that minimize the maximum value yield the most accurate solution.

11.3.5 Communication Complexity

As we have seen with active networking in Section 11.2.3, one of the great, unsettled debates has been the relation between communication and computation, including when, where, and how to place computation in a communication system and when, where, and how communication is best utilized in computation. Kolmogorov complexity addressed the notion of complexity of information as the smallest program that can generate the information. Communication complexity does nearly the opposite: it considers what is the smallest amount of communication required to perform a computation. Ideally, communication complexity would yield the design of a system that accomplishes its goal with the least amount of communication necessary. Unfortunately, like Kolmogorov complexity, communication complexity is difficult to apply to general-purpose computation in real systems. However, it does provide insight for simple, easily-analyzed scenarios.

The basic model for communication complexity may be viewed quite simply as two entities, often called Alice and Bob, who each have information known only to themselves. The goal is for Alice and Bob to cooperatively compute a function that requires all of their information as input while using as little communication as possible. The function can be defined as $f : X \times Y \to Z$, where X is the information that Alice has and Y is the information that Bob has. Since the goal is to focus on minimizing the total amount of communication required, both Alice and Bob are assumed to have infinite computing capacity at their disposal.

Some simple examples serve to show how the concept works. Suppose the function is to determine whether the condition of equality holds for the information that Alice and Bob have. In other words, is Alice's string of bits equal to Bob's string of bits? This is shown in Equation 11.25, where the function is shown in Equation 11.26:

$$EQ_n : \{0, 1\}^n \times \{0, 1\}^n \to \{0, 1\}, \tag{11.25}$$

$$EQ_n(x, y) = \begin{cases} 1 & \text{if } x = y \\ 0 & \text{otherwise} \end{cases}. \tag{11.26}$$

The easiest solution to this (or almost any) function is for one party to send all its information to the other. Thus, Alice could send all her bits to Bob for a total communication cost of $n + 1$ bits, n for Alice's transmission of her information to Bob, and then one more bit for Bob to respond with the result. Of course, such simple solutions are not really the goal of communication complexity. The purpose is to find the absolute minimum amount of communication required considering the space of all possible solutions.

Determining parity is a slightly more interesting example. This is shown in

$$PARITY_n(x, y) = \oplus_{i=1}^{n}(x_i \oplus y_i) = \begin{cases} 1 & \text{if } \sum_{i=1}^{n}(x_i + y_i) \equiv 1(\text{mod } 2) \\ 0 & \text{otherwise} \end{cases}. \tag{11.27}$$

Here, Alice could send all her bits to Bob as before. However, this would be costly from a communication perspective when alternative approaches would require much less communication. Instead, Alice could send $\oplus_{i=1}^{n} x_i$; that is, just send the single bit that represents the exclusive-or of her bits. This will work because $\oplus_{i=1}^{n}(x_i \oplus y_i) = (\oplus_{i=1}^{n} x_i t) \oplus (\oplus_{i=1}^{n} y_i)$. Thus, the total communication cost is just two bits: the single exclusive-or bit of Alice's bits to Bob and Bob's single bit result back to Alice.

Hopefully these examples provide simple, introductory insight into the nature of communication complexity. There are many more clever examples of minimizing communication cost for specific functions as well as general bounds on communication complexity for some problems. For example, communication between Alice and Bob in computing a function can be represented by a tree. The leaves of the tree are the results of the interaction, and the height of the tree represents the cost of communication. However, like many forms of complexity, a single, easily generalizable approach for computing communication complexity for any complex function is lacking and hampers the widespread use of this potentially useful concept.

Communication complexity would help to understand the trade-offs in computation versus communication in the power grid, particularly for distributed control and protection systems that require low latency but also require analysis of information from widely distributed sources throughout the grid.

11.4 Machine Intelligence in the Grid

Regardless of the amount of communication required, machine intelligence will be a distinguishing feature of the smart grid. This section provides an introduction into some typical machine-learning approaches. We begin with neural networks. In all of these approaches, keep in mind how they might be distributed in the smart grid and what the communication requirements might be.

11.4.1 Neural Networks

A well-known technique derived from artificial intelligence is artificial neural networks. The artificial neural network had it origins in 1957 with the perceptron, an algorithm for supervised classification of an input into one of two possible outputs. It is a classification algorithm that predicts based upon a linear predictor combining a set of weights with the feature vector describing a given input. It was shown later that the perceptron was unable to learn to classify many types of problems. However, the perceptron was only a single-layer, feed-forward network structure. The artificial neural network was inspired by the structure and presumed operation of the central nervous system of living organisms, with many layers of complex interconnected neurons. The artificial neural network has a learning phase in which it is able to adjust its structure. Typically, the structure adjusts by changing the weights of its connections as it learns to properly classify information. Artificial neural networks have been researched extensively for use throughout power systems. The hope is that artificial neural networks can learn to extract meaning from data that is complex, imprecise, or missing values. In fact, they should be able to do this faster and better than a human or other proposed machine-learning technique.

The "normal" (that is, non-learning) mode of operation of an artificial neural network is relatively simple. Each node in a layer in the network receives input from all the nodes in a preceding layer. Let X_j be the output of node j in the current layer. Let y_i be the input to node j in the current layer from node i in the previous layer (or directly from the input itself). The key to obtaining the correct result is that each input is weighted by a value W_{ij}. This represents the "strength" of the neural connection feeding into the current node j of the current layer. All weighted inputs are summed as shown in Equation 11.28:

$$X_j = \sum y_i W_{ij}. \tag{11.28}$$

The output of node j is y_j; it is typically a function of the weighted summation X_j. The sigmoid function is routinely used, as shown in

$$y_j = \frac{1}{1 + e^{-X_j}}. \tag{11.29}$$

If the weights W_{ij} are set correctly, then the output from each node of the final layer should correctly classify the input sequence. However, setting the weights is the next challenge to consider. This can be done automatically through a training process known as back-propagation. In this process, a sequence of inputs is provided for which the results are known. This way, the error can be computed and fed back into the network to properly adjust the weights. We know that y_i is the output of the final layer for node i. Since the correct result is known, its value is d_i for node i. Thus, the error E is

$$E = \frac{1}{2}\sum_i (y_i - d_i)^2. \tag{11.30}$$

The error derivative is a measure of how fast the error changes as an output unit changes. The error derivative EA_j is

$$\mathrm{EA}_j = \frac{\partial E}{\partial y_j} = y_i - d_i. \tag{11.31}$$

Next, determine the rate of change of error with change in input. This value is EI_j:

$$\mathrm{EI}_j = \frac{\partial E}{\partial x_j} = \frac{\partial E}{\partial y_j} \times \frac{\partial y_j}{\partial x_j} = \mathrm{EA}_j y_j (1 - y_i). \tag{11.32}$$

Next, we can determine the sensitivity of the error on the change of individual weights EW_{ij}:

$$\mathrm{EW}_{ij} = \frac{\partial E}{\partial W_{ij}} = \frac{\partial E}{\partial x_j} \times \frac{\partial x_j}{\partial W_{ij}} = \mathrm{EI}_j y_j. \tag{11.33}$$

Finally, we can determine how fast the error changes as a particular node in the previous layer changes:

$$\mathrm{EA}_i = \frac{\partial E}{\partial y_i} = \sum_j \frac{\partial E}{\partial x_j} \times \frac{\partial x_j}{\partial y_j} = \sum_j \mathrm{EI}_j W_{ij}. \tag{11.34}$$

Equations 11.31–11.34 enable us to see the impact of a neural weight on the error and allow us to properly adjust the weight to minimize the error. In summary, the steps for back-propagation are first to perform normal, forward-propagation using a training input sequence; that is, apply the training input and read the artificial neural network output and determine the amount of error. Then apply back-propagation of the error to adjust the weights. We want to adjust the weights in such a manner as to minimize each node's contribution to the error. Since a training input is used that has a known output, the amount of error, or output delta, can be determined. We can multiply the output delta and the input activation to the gradient of the weight. Thus, we need to move the weight in the opposite direction of the error gradient by subtracting a proportion of it from the weight. The proportion to subtract influences the speed and quality of the training process and is called the learning rate. The process is performed for each node at each layer of the network. The training sequence is repeatedly applied and the back-propagation process occurs until error is minimized.

11.4.2 Expert Systems

Expert systems are a well-known approach developed from artificial intelligence that have been applied to many aspects of the power grid. The first expert systems were created in the

1970s and were widely utilized in the 1980s. An expert system uses rules that attempt to emulate the ability of a human expert to make a decision. The rules are developed and encoded by a knowledge engineer; however, the specific rules used and the order in which they are used is entirely determined by the current scenario and the problem to be solved. An expert system has two parts: (1) the inference engine, which is fixed, regardless of the particular application, and (2) the knowledge base, which is comprised of the set of rules specific to a particular application. The inference engine uses the rules to reason about a particular problem, attempting to emulate the manner in which a human might recall and assemble rules to reason about a problem. Unlike a neural network, expert systems have the capability to be queried about how they reached their decisions.

Rules are manipulated using forms of logic. Propositional logic, predicates, epistemic logic, model logic, temporal logic, fuzzy logic (discussed next), are some of the many possible forms of logic. The use of predicates and propositional logic are the simplest to understand, taking the form of if–then statements. Using the various forms of logic, the expert system is able to generate new information from the knowledge base and the input data requested to solve a particular problem.

The expert system can use forward-chaining to ask fundamental questions of the user as it proceeds to build a solution. In backward-chaining, the expert system assumes a solution and works backward, attempting to verify it. Mixed-chaining is also possible, in which a combination of forward- and backward-chaining is utilized. Advanced expert systems are capable of determining contradictions in either user input or in the knowledge base and can alert the user with a clear explanation of the contradiction. Also, as previously mentioned, it is possible to ask the expert system either how it arrived at a particular result or why it is asking for a particular piece of data.

One disadvantage of expert systems is the fact that rules for a specific application (that is, the knowledge base) have to be encoded more or less manually, by a knowledge engineer. The knowledge base can become extremely large and unmanageable, leading to errors and inconsistent rules. While expert systems have advanced to the level of helping the knowledge engineer collect and manage rules, the process tends to be tedious and time consuming.

11.4.3 Fuzzy Logic

Fuzzy logic enables one to move from certain, clear, crisp data (for example, either yes–no or true–false), to uncertain, ambiguous, fuzzy data or results. However, it is important to make a subtle distinction between "degree of truth" and "likelihood," which is related to the interpretation of probability. Probability can refer to physical outcomes, in which all possible outcomes are known, but the likelihood of one of the outcomes can only be estimated. However, probability can also refer to the weight of evidence, or the belief that something is true. It is this latter interpretation to which fuzzy logic applies. Note, though, that both interpretations of probability operate in a similar manner, utilizing probabilities between zero and one. In other words, fuzzy logic deals with the degree of truth obscured by the vagueness of a phenomenon; physical probability is a model of ignorance of the outcome.

Figure 11.1 shows the high-level architecture of a fuzzy logic system. Crisp inputs pass through a fuzzifier where fuzzy values are returned. Fuzzy rules are applied via a fuzzy logic

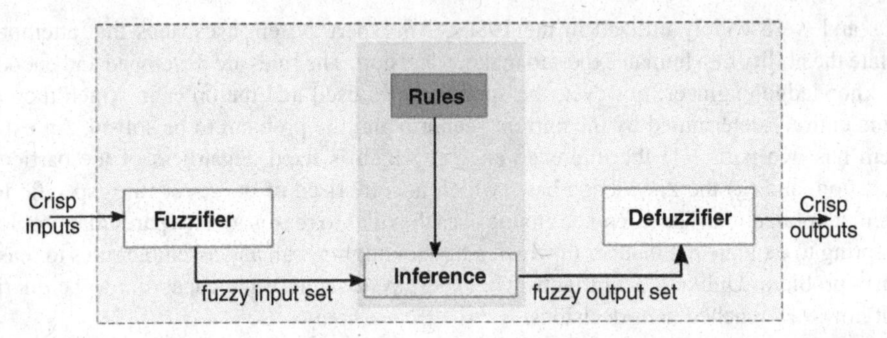

Figure 11.1 A fuzzy logic system converts normal values, known as crisp data, into fuzzy data. Rules that implement fuzzy logic are applied the data. The fuzzy result is converted into crisp data and is output as the final result. Source: Mendel, 1995. Reproduced by permission of IEEE.

inference engine that result in fuzzy output. The fuzzy output passes through a defuzzifier, yielding crisp output.

Keep in mind that this is only a brief introduction; the goal is to provide a quick, intuitive feel for how the technique works. Fuzzy logic has been applied to many different applications, including throughout power systems and control systems in particular. The key to fuzzy logic is the membership function $\mu_A(x)$, which defines the degree of membership of x in A. It is possible for an element to have a degree of membership in more than one set simultaneously. Operations take place on membership functions. An intuitive example is shown in Equation 11.35, where the complement, or NOT operation, is shown:

$$\mu_{\bar{A}}(x) = 1 - \mu_A(x). \tag{11.35}$$

Following our intuition:

$$\begin{array}{rl} \text{MAX} & \max\{\mu_A(x), \mu_B(x)\} \\ \text{ACCUMULATED SUM} & \mu_A(x) + \mu_B(x) - \mu_A(x)\mu_B(x) \\ \text{BOUNDED SUM} & \min\{1, \mu_A(x) + \mu_B(x)\} \end{array} \tag{11.36}$$

and

$$\begin{array}{rl} \text{MIN} & \min\{\mu_A(x), \mu_B(x)\} \\ \text{PROD} & \mu_A(x)\mu_B(x) \\ \text{BOUNDED DIFFERENCE} & \max\{0, \mu_A(x) + \mu_B(x) - 1\} \end{array} \tag{11.37}$$

show how the maximum and minimum operations can be used to implement a fuzzy logic OR and AND operation, respectively. However, note that there are other possible ways of defining logical operations as well.

It is possible that multiple results need to be combined in some manner in order to infer a result. Equation 11.38 shows examples of operations to perform such inference:

$$\begin{aligned} &\text{MAXIMUM} && \max\{\mu_A(x), \mu_B(x)\}, \\ &\text{BOUNDED SUM} && \min\{1, \mu_A(x) + \mu_B(x)\}, \\ &\text{NORMALIZED SUM} && \frac{\mu_A(x) + \mu_B(x)}{\max\{1, \max\{\mu_A(x'), \mu_B(x')\}\}}. \end{aligned} \tag{11.38}$$

Finally, a result typically needs to be converted back into a crisp logical answer, which we are used to dealing with. Equation 11.39 shows various ways of converting fuzzy answers back into crisp results, where U is the result of defuzzification, u is the output variable, p is the number of singletons, and μ represents the membership function after accumulation:

$$\begin{aligned} &\text{Center of gravity} && U = \frac{\int_{\min}^{\max} u\mu(u)\,\mathrm{d}u}{\int_{\min}^{\max} \mu(u)\,\mathrm{d}u}, \\ &\text{Center of gravity for singletons} && \frac{\sum_{i=1}^{p}[u_i\mu_i]}{\sum_{i=1}^{p}[\mu_i]}, \\ &\text{Leftmost maximum} && U = \inf(u'), \mu(u') = \sup(\mu(u)), \\ &\text{Rightmost maximum} && U = \sup(u'), \mu(u') = \sup(\mu(u)). \end{aligned} \tag{11.39}$$

11.4.4 Evolutionary Computation

There is a class of artificial intelligence known as evolutionary computation, with numerous approaches to solving optimization problems. The common element in these approaches is that they involve a population that attempts to grow, or self-organize, toward an optimal solution. The idea is that the population represents a large number of adaptable, potential solutions working together to explore the entire space of possible solutions; the manner in which nature has utilized such techniques often serves as an inspiration. Examples include genetic algorithms, ant colony optimization (ACO), and PSO. The interesting aspect to consider regarding communications is that, while all of these approaches are easily distributed or parallelized, there can be large amounts of communication required among the individual elements depending upon the optimization problem; variations of these techniques often attempt to limit the amount of interaction among the elements in order to achieve efficiency.

11.4.4.1 Genetic Algorithms

A genetic algorithm, as the name implies, attempts to mimic the genetic adaptation of the population of a species to optimize its chances of survival in a changing environment. In a genetic algorithm, there is a population of chromosomes, each encoding a potential solution to the optimization problem. The chromosomes exchange genetic material with one another, attempting to improve their fitness. Note that fitness needs to be specified by the user for the particular optimization problem. The goal is for the population of chromosomes to adapt and evolve to improve their overall fitness. There is a distinction between the chromosomes and its "genetic" material, known as the genotype, from the encoded solution that it expresses, which is known as the phenotype.

Evolution typically starts from a population of randomly encoded chromosomes in order to avoid falling into local optima. The process is divided into time periods known as generations, where the most-fit chromosomes from each each generation are selected to undergo potential genetic operations. These can include recombination (that is, the random exchange of genetic material), or crossover, and mutation. The result of the operations forms the next generation and the process is repeated. This process continues until either a predefined number of generations has passed or a target fitness level has been reached.

11.4.4.2 Particle Swarm Optimization

PSO is another evolutionary computing technique that uses a population of possible solutions, known as particles. The particles can move, changing both their position and velocity, as they "swarm" through the space of possible solutions. There is no exchange of genetic material as in genetic algorithms; instead, there is an exchange of information among particles. The particles are guided by their own past knowledge of local best positions as well as the global best position of all the particles in the swarm. Thus, instead of genetic evolution toward a more fit society, there is a swarming behavior toward more-fit locations in the search space, analogous to birds moving in a flock or fish swimming in a school. PSO has been applied extensively to a variety of applications, including power systems (Del Valle *et al.*, 2008) and communications (Kulkarni and Venayagamoorthy, 2011). Like many evolutionary algorithms, there is no guarantee that the optimal solution will be found as in explicit mathematical solutions that utilize a gradient to guarantee movement toward an optimal solution. The benefit of PSO is that the problem need not be continuous or differentiable, allowing for potential solutions to otherwise intractable problems. Thus, it is useful for problems that are irregular, noisy, or change over time.

Another problem with many evolutionary computation techniques is becoming stuck in local minima. Randomization or random mutations are often used to avoid such minima. In PSO, local minima are also avoided by ignoring the global best position and using a function of position of a subswarm; that is, a subset of the particles. If the particles are thought of as communicating with one another, then the hypothetical network formed by such communication is known as the topology of the PSO. Variations of the PSO technique may utilize different topologies in determining the best movement patterns.

More specifically, the algorithm begins by creating a population of randomly placed particles with random velocities within the search space. Just as in genetic algorithms, a fitness of each particle must be determined based upon the specific optimization objective. Each particle i keeps track of the best position in its history pbest_i and its current position is always p_i. The best particle g in the swarm is determined. Then, all other particles update their velocity and position as shown in

$$\begin{cases} \vec{v}_i \leftarrow \vec{v}_i \vec{U}(0, \phi_1) \otimes (\vec{p}_i - \vec{x}_i) + \vec{U}(0, \phi_2) \otimes (\vec{p}_g - \vec{x}_i) \\ \vec{x}_i \leftarrow \vec{x}_i + \vec{v}_i \end{cases} . \tag{11.40}$$

$\vec{U}(0, \phi_i)$ is simply a vector of uniformly distributed random numbers $[0, \phi]$ and $\otimes$ simply represents a component-wise multiplication. Each particle's velocity is maintained within the range $[-V_{max}, +V_{max}]$. Thus, we can see that each particle will tend to move stochastically toward a combination of its best past position and the globally best position. Just as in a genetic algorithm, the process is repeated for a given number of iterations or until the result has come sufficiently close to an optimal value.

It was found that velocity played a crucial role in the efficiency of convergence toward an optimal solution. If too large, the swarm could overshoot a potential solution and perhaps never converge; however, if too small, the swarm could take a significant amount of time to arrive at a solution. A variant of PSO uses an adaptive velocity, as shown in

$$\begin{cases} \vec{v}_i \leftarrow \omega \vec{v}_i \vec{U}(0, \phi_1) \otimes (\vec{p}_i - \vec{x}_i) + \vec{U}(0, \phi_2) \otimes (\vec{p}_g - \vec{x}_i) \\ \vec{x}_i \leftarrow \vec{x}_i + \vec{v}_i \end{cases} . \tag{11.41}$$

Here, the velocity is modified by ω, which may begin high, allowing an initial rapid search, and then decrease over time, allowing the swarm to more easily converge upon the optimum. The PSO concept is a fertile area for tuning algorithms, so there are, of course, many other variations.

11.4.4.3 Ant Colony Optimization

ACO is yet another variation on the evolutionary computing concept. As the name implies, it receives inspiration from the movement of large numbers of ants as they appear to self-organize along trails in their search for food. Thus, the concept applies well to finding paths along graphs and is a natural application to communication networks and routing, as well as for traveling salesman-type problems.

Ants appear to initially roam randomly until they find food. As they bring the food back to their colony, they lay down a pheromone trail that other ants can follow. As other ants follow the trail and are rewarded by finding food, they too return to the colony and reinforce the trail by depositing more pheromone. To counteract the continued deposition of ever-more pheromone, pheromone evaporates over time. Thus, trails can change and adapt over time. The dissipation of the pheromone trail is important because, while it attracts ants in a certain direction, it also weakens enough that they can loose the scent and travel randomly as well, thus exploring more space and avoiding a potential local minima. Also, longer trails with low traffic will tend to dissipate pheromone more quickly than short trails that are more heavily traveled. Thus, the longer, less optimal paths are forgotten and the shorter, optimal paths are strengthened, leading to an optimal, shortest path solution to a network problem. The idea of using the environment to store and communicate information (for example, using pheromones) is called stigmergy.

11.4.5 Adaptive Dynamic Programming

Notice that the computational and machine-learning techniques we are reviewing are related to solving optimization problems. This is important for the smart grid because communication will be used to carry information that is ultimately involved in controlling the power grid, either to collect information for state estimation or to issue control commands. Stability of the control system is important; an unstable control system could oscillate wildly or never converge toward a proper response. While stability is essential, a stricter requirement is optimal control; that is, the ability for the system to respond in such a manner as to achieve the required control action while also meeting constraints that make the actions as efficient as possible. In other words, achieving control in such a manner as to also meet requirements of an optimization's objective function, such as minimizing cost or maximizing a given utility. Thus, an optimal control problem includes cost as a function of the state and control variables. In optimal control, the control response must ensure that the state and control variables minimize cost while achieving the desired control over the system. A simple example is the automatic speed control in a vehicle. The control system can achieve the required speed under different conditions. For example, it can achieve the desired speed while minimizing fuel consumption or it can simply achieve the desired speed as quickly as possible, or it can achieve the desired speed while minimizing acceleration and jerk experienced by mechanical systems and passengers, saving wear and tear on both. In all cases, the control system is attempting to maintain the same speed, but the manner in which it does so could differ, choosing to accelerate or decelerate at different times under different conditions. An important point is that an optimal control system is guaranteed to be stable.

Adaptive dynamic programming ADP techniques allow for efficient forms of optimal, and thus stable, control. Let us run through a quick and intuitive introduction of this concept. Recall the principle of optimality, that remaining decisions must be optimal with regard to the state resulting from the first decision.

Equation 11.42 represents a discrete-time nonlinear system, in which x is the state vector, u is the control action, and F is the system function:

$$x(k+1) = F[x(k), u(k), k], \quad k = 0, 1, \ldots \tag{11.42}$$

As previously mentioned, we can associate a cost with the control actions, as shown in

$$J[x(i), i] = \sum_{k=i}^{\infty} \gamma^{k-i} U[x(k), u(k), kt]. \tag{11.43}$$

Here, U is a utility function and γ is the discount factor; that is, the cost of waiting longer for a result that is closer to optimal.

Thus, optimal control seeks to find the sequence $u(k)$ for time $k = i, i+1, \ldots$ such that cost J is minimized. We know from earlier in the chapter that we can apply the Bellman equation, which would be of the following form:

$$J^*(x(k)) = \min_{u(k)} t\{U(x(k), u(k)) + \gamma J^*(x(k+1))\} \tag{11.44}$$

Here, the optimal control decision $u^*(k)$ is the one that minimizes Equation 11.44, as shown in

$$u^*(k) = \arg\min_{u(k)}\{U(x(k), u(k)) + \gamma J^*(x(k+1))\}. \tag{11.45}$$

The continuous version of this problem is shown in Equation 11.46 and the corresponding cost is shown in Equation 11.47:

$$\dot{x}(t) = F[x(t), u(t), t], \quad t \geq t_0, \tag{11.46}$$

$$J(x(t)) = \int_t^{\infty} U(x(\tau), u(\tau))\, \mathrm{d}\tau. \tag{11.47}$$

Bellman's principle of optimality can be applied in a similar manner to this continuous-time problem. Here, the optimal cost $J^*(x_0) = \min J(x_0, u(t))$ takes the form of the Hamilton–Jacobi–Bellman equation shown in Equation 11.48.

$$\begin{aligned} -\frac{\partial J^*(x(t))}{\partial t} &= \min_{u \in U} \left\{ U(x(t), u(t), t) + \left(\frac{\partial J^*(x(t))}{\partial x(t)}\right)^{\mathrm{T}} \times F(x(t), u(t), t)) \right\} \\ &= U(x(t), u^*(t), t) + \left(\frac{\partial J^*(x(t))}{\partial x(t)}\right)^{\mathrm{T}} \times F(x(t), u(t), t)). \end{aligned} \tag{11.48}$$

As a side note, if the system is linear and the cost function is quadratic, then optimal control takes the form of a linear feedback control system where the gain can be determined using the Riccati equation. However, in general, control systems do not meet these conditions and become intractable to solve using the standard dynamic programming approach just described. As an alternative, artificial-intelligence-based approaches are being recruited to approximate optimal control solutions. One approach is to use an artificial neural network to approximate the cost function, called a "critic." There are many different names for this concept, including adaptive critic designs, approximate dynamic programming, asymptotic dynamic programming, ADP, heuristic dynamic programming, neuro-dynamic programming, neural-dynamic programming, and reinforcement learning.

Two specific implementations of ADP are heuristic dynamic programming (HDP) and dual heuristic programming dielectric (DHP). The focus of ADP is on estimating the cost function. However, finding optimal control decisions requires understanding the environment as well.

HDP finds an estimate of the cost $\hat{J}$, an estimate of J from Equation 11.43, by minimizing the error thus:

$$\|E_{\mathrm{h}}\| = \sum_k E_{\mathrm{h}}(k) = \frac{1}{2}\sum_k [\hat{J}(k) - U(k) - \gamma \hat{J}(k+1)]^2. \tag{11.49}$$

This is a discrete-time formulation, so $\hat{J}(k) = \hat{j}[x(k), u(k), k, W_{\mathrm{C}}]$ and W_{C} holds the parameters of the critic's artificial neural network. Once error is driven to zero for all k, Equation 11.50 holds:

$$\hat{J}(k) = \sum_k [U(k) + \gamma \hat{J}(k+1)t]. \tag{11.50}$$

A little rearrangement yields Equation 11.51, which is the same as Equation 11.43:

$$\hat{J}(k) = \sum_{i=k}^{\infty} \gamma^{i-k} U(i). \tag{11.51}$$

DHP estimates the derivative, or gradient, of the cost function. Notice that minimizing the cost function has some similarity to minimizing error in artificial neural network backpropagation. In order to minimize the gradient of the cost function, the same approach is used in forming an error measure as defined in Equation 11.49. The resulting error function for the gradient is

$$\|E_D\| = \sum_k E_D(k) = \frac{1}{2}\sum_k \left[\frac{\partial \hat{J}(k)}{\partial x(k)} - \frac{\partial U(k)}{\partial x(k)} - \gamma\frac{\partial \hat{J}(k+1)}{\partial x(k)}\right]^2. \tag{11.52}$$

Similar to the error measure definition before, $\frac{\partial \hat{J}(k)}{\partial x(k)} = \frac{\partial \hat{J}[x(k),u(k),k,W_C]}{\partial x(k)}$ and W_C again represents the parameters of the critic artificial neural network. When the error is driven to zero, Equation 11.53 holds, as can be seen from Equation 11.52:

$$\frac{\partial \hat{J}(k)}{\partial x(k)} = \frac{\partial U(k)}{\partial x(k)} + \gamma\frac{\partial J(k+1)}{\partial x(k)}. \tag{11.53}$$

11.4.6 Q-Learning

Q-learning is a type of reinforcement learning. Reinforcement learning combines dynamic programming with supervised learning in an attempt to achieve a "smarter" form of control: a control algorithm that can learn. Dynamic programming, discussed earlier in this chapter, is limited in the size and complexity of the types of problems it can handle. Supervised learning, on the other hand, requires training samples from which to learn. Reinforcement learning attempts to combine the efficiency of dynamic programming with the ability to learn that supervised learning enables. Ideally, the goal is to allow an agent to learn a control algorithm by interacting with the environment and learning through trial and error. However, it would rarely be practical to allow a real control system to literally learn from experience; for example, the actions that cause a plane to crash by actually crashing a plane. Instead, the environment is simulated. There are three parts to a reinforcement learning problem in general: the environment, the reinforcement function, and the value function. As mentioned, the environment is typically simulated. The reinforcement function is the function that the reinforcement agent seeks to maximize. Definition of the reinforcement function defines what the learning system is to achieve. It is assumed that the system is defined by states and actions that lead from one state to another. The value of a particular state is the sum of the reinforcements (rewards) received by starting in that state and following a given policy to a final state. The value function is the mapping from state to value using any type of function approximation.

It turns out that dynamic programming is a good way to represent the value function. Consider a more detailed explanation as follows. Assume an agent, states S, and a set of actions for each state A. By performing an action $a \in A$, the agent can move from state to state. As mentioned, each state provides the agent a reward, represented by a scalar value. Of course, the agent learns the actions that are optimal for each state in order to maximize its total reward. The quality (thus the letter "Q" in Q-learning) is a state–action combination, as shown in

$$Q : S \times A \rightarrow \mathbb{R}. \tag{11.54}$$

Values for Q are set by the developer. The rest is very similar to dynamic programming. It assumes the current value and makes a correction based on new information, as shown in

$$Q(s_t, a_t) \leftarrow \underbrace{Q(s_t, a_t)}_{\text{old value}} + \underbrace{\alpha_t(s_t, a_t)}_{\text{learning rate}} \times \left[\overbrace{\underbrace{R_{t+1}}_{\text{reward}} + \underbrace{\gamma}_{\text{discount factor}} \underbrace{\max_{a_{t+1}} Q(s_{t+1}, a_{t+1})}_{\text{max future value}}}^{\text{learned value}} - \underbrace{Q(s_t, a_t)}_{\text{old value}} \right], \tag{11.55}$$

where R_{t+1} is the reward observed after performing a_t in s_t and $\alpha_t(s, a)$ $(0 < \alpha \leq 1)$ is the learning rate. The discount factor γ is assumed to be within the range $0 \leq \gamma < 1$. Next, we turn our attention to communication concepts that have claimed to incorporate machine intelligence and see how they relate to the smart grid.

11.5 Machine-to-Machine Communication in Smart Grid

The term "Internet of things" (IoT) appears to have first been coined with the development of RFID and the notion of a network of RFID devices having the capability of uniquely identifying any object. The term is ascribed to Kevin Ashton's first use in 1999. In one opinion, the IoT envisions machine perception of the physical world and direct interaction with it. Thus, it would require identifiable objects and their virtual representations in an Internet-like form. However, as is often the case with imprecise terminology, it will likely gain a life of its own and be interpreted many different ways and for different applications.

M2M communication appears to have a similar connotation, although its origin appears to be less clear. An early example of M2M communication is the well-known OnStar system, a product that enables automatic vehicle-to-infrastructure communication. A key feature of this system is that the vehicle may initiate communication on the users' behalf. The thread behind both M2M and IoT is the lack of human intervention: devices know the devices with whom they need to communicate and handle communication automatically. While both M2M and IoT are most likely used to connote the same thing, IoT, because it contains the word "Internet," tends to imply network-layer aspects of interdevice communication, while M2M tends to indicate lower, physical aspects of such communication.

Clearly, the evolving power grid has almost always had devices communicating directly with devices; so, when people use the terms M2M and IoT, it is important to understand whether M2M and IoT provide any unique new idea or benefit to the power grid or are simply academic terms. In the discussions that follow, the term M2M will be assumed to include IoT as well.

Because M2M implies machine-directed communication without a human directly in the loop, participating machines need some level of "intelligence" in order to: (a) "wish" to communication, (b) "decide" with whom to communicate, and (c) initiate the communication. Thus, it is natural that machine intelligence and machine learning will play a larger role in M2M communication, as we discuss next.

11.5.1 Semantic Web

The terms "M2M communication" or "IoT" are unfortunate. They simply convey the notion of machines communicating data to one another; something machines and "things" have been designed to do for decades. The unarticulated concept behind these terms may instead be communication that is entirely conceived, initiated, and controlled by machine. In other words, machines communicate to one another of their own accord. This touches upon many subtle points. For example, what is the definition of a machine and at what point is it considered to be initiating its own communication? Suppose one sets up two "machines" on two ends of a wire that each measure current, the current entering the wire and the current leaving the wire. They communicate to compare current measurements and they break the circuit when there is mismatch between the current measurements that is large enough to be considered an indication of a fault. Are these "machines" or simply sensors and actuators? Are the machines initiating their own communication or is the scenario too simple to be considered machine-initiated communication? Note that the scenario just described is a common one in power protection and has been used since the early 19th century.

One might imagine more complex scenarios in which machines have "semantic" understanding of other machines and their components and begin communicating in order to accomplish objectives in real time in a manner that a human would not have considered. Certainly, that might be closer to the intent behind M2M communication and the IoT. This would imply that machines have some form of reasoning capability. Research in this area is being addressed by projects that associate themselves with the colloquially termed "semantic web."

The terms "M2M" and "IoT" are still ill-defined concepts; different people often imagine different and evolving ideas for what these terms mean. The fact that they are ill-defined is apparent in the fact that the two terms are often used interchangeably for the same concept. But what precisely is the concept? Certainly it involves devices communicating autonomously with other devices. A simple intuitive scenario is the refrigerator that scans packaging to determine what items are missing or in short supply and automatically places an order for needed items. The power grid has had devices that communicate with one another for nearly a century, so the concept is not new for smart grid. As one specific example, differential fault detection requires current transformers on each end of a line to communicate with one another and actuate a recloser or breaker if there is a significant deviation in current flow. In fact, the power grid has been, to a large extent, an M2M communication environment. So the question is what is unique about the M2M communication or the IoT concept? For researchers in traditional telecommunications, who have spent a lifetime working on voice and video applications, M2M communication may seem novel, so we should try to identify what, if anything, is really unique or novel about it and examine what is real and what is perhaps myth.

One aspect sometimes proposed as being unique is the large number of devices in M2M communication versus human-to-human communication. Of course, this has been anticipated for a long time in the development of sensor networking. The concept of innumerable sensors communicating with one another and its anticipated problems has been the focus of research almost since the beginning of sensor networking. Another proposed unique aspect is that traffic patterns in M2M networks will be bursts of machine-readable data rather than longer streams of human-interpretable information, such as text, voice, or video. Sensor network design assumes transmission of sensor data in machine format usually transmitted as sparsely as possible to save battery power. Finally, M2M devices are assumed to be deployable without human intervention, self-configuring, and operate autonomously. Again, sensor networking technology has been advancing along these lines and with these goals in mind.

However, for the power grid, we need to keep in mind that there are two operational aspects that coincide: the M2M communication system and the power system. Power system self-configuration and autonomous operation require a bit more care because mistakes are more catastrophic.

11.5.2 Cognitive Radio

Software-defined radio (SDR) is a radio communication system in which components that have been typically implemented in hardware (for example, mixers, filters, amplifiers, modulators/demodulators, detectors) are instead implemented by means of software on a personal computer or embedded computing device. In theory, this should provide greater flexibility and opportunity to embed intelligence behavior into the device.

Cognitive radio builds upon the notion of SDR. It is a paradigm for wireless communication in which either a network or a wireless node changes its transmission or reception parameters to communicate efficiently, dynamically avoiding interference with licensed or unlicensed users. This alteration of parameters is based on active monitoring of several factors in the external and internal radio environment, such as RF spectrum, user behavior, and network state. SDR and cognitive radio take the active network concept from the network into hardware and seek to increase network flexibility and add the appearance of intelligence to the communication system. The opportunities for SDR, cognitive radios, and cognitive networks applied to a smart grid derive from the same fundamental concepts as active networks discussed in Section 6.5.2.

11.6 Summary

Communication and machine intelligence are intimately related. In fact, intelligence has been defined as the ability to compress information; this makes information theory, compression, and intelligence nearly synonymous. It also suggests that a disorganized, ad hoc collection of power system components, communication, and computation will not lead to a "smart" grid. Communication technology has been enamored with intelligence; terms such as "cognitive" radios and "cognitive" networks express the desire for communication to be seen as somehow "smart" or self-aware. Active networking was discussed as a prime example of enabling embedded intelligence within the network. However, going back to the fact that there is no clear or computable definition of intelligence, the claim that these network concepts are smart, intelligent, or even beneficial has been highly debatable. The same will doubtlessly be true of the "smart" grid as well; there are good lessons to be learned from communication and networking in this respect. If added intelligence is not directly supporting the system's main function, it may be wise to reconsider whether the cost of intelligence may outweigh its benefit.

Clearly, there are many different computational models applicable to power grid applications. Most of these are decision-support and optimization-related algorithms that find general use in both communication and power systems. However, the lingering question remains as to when and where to use such algorithms. The discussion on communication complexity helps to at least frame the problem in an interesting manner. Then the chapter reviewed common techniques that derived from the old field of artificial intelligence.

The chapter ended on three related topics: M2M communication, the semantic web, and cognitive radio. These topics are all direct attempts to embed intelligence within communication at various levels. M2M communication assumes that machines are smart enough to have direct, meaningful communication with one another. Of course, this is not a new idea in the power grid, as SCADA and process control systems have long interacted with one another in the power grid. The semantic web has experimented with embedding semantic information as a standard part of the Internet with the goal of enabling reasoning to be more easily implemented. Some smart grid standards define objects in a manner similar to the semantic web standards. Finally, cognitive radios are a direct attempt at embedding cognitive capability in a communication component. Whether the smart grid needs cognitive transformers, cognitive capacitor banks, and so on is left as an exercise for the reader.

Chapter 12 builds upon some of the optimization techniques introduced in this chapter to implement state estimation and stability in the smart grid. Observability and stability are key aspects of control. It is impossible to operate and manage the power grid efficiently without knowing its current state. How does smart grid address this problem? Keeping the power grid stable is, or course, a key requirement. Stability in the power grid broader than keeping generators stable, so it's important to better define stability in its broadest aspect as it relates to the power grid. How will new smart grid techniques help with monitoring and controlling stability? In addition, networked control, introduced in earlier chapters, is explained in more detail in the next chapter. State estimation and stability are fundamental components of power grid control and are also used in communication systems; both topics are reviewed with particular emphasis on application in power systems. Networked control extends control of the power grid over communication networks and must overcome challenges that variable performance and reliability of the communication network impose upon the control system. While state estimation and stability are well-understood when implemented via centralized processing, scale and reliability requirements of the power grid requires that they be implemented using distributed processing. Thus, distributed state estimation is explained as a more scalable and fault-tolerant approach. The need for stability occurs in many forms throughout the power grid. One aspect of stability is related to the high penetration of distributed generation sources. Many moderate or low-power generators will lack the stable momentum of a single, large, centralized generator. Solutions to stability and the corresponding communication requirements are discussed.

11.7 Exercises

Exercise 11.1 Machine Intelligence

1. What is Kolmogorov complexity and how is it used as a measure of machine intelligence?
2. How is Kolmogorov complexity related to information entropy?
3. What is the difference between Kolmogorov complexity and Levin's *Kt* complexity? Why is Levin's *Kt* complexity useful?

Exercise 11.2 Machine Intelligence and its Impact on Communication

1. As machine intelligence improves and becomes more embedded within devices and throughout the power grid, what impact could that have upon the nature of the traffic that communication is required to transport?
2. What impact could machine intelligence have upon communication?

Exercise 11.3 Machine Learning

1. What precisely is machine learning?
2. What are the three main components of machine-learning algorithms?
3. How does machine learning relate to algorithmic information theory, specifically minimum message length?

Exercise 11.4 Machine Learning and Generalization

1. What is overfitting in machine learning?
2. What is the "no free lunch" theorem and how does it relate to generalization?
3. What could happen to a smart grid machine-learning application if it failed to generalize properly?

Exercise 11.5 Machine Learning in Communications

1. What is the end-to-end principle in communication networking?
2. What is an active communication network? How does it relate to the end-to-end principle?
3. What are the implications of the end-to-end principle with respect to intelligence within the network?
4. How does a cognitive radio attempt to utilize machine intelligence?

Exercise 11.6 Subjective Logic

1. How does subjective logic extend the notion of belief and probability?
2. What are the four quantitative parameters that define an opinion in subjective logic?
3. How does subjective logic help with operating in an uncertain environment such as the power grid?
4. Describe how subjective logic might be used in power grid state estimation?
5. How could subjective logic help with unreliable or high-latency communication?

Exercise 11.7 Machine-Learning Smart Grid Architecture

1. Consider the definition of the smart grid. Where is machine intelligence likely to be applied within the power grid to meet the criteria of being a smart grid?
2. What implication does the above have upon supporting communication?

Exercise 11.8 Dynamic Programming

1. What is the key concept behind dynamic programming?
2. What is the Bellman equation?
3. What is the principle of optimality?

4. How does the Bellman–Ford algorithm for communication network routing utilize dynamic programming?
5. How does dynamic programming solve the problem of optimizing a complex system?
6. How does dynamic programming allow a trade-off between speed and accuracy?

Exercise 11.9 Stochastic Programming

1. What are the key concepts behind stochastic programming?
2. How are random events handled in stochastic programming?
3. How does stochastic programming relate to forward pricing and option pricing from Section 7.3?

Exercise 11.10 Lagrangian Relaxation

1. How does Lagrangian relaxation address the problem of simplifying a complex optimization problem?

Exercise 11.11 Communication Complexity

1. What is communication complexity?
2. How does communication complexity relate to algorithmic information theory?
3. How does communication complexity relate to distributed computation in the power grid?

Exercise 11.12 Particle Swarm Optimization

1. What are advantages and disadvantages of PSO?
2. What are examples of power grid applications utilizing PSO?
3. How can PSO be used for communication in the power grid?
4. What is distributed PSO?
5. Could distributed PSO be run over a network in the grid?

Exercise 11.13 Cognitive Networks

1. What are the difference between cognitive radios and cognitive networks?
2. How are cognitive networks similar to a smart grid?

Exercise 11.14 Neural Networks

1. What are the advantages and disadvantages of neural networks?
2. What are some example power grid applications?
3. How can they be used for/with communication in the power grid?

12

State Estimation and Stability

The killer app that got the world ready for appliances was the light bulb. So the light bulb is what wired the world. And they weren't thinking about appliances when they wired the world. They were really thinking about – they weren't putting electricity into the home. They were putting lighting into the home.

—Jeff Bezos

12.1 Introduction

This is the second chapter of Part Three, which focuses upon state estimation and stability within the power system. Both of these aspects fall under the concept of controls; namely, correctly estimating the state of the system being controlled and making sure that the controller is programmed to respond in a manner that keeps the output of the system at the desired value when perturbations occur. These same aspects of control theory also apply in communication systems; for example, state estimation is used in estimating the quality of a communication channel and stability is critical in keeping traffic flowing smoothly via access control mechanisms. A significant component of the smart grid is the integration of communication. A particularly challenging aspect is using communication for control within the power grid. This is known as networked control. It is challenging because remotely controlling a system using a communication network can introduce variable delay into the control system, thus making it challenging to respond appropriately to a rapidly changing physical system such as the power grid. We introduce how to analyze latency in control systems and how networked control systems have attempted to meet the challenge of variable latency in the network. State estimation begins with monitoring the power grid for the appropriate information. Thus, we discuss monitoring and sensor networking in the power grid. Next, the concept of state estimation is introduced. State estimation is used in communications, for example, in order to estimate the characteristics of a communication channel, known as channel estimation, or to determine a received signal in the presence of noise. State estimation plays a significant role in observing the state of the power grid so that control systems can respond quickly and accurately. Next, we review distributed state estimation. As communication becomes an integral part of the power grid, it opens up the possibility for more processes to be distributed. Distributed systems may

---

Smart Grid: Communication-Enabled Intelligence for the Electric Power Grid, First Edition. Stephen F. Bush.

be more fault tolerant and efficient. Then we focus on the topic of stability. The need for stability occurs in many forms throughout the power grid. An entirely separate section is devoted to stability issues involved with a high penetration of DG. A concern is that many moderate or low-power generators will lack the stable momentum of a single, large, centralized generator. The chapter ends with a brief summary and exercises for the reader.

This chapter explores the changes that the smart grid will bring to two fundamental and interrelated aspects of maintaining an operational power grid: accurate and efficient estimation of state and maintaining stability. Maintaining stable, healthy, and efficient operation benefits from rapid and accurate state estimation. In fact, much like multimedia quality in communications, there is a conceptual rate–distortion curve defining the accuracy and bandwidth required for estimating state. In the ideal case, state estimation should take place continuously, providing a perfect reflection of the actual state of the power grid. However, communication and processing latency make this impossible; instead, estimation runs, at most, every few minutes or is triggered by a major event.

State estimation is required because the power grid has many interacting, complex components. Power generation, particularly smaller, distributed and renewable generators have random fluctuations in power output, loads constantly vary in the amount of power they consume, switches may be in an unknown configuration, and power transport may experience random effects of weather and the environment. Unless all locations in the power grid were instantaneously observable via sensors that were collecting all conceivable properties of electric power, the exact state of the power grid would not be known with certainty. Clearly, this is infeasible. This is where estimation theory becomes useful. Estimation theory focuses upon determining the value of parameters based on either measured or empirical data that has a random component. The parameters are assumed to describe the underlying physics of a system in such a way that their value affects the distribution of the measured data. An estimator attempts to approximate the unknown parameters using measurement to reveal an estimate of the current state.

Let us briefly review the role of state estimation in the power grid. Power generation and transmission must attempt to meet customer demand while minimizing cost and maintaining security; that is, maintaining stability and reliability given likely, adverse scenarios. As introduced earlier, the EMS is responsible for meeting these goals. The power grid is a complex system whose state is continuously changing to meet varying power demand. In order to effectively and efficiently control the system, its current state must be known. In the past, because of its large size, complexity, spatial extent, and the cost of sensors and poor performance of communication, the state of the power grid has been approximated from a relatively few, key measurements and has been determined relatively infrequently.

However, after deregulation of utilities and with the impending, widespread implementation of DR, the so-called smart grid will be making significant demands upon state estimation from at least two fronts. First, the state of the system is becoming more complex, with more sophisticated control devices that we will discuss in more detail in the following chapters. Second, business transactions, such as those enabled by DR, will potentially require faster control responses and more detailed and accurate state information. It may no longer be possible to rely on quasi-static state approximations. Instead, state estimation will have to become more accurate (that is, more closely reflect the actual state) and do it with greater precision. This will have to be accomplished at a rate that more closely approaches real time.

The hope is that new components of the smart grid will help meet the greater requirements on state estimation. First, by utilizing the latest in communication technology and by utilizing more precise, widely deployed sensing mechanisms, more demanding state estimation challenges can be overcome. The rising star and distinguishing feature of the smart grid era for sensing has been the PMU and the use of synchrophasors. Synchrophasors will be discussed in more detail in Chapter 13.

The process of state estimation is comprised of two parts: determining the topology and estimating analog and discrete values based upon the topology. Topology is determined by the initial design of the power grid and by the state of all breakers and switches; that is, whether interconnections in the power network are opened or closed. The main values of interest are typically voltage and angle at all buses. However, there may be inaccurate data regarding the state of many required grid parameters, such as the position of the tap in tap changing transformers, shunt capacitors, and breakers. Traditionally, a weighted least-squares technique has been applied to estimate the most likely, true state and to filter potentially bad data. However, there are problems that can be exacerbated by the increasingly difficult challenges required by more accurate state estimation in the smart grid. One problem is that, as the grid becomes more complex and changes occur more frequently, the state estimation process may not converge, particularly if the actual state within the power grid changes faster than the data reported by the communication system. Also, if critical data are not communicated, for whatever reason, state estimation may either fail to converge or suffer from accuracy that is too poor to be useful. However, if there is enough redundancy in state estimation information, the impact of such problems can be reduced.

Only a finite portion of the power grid can be instrumented with sensors; this is known as the "observable" portion of the power grid. Portions that are not instrumented are "unobservable"; thus, there are islands of observability. The configuration of the unobservable portions of the grid must be modeled and inferred from those portions that are observable. "Generalized" state estimation refers not only to estimating state values of interest, but also to inferring, or estimating, the configuration of the topology and its associated parameters in unobservable portions of the grid.

The distinction between observable and unobservable portions of the grid is important; the larger the observable portion of the power grid, the greater the accuracy of the estimated state. If the configurations of bus sections and switches are known, the portion of the grid considered to be observable can be extended through these devices. The state estimation algorithm can be fed with what are known as "pseudo-measurements." Quite simply, if a switch is closed, but has an unknown impedance, then the voltage difference across the switch is assumed to be zero and the angle is also assumed to be zero when performing state estimation. This is an example of a pseudo-measurement. On the other hand, if the switch is open, the impedance is assumed to be infinite and the state estimation process is informed of zero active and reactive power flow. A more critical concern is that of a switch in an unknown state. In this case, the switch state is left as unknown with the hope that enough redundant information has been supplied such that the state estimation process will yield the proper result, as explained below.

It is possible to further extend the idea of feeding pseudo-measurements into the state estimation process through branches with unknown impedances. Assuming there is a branch impedance z_{km} from node k to node m, we want to find the state variables $V_k \, e^{j\theta_k}$ and $V_m \, e^{j\theta_m}$. The branch complex power flows are $P_{km} + jQ_{km}$ and $P_{mk} + jQ_{mk}$ and we also know that current is conserved through the branch connection, $I_{km} + I_{mk} = 0$. The following pseudo-measurements can then be used:

$$\begin{aligned} P_{km}V_m + (P_m k \cos\theta_{km} - Q_m k \sin\theta_{km} V_k) = 0, \\ Q_{km}V_m + (P_m k \sin\theta_{km} - Q_m k \sin\theta_{km} V_k) = 0. \end{aligned} \tag{12.1}$$

Consider z_j as the jth measurement. Let $\vec{x}$ be the actual, true vector of states. Also, let $h_j(\cdot)$ be a nonlinear, scalar function that transforms the jth measurement to states. Since the purpose of the state estimator is to handle measurement error, let e_j represent the measurement error. An assumption in this process is that error will have zero mean and variance σ^2. Finally, there are m measurements and n state variables. Clearly, m must be greater then n in order to have measured the n states and have additional information to provide redundancy, $m > n$. These values are related thus:

$$z_j = h_j(\vec{x}) + e_j. \tag{12.2}$$

Simply put, z_j represents actual, measured values with the addition of measurement or communication error. It is important to note that the error, also known as the residual, is $z_j - h_j(\vec{x})$.

The key to understanding the state estimation process is that we want to find a solution that minimizes the error, or residual. This is precisely what is done in Equation 12.3:

$$\begin{aligned} &\text{minimize} \quad J(\vec{x}) = \frac{1}{2}\sum_{j=1}^{m}\frac{r_j^2}{\sigma_j^2} \\ &\text{subject to} \quad g_i(\vec{x}) = 0 \text{ and } i = 1, n_g \\ &\qquad\qquad\quad c_i(\vec{x}) \leq 0 \text{ and } i = 1, n_c. \end{aligned} \tag{12.3}$$

$J(\vec{x})$ is the sum of the residuals over all measurements m, normalized by the error variance. The $g_i(\vec{x})$ and $c_i(\vec{x})$ are typically constraints imposed by the physics of power flow. The equalities are target values and the inequalities are limits within the unobservable portions of the network.

Now let us consider specific types of state and measurement values. The state $\vec{x}$ (that is, the values we wish to estimate) are typically node values such as voltage, including the magnitude V_k and angle θ_k. The state also includes transformer turn-ratios including magnitude t_{km} and phase shift angle ϕ_{km}. Finally, the key types of state values to estimate are the power flows, namely, complex power flow P_{km} and P_{mk} and reactive power flow Q_{km} and Q_{mk}.

Referring back to Equation 12.3, the equality constraints $g_i(\vec{x})$ can extend observability by making simplifying assumptions, such as the voltage difference across a closed switch is zero. The inequality constraints $c_i(\vec{x})$ can be limits such as a VAr limit $Q_k^{\lim}$, a tap transformer limit t_{km}^{lim}, or a phase shift limit $\phi_{km}^{\lim}$. The limits, although not precise values, provide additional information that helps guide optimization toward a feasible solution.

The objective function is transformed to matrix form

$$J(\vec{x}) = \frac{1}{2}\sum_{j=1}^{m}\frac{r_j^2}{\sigma_j^2} = \frac{1}{2}\vec{r}' R_z^{-1}\vec{r} = \frac{1}{2}\vec{r}'\vec{r}, \tag{12.4}$$

where $\vec{R}_z$ is a diagonal matrix of measurement variances and $\vec{\tilde{r}} = R_z^{-1/2}\vec{r}$ is a vector of weighted errors, or residuals.

The optimum value must be on a critical point, where a critical point is the location where the first derivative is zero. This condition is reflected in

$$\frac{\partial J(\vec{x})}{\partial(\vec{x})} = -\sum_{j=1}^{m} \frac{r_j}{\sigma_j^2} \frac{\partial h_j}{\partial(\vec{x})} = 0. \tag{12.5}$$

Next, we introduce the Hessian. Note that Hessian matrices are often found in large-scale optimization problems because they are the coefficient of the quadratic term of a local Taylor expansion of a function. That is, $y = f(\mathbf{x} + \Delta\mathbf{x}) \approx f(\mathbf{x}) + J(\mathbf{x})\Delta\mathbf{x} + \frac{1}{2}\Delta\mathbf{x}^{\mathrm{T}}H(\mathbf{x})\Delta\mathbf{x}$, where J is the Jacobian matrix, which for scalar-valued functions is a vector; namely, the gradient. Recall that $h_z(\cdot)$ was the function that transforms a state value to a measurement. In Equation 12.6, $\vec{H}$ represents a matrix of m and n columns such that the $\frac{\partial h_j{}'}{\partial(\vec{x})}$ is the jth row of the matrix:

$$-\sum_{j=1}^{m} \frac{r_j}{\sigma_j^2} \frac{\partial h_j}{\partial(\vec{x})} = \vec{H}'\vec{R}_z^{-1}\vec{r} = 0. \tag{12.6}$$

Equation 12.7 is the Taylor approximation applied to J:

$$\left.\frac{\partial J}{\partial(\vec{x})}\right|_{(x+\Delta x)} \approx \frac{\partial J}{\partial(\vec{x})}\Big|_{(x)} + \frac{\partial^2 J}{\partial(\vec{x}^2)}\Big|_{(x)}\Delta\vec{x}. \tag{12.7}$$

Equation 12.8 is the Hessian matrix:

$$\left.\frac{\partial^2 J}{\partial(\vec{x}^2)}\right|_{(x)} \Delta\vec{x} = \sum_{j=1}^{m} \sigma_j^{-2} \frac{\partial h_j}{\partial\vec{x}} \frac{\partial h_j{}'}{\partial\vec{x}} - \sum_{j=1}^{m} \frac{r_j}{\sigma_j^2} \frac{\partial^2 h_j}{\partial\vec{x}^2}. \tag{12.8}$$

Equation 12.9 implements the iterative process that (hopefully) converges to the approximate state:

$$\begin{aligned} \vec{G}(\vec{x}^{\nu})\Delta x^{\nu} &= \vec{H}'(\vec{x}^{\nu})\vec{R}_z^{-1}\vec{r}(\vec{x}), \\ \vec{x}^{\nu+1} &= \vec{x}^{\nu} + \Delta\vec{x}^{\nu}. \end{aligned} \tag{12.9}$$

An initial value is chosen for the state estimate to begin the process. A $\Delta\vec{x}$ value will be returned indicating the change necessary to the original approximation to achieve a better approximation. The new, updated approximation is then input and a new $\Delta\vec{x}$ is returned. This process continues until $\Delta\vec{x}$ becomes smaller than a specified value or it is determined that the process is not converging to smaller $\Delta\vec{x}$ values. $\vec{G} = \frac{\partial^2 J}{\vec{x}^2}$ is a gain matrix. In the Gauss–Newton method, $\vec{G} = \vec{H}'R_z^{-1}\vec{H}$. This introduces the basic algorithm; much research has been done to find ways to tweak the equations and approaches to improve performance. However, let us return to the issue of the smart grid. As previously mentioned, synchrophasors have only relatively recently been deployed into normal operation within the power grid. Synchrophasors are discussed in more detail in Section 13.2; however, for the purpose of this chapter, consider synchrophasors as simply measures of current and voltage using a phasor representation and sampled such that all values are synchronized with an absolute timestamp.

PMUs measure current and voltage phasors assuming that the GPS or other accurate timing is available to incorporate timestamps with the measurements. The idea is that the timestamp allows for a precise comparison of magnitudes and phases that are spatially separated. Communication does not necessarily have to be low latency because timestamps allow for comparison after any arbitrary communication delay. A PMU may utilize multiple channels in order to measure each of the phases of a three-phase system. The PMU measures the positive sequence voltage and current phasors. Essentially, PMUs allow for the measurement of voltage and current phase angles that could never have been accurately and directly measured before the deployment of the GPS satellite constellation.

However, incorporating synchrophasors into state estimation may not be entirely straightforward. The traditional state estimation process assumes only one arbitrarily selected bus has an a priori known and fixed phase angle, which is assumed to be zero (recall Section 3.3.4). All other phase angles are referenced by differences from this fixed angle. The arbitrarily selected bus's phase angle is left out of the state vector and the resulting column of the Jacobian. The introduction of synchrophasors results in a precisely known magnitude and phase angle, perhaps for all buses if PMUs are widely deployed. A reference phase angle is no longer left open to be fixed later and set all the other phase angles with reference to it. One solution is to reformulate the state estimation problem without using a reference bus and explicitly include the phase angle for all measurements in the Jacobian, which will increase the state vector to twice the number of buses (Zhu and Abur, 2007). Thus, state estimation may take place without a reference and with larger matrices as a result. The corresponding amount of additional communication and computation will be significant.

However, instead of placing PMUs everywhere, a related problem considers a limited set of the best locations to place PMUs. This approach considers that PMUs have an associated cost, both to purchase and install. Although they may be placed ubiquitously in the future, one can assume that there are only a finite number available due to cost constraints and that it would be ideal to determine the most effective use at minimal cost (Abur, 2009). This has significant implications for communication and computation as well. In Section 15.5, we discuss quantum computing that leverages superconducting components within the power grid to help solve the computational problems. In Section 13.4, the capacity required for communication of synchrophasors is discussed.

Suppose that there are N buses and L branches in the power system whose state is to be estimated. One could place a PMU on each branch. However, in this optimization problem, the goal is to minimize the cost of purchasing and installing PMUs. Then let the location of PMUs be represented by a vector X, where x_i is 1 if there is a PMU on the branch and 0 if no PMU is installed on branch i. Let ρ_i be the cost of installing a PMU on branch i, where cost may include not only purchase and installation cost, but could also include communication cost. For example, it could be the cost of maintaining a channel for a PMU on that branch, in which case the optimization results in finding the best placement of PMUs for communication purposes. Finally, an auxiliary variable T_{ij} represents whether branch j is incident to branch i; a 1 indicates the branches are incident, and a 0 indicates that they are not incident to one another. Note that this is a variant of a simple graph incidence matrix. Another auxiliary variable is $\vec{1}$, which is the all-1s vector transpose, $\vec{1} = [1 \ldots 1]^T$. Putting all these values together, the PMU placement optimization problem for state estimation is

$$\begin{aligned} \text{minimize} \quad & \sum_{i}^{L} \rho_i \cdot x_i, \\ \text{such that} \quad & T \cdot X \geq \vec{1}. \end{aligned} \tag{12.10}$$

The optimization represents the minimization of total cost, while the constraint assumes that a PMU can measure every branch incident to a bus. The result is that all branches will be measured at the lowest cost.

However, consideration is given to the fact that there are different PMUs created by different manufacturers that have varying measurement and communication capabilities. In particular, there may be a limit to the number of channels that a PMU has for independent measurement. If the number of channels per PMU L is large enough to measure the number neighbors N_k of bus k, then a single PMU placed at the bus will be enough to cover that bus. However, if L is smaller than N_k, then there will be r_k combinations of possible channels to branches incident at bus k, where r_k is the number of combinations of L chosen from N_k. The total number of possible choices is thus $n = \sum r_k$ and the optimization problem is modified thus:

$$\begin{aligned} \text{minimize} \quad & \sum_{i}^{n} \rho_i x_i, \\ \text{such that} \quad & T \cdot X \geq \vec{1}. \end{aligned} \tag{12.11}$$

Here, r_k is shown explicitly in Equation 12.12, where ${}^{N_k}C_L = \frac{N_k!}{(N_k-L)!L!}$:

$$r_k = \begin{cases} {}^{N_k}C_L & L < N \\ 1 & L \geq N_k \end{cases}. \tag{12.12}$$

As mentioned, L is the number of channels available per PMU, N_k is the number of neighbors of bus k, X is a vector of size n whose ith element is x_i, and x_i is defined in

$$x_i = \begin{cases} 1 & i\text{th PMU chosen for placement} \\ 0 & \text{otherwise.} \end{cases} \tag{12.13}$$

As before, ρ_i is the cost of installation of the ith PMU. T serves a similar purpose as in the previous optimization problem and is a binary connectivity matrix. However, it now has a total of n rows, where each bus k contributes r_k rows and each row contains $L + 1$ nonzero values. As before $\vec{1} = [1 \ldots 1]^T$ is the 1s-vector transpose. Given that the node degree distribution of a power grid network tends to be relatively small, an exhaustive search to solve this optimization problem may be potentially feasible. Now that we have discussed observability, let us consider networked control again before discussing state estimation.

12.2 Networked Control

Networked control was introduced in Section 8.6. Communication for control in the power grid is an interesting challenge. In this section, the topic is developed further with the goal of applying it to wide-area applications.

Figure 12.1 illustrates a wide-area networked control system. Measurements are taken from a wide-area power system in the box labeled "Sampling." The controller is spatially remote from the location where sampling takes place, and sampling and control take place over a communication network. The communication network can introduce variable delay as well as drop or reorder packets. In this application, a zero-order hold is used to hold values before communication failure occurs with the goal of helping to maintain stability when communication problems occur. Note that the sampled devices may include PMUs.

The system is abstractly modeled thus:

$$\begin{cases} \dot{x} = f[x(t), y(t), u(t)] \\ 0 = g[x(t), y(t)] \end{cases}. \tag{12.14}$$

where x is a vector of power system state variables that could include any of the typical state variables, such as voltage magnitudes or angles, y is a vector of power system algebraic variables, and u is a vector of power system control variables.

A simplifying approach is to create a linear approximation of the dynamic behavior as shown in Equation 12.15, which takes place around an assumed operating point (x_0, y_0, u_0):

$$\begin{aligned} A_f &= \left.\frac{\partial f}{\partial x}\right|_{x,y_0,u_0} & A_g &= \left.\frac{\partial g}{\partial x}\right|_{x,y_0} \\ B_f &= \left.\frac{\partial f}{\partial u}\right|_{x,y_0,u_0} & C_g &= \left.\frac{\partial g}{\partial y}\right|_{x,y_0} \\ C_f &= \left.\frac{\partial f}{\partial y}\right|_{x,y_0,u_0} \end{aligned} \tag{12.15}$$

The linear approximation of Equation 12.14 is

$$\begin{cases} \dot{x} = A_f\Delta x + C_f\Delta y + B_f\Delta u \\ A_g\Delta x + C_g\Delta y = 0 \end{cases}, \tag{12.16}$$

where $\Delta x(t) = x(t) - x_0$, $\Delta y(t) = y(t) - y_0$, and $\Delta u(t) = u(t) - u_0$.

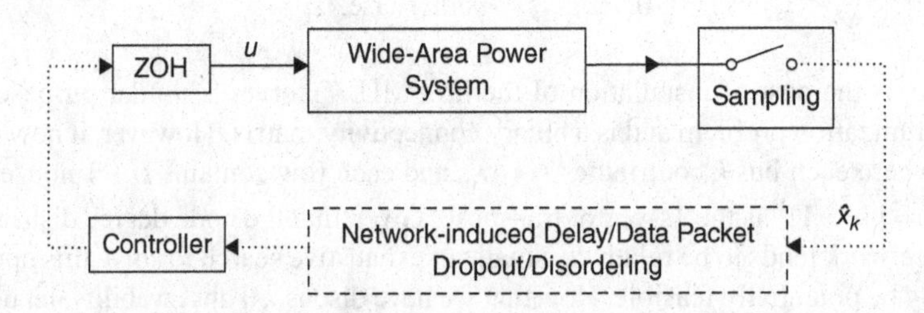

Figure 12.1 A wide-area networked control system diagram using a zero-order hold mitigates network delay by holding the last sampled value for the duration of the delay. Source: Wang, *et al.*, 2012. Reproduced by permission of IEEE.

If we ignore the impact of the network for now, the control input is $\Delta u(t) = K\Delta x_k = K(x_k - x_0)$, where x_k is the system state variable at time $t = t_k$. The zero-order hold causes $\Delta u(t) = K\Delta x_k$ to be held until the next data packet arrives. This time period is defined as

$$\tau = t - t_k \geq 0. \tag{12.17}$$

Utilizing Equation 12.17, Δx_k can be defined as

$$\Delta x_k = \Delta x(t_k) = \Delta x(t - \tau). \tag{12.18}$$

Note that τ is a function of t and $d\tau/dt = 1$.

$$\Delta x_k = \Delta x(t_k) = \Delta x(t - \tau) \tag{12.19}$$

Referring to the linear equations in Equation 12.16, Δy can be eliminated and $\Delta u = K\Delta x(t - \tau)$ can be substituted, resulting in

$$\Delta \dot{x} = (A_f - C_f C_g^{-1} A_g)\Delta x + B_f K \Delta x(t - \tau). \tag{12.20}$$

Substituting $A_0 = A_f - C_f C_g^{-1} A_g$ and $A_\tau = B_f K$ into Equation 12.20 yields

$$\Delta \dot{x} = A_0 \Delta x + A_\tau \Delta x(t - \tau), \tag{12.21}$$

which shows how the linearized x changes as a function of both the value of the linearized x and the zero-order hold.

Remember that x is a vector of power system state variables and u is a vector of control variables. The values in x may be delayed, dropped, or disordered by network effects. The question to be resolved is how these network effects impact the initial system description assumed in Equation 12.14 given the zero-order hold. If a packet x_{k+1} is dropped, then the value from packet x_k will be held until packet x_{k+2} arrives. If packets are disordered and value x_{k+2} arrives before x_{k+1}, then x_{k+2} will be used instead of x_{k+1}. While beyond the scope of this book, it is possible to determine bounds on τ for which the system can be guaranteed to remain stable (Wang *et al.*, 2012). Now let us consider power system state estimation in more detail.

12.3 State Estimation

The control center, being the nerve center for the power grid, needs to have as much knowledge as possible of the current state of the power system in order to make intelligent control decisions. There have been three general types of knowledge available at the control center. The first is analog measurement of values (such as real and reactive power flow), real and reactive power injection (which include generation or demand at buses), and bus voltage magnitudes. The second is logical measurement such as the status of switches and breakers, as well as load tap changing transformer positions. Finally, there are pseudo-measurements such as predicted bus load and generation.

State estimation is the mathematical process that yields a description of the power system state. It computes the best estimate of the power system state variables such as bus voltages and angles of the power system based upon noisy data. Once the state has been estimated, enough information should be available to derive all relevant aspects of the power system such as power line flows. State estimation has traditionally had to emphasize "estimation" or approximation. Equipment specification had to be used as an approximation for actual measurements and many variables had to be inferred because real-time measurements were simply unobtainable, particularly in real time. However, one of the more obvious goals of the smart grid is to shift the emphasis from "estimation" to actually sensing state through extensive sensor networking. However, it is difficult to appreciate the benefits of communication without first understanding traditional state estimation.

Assume there are n state variables, $i = 1 \ldots n$, and that m measurements are taken. Let z be the measurement vector with the m measurements. The state vector is x and v is the noise or approximation in the measurements. Finally, let $h()$ represent a mapping between measurement and state. The resulting equation for the measurements is

$$z_i = h_i(x) + v_i. \tag{12.22}$$

If h is assumed to be linear – that is, a linear combination of the state – then it can be represented as a matrix:

$$z = Hx + v. \tag{12.23}$$

H is more formally called the measurement matrix. There are many approaches to finding the best estimate for x, which is indicated by $\hat{x}$. A simple and widely used approach is weighted least squares (WLS). The goal in this approach is to minimize the difference between the measurements and the corresponding mapping to state values through the measurement matrix. This corresponds to minimizing the objective function

$$J(x) = \sum_{1}^{m} k_i[z_i - h_i(x)]^2, \tag{12.24}$$

where k_i is a linear constant.

The goal is to solve for x in $Hx = z$. To do this, multiply both sides by H^T yielding $H^T Hx = H^T z$. Then multiply both sides by $[H^T H]^{-1}$, yielding $x = [H^T H]^{-1} H^T z$. Equation 12.25 shows the recursive process:

$$x_{k+1} = x_k + [H_k^T W H_k]^{-1} H_k^T W[z - h(x_k)]. \tag{12.25}$$

Notice that this equation is setting the updated state value to the previous value plus a ratio. The lower portion of the ratio will collapse to a scalar based upon the weights and the linear mapping H or measurement matrix. The upper portion of the ratio is similar to the lower except that it includes the difference between the measurements and the mapping to state values. As the recursion progresses, the ratio should become smaller as the error becomes smaller.

In the general case, the weights W can be set to the inverse of the variance in measurement; this assumes that a greater measurement variance indicates less certainty, or confidence, in that value. However, these weights can be explicitly set by an expert who knows what the confidence should be. In the ideal case, the equation will converge to a solution. If this happens, the system is said to be observable. This requires that the number of measurements is sufficient and that they are uniformly distributed throughout the network. Before state estimation is undertaken, a study should be done to determine the likelihood that the system will be observable. If it is not likely to be observable, then more meters or sensors need to be added or repositioned and more pseudo-measurements may be added. Next, we have to address the challenge that the power grid is too large to implement a single, centralized state estimation system. Instead, the grid must be managed in parts and the approach distributed throughout the power grid. This is where communication plays a critical role and this is the topic of the next section.

12.4 Distributed State Estimation

As the power grid grows larger and more complex, state estimation will become more challenging. Hope for a solution lies in the evolution toward the smart grid; communication plays a significant role. The state estimation process can be parallelized and distributed, allowing computation to more closely match the distributed nature of the evolving power grid. However, as computation becomes more distributed, communication among distributed processes becomes more critical to reducing the computation time required to obtain a solution. Distributed processors can focus on processing only the information in their immediate, local area, thus efficiently dividing the work and improving speed toward a global solution. However, processors must also wait for neighboring processors to share boundary information, which will be described in more detail shortly. Communication system design needs to be aware of the computational aspects and architecture of the power grid.

The parallel and distributed processing approaches are similar. In parallel processing, all information is communicated to a central computer that virtually partitions the power grid into subareas, which are then assigned to each processor within the centralized computer. The individual processors share boundary information among processors using internal interprocessor communication, which is assumed to be relatively fast. In the distributed processing approach, processors are spatially distributed throughout the power grid. These processors perform state estimation on their local subareas only, sharing boundary information as needed with neighboring processors. In this case, the interprocess communication latency is significantly longer and more variable. Of course, a design goal in partitioning the power grid for state estimation is to reduce the amount of interprocess communication required. We can think of the parallel approach as a "virtual" partition of the power grid for computational purposes, while the distributed approach is an actual partition of the grid into processing areas. Finally, keep in mind that the parallel and distributed techniques discussed here are relatively incremental extensions of well known, mature state estimation algorithms; entirely different machine-learning algorithms that may be applied instead were discussed in Chapter 11.

Recall Equation 12.25 with n measurements and the Jacobian $H = \partial x / x_i$. In distributed and parallel computation of state estimation, H is partitioned into multiple subareas H_i each with

n_i measurements. The gain matrix of the ith subarea is

$$G_i(x) = H_i^{\mathrm{T}}(x)W_iH(x), \quad i = 1, \ldots, N. \tag{12.26}$$

Each processor can perform the iteration shown in Equation 12.27. Note that x represents the state vector for the entire system; for a specific subarea, only the x_i are required.

$$G_i(x)\Delta x_i = H_i^{\mathrm{T}}(x)\Delta z_i(x) \quad i = 1, \ldots, N. \tag{12.27}$$

Equation 12.28 shows the gain matrix relevant to a specific subarea in which x_i represents state variables internal to the subarea and $x_{i,b}$ represents a state variable at the boundary, between a subarea and an adjacent subarea:

$$G_i(x, x_{i,b}) = H_i^{\mathrm{T}}(x_i, c_{i,b})W_iH(x_i, x_{i,b}) \quad i = 1, \ldots, N. \tag{12.28}$$

Thus, Equation 12.29 shows the detailed state equation for a single subarea:

$$G_i(x_i, x_{i,b})\Delta x_i = H_i^{\mathrm{T}}(x_i, x_{i,b})\Delta z_i(x_i, x_{i,b}) \quad i = 1, \ldots, N. \tag{12.29}$$

Here we see that each subarea will need to wait for the transfer of information from any processors handling adjacent subareas in order for the $x_{i,b}$ to be updated appropriately. If processing in the individual areas is not balanced, then a significant amount of time will be spent waiting to synchronize with adjacent subareas.

All forms of state estimation, centralized, parallel, or distributed, require checking for, and handling, bad data. The simplest and most direct approach is to examine the residuals, or total error, that we wish to minimize in the form of the objective function $J(x)$. If $J(x)$ cannot be minimized below a given threshold, then that is an indication of bad data that needs to be handled. For parallel and distributed approaches, this is relatively straightforward. Each subarea can compute its portion of the objective function as shown in Equation 12.30 and totaled in Equation 12.31. However, note that this generally assumes that one particular processor will be the "coordinator" it will receive and total values from all other processors.

$$J_i(x) = [z_i - h_i(x)]^{\mathrm{T}} W_i[z_i - h_i(x)] \quad i = 1, \ldots, N \tag{12.30}$$

$$J_i(x) = \sum_{i=1}^{N} J_i(x). \tag{12.31}$$

Another relatively simple approach is to check the residuals for each measurement, as shown in Equation 12.32. The larger the residual, the more likely it is to be an erroneous or inconsistent value.

$$\Delta z_i = z_i - h_i(x) \quad i = 1, \ldots, N. \tag{12.32}$$

All subareas can transmit their residuals to the coordinating processor, which can form the residual vector shown in Equation 12.33; recall that N is the number of subareas:

$$\Delta z_i = \begin{bmatrix} \Delta z_1 \\ \vdots \\ \Delta z_N \end{bmatrix}. \tag{12.33}$$

A normalizing value, $d = \sqrt{(\Delta z_1)^2 + \cdots + (\Delta z_N)^2}$, can be used to obtain the result shown in Equation 12.34:

$$\Delta z = \begin{bmatrix} \Delta z_1/d \\ \vdots \\ \Delta z_N/d \end{bmatrix}. \tag{12.34}$$

Normalizing over all subareas allows for a fair contribution of the residual from each subarea. This result allows for each piece of data to be individually checked to determine whether it is considered bad data.

Now we move on to distributed state estimation, the approach in which processors that are dispersed in different parts of the power grid work collaboratively to perform state estimation. Here, subareas are not conceptually divided as in parallel state estimation; rather, they are actually, physically divided in space, requiring time for any communication between subareas to take place. Subareas may be divided following organizational boundaries, such as ISO and RTO boundaries, control centers, or by areas with DG capability. Each area may have its own portion of the communication network or SCADA system under its control. Remember that each SCADA system typically scans on a cyclical basis for updates, so information is near-real time.

There is an observability issue to be considered. As discussed for parallel state estimation, buses or tie-lines that interconnect subareas play a critical role. If such lines are not observable, the system may be split into observable islands or rendered unobservable, even though the system may have been observable for sequential (that is, nonparallel), nondistributed, state estimation. In fact, techniques used in network communications apply to determining observability. If there are N subareas and $N-1$ tie-lines interconnecting the subareas, then a minimum spanning tree will be formed, allowing each subarea to receive information from all other subareas, particularly the subarea in which the reference bus resides for flow analysis. This is a sufficient, but not necessary, condition for total system observability. The necessary and sufficient condition is simply that all subareas must be able to either directly or indirectly receive reference information through the exchange of information across boundary connections, such as boundary buses or tie-lines.

As in the parallel approach, the Jacobian H is partitioned into H_i where there are i subareas. This is done in such a manner that all state variables in areas outside H_i are assumed to be constant and thus their Jacobian values are zero. Now we need notation to take communication delay into account by means of the following: $x^i(t) = (x_1(\tau_1^i(t)), \ldots, x_N(\tau_N^i(t))$, where $x_j(\tau_j^i(t))$ represents state information x_i at time t sent by processor j to processor i at time $\tau_j^i(t)$. In other words, $\tau_j^i(t)$ represents the time when the jth component becomes available at the updated $x_i(t)$ at time t. The time difference, $t - \tau_j^i(t)$, is the communication delay. Then each processor performs the algorithm by iterating over

$$G_i(x^i(t))\Delta x_i = H_i^{\mathrm{T}}(x^i(t))\Delta z_i(x^i(t)) \quad i = 1, \ldots, N. \tag{12.35}$$

Thus, the variable communication latency is taken into account in Equation 12.35. Only state variables of boundary power flows need to be exchanged. Bad data processing still requires a central computer capable of comparing all subarea results using the technique previously explained for parallel bad data processing beginning in Equation 12.30. Now the we have covered state estimation, we consider the issue of using the estimated state to maintain stability in the power grid and how smart grid communication will help maintain a stable power system.

12.5 Stability

In the general terms, stability of the power grid is its ability to reestablish its initial state after any disturbance or interruption causing a deviation from the initial parameter values. A distinction is made between static and dynamic stability; that is, between the ability to reestablish the initial state after small and large disturbances. Clearly, stability is necessary for reliable operation of the power grid. In steady state, the power coming into the system from the outside is expended by the load W_1, which offsets power losses ΔW. The occurrence of a disturbance in the system produces a deviation in system parameters P. If the power being expended $W_l + \Delta W = \phi(P)$ after the disturbance is greater than the maximum power $\Delta W_g = f(P)$ that the outside supply can replace, then it is necessary to restore the system to its former state. A system that can naturally return to its initial operating condition after such a disturbance is said to be stable.

Figure 12.2 classifies different types of stability. The main types of stability are related to maintaining generator rotor synchronization, maintaining a constant, nominal alternating-current frequency, and maintaining voltage within a constant, correct range. Each of these

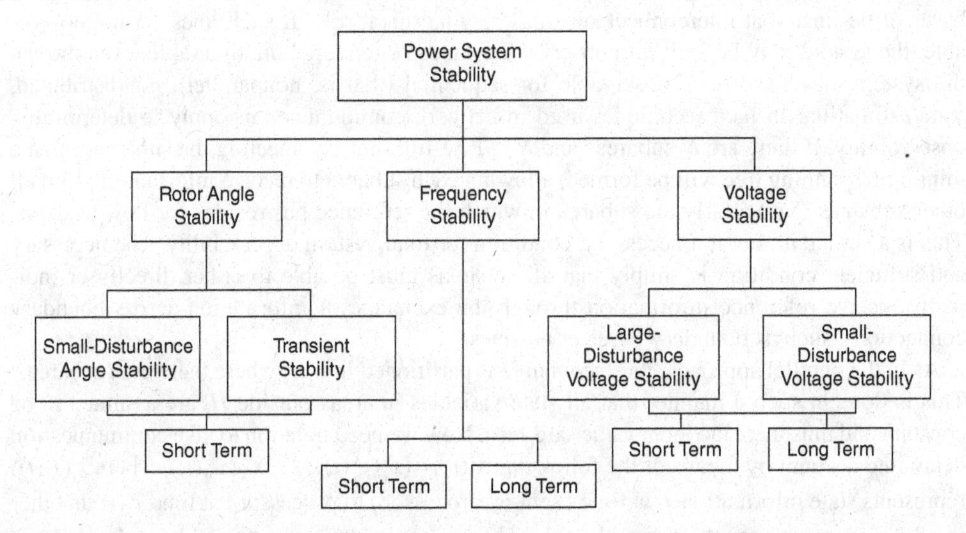

Figure 12.2 A classification of stability into three major types: rotor, frequency, and voltage. Source: Kundur *et al.*, 2004. Reproduced with permission of IEEE.

types of stability has a short-term and long-term incarnation. Clearly, short-term instability is dampened and dies out rapidly, while long-term instability continues to oscillate.

12.5.1 System Performance

Next we consider performance of the power grid as a whole. Reliability is one measure of performance; here, it used as a measure of the ability of the system to deliver power to consumers when and where they need it. Traditionally, reliability had focused upon the ability of generation to meet demand because transmission and distribution had been well provisioned, if not overprovisioned. However, this is no longer the case; transmission and distribution systems are reaching their capacity limits.

From a generation standpoint, reliability can be assured by having a *reserve margin* of generating capacity above and beyond the maximum anticipated load. Before the 1970s, a reserve margin of up to 25% was not uncommon. However, economic conditions no longer allow such large reserve margins. The system must be operated closer to its limits. In fact, this is a theme throughout power systems today and for the smart grid: How close to system limits we can operate while safely squeezing more from the existing system?

Another measure is the *loss-of-load probability*. This is the probability that, over a given interval, the system will fall short of meeting demand. This is often expressed as the estimated number of days over a relatively long period, such as 10 years or the life of the system. This value is computed by summing the probabilities of all events in which load fails to meet demand. Another measure is the *loss-of-load expectation*. This is the expected number of days that the peak demand exceeds generation capacity. There is also the *expected unserved energy*. This can be computed by simply multiplying the probability of loss of load with the amount of load that exceeds the generation capacity in megawatts. This is the expected amount of load that would need to be shed in order to keep the system operating.

As mentioned, a growing cause of power loss is not the failure of generation capacity, but the failure of the large and aging transmission and distribution system. Reliability is highly variable, depending upon terrain, weather, and population density in the local region. There are a number of different reliability measures that seem to reflect the different philosophies of what constitutes an outage and its impact. One could perhaps argue that power quality, rather than a complete loss of power, is an "outage." However, assuming complete loss of power constitutes an outage, there are several measures that take into account the frequency, duration, and number of customers affected. These are measures by which the overall impact of a "smart grid" may be rated. In particular, an interesting question will be to what extent communication has served to either increase or decrease these metrics. A description of each metric follows.

The SAIDI is commonly used as a reliability indicator and represents the average outage duration for each customer served:

$$\text{SAIDI} = \frac{\sum \text{Individual outage durations}}{\text{Total customers served}}. \tag{12.36}$$

The SAIFI is the average number of interruptions that a customer would experience and is defined as

$$\text{SAIFI} = \frac{\sum \text{Customers interrupted}}{\text{Total customers served}}. \tag{12.37}$$

The customer average interruption frequency index (CAIFI) is designed to show trends in customers whose power is interrupted and helps to show the number of customers affected as a proportion of the entire customer base:

$$\text{CAIFI} = \frac{\sum \text{Customer interruption duration}}{\text{Total number of customers interrupted}}. \tag{12.38}$$

The customer total average interruption duration index (CTAIDI) is

$$\text{CTAIDI} = \frac{\sum \text{Customer interruption duration}}{\text{Total number of customers interrupted}} = \frac{\sum r_i N_i}{\text{CN}}, \tag{12.39}$$

where CN is the total number of customers who have experienced a least one sustained interruption. This represents the total average time that customers who experienced an interruption were without power. Note that this is a hybrid of customer average interruption duration index (CAIDI) and is calculated similarly, except that customers with multiple interruptions are counted only once. The CAIDI gives the average outage duration that any given customer would experience and may also be viewed as the average restoration time and is given by

$$\text{CAIDI} = \frac{\sum \text{Total number of customers interrupted}}{\text{Total number of customers interrupted}} = \frac{\sum N_i}{\text{CN}}. \tag{12.40}$$

The MAIFI is

$$\text{MAIFI} = \frac{\sum \text{Total number of customer momentary interruptions}}{\text{Total number of customers served}} \tag{12.41}$$

It represents the average number of momentary interruptions that a customer would experience during a given period, typically a year. Utilities may define momentary interruption differently. Some may consider a momentary interruption to be an outage of less than 1 min in duration, while others may consider a momentary interruption to be an outage of less than 5 min in duration.

The momentary average interruption event frequency index (MAIFI_E) indicates the average frequency of momentary interruption events. This index does not include events immediately preceding a recloser lockout. It is defined as

$$\text{MAIFI}_\text{E} = \frac{\sum \text{Total number of customer momentary interruption events}}{\text{Total number of customers served}}. \tag{12.42}$$

Customers experiencing multiple sustained interruptions and momentary interruption events (CEMSMI_n) is defined as

$$\text{CEMSMI}_n = \frac{\sum \text{Total number of customers experiencing more than } n \text{ interruptions}}{\text{Total number of customers served}} \tag{12.43}$$

It is the ratio of individual customers experiencing more than n sustained interruptions and momentary interruption events to the total customers served. The purpose is to help identify issues that cannot be observed using averages.

Some loads are more critical than others; some loads may be air-conditioning for an elderly or sick person on an exceptionally hot day, traffic signals at a dangerous intersection, or crucial medical equipment required to keep someone alive. It has been impossible for utilities to discriminate among loads; every load has had to be treated as equally critical and we have become used to assuming power should always be available. Most electronic appliances are designed assuming power will be continuously available. Some have considered that utilities have gone overboard in spending in order to provide such high reliability. Historically, utilities were easily reimbursed by customers for providing this level of service. A key element has been to distinguish what individual customers would be willing to pay and for what level of service. This has been attempted in different ways in the past; however, with a smart grid, the hope is that DR mechanisms will allow the utility to make this distinction among customers and to allow customers a high-resolution scale of service options.

Security, not to be confused with cybersecurity, is a power systems term related to reliability and system performance that quantifies the amount of "room" remaining for alternatives to compensate for failures before the limit is reached and the grid fails. Considered from the opposite viewpoint, power systems security is the number of *contingencies*, or disruptive events, that the system can sustain before an outage occurs or equipment is damaged. It is important for people in the communications field to realize that security in power systems has had little to do with cybersecurity, although it is possible that cybersecurity events could certainly lead to damage of the grid and reduce overall security from a power systems standpoint. The simplest example of a contingency is a power line going down; security is then related to the number of alternative configurations, or paths, available to continue to supply power. Thus, security is the ability of the power system to reconfigure after an event and still supply the power demanded, where the event could be anything from loss of a generator to a fallen power line. Security is a high-level term used to describe the number of discrete, alternative configurations available to work around problems in the power system. However, the actual transition from one configuration to another may involve sudden changes in the system. These sudden changes are closely related to the topic of stability.

Reliability, security, and stability may all seem to the reader like different words describing the same thing. However, there are clear differences. Reliability refers to the *probability* that the power system will continue to supply power over the long term. Security describes the amount of risk in the power system's ability to survive perturbations, failures, or contingencies without interrupting the flow of power to consumers. Stability refers to the ability of the system to safely transition from one configuration to another immediately after an event. To be reliable, the system must be secure; security implies having enough redundancy to increase reliability. To be secure, the system must not only have alternatives to keep power flowing, but also have stability when transitioning among alternatives. To further distinguish these terms, a system may be stable, having successfully transitioned to a new configuration in order to avoid an outage; however, it may have reduced security because there are fewer backup configurations available.

Security has been increased over the years by having highly interconnected power systems. A very simple example goes back to the radial distribution system. A single fault anywhere on one radial power line will cut off power to all remaining customers on the line downstream of the fault from the feeder. A loop distribution system enables two or more radial systems to connect, usually at their end points, when a fault occurs so that one radial system can connect to the faulted system and supply power to its customers who have lost power due to the fault. A more highly meshed distribution system will allow multiple points of connection, through tie-switches that are normally open, allowing a higher degree of redundancy. In this case, security is related to the number of redundant paths available. Clearly, this is similar to routing in communication networks; an ad hoc or mesh communication network has many possible redundant communication paths. Highly interconnected power systems also allow for a greater degree of pooling among generators. If one generator goes down, it is possible to meet demand by tapping into another, connected generator. This is similar to either having a print server directly connected to your computer or being on a communication network and sharing from among many such servers.

In order to determine the degree of security a CA is performed. The "normal-minus-one" (N-1) CA is fairly standard and requires that the system should remain operational after one contingency. One example of security analysis is the analysis of line flow limits. Recall that line flow limits are based upon both their thermal limits (power capacity) and stability limits. A line flow limit analysis will compute the ability of the network to deliver power as lines go down and load is shifted to other lines. Operators know what the most likely contingencies will be from experience and can run through scenarios that not only include power line limits but also peak loads and factors influencing generation capability. The result is not only an understanding of power system security but also of where additional capacity and redundancy are needed for planning purposes.

While security deals with discrete states of the power system, stability is focused upon the continuous change between states and making sure that the system does not become unstable when those changes occur. Stability quantifies the ability of the system to remain synchronous and balanced. Angle stability refers to the ability of the system to keep all components, both electrical and mechanical, running in step with one another at the same frequency. Voltage stability is another related aspect of stability and refers to keeping the magnitude of the voltage constant. We first focus on angle stability and then cover voltage stability.

12.5.1.1 Angle Stability

Stability in power systems retains its meaning from physics; namely, it refers to the ability of the system to return to its desired operating state after a perturbation. Conceptually, a marble in a rounded bowl will return to the bottom center of the bowl no matter where it is released or how it has been perturbed within the bowl. This is an example of a stable system with the interplay between mass, gravity, and the concave shape of the bowl yielding a stable system for the marble. However, if the rounded bowl is flipped upside down, and the marble placed on top of the now inverted bowl, unless the marble is placed and balanced extremely carefully, it will fall off the side of the inverted bowl. This is an unstable system due to the convex nature of the terrain upon which the marble is placed. The analog of the marble in power systems is the voltage angle or power angle δ. Recall from earlier chapters that the power angle may be different in different parts of the system depending upon how "hard" generators are working relative to one another. A generator that is working harder will have a leading power angle; a circulating current will flow among the generators as the harder working generator takes more of the load and produces more power relative to the others. The power angle also indicates how much real power is being injected or withdrawn from the system. Stability is involved with understanding the continuous change in the power angle as a dynamically changing variable.

A stable power system is one in which generators are operating synchronously; their voltage waveforms rise and fall at the same time, they operate at the same frequency and phase. If the generators fall out of synchrony, one of the results will be large circulating currents, enough to trip breakers or destroy power lines. As long as there is a natural restoring force to bring the generators back into synchronization, the system is stable. This restoring force is the analog of gravity in our marble-in-the-bowl analogy. It is electromagnetic reaction that increases the force against a faster generator and reduces the force against slower generators, causing them all to converge to the same speed. Consider the analogy of stability with balancing a long stick in the palm of your hand. A long stick is easier to balance because there is more time to react to changes in the angle of the stick to keep it upright. Keeping the stick upright is maintaining balance among generators. A small difference in power angle is analogous to a long stick, allowing relative ease in maintaining synchrony. However, a long transmission line between generators, one that approaches its stability limit, would be like a short stick. It becomes more difficult to maintain the stick upright in the palm of the hand; sudden, larger motions are required to maintain stability of the stick. As the stability limit is exceeded, it no longer becomes possible to keep the stick upright. The system is no longer stable and will collapse at the slightest perturbation.

Consider a power line with generators and loads at both ends, where the ends are designated by the subscripts 1 and 2. The power angle δ is the angle formed by the relative difference in power between the end of the line and the generator at that end. δ_{12} is the difference between the power angles at each end of the line. The more power that is transmitted through the line, the larger the difference. Equation 12.44 shows the real power transmitted across the line:

$$P = \frac{V_1 V_2}{X} \sin \delta_{12}. \tag{12.44}$$

Note that a simplifying assumption is that the line is lossless. That is, it is a power line in which resistive losses are zero; there is only reactance. This assumption also allows real and reactive power to be decoupled and treated separately in power flow analysis, as discussed in previous chapters.

Assuming voltage magnitudes and the line impedance are fixed, the only way to transmit more power is to have a larger power angle. On the other hand, if power angles on either side of the transmission line are equal, no power will be transmitted, regardless of how large the voltage magnitudes are or how small the line impedance becomes. The largest possible difference in power angle would be 90°, yielding a sine of one. However, a power angle difference this large would be unstable. As the power angle increases, feedback, in terms of circulating current among the generators, decreases.

Examining Equation 12.44, if all parameters except the power angle are fixed, then the real power versus power angle plot is simply a sine function, with a maximum value at 90°. The problem occurs if the power angle were to accidentally exceed 90°, even by a small amount. Beyond 90°, the sine function begins to decrease back toward zero, which means that power decreases as the power angle continues to increase. Decreasing power results in decreasing circulating current feedback among the generators and the faster generator continues to run even faster, causing the power angle to increase even further, causing a further reduction in feedback. This vicious cycle is clearly the result of the system becoming unstable. Also, looking at the sine wave, one can see that at small power angles the curve is steep; small changes in power angle result in large changes in real power and a stronger restoring force. As the power angle approaches the top of the sine wave the slope is flatter; there is less feedback power corresponding to changes in the power angle at higher differences in the angles.

Another way to visualize the ineffectiveness of large power angles on maintaining stability is to imagine two, interconnected generators, first with a relatively-small power angle. Because the power angle is small the voltage of each generator will be shifted by only a small amount. The voltage difference, which drives the circulating current, will appear 90° ahead of the system voltage as seen by the leading generator. This is because we assume the voltages are roughly the same magnitude and only slightly shifted, so their difference cancels near their maximum and minimum and is greatest while they are near the zero crossing. Because the system in which the circulating current resides is almost entirely inductive, the current lags the voltage by 90°. This actually shifts the current into phase with the slower generator and creates additional power output. It is out of phase with the faster generator and reduces its output, serving to lead both into synchrony. However, as the power angle increases, and the voltage from each generator shifts more out of phase with one another, the circulating current phase tends to shift between the voltage phases of both generators. This results in power oscillating between the generators, rather than flowing in only one direction, and the stabilizing force is reduced. Such oscillation is also damaging to the equipment, creating eddy currents and causing abnormal heating. The only practical way to transmit more power while keeping the power angle low is to decrease the reactance X of the line.

The previous discussion was about steady-state stability; that is, how well the system is designed to return to a stable state. Returning to the analogy of balancing a stick in the palm of your hand, steady-state stability is about the length of the stick. Dynamic stability, on the other hand, is about how quickly we can react to a sudden change – for example, a gust of wind that blows the stick out of position – and still maintain stability. We can take the actual analysis right to the heart of the generator; namely, the rotor and the transformation of mechanical power into electrical power. As always, conservation of energy applies. Thus, mechanical power applied to the input of the rotor must equal the electrical power generated by the rotor, minus any loss due to inefficiency. More specifically, the mechanical torque applied to the rotor equals the electrical power generated from the armature windings plus damping power. Damping power comes from both friction and the exciter winding. Equation 12.45 shows the power balance as a function of the power angle δ:

$$M\ddot{\delta} + D\dot{\delta} + P_G(\delta) = P_M^0. \tag{12.45}$$

P_M^0 is the mechanical power, where the superscript zero indicates the system is in equilibrium. $P_G(\delta)$ is the electrical power that is output, which is clearly a function of the power angle. Recall that the power angle describes the difference in angle between the internal generator voltage waveform and the waveform of the power system as whole. Since the waveforms are sinusoidal, then as the power angle increases, power output will increase, but only up to a point. The increasing power angle is really describing shifting voltage phases; beyond a maximum point, the phase difference, or difference in rotation of the rotors, will be so great that the leading rotor will actually appear to be behind the other rotors. The result is loss in power. The power angle versus power curve forms a nonlinear convex curve. In Equation 12.45, $P_G(\delta)$ is the magnetic force that pushes back on the rotor when it attempts to move faster than its initial equilibrium speed. If it moves too far ahead, the restoring force will pass its maximum value and decrease, becoming zero, or even become negative if it continues to move too far. D is the damping force due to friction and $D\dot{\delta}$ is the power lost due to friction. M, for momentum, is the rotor's inertia, its tendency to resist change in speed. $M\ddot{\delta}$ is the power lost in accelerating or decelerating the rotor.

Recall that the superscript zero indicates an equilibrium condition. When the rotor is running at equilibrium, this implies that there is no change in δ; that is, the rate of change, or derivative, is zero. In this case, the damping and inertia terms in Equation 12.45 are zero and the equation simplifies to $P_G(\delta) = P_M^0$. Recall that the goal of dynamic stability is to understand what happens when there is a perturbation in δ, just as there would be a perturbation in the position of the stick balanced in the palm of the hand after it was suddenly blown by a wind gust. The question is whether the system will be able to stabilize after the perturbation, or whether we can continue to balance the stick upright in our analogy, and how much perturbation can be tolerated in both cases.

Consider what happens when a switching event occurs and the load suddenly changes. Specifically, consider a transmission line that becomes disconnected along with all its load for a short time interval. The mechanical power P_M^0 continues to be applied to the rotor as though the rotor were still supplying the original load. However, with the reduction in load due to the transmission line being disconnected, the same mechanical power now causes the rotor to accelerate. The rotation rate is no longer constant, the power angle δ increases, and its derivatives are no longer zero. The frequency of the rotor increases. The damping and inertial terms in Equation 12.45 are now no longer zero as long as the transmission line remains disconnected. These terms represent the excess energy that was accumulated during the transient transmission line outage. If the disconnect time was short, then the excess power can be absorbed by the generator and the system will return to normal after the line is reconnected. However, beyond a certain outage time the system will not be able to absorb the excess energy after the transmission line reconnects, and the system will become unstable.

If the transmission line outage is of relatively short duration, then the power angle δ will move ahead of its equilibrium position, then fall behind it, then ahead, oscillating back and forth as the magnitude of the oscillation decays over time due to damping forces, eventually returning to its equilibrium position. This is similar to our analogy of the balanced stick after a wind gust; to maintain balance, one will attempt to move one's hand in one direction to compensate, but will usually overcompensate, requiring a motion back toward the original position. These oscillations will continue, becoming smaller, until the stick is balanced again.

Inside the generator, once the transmission line and its load have reconnected, the magnetic power $P_G(\delta)$ applied against the rotor will be large due to the sudden increase in load and the large value of δ. This will cause the rotor to decelerate. However, owing to the inertia of the rotor, deceleration will take time as the rotor continues to move ahead a bit. Eventually the rotor begins to decelerate toward δ^0 and eventually to a point just slower than δ^0. At this point, the mechanical power P_M^0 causes rotation to accelerate toward δ^0, again potentially overshooting it slightly. This cycle of deceleration and acceleration continues to oscillate in a decaying manner until the rotor comes to its steady-state equilibrium speed again.

On the other hand, if the transmission line outage lasts too long, δ could become too large and, as discussed, the restoring power becomes weaker at larger values of δ. In our analogy, this is similar to the wind gust being significantly stronger. It is likely that stability will be lost because the person will not be able to maintain balance due to the increasing speed and rate of motion required to keep the stick balanced upright.

The transient stability example just described can be seen in terms of potential and kinetic energy. Suppose, for simplicity, that there are only two generators and one goes offline for only a short time. The offline generator builds up kinetic energy through mechanical power applied to its rotor. Its rotor speeds up. The other generator must now supply power to all the loads and its rotor slows down. When the offline generator is brought back online, there will be a temporary imbalance, with one generator going too fast and the other too slow. This will lead to an oscillation in energy between the generators that takes place through the circulating current between the generators until both generators reach equilibrium again.

12.5.1.2 Voltage Stability

Now we consider the issue of voltage stability. Voltage magnitude and (power) voltage angle are related; however, as we discussed in the decoupled power flow analysis, they can be considered separately. Voltage magnitude was found to be associated with reactive power and angle with real power. Recall that when a load's impedance goes down, it draws more current and increases its power consumption. However, this can only continue up to a limit; after a certain point, the power drawn is so large that voltage cannot be sustained. In order to keep voltage from dropping, either reactive power or more voltage support must be added, allowing it to draw more power.

12.6 Stability and High-Penetration Distributed Generation

This section covers what could happen to the power grid when centralized generation is no longer dominant and providing a common frequency. In a nutshell, isochronous speed control refers to the prime mover governor speed control mode that controls the frequency (speed) of an alternating current generator (alternator), and droop speed control refers to the prime mover governor speed control mode that allows multiple alternating current generators (alternators) to be operated in parallel with one another to power large electric loads, or to "share" load.

The frequency of a synchronous alternating current generator (the type most commonly used in alternating current power generation) is directly proportional to the speed of the rotating electrical field(s) $F = PN/120$, where F (Hz) is frequency, P is the number of poles of the rotating electrical field, and N is the speed of the rotating electrical field in rotations per minute (RPM).

In isochronous speed control mode, energy being admitted to the prime mover is regulated very tightly in response to changes in load, which would tend to cause changes in frequency (speed). Any increase in load would tend to cause the frequency to decrease, but energy is quickly admitted to the prime mover to maintain the frequency at the setpoint. Any decrease in load would tend to cause the frequency to increase, but energy is quickly reduced to the prime mover to maintain the frequency at the setpoint.

In droop speed control mode, the governor of the prime mover is not attempting to control the frequency (speed) of the alternating current generator. The term "share the load" causes much confusion, but it simply refers to the ability of the prime movers of alternating current generators to smoothly control the production of torque when connected in parallel with other generators supplying an electric load.

Droop speed control, refers to the fact that the energy being admitted to the prime mover of the alternating current generator is being controlled in response to the difference between a speed (frequency) setpoint and the actual speed (frequency) of the prime mover. To increase power output of the generator, the operator increases the speed setpoint of the prime mover, but since the speed cannot change (it is fixed by the frequency of the grid to which the generator is connected) the error, or difference, is used to increase the energy being admitted to the prime mover. So, the actual speed is being "allowed" to "droop" below its setpoint.

On a small electric grid, one machine is usually operated in isochronous speed control mode, and any other, usually smaller, generators that are connected to the grid are operated in droop speed control mode. If two prime movers operating in isochronous speed control mode are connected to the same electric grid, they will usually "fight" to control the frequency, and wild oscillations of the grid frequency usually result. Only one machine can have its governor operating in isochronous speed control mode for stable grid frequency control when multiple units are operated in parallel. There are isochronous load sharing schemes in use in various places around the world, but they are not common.

On very large electrical grids – commonly referred to as "infinite" electric grids – there is no single machine operating in isochronous speed control mode which is capable of controlling the grid frequency; all the prime movers are being operated in droop speed control mode. But there are so many of them and the electric grid is so large that no single unit can cause grid frequency to increase or decrease by more than a few hundredths of a percent as load varies.

Very large electric grids require system operators to quickly respond to changes in load in order to control grid frequency properly since there is no isochronous machine doing so. Usually, when things are operating normally, changes in load can be anticipated and additional generation can be added or subtracted in order to maintain tight frequency control.

One method many electrical grid operators use to control grid frequency is called AGC. Units being operated in AGC have their droop speed control setpoints adjusted remotely in response to commands from the system operator to maintain grid frequency. Since this remote control is taking place over a communication network, we have networked control as discussed earlier in this chapter.

12.7 Summary

State estimation and stability are significant components of power grid management. Operational control requires knowledge of the state of the grid, and control is required to maintain stability. The power grid is evolving toward incorporating a more widely deployed collection of sensors, including phasor measurement units. Thus, state estimation, from a smart grid perspective, involves determining how best to utilize information collected from new and more widely deployed types of sensors. In particular, this chapter looked at how to handle synchrophasor data in state estimation as well as how to optimize deployment of PMUs. How data are utilized and how sensors are deployed will greatly impact communication requirements.

Next, the chapter picked up the recurring topic of networked control. This time, the use of zero-order hold was considered as a solution to handling the impact of message disruption due to the communication network. Next, the typical WLS approach to state estimation was explained in detail. The approach takes the form of an optimization problem; namely, minimizing error. An iterative approach to accomplish this was explained. One approach to handling the growing computational burden of state estimation is to distribute the process. This is explained along with the impact it could have for communication to handle the interprocess communication among the distributed components. Stability appears in several different forms in the power grid. While it appears to be most commonly associated with generator rotor angle stability, there are several other dynamic properties that require stable operation. Of course, the issue of a power grid in which centralized generation is no longer dominant is a common area of concern and ends the chapter.

Chapter 13 introduces synchrophasors and their potential applications in detail. This includes synchrophasor compression and communication networking. Synchrophasors grew from success of symmetrical components used to implement protection mechanisms and accurate timing provided by GPS. Phasors, symmetrical components, and GPS are explained. Understanding synchrophasors and their applications allows their communication and networking requirements to be better understood and their properties to be exploited to improve communication. Synchrophasor applications are numerous; the distance, frequency, delay tolerance, and the number of synchrophasors required may vary considerably from one application to another. Phasor measurement units and phasor data concentrators are significant components of synchrophasor communication and are explained. Compression of synchrophasors, by both passive (classical source coding) and active (active networking techniques) means is also explained. In order for applications to use synchrophasors, they must be transported in a standard manner. Thus, synchrophasor standards are reviewed.

12.8 Exercises

Exercise 12.1 Basics of State Estimation

1. Define state estimation as it is used in the power grid. Why is state estimation required? What problems does it address?
2. What are the two general parts of state estimation in the power grid?
3. What is a pseudo-measurement?
4. How is minimum description length similar to state estimation?

Exercise 12.2 Least Squares

Variations of the least-squares method are often used for power grid state estimation.

1. What is the key concept behind least-squares state estimation?
2. Least squares assumes that the system is overdetermined. Explain precisely what this means with regard to state estimation in the power grid, number of sensors required, and communication required.
3. What is the assumption regarding the type of statistical distribution that errors exhibit for least squares to be valid?

Exercise 12.3 Gauss–Newton

1. Explain how iteration works in the Gauss–Newton method.

Exercise 12.4 Phasor Measurement Units and Estimation

Synchrophasors have received a lot of attention as part of smart grid. Consider how they relate to state estimation.

1. The trade-off between minimizing cost and maximizing observability using PMUs can be represented mathematically. Derive an optimization problem for this trade-off.

Exercise 12.5 Communication and Estimation

1. Can state estimation as implemented in the power grid be used to compensate for communication errors?
2. How is state estimation used in communications to mitigate errors?

Exercise 12.6 Distributed State Estimation

1. What is the difference between parallel processing and distributed processing of state estimation?
2. What modifications to state estimation are required to enable distributed state estimation?

Exercise 12.7 Definition of Stability

1. What is power system stability in the most general sense of the definition?
2. How does stability relate to reliability? List several measures of stability.

Exercise 12.8 Fault Detection, Isolation, and Restoration Metrics

1. What is the system average interruption duration index (SAIDI), and how is it defined?
2. What is the system average interruption frequency index (SAIFI), and how is it defined?
3. What is the momentary average interruption frequency index (MAIFI), and how is it defined?
4. How might these indices change by the addition of improved communication and control? How do these power indices compare and contrast with communication reliability measures?

Exercise 12.9 Stability and Power Angle

1. Define "power angle."
2. How do differences in power angle impact stability?

Exercise 12.10 Types of Stability

1. What is the difference between steady-state stability and dynamic stability?
2. What is "damping power" and how does it impact power balance?
3. How can a sudden fault create instability? What conditions increase the impact of such a fault on stability?
4. What techniques are used to measure stability in the power grid?

5. What techniques are used to ensure stability in the power grid?
6. In general, how does power grid stability relate to the communication requirements necessary to maintain stability?
7. Define voltage stability? How does it differ from power stability?

Exercise 12.11 Stability and DG

1. What will be the impact of large amounts of DG upon stability? Will more DG require more communication to maintain stability? Why or why not?
2. How do microgrids address the problem of DG and stability?

13

Synchrophasor Applications

Those who say it cannot be done should not interfere with those of us who are doing it.

—S. Hickman

13.1 Introduction

This is the third chapter of Part Three on embedded and distributed intelligence for a smarter grid. In this chapter, the focus is on synchrophasors, their potential application within the power grid, and their interaction with communication. While phasors were introduced earlier, we begin by explaining precisely what synchrophasors are and how they can be used with an emphasis on communication. This includes a discussion of compression of synchrophasors. There is much anticipation in some quarters about the impending widespread use of synchrophasors, including new applications that may require their use in, or near, real time. For such applications, fast and efficient transport is a *sine quo non*. Phasor measurement units are where initial measurements are taken and synchrophasors are constructed, so there is a brief explanation of these devices. Once constructed, synchrophasors must be transmitted to the applications that will utilize them. Synchrophasor applications are numerous, and distance, frequency, delay tolerance, and number and rate of synchrophasors required may vary considerably from one application to another. This section reviews such communication networking considerations. In order for there to be widespread use of synchrophasor applications, synchrophasors must be transported in a standard manner. Thus, synchrophasor standards are reviewed. Next we consider a range of possible applications, which includes nearly all aspects of the power grid. An exhaustive list would be intractable; however, a few representative applications from different aspects of power grid operation are discussed. Two applications of synchrophasors are significant enough that they each have their own section. First, synchrophasors are a natural metric for estimating the state of the power grid. Thus, state estimation that incorporates synchrophasors is considered within its own section. Also, power system protection is another wide area of application and also has a section devoted to it. The chapter ends with a brief summary and exercises.

Smart Grid: Communication-Enabled Intelligence for the Electric Power Grid, First Edition. Stephen F. Bush.

13.2 Synchrophasors

Simply stated, synchrophasors place measurements from any portion of the power grid onto the same, absolute time base, allowing measurements to be meaningfully compared. This is done by including a timestamp with measurements, providing a method of correlating values from different locations that take different amounts of time to arrive at a common collection point. This simple technique provides a tool to view the power system more holistically and to precisely compare measurements from different points in time. The fact that these values can be synchronized with one another explains the derivation of "synchro" in the term "synchrophasor." Achieving time accuracy over wide areas of the power grid had been the challenge keeping this concept from being developed earlier. The advent of the GPS and the precise timing that it provides has been the solution allowing synchrophasors to become widely deployed and more fully developed. This is another example of communication technology, which is intimately related with timing and information transmission, helping to advance power systems for use in the power grid.

13.2.1 Phasors

The second portion of the term "synchrophasor" is "phasor." A phasor, as discussed earlier, has its origin in the term "phase vector." It is a constant vector representation of a sinusoidal function. It is assumed that the amplitude, frequency, and phase remain constant and thus can be represented as a vector. It is then possible to easily manipulate the vectors, utilizing linear combinations of the vectors, to solve problems that would otherwise be more complex, involving trigonometry and linear differential equations. An example of the sum of phasors is shown in Figure 13.1.

To be more specific, the development of synchrophasors grew from the convergence of two technologies: (1) symmetrical components used to implement protection mechanisms and (2) the accurate timing provided by GPS (Phadke and Thorp, 2006). The method of symmetrical components was discovered in 1913 by Charles Legeyt Fortescue; this method has been used as the basis for many applications within power systems ever since. The technique motivates the use of a phasor representation, as we will discuss shortly. The use of symmetrical components for protection mechanisms and their early implementation in digital microprocessors led directly to the development of the synchrophasor; thus, the synchrophasor's native application is power protection. However, it can be, and has been, applied to many other applications.

Let us first briefly review the concept of symmetrical components on our way to understanding the synchrophasor. The main idea of Fortescue's symmetrical components is that any set of unbalanced, three-phase electrical values can be expressed as the sum of three balanced vector components. This simplifies the analysis of three-phase power systems and, in particular, allows unbalanced faults to be more effectively detected and analyzed.

Consider either current or voltage represented by a phasor, which has magnitude and phase in the complex plane as shown in Figure 13.2. The magnitude is the RMS of the corresponding sinusoidal waveform, and the phase is the offset of the waveform, typically considered the distance between the peak and a reference point.

Clearly, in a correctly operating three-phase system, peaks should be 120° from one another, as the output of a properly operating generator. However, just as in communication systems,

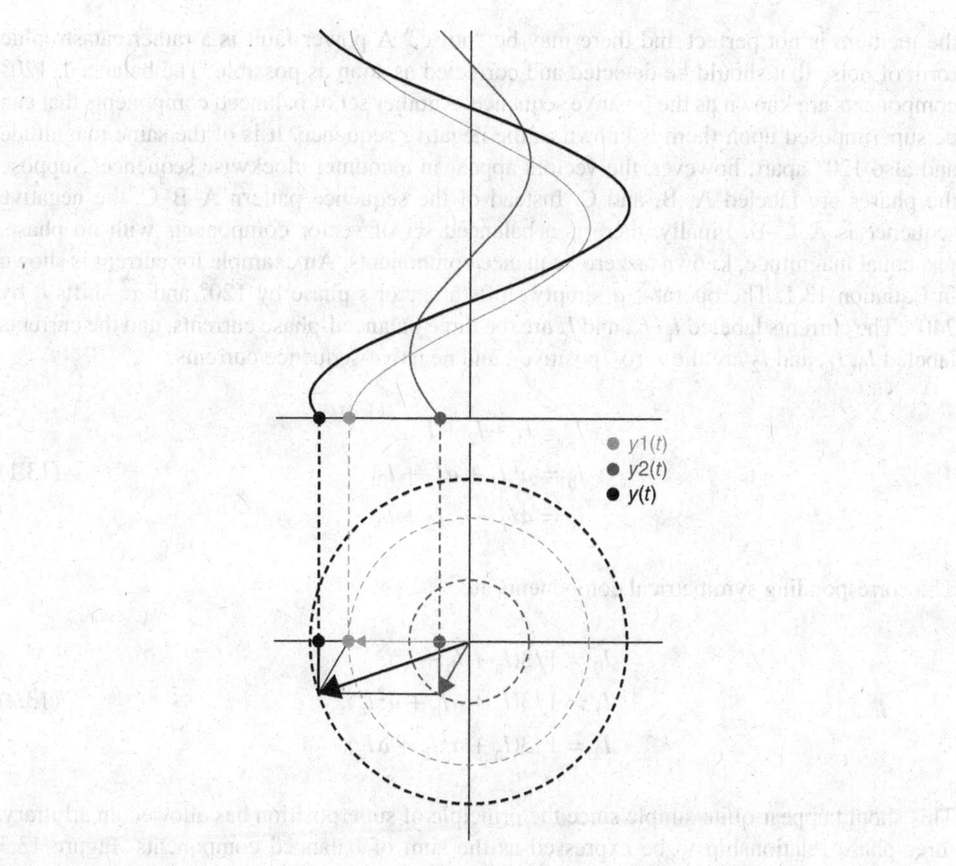

Figure 13.1 The sum of two sinusoids, y_1 and y_2, is shown in relation to their phasor representations. The lower portion of the figure shows how the vectors that represent the sinusoids add to produce the vector sum. The sinusoid of the vector sum is shown in the upper portion of the figure Source: By Gonfer (en wikipedia) [GFDL (http://www.gnu.org/copyleft/fdl.html) or CC-BY-SA-3.0 (http://creativecommons.org/licenses/by-sa/3.0)], via Wikimedia Commons.

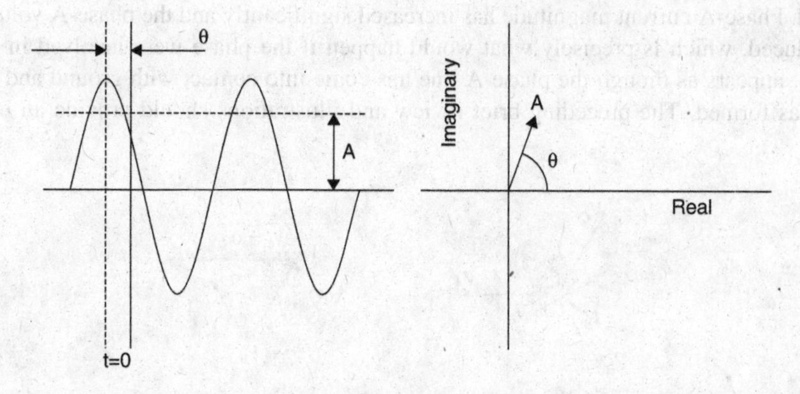

Figure 13.2 This cosine is offset to the left by θ and has an RMS magnitude of A. The corresponding phasor is shown on the right side of the figure. Source: Martin and Carroll, 2008. Reproduced with permission of IEEE.

the medium is not perfect and there may be "noise." A power fault is a rather catastrophic form of noise that should be detected and corrected as soon as possible. The balanced, 120° components are known as the positive sequence. Another set of balanced components that can be superimposed upon them is known as the negative sequence. It is of the same magnitude and also 120° apart; however, the vectors appear in a counter-clockwise sequence. Suppose the phases are labeled A, B, and C. Instead of the sequence pattern A–B–C, the negative sequence is A–C–B. Finally, there is a balanced set of vector components with no phase, just equal magnitude, known as zero-sequence components. An example for current is shown in Equation 13.1. The operator α simply shifts a vector's phase by 120° and α^2 shifts it by 240°. The currents labeled I_a, I_b, and I_c are the three balanced-phase currents, and the currents labeled I_0, I_1, and I_2 are the zero-, positive-, and negative-sequence currents.

$$\begin{aligned} I_a &= I_1 + I_2 + I_0, \\ I_b &= \alpha^2 I_1 + \alpha I_2 + I_0, \\ I_c &= \alpha I_1 + \alpha^2 I_2 + I_0. \end{aligned} \tag{13.1}$$

The corresponding symmetrical components are

$$\begin{aligned} I_0 &= 1/3(I_a + I_b + I_c), \\ I_1 &= 1/3(I_a + \alpha I_b + \alpha^2 I_c), \\ I_2 &= 1/3(I_a + \alpha^2 I_b + \alpha I_c). \end{aligned} \tag{13.2}$$

This should appear quite simple since the principle of superposition has allowed an arbitrary, three-phase relationship to be expressed as the sum of balanced components. Figure 13.3 shows a balanced system.

Figure 13.4 shows an unbalanced system caused by an open phase. The magnitude of the current on phase A has been greatly reduced, while the voltages appear to remain normal. It appears as though the line for phase A has been cut.

Figure 13.5 shows an unbalanced system in which a single phase-to-ground fault has occurred. Phase-A current magnitude has increased significantly and the phase-A voltage has been reduced, which is precisely what would happen if the phase were involved in a short circuit. It appears as though the phase-A line has come into contact with ground and a short circuit has formed. The preceding brief review and illustrations should provide an intuitive

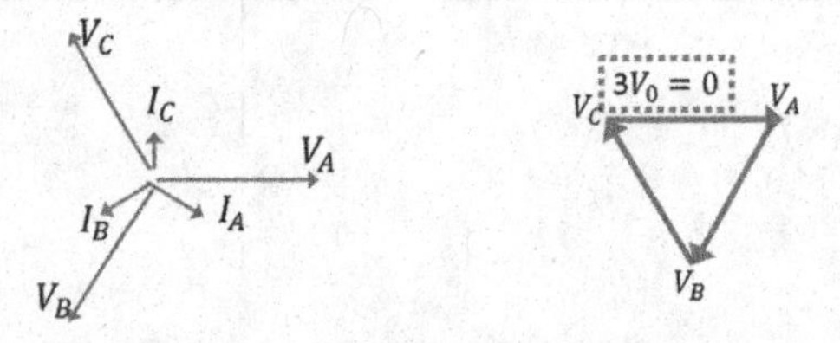

Figure 13.3 Symmetric components are shown within a balanced system. Fortescue determined that a balanced system is comprised of only positive-sequence currents and voltages.

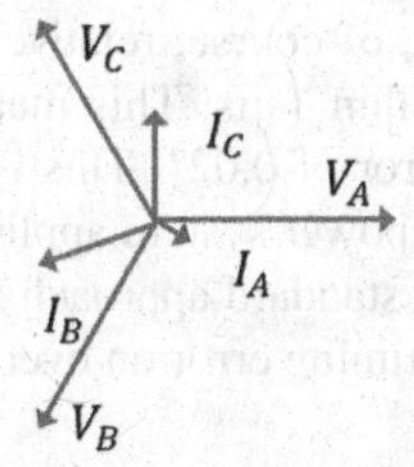

Figure 13.4 An unbalanced system is shown represented by symmetric components. Voltage and current are illustrated as solid vectors. All sequence components are nonzero. Voltages are balanced and appear normal; however, the current is abnormal. This looks like an open circuit on the B phase.

feel for how phasors have been used in protection systems long before the synchrophasor was developed.

13.2.2 Timing and Synchronization

Synchrophasors are, quite simply, synchronized phasors. The topic of phasors was covered in the previous section. Now the topic of timing and synchronization is discussed in order to complete the notion of a synchrophasor. As mentioned, the technology that is most often relied upon to provide the timing, and thus the synchronization capability, is the GPS.

The GPS was deployed in the early 1980s; it was quickly recognized that the timing provided by such a system would be both ubiquitous and precise enough to allow phasor references any time and across large distances, not just local comparison with one another at a single time instant. PMUs were initially expensive, in part due to the incomplete installation of all satellites within the GPS constellation and the need for very accurate clocks to maintain time until the next satellite came within view. However, with the completion of the GPS

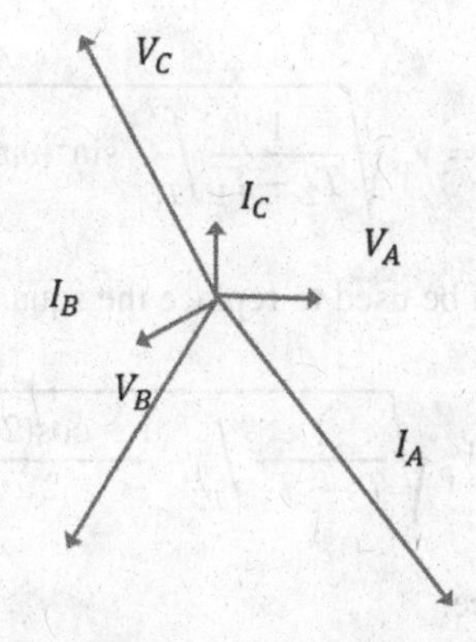

Figure 13.5 The voltage and current symmetric representations appear to be abnormal. Voltage and current are illustrated as solid vectors. There are negative-sequence voltages and currents. A single phase-to-ground fault has occurred in which voltage on phase A has been reduced and the current is excessive.

constellation, accurate timing is now, of course, relatively cheap and widely available. The accuracy of a GPS time signal is within 1 μs. This means that for a 60 Hz power system there would be a maximum phase error of 0.02°. This is well within the tolerance required to properly analyze phasors for most power system applications. Next, consider the standard definition of a synchrophasor and the standard approach for measuring error. That will allow us to better understand the impact of timing error on overall phasor accuracy.

Let us begin by defining a phasor:

$$x(t) = x_{\mathrm{m}} \cos(\omega t + \phi), \tag{13.3}$$

where X_{m} is the magnitude of the sinusoidal waveform, $\omega = 2\pi f$, where f is the instantaneous frequency, and ϕ is the angular starting location of the waveform. The point at which the waveform starts is important in comparing waveforms. Thus, ϕ defines the phase of the wave. The synchrophasor is defined relative to the cosine function, which has a value of one at 0°. This clearly defines the starting point of the waveform. Phasor notation is shown in Equation 13.4:

$$X = X_{\mathrm{m}} \angle \phi. \tag{13.4}$$

The magnitude X_{m} is the peak value. However, periodic signals often have values measured using RMS. To see how this impacts the magnitude, consider the RMS value of a sinusoidal voltage:

$$V_{\mathrm{RMS}} = \sqrt{\frac{1}{T_2 - T_1} \int_{T_1}^{T_2} (V_{\mathrm{p}} \sin(\omega t))^2 \, \mathrm{d}t}, \tag{13.5}$$

where t is time, ω is the angular frequency, and $\omega = 2\pi/T$, where T is the period of the wave. Since V_{p} is a positive constant it can be removed from the integration:

$$V_{\mathrm{RMS}} = V_{\mathrm{p}} \sqrt{\frac{1}{T_2 - T_1} \int_{T_1}^{T_2} \sin^2(\omega t) \, \mathrm{d}t}. \tag{13.6}$$

Next, a trigonometric identity can be used to replace the squared-sine term:

$$V_{\mathrm{RMS}} = V_{\mathrm{p}} \sqrt{\frac{1}{T_2 - T_1} \int_{T_1}^{T_2} \frac{1 - \cos(2\omega t)}{2} \, \mathrm{d}t}. \tag{13.7}$$

Then the integration can take place:

$$V_{\mathrm{RMS}} = I_{\mathrm{p}} \sqrt{\frac{1}{T_2 - T_1} \left[\frac{t}{2} - \frac{\sin(2\omega t)}{4\omega} \right]_{T_1}^{T_2}}. \tag{13.8}$$

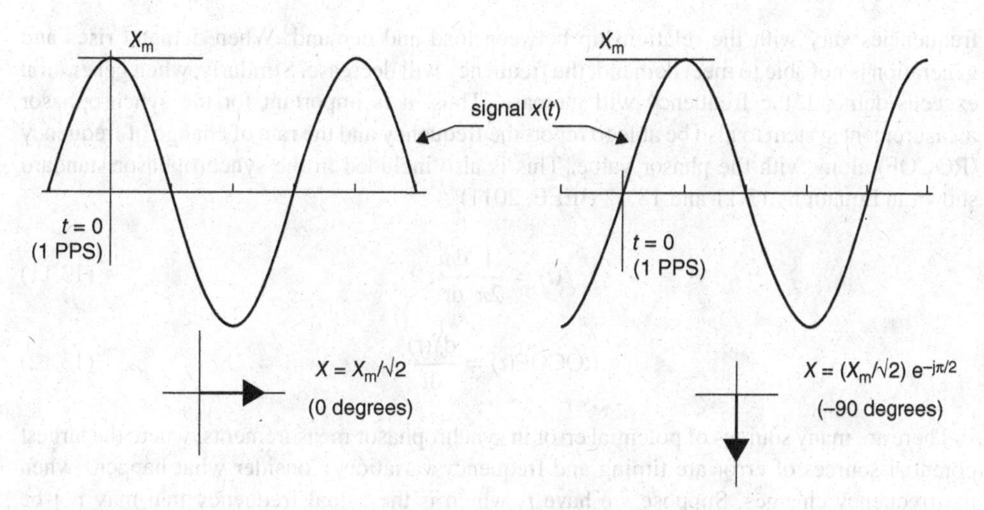

Figure 13.6 The synchrophasor representation is shown below each waveform assuming one pulse per second (PPS). Source: IEEE, 2011. Reproduced with permission of IEEE.

However, notice that the sine terms cancel because they are a whole number of full cycles:

$$V_{\mathrm{RMS}} = V_{\mathrm{p}}\sqrt{\frac{1}{T_2 - T_1}\left[\frac{t}{2}\right]_{T_1}^{T_2}} = V_{\mathrm{p}}\sqrt{\frac{1}{T_2 - T_1}\frac{T_2 - T_1}{2}} = \frac{V_{\mathrm{p}}}{\sqrt{2}}. \tag{13.9}$$

The result is that the voltage peak magnitude V_{p} has an RMS value of $V_{\mathrm{p}}/\sqrt{2}$. Thus, adding $1/\sqrt{2}$ allows equivalence between the peak and RMS representation. Thus, the synchrophasor definition is

$$X = \frac{X_{\mathrm{m}}}{\sqrt{2}}\angle\phi. \tag{13.10}$$

Given absolute time, a synchrophasor is defined as the magnitude and angle of a cosine as referenced to an absolute point in time. Two examples are shown in Figure 13.6 to illustrate the points discussed so far. The synchrophasors shown in the figure are at 0° and 90° degrees. The 0° point starts when the maximum of $x(t)$ occurs at the universal time coordinate (UTC) second rollover. Note that UTC is used in the standard for the official time reference. UTC was defined in 1963 by the International Radio Consultative Committee in Recommendation 374 and is based upon international atomic time (TAI for temps atomique international) with leap-seconds added at irregular intervals to compensate for the slowing of the Earth's rotation. The synchrophasor standard refers to a one-pulse-per-second time signal. Finally, in the second example in Figure 13.6, notice that the −90° phase occurs at the positive zero crossing of the UTC-second rollover.

It is important to emphasize that the synchrophasor is only meaningful as long as the frequency of the values under measurement is constant. However, we know that power-related frequencies vary with the relationship between load and demand. When demand rises and generation is not able to meet demand, the frequency will decrease. Similarly, when generation exceeds demand, the frequency will increase. Thus, it is important for the synchrophasor measurement system to also be able to report the frequency and the rate of change of frequency (ROCOF) along with the phasor value. This is also included in the synchrophasor standard shown in Equations 13.11 and 13.12 (IEEE, 2011):.

$$f(t) = \frac{1}{2\pi}\frac{\mathrm{d}\phi}{\mathrm{d}t}, \tag{13.11}$$

$$\mathrm{ROCOF}(t) = \frac{\mathrm{d}f(t)}{\mathrm{d}t}. \tag{13.12}$$

There are many sources of potential error in synchrophasor measurements, where the largest potential sources of error are timing and frequency variation. Consider what happens when the frequency changes. Suppose we have f, which is the actual frequency that may not be constant, and f_{nominal}, which is the assumed, correct constant, operating frequency. Let $\bar{X}$ be the actual magnitude of the phasor value. Let δt be $1/(Nf_{\mathrm{nominal}})$, where there are N samples per cycle, and δt be the time between samples. The value of $1/2$ is added to k to allow for an offset necessary to achieve a perfect centering around the sample time. The resulting phasor under these conditions from a single frequency component is

$$x(\delta t(k+1/2)) = \sqrt{2}\,\mathrm{Real}[\bar{X}\,\mathrm{e}^{\mathrm{j}(k+1/2)(2\pi/N)(f/f_{\mathrm{nominal}})}]. \tag{13.13}$$

It is possible to transform Equation 13.13 into form that looks like

$$\hat{X} = A \cdot \bar{X} + B \cdot \bar{X}^{*}, \tag{13.14}$$

where $\bar{X}^{*}$ is the complex conjugate of $\bar{X}$. Equation 13.15 shows the values of the magnitudes:

$$A = \frac{\sin\left[\pi\left(\frac{f}{f_{\mathrm{nominal}}} - 1\right)\right]}{N\sin\left[\frac{\pi}{N}\left(\frac{f}{f_{\mathrm{nominal}}} - 1\right)\right]},$$
$$B = \frac{\sin\left[\pi\left(\frac{f}{f_{\mathrm{nominal}}} - 1\right)\right]}{N\sin\left[\frac{2\pi}{N} + \frac{\pi}{N}\left(\frac{f}{f_{\mathrm{nominal}}} - 1\right)\right]}. \tag{13.15}$$

The value of this exercise is that $A \cdot \bar{X}$ and $B \cdot \bar{X}^{*}$ indicate the amount of error when the frequency changes from its nominal value. When the frequency is correct (that is, constant at its nominal value), then $A = 1$ and $B = 0$ and $\hat{X} = \bar{X}$; that is, there is no error or distortion due to a change in frequency. However, as the frequency changes from its nominal value, A changes from one and B from zero. B will have a positive value for a frequency increase and negative one for a frequency decrease. It is possible to manipulate Equation 13.14 in various ways to form measures of error as well as visualize the distortion caused by changing frequency.

The standard performance criteria for a synchrophasor is called the total vector error (TVE) shown in Equation 13.16. Here X_r and X_i are the real and imaginary parts of the theoretical, exact phasor and $X_r(n)$ and $X_i(n)$ are real and imaginary parts of the approximated phasor. Note that the TVE is not simply the magnitude of the error, but rather the magnitude normalized to the theoretically correct phasor. To provide some perspective, a 1% TVE is required in the most demanding case by the standard.

$$\text{TVE} = \frac{\sqrt{(X_r(n) - X_r)^2 + (X_i(n) - X_i)^2}}{X_r^2 + X_i^2} \times 100. \tag{13.16}$$

13.2.3 Synchrophasor Compression

The IEEE C37.118.2-2011 standard specifies synchrophasor reporting rates of from 10 to 60 frames-per-second, where a frame is defined in IEEE C37.118.2-2011. The frame typically ranges from 40 to 70 bytes-per-phasor measurement unit. Next, consider the size of the power grid and all the locations where it would be useful to place a PMU. It is easy to see the potential for a flood of information just from synchrophasors alone. Thus, development is required not only of efficient synchrophasor communication, but also of efficient synchrophasor representation and storage. As one example, the super phasor data concentrator (SuperPDC) developed by the Tennessee Valley Authority (TVA), as of 2008, had 1 billion syncrophasor measurement data points daily from the PMUs in the Eastern Interconnect (Carroll *et al.*, 2008) and 3.6 billion per day by 2012. The amount is expected to grow rapidly as the use of synchrophasors becomes more widespread.

One way to help alleviate the synchrophasor data and communication problem is to compress the synchrophasor samples. There are two general approaches to achieve compression of information: (1) lossless – that is, reduce the size of information by removing redundancy, while ensuring that no information is lost; (2) lossy – that is, reduce the size of information by removing information that appears to be of least value. The argument for lossy compression is a practical one. The assumption is that we know exactly what accuracy is required; any additional precision or accuracy adds no additional value to the information. However, the argument for lossless compression is that we may not know what applications may be developed in the future; thus, it is important to preserve as much accuracy and precision as possible. One may argue either way regarding the fact the measurements are approximations anyway. Synchrophasor samples are not perfect measurements due to varying frequency, sampling error, and other sources of noise during measurement. Since the measurements are not perfect, lossy compression may be no more noisy than the existing noise, or that we should not need to add any more noise than already exists in the samples. We can expect that both lossy and lossless approaches will be used depending upon the mindset of the developers and the particular applications developed in the future, some of which we will discuss later in this chapter.

A straightforward approach to lossy synchrophasor compression involves leveraging the well-known technique of principal component analysis (PCA) (Dahal *et al.*, 2012). The difference between the typical implementation of PCA, which assumes all data are immediately available, is that the synchrophasor data are assumed to be so large that it must be evaluated on-the-fly; that is, as the synchrophasor data stream into the system. Thus, the PCA algorithm must be able to adapt as synchrophasor data arrive.

First, we begin with a basic review of PCA. Consider data defined in a matrix $\mathbf{X}^{\mathrm{T}}$ with zero mean and where there are n rows representing a distinct set of measurements and there are m columns where each column is data from a different synchrophasor. Now note that the singular value decomposition (SVD) of $\mathbf{X}$ is $\mathbf{X} = \mathbf{WSV}^{\mathrm{T}}$, where matrix $\mathbf{W}$ size $m \times m$ contains the eigenvectors of the covariance matrix $\mathbf{XX}^{\mathrm{T}}$, the matrix $\mathbf{S}$ is an $m \times n$ rectangular diagonal matrix with nonnegative real numbers on the diagonal, and the $n \times n$ matrix $\mathbf{V}$ is the matrix of eigenvectors of $\mathbf{X}^{\mathrm{T}}\mathbf{X}$. Now consider that we know $\mathbf{X}^{\mathrm{T}}$; that is, the data from the synchrophasors. The matrix $\mathbf{W}$ contains the eigenvectors of $\mathbf{XX}^{\mathrm{T}}$, so that is known as well. Then in Equation 13.17, the SVD can be applied to obtain $\mathbf{Y}$:

$$\begin{aligned} \mathbf{Y}^{\mathrm{T}} &= \mathbf{X}^{\mathrm{T}}\mathbf{W} \\ &= \mathbf{V}\boldsymbol{\Sigma}^{\mathrm{T}}\mathbf{W}^{\mathrm{T}}\mathbf{W} \\ &= \mathbf{V}\boldsymbol{\Sigma}^{\mathrm{T}}. \end{aligned} \tag{13.17}$$

The first column of $\mathbf{Y}^{\mathrm{T}}$ results in a "score" of the fit to the first "principal" component, the next column contains the scores with respect to the "second principal" component, and so on. Note that some researchers refer to these "scores" as "energy" and also view the dimensions as "hidden variables." The number of principal components will be equal to or less than the number of variables that describe the data. The first principal component has the largest possible variance, so that it accounts for as much of the variability in the data as possible. Each succeeding component has the highest variance possible under the constraint that it is orthogonal to, and thus uncorrelated with, the preceding components. Lossy compression is implemented by removing the lowest components; that is, the components that have the least impact upon the data. Suppose that only the first L components are desired to be retained. This means that the data $\mathbf{X}$ will be projected onto the reduced space defined by only the first L singular vectors $\mathbf{W}_L$ as shown in

$$\mathbf{Y} = \mathbf{W}_L^{\mathrm{T}}\mathbf{X} = \boldsymbol{\Sigma}_L\mathbf{V}^{\mathrm{T}}. \tag{13.18}$$

In this equation, $\boldsymbol{\Sigma}_L = \mathbf{I}_{L \times m}\boldsymbol{\Sigma}$, where $\mathbf{I}_{L \times m}$ is the $L \times m$ rectangular identity matrix.

The matrix $\mathbf{W}$ of singular vectors of $\mathbf{X}$ is equivalently the matrix $\mathbf{W}$ of eigenvectors of the matrix of observed covariances $\mathbf{C} = \mathbf{XX}^{\mathrm{T}}$:

$$\mathbf{XX}^{\mathrm{T}} = \mathbf{W}\boldsymbol{\Sigma}\boldsymbol{\Sigma}^{\mathrm{T}}\mathbf{W}^{\mathrm{T}}. \tag{13.19}$$

Given a set of points in Euclidean space, the first principal component corresponds to a line that passes through the multidimensional mean and minimizes the sum of squares of the distances of the points from the line. The second principal component corresponds to the same measure after all correlation with the first principal component has been subtracted from the points. The singular values (in $\mathbf{S}$) are the square roots of the eigenvalues of the matrix $\mathbf{XX}^{\mathrm{T}}$. Each eigenvalue is proportional to the portion of the "variance" (more correctly of the sum of the squared distances of the points from their multidimensional mean) that is correlated with each eigenvector. The sum of all the eigenvalues is equal to the sum of the

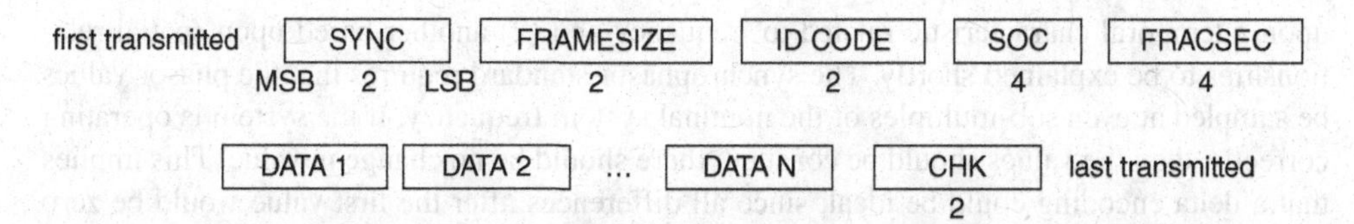

Figure 13.7 The IEEE standard for synchrophasors (IEEE Std C37.118-2005) defines a frame comprised of a frame sync word (SYNC), bytes in the frame (FRAMESIZE), PMU ID number (ID CODE), timestamp (SOC), fraction-of-second and time quality (FRACSEC), the actual data (DATA 1, ..., DATA N), and 16-bit CRC (CHK). Source: Martin, 2006. Reproduced with permission of IEEE.

squared distances of the points from their multidimensional mean. PCA essentially rotates the set of points around their mean in order to align with the principal components. This moves as much of the variance as possible (using an orthogonal transformation) into the first few dimensions. The values in the remaining dimensions, therefore, tend to be small and may be dropped with minimal loss of information. PCA is often used in this manner for dimensionality reduction. PCA has the distinction of being the optimal orthogonal transformation for keeping the subspace that has largest "variance," as defined above. This advantage, however, comes at the cost of greater computational requirements if compared, for example, with the discrete cosine transform (DCT), and in particular to the DCT-II, which is simply known as the "DCT."

PCA is sensitive to the scaling of the variables. If we have just two variables and they have the same sample variance and are positively correlated, then the PCA will entail a rotation by 45° and the "loadings" for the two variables with respect to the principal component will be equal. But if we multiply all values of the first variable by 100, then the principal component will be almost the same as that variable, with a small contribution from the other variable, whereas the second component will be almost aligned with the second original variable. This means that whenever the different variables have different units (like temperature and mass), PCA is a somewhat arbitrary method of analysis.

A typical standard synchrophasor frame is shown in Figure 13.7 in which there is a frame sync word, bytes in frame, PMU ID number, timestamp (Second-of-century), fraction-of-second, and time quality (clock accuracy), and a 16-bit CRC.

A lossless synchrophasor compression technique, taken from image compression, is explored in Klump *et al.* (2010). Compression of arbitrary bit-sequences is a mature topic and there are many well-known techniques on this topic. The benefit to synchrophasor compression is determining correlation that will help improve compression for the specific case of synchrophasors. The technique to accomplish this is called Slack Reference Encoding (SRE) (Klump *et al.*, 2010).

Suppose there is a list of measurements M_i as shown in

$$M_i = [M_{i,1} \cdots M_{i,q}], \tag{13.20}$$

where each M_i is a vector of length q of measurements. The goal is to understand what characteristics of M, with regard to synchrophasors, can increase compression of the measurements. There are two specific characteristics capitalized upon to increase compression: (1) one based upon a temporal characteristic related to frequency and (2) another based upon spatial relationships to be explained shortly. The synchrophasor standard requires that the phasor values be sampled at even sub-multiples of the nominal system frequency. If the system is operating correctly, then the values should be constant; there should be no change in value. This implies that a delta encoding could be ideal, since all differences after the first value would be zero in the ideal case. Of course, as we discussed, there are numerous sources of noise; the values will not be precisely zero in a real system.

The second characteristic was spatial, which relates to how electrically close the measurements are to one another. Furthermore, in an ideal system, if generation, topology, and load remain constant, then the angles, although they will differ from one another on various electrical buses, will remain constant. This points toward a differencing algorithm in which only the relative angle differences are required and the rest of the angle differences will be zero. But no system is ideal, and variations will exist in a real system. Note that the notion of choosing a slack bus in power flow analysis is commonly used as discussed in a previous chapter.

The SRE algorithm is quite simple; it combines the two characteristics just described into one algorithm by choosing a reference signal and taking the difference between the last measurement and the current measurement for all values and taking the difference between that value and the difference between the last and current measurement of the reference signal. Thus, there are two differences being computed: the difference in time between the current and last measurement, and the difference in electrical distance between the current signal and the difference in time of the reference signal.

13.3 Phasor Measurement Unit

The PMU is the sensor used to detect, create, and report a synchrophasor value. The main hardware components are current and voltage sensors to sample both current and voltage, a unit to supply accurate, absolute timing, and a processing unit to provide a means of creating and operating upon the data. As we have seen, the primary purpose of the PMU is to measure a waveform with reference to an absolute time so that the waveform can be compared, in a meaningful way, with other waveforms. A PMU may be a standalone device or integrated with other power system devices. As technology progresses, it is likely that the PMU will become a standard part of many power grid devices. The PMU samples power at the nominal frequency of 50 or 60 Hz and reports on current and voltage at a typical rate of 48 samples-per-cycle with 1 μs accuracy. The output of the PMU is a stream of time-tagged synchrophasors at a rate of up to 60 samples-per-second. The synchrophasor samples may be used locally or transmitted to a central location for a particular region of the power grid. The value of a synchrophasor becomes evident over a wide area. This is because the same result could be obtained with local-area measurements without synchrophasors. In other words, waveforms that can be sampled simultaneously and compared locally do not require a common time. However, when such information is used over a wide area, the absolute timing of synchrophasors becomes a necessity, assuming that values will experience variable delay in transmission. In some sense, synchrophasors are a solution to the communication problem of variable latency. As the PMU becomes ubiquitous, while also transmitting samples at high rates, there is a growing concern about overloading the communication network, particularly in applications that rely upon real-time or near-real-time control using synchrophasors.

13.3.1 Phasor Data Concentrator

In the previous section, the means of creating synchrophasors via a PMU was explained. A single synchrophasor value is a snapshot of the current or voltage at a specific point in space and time; it is a single value from a specific point in the power grid at a specific time. Often, a value only becomes meaningful when compared with other synchrophasor values over space and time; that is, as part of a trend from the same location or in comparison with synchrophasor values from other locations. Aggregation and comparison of synchrophasors takes place in another hardware unit known as a PDC. It should be noted that the PDC is not necessarily a standalone device or software, but rather is functionality that may be integrated wherever required.

One or more PMU devices feed synchrophasor data into a PDC and multiple PDC units may be interconnected (IEEE, 2011). The concept behind the PDC is to correlate multiple streams of synchrophasor input to create a single, aggregate, output that can then be fed directly to applications or may be fed to other PDC devices for further aggregation with other streams. Since the output streams are time-aligned, aggregation of each input stream adds a broader view of the power grid at each point in time. There may be a hierarchy of PDC units, with local PDC units aggregating streams directly from power grid devices, mid-level PDC units aggregating streams from local PDC units within a region, and higher level PDC units aggregating regional streams. The high-level PDC units are called "SuperPDCs" and can include synchrophasor information from an entire power grid interconnection. Each layer in the hierarchy, since it encompasses an area of different size, may have different expectations regarding latency, coverage, and quality of the synchrophasor data.

13.4 Networking Synchrophasor Information

Various network architectures for the communication network for synchrophasor data have been explored. The communication network must provide the interconnection for the flow of synchrophasor data from PMUs to PDCs and synchrophasor applications. There have been a few general approaches. One approach has been to develop middleware, an intermediate, software layer built upon the communication network to specifically handle the unique needs of synchrophasor traffic. Another approach has included the concept of "stacking" or "chaining" PDC units within the network. In this approach, the PDC is treated as part of the network and synchrophasor traffic passes through PDC units on the way to its destination, which should call to mind active networking introduced in Section 6.5.2. Another approach is to utilize the basic IP and Internet multicast in particular (Myrda *et al.*, 2012). IP multicast is a subscription-based approach in which nodes that wish to receive information from a multicast source must subscribe to the source node. The multicast protocol is designed to run over IP and efficiently scale with the number of subscribers and efficiently handle errors in the multicast transmission.

From the discussion in Section 10.6 which covers synchrophasor standards in more detail, there are three general message formats designed to handle synchrophasor traffic: (1) IEC 61850 GOOSE (2) C37.118-2005, and (3) IEC 61850-90-5 Routable GOOSE. The first, IEC 61850 GOOSE messages, are typically placed directly within Ethernet frames. Thus, they are restricted to Ethernet LANs or potentially Ethernet virtual LANs. They lack the ability for general IP routing and so cannot be transmitted throughout the Internet. The latter two message formats, C37.118-2005 and IEC 61850-90-5 Routable GOOSE, may be encapsulated within IP packets and used in IP multicast transport.

Typically, synchrophasor data in IP-routable message formats use either UDP or TCP for transport. These protocols are predefined, point-to-point, transport protocols. Thus, they require that messages are sent to only one destination node. Then, as we have seen, the destination PDC must aggregate the synchrophasor data from multiple incoming streams. One challenge with this operation is that the PDC must correlate the synchrophasor data by time from each of the multiple, incoming streams. This means that if any of the streams are slow in delivering their messages, the PDC must delay processing until the slowest message arrives. Because each IEEE C37.118 message has a single timestamp for all the data in the message, the entire message must arrive before processing of the message can begin. This also can cause a delay if information from part of a message needs to processed and transmitted to another receiver. In other words, the message must be completely parsed and broken down in order to obtain the timestamp required for each element of the message. These delays, both the need to wait for the slowest input and the need to completely parse a message and find its timestamp, can cause highly variable latency in synchrophasor traffic flow through the network.

Thus, stacking or daisy-chaining of PDC units as well as forwarding delays in middleware approaches all add to overall delay, stochastic latency, and jitter which could easily make a control system based upon synchrophasors unstable. An alternative architecture is to utilize the IP multicast as it would normally be used in the Internet. That is, move as many processing elements as possible, such as PDC units and synchrophasor applications, to the edge of the network and place synchrophasor messages within IP multicast packets destined for each of the processing elements. In this case, each PMU becomes a multicast source. IP multicast handles the formation of a tree structure that is created as each subscriber, such as a PDC, subscribes to messages from the PMUs that should stream into it.

The ongoing development of the synchrophasor network echoes the often-repeated research involved in the discussion on intelligence in networking versus the end-to-end principle from Section 11.2.3. The approach just discussed is an example of the relatively simple network and complex edge approach. An example of the opposite approach can be seen in the next architecture (Arya *et al.*, 2011). In this approach, intelligence is added to the network such that aggregation is specifically performed within the network and a new, three-layered network architecture is proposed. The bottom layer provides QoS, the middle layer provides a wide-area publish–subscribe mechanism, and the top layer provides a distributed processing environment for application-specific, in-network aggregation. One advantage of this approach is that careful consideration is given to how network congestion should be handled. In this case, it is handled in an application-specific manner, using techniques leveraged from video quality. Namely, a means is provided so that when congestion occurs, data of least value can be dropped so that impact on quality of the overall system is minimized.

An example to illustrate the concept comes from maintaining voltage stability, although it is important to keep in mind that any power application could be implemented in a similar manner. Voltage stability involves monitoring the power grid in order to determine what needs to be done to keep voltage at a healthy level throughout the grid. As we have seen in other chapters, voltage and reactive power are tightly coupled, and often the loss of reactive power can lead to voltage collapse. There are many indices and algorithms for attempting to estimate voltage stability. Also, it should be noted that there are several types of stability in the power grid; voltage stability should not be confused with power flow stability, for example. One form of the voltage stability index (VSI) is defined in Equation 13.21 for bus i:

$$\mathrm{VSI}_i = \frac{\partial P_i / \partial \delta_i}{\sum_{j=1,j!=i}^{n} B_{ij} V_j}. \tag{13.21}$$

In this definition, n is the number of buses, P_i is the real power injected at bus i, V_j is the magnitude of the voltage at bus j, δ_i is the phase angle of the voltage at bus i, and B_{ij} are elements of the network admittance matrix. The basic idea is that if real power is decreasing as a function of voltage phase angle relative to all the other buses in the system, then it is likely that the voltage at that bus is becoming unstable. Thus, a low value of the VSI indicates greater potential instability. The VSI should be checked for each bus in the system and the bus with the lowest value is the VSI for the system. A VSI greater than 0.5 is considered stable, while a value below 0.5 indicates a possible voltage collapse.

Information required to compute the VSI, as defined above, can be obtained from a PMU. The simple network approach is to design the network such that it is, at most, capable of transmitting or multicasting each message to an end-point or edge node of the network, letting the edge node perform the computation needed to compute the solution. The alternative is to provide some knowledge of the application to the network and enable the network to make decisions about how the information should best be handled. For example, QoS might be improved by allowing for message prioritization of the data from the PMUs. This is relatively simple and can be based upon the voltage magnitude. Voltage magnitudes that are very low or that originate from locations that are known to be sensitive indicators of voltage collapse should have priority in transport through the network. If the network is becoming highly congested, values with little change from the past value and little impact on sensitivity of the calculation can be dropped. In essence, the network is implementing a form of source coding or lossy compression on the information.

A more interesting and sophisticated technique comes from the active networking concept. This involves allowing partial computation within the network. For example, instead of sending only raw data, the PMU or other computational elements within the network perform partial computation on the data before forwarding it. This involves being able to decompose the primary function into smaller functions that can be computed in a distributed manner. For example, Equation 13.22 shows a partial summation for bus i:

$$\frac{1}{\mathrm{VSI}_i} = \frac{\sum_{j=1,j!=i,j\in A}^{n} B_{ij} V_j}{\partial P_i / \partial \delta_i} + \frac{\sum_{j=1,j!=i,j\notin A}^{n} B_{ij} V_j}{\partial P_i / \partial \delta_i}, \tag{13.22}$$

where A is a partition within the bus network. The function can be easily distributed by partitioning the power grid network into multiple sections and computing the partial function within each section. Partial values can then be summed as they are transported on their way to their final destination within the network.

13.5 Synchrophasor Applications

This section provides an overview of applications of synchrophasors. There are numerous applications, including substation voltage and current measurement, SCADA verification and backup, wide-area frequency monitoring, improved state estimation, wide-area disturbance recording, DG control, system black start, and protection. An exhaustive list of synchrophasor applications would be intractable since synchrophasors have been suggested for use in almost all grid applications. The analogy has been made that synchrophasors are "like the MRI of the bulk power system" (Schweitzer *et al.*, 2010). The goal of this section is to review a few selected applications in order to provide insight into how they can be used. Synchrophasors were initially suggested for use in non-real-time applications; that is, where synchrophasor data are collected from a wide area and then processed for post-event analysis and visualization. With the advent of more efficient power grid communication, it will become possible to utilize synchrophasors in real-time, or near-real-time, applications.

Continuing with the application of voltage stability, since voltage is closely related to reactive power, a simple indicator of voltage stability is a direct measure of the amount reactive power in each bus, normalized in some manner or determined as a marginal value required for a specific system. An example of this is the incremental reactive power cost IRPC:

$$\mathrm{IRPC}_j = \sum_{k=1}^{n} \frac{\Delta Q_{\mathrm{gen}_k}}{\Delta Q_{\mathrm{bus}_j}} \tag{13.23}$$

Here, IRPC_j is the value of the voltage stability metric for bus j, where n is the number of reactive power sources, $\Delta Q_{\mathrm{gen}_k}$ is the change in the kth generator reactive power output for a small change in reactive load at the bus, and $\Delta Q_{\mathrm{bus}_j}$ is the change in reactive power load at the bus. The summation is over all the reactive power generators. The result is the marginal reactive power needed to feed the bus. As IRPC_j becomes large for a particular bus, it is likely to be approaching a voltage collapse. Again, all this information can be collected from synchrophasor data at a PMU and sent to a control center where the metric can be computed and action taken if necessary.

Power stability, another application, can be easily determined from synchrophasor data because it typically involves looking at the difference in power angles throughout the power grid and determining partitions of the grid where the angles may be so far apart as to cause portions of the grid to become unstable. A similar approach, applied on a smaller scale, within the distribution system can help solve the problem of islanding. Given the growing use of DG within the distribution system, it is important to determine whether an isolated portion of the grid has become disconnected. This may not be obvious, since the island is providing its own power and devices within the island may not recognize that a disconnection from the main power grid has occurred. One technique to detect this is to utilize the synchrophasor angle measurement. Let δ_k be the voltage angle at time k. These values will be measured by PMUs located at the distributed generators within the potentially islanded area and at the interface with the main power grid. In this case, consider two voltage angle measurement locations: $\angle V_k^{(1)}$ can be taken at a recloser or relay located near a DG source and $\angle V_k^{(2)}$ can be taken near the main power grid at the interface to the potentially islanded area. Assume that the synchrophasors are being sampled at rate MRATE. Then the voltage difference between locations is

$$\delta_k = \angle V_k^{(1)} - \angle V_k^{(2)}, \tag{13.24}$$

the slip in frequency S_k is

$$S_k = (\delta_k - \delta_{k-1})\text{MRATE}, \tag{13.25}$$

and the acceleration A_k is

$$A_k = (S_k - S_{k-1})\text{MRATE}. \tag{13.26}$$

Thus, the idea is to simply measure the rate and acceleration of change in the difference of the voltage angles assuming that a significant change in difference indicates a separation in control. When the island is connected to the power grid and everything is working perfectly, there will be little or no voltage difference and the slip frequency and acceleration with both will be zero. However, it is possible for there to be a moderate amount of positive slip frequency as long as there is negative acceleration to bring the slip frequency back toward zero. Similarly, there may be a negative slip frequency as long as there is positive acceleration to bring the slip frequency back toward zero. This slack in the relationship between slip frequency and acceleration is required in order to reduce false islanding alarms. However, if either the slip frequency, acceleration, or both, become too large, then that is a clear indication of islanding.

An indirect approach is shown in Equation 13.27 where change in frequency from its nominal value f_{NOM} is used as an indicator:

$$\text{TE}_k = \text{TE}_{k-1} + \frac{1}{f_{\text{NOM}}}(f_k - f_{\text{NOM}})\Delta t, \tag{13.27}$$

where TE is called the time error measurement. Since frequency is the derivative of phase, this technique is indirectly measuring phase difference.

As mentioned, synchrophasors are enabling measurements to be accurately made over wider areas than were possible before GPS timing became available. It should be clear that synchrophasors can and will be used in many different applications throughout the power grid. The concern will be the ability to rely upon communication to have the availability, accuracy, and low latency to directly utilize synchrophasor results for critical control situations.

13.6 Summary

The unique aspect of synchrophasors is their ability to be referenced to a common time. This allows phasor readings from across the power grid to be directly compared and analyzed with one another. Because this had not been possible in the past, applications typically left phasor angles open or as slack variables to be filled in with an arbitrary reference value. Thus, synchrophasor application developers are seeking to develop applications that take advantage of wide-area communication to bring together synchrophasor readings from spatially diverse areas of the power grid. Synchrophasor applications are wide ranging; a few representative example applications were discussed in this chapter.

Given the expectation for a large number of synchrophasor applications, some of them real time, synchrophasor communication can become a challenging problem. Numerous aspects of synchrophasor communication were considered, including phasor data concentration, compression, and active networking techniques that process synchrophasors as they travel through the network.

Chapter 14 introduces power system electronics. The bottom line is that the smart grid will only be as good as its underlying components, and those components are largely implemented with high-power electronics. As will become apparent, these electronic technologies are transitioning power from an analog to a digital product. This is having a significant impact on the operation of the power grid and on its communications infrastructure. Advances in power electronics will impact both the smart grid and its supporting communication. Advances in high-power solid-state electronics along with communication will enable the power grid to operate in a more efficient and flexible manner. A comprehensive review of power electronic advances in flexible alternating-current transmission systems and the solid-state transformer is given. The ability of advanced power electronics to increase controllability and provide advanced features is highlighted. The manner in which the field of high power electronics will impact communication is discussed, both from the perspective of their communication requirements and their potential to support new forms of communication. Superconducting technology offers new capability to power grid components that may be leveraged not only for power system efficiency but also for new communication and computation technologies such as quantum computation within the power grid. Thus, Chapter 14 provides a look into the future of power systems and smart grid.

13.7 Exercises

Exercise 13.1 Phasor

1. What is a phasor?
2. What three components of a sinusoid are assumed to remain time invariant in a phasor?
3. Does the standard definition of phasor use peak amplitude or RMS?

Exercise 13.2 Synchrophasor

1. What is a synchrophasor?
2. What is the official absolute time to which all synchrophasors are referenced?
3. Precisely when is the synchrophasor angle defined to be 0°?
4. To which sinusoidal function is the standard synchrophasor referenced?

Exercise 13.3 Phasor Measurement Unit

1. What are the main components of a PMU?
2. How is the sliding window approach used to help keep phasor measurement stable?
3. What are the typical reporting rates of a PMU according to the standard for 60 Hz systems?
4. What is a typical frame size?

Exercise 13.4 Error Measurement

1. How is synchrophasor error measured according to the standard?
2. What is the sensitivity of the synchrophasor to GPS timing error using the standard measure of error?

Exercise 13.5 Synchrophasor Compression

1. What is the general two-step approach to synchrophasor compression discussed in this chapter? How is the first step designed to aid in compression?
2. How can temporal and spatial correlation be used to aid in synchrophasor compression?

Exercise 13.6 Phasor Data Concentrator

1. What is a PDC?
2. How does a PDC differ from a PMU?
3. How can a PDC reduce the network synchrophasor traffic load?

Exercise 13.7 Real-Time Synchrophasor Networking

While much can be done with synchrophasors offline, consider the following questions in the context of real-time synchrophasor applications.

1. Latency is a significant issue in operating a real-time synchrophasor network. What are some of the main causes of variation of latency in communication networks?

Exercise 13.8 Synchrophasor Timing

Along with latency, accurate timing is a critical requirement for real-time synchrophasor networking.

1. Explain the operation of the IEEE 1588 time distribution protocol.
2. Does IEEE 1588 solve the problem of asymmetric channels?

Exercise 13.9 Synchrophasor Network Architecture

1. Why does synchrophasor message aggregation require a PDC instead of applying simple message aggregation within the network?
2. What are the output formats of a PMU?
3. What are the benefits and trade-offs of PDCs being within the network instead of edge nodes? How does this relate the active network architecture versus the end-to-end network principle discussed in Section 11.2.3?

Exercise 13.10 Synchrophasors and Stability

1. What assumption regarding compression is made when monitoring rotor angle using synchrophasors for stability? What does the synchrophasor measure? What generator component is involved in stability?

2. Using the synchrophasor stability algorithm discussed in the text, what impacts the synchrophasor bandwidth requirements?
3. What are the actions taken to restore stability?

Exercise 13.11 Synchrophasors and State Estimation

1. What values are used in conventional (non-synchrophasor) state estimation?
2. What value do synchrophasors add to conventional state estimation?
3. What is a pseudo-measurement?
4. Explain a simple way to include synchrophasors into state estimation.

14

Power System Electronics

Ampère was the Newton of Electricity.

—James C. Maxwell

14.1 Introduction

This is the fourth chapter in Part Three on embedded and distributed intelligence for the smart grid and introduces advances in power electronics that will impact both the smart grid and its supporting communication. Adding communication to the existing power grid can only yield a limited amount of efficiency because the operational components of the grid are themselves inflexible and of limited efficiency; advances in high-power, solid-state electronics along with communication will enable the power grid to operate in a more efficient and flexible manner. As a simple example, large, inductive losses within the coil of an old-fashioned transformer can be significantly reduced by power electronics and power electronics allow renewable generators and the power grid to adapt to one another rapidly.

The chapter begins with a brief introduction to power electronics assuming no prior knowledge. In Section 14.2, the first application of high-power, solid-state electronics discussed is FACTS. This system allows power flow over individual transmission lines to be controlled, which enables efficient and adaptable power transport. This is followed in Section 14.3 by the solid-state transformer. As previously mentioned, the solid-state version of the transformer removes many of the worst inefficiencies. In a more general sense, the solid-state transformer can be viewed as a power router, connecting the medium-voltage grid system with many lower voltage customers. However, the solid-state transformer contains an internal direct-current bus that can directly connect to renewable power generators, such as PV systems. The inverter within the solid-state transformer can convert direct current from customer renewable sources into alternating current for other customers. Multiple, interconnected solid-state transformers can form an electric energy routing analogous to the way an IP router transports protocol data units. Solid-state power electronic devices operate in a significantly different manner from classical inductors and capacitors. Thus, communication through and with such devices may reflect this change in operation, perhaps better utilizing their native characteristics. An example of this is using pulse-width modulation not just for shaping waveforms, but also simultaneously for communication. The phasor measurement unit was discussed in Chapter 13; however, here we have a section devoted briefly to solid-state approaches. Protection mechanisms will also benefit from advances in solid-state electronics. Section 14.4 focuses on current limiting in particular. The idea behind current limiting is to protect power systems from electrical faults without completely disengaging power to the circuit and disrupting the entire system. In other words, current flow is simply limited to a safe value. The final topic in this chapter discusses superconducting power cables. These are very low-loss power transmission cables. This section begins to touch upon utilizing unique quantum effects for the power grid. This general topic is picked up in more detail in Chapter 15. The chapter ends with a brief summary of the main points and exercises for the reader.

Because the focus of this book is on communication for the power grid, the reader may be wondering why there is a chapter on power devices and advances in power electronics. The reason is that communication is one part of the evolving power grid and must interact with all parts of the power grid as they evolve. The physics and principles of operation of the power grid will also be advancing and changing along with communication. Thus, in order to remain relevant and valuable, communication experts need to anticipate advances in power electronics that are likely to occur within the next 10–20 years. The fundamental, operational paradigm in which power electronics directly handles power is moving from analog to digital; power flow itself is moving towards switched, high-speed, on–off operation, analogous to digital bits. Understanding this trend will enable communication to fit more efficiently with and into the components of these evolving devices. The goal of this chapter is to provide a very brief, simple introduction to power electronics such that a communication engineer can anticipate how to manage coming advances in the power grid.

Recall that, long before the notion of the smart grid, communication was always implicit in power systems, even when no communication equipment was explicitly used; the power grid physics itself was used as a communication medium. Devices have been configured to detect and react to changes sensed locally, although initially induced from spatially distant locations. As we have already seen in preceding chapters, more communication is being used to explicitly aid the operation of the power grid, allowing changes in one location of the grid to be detected and sent to other locations before a noticeable physical perturbation propagates through power components. Advances in power electronics are moving from passive capacitive and inductive power components to more active operation, reminiscent of active networking in communications. This changes the nature of monitoring and control of information and may perhaps open the door to new types of communication in the power grid. One possibility is leveraging the connection between superconducting power cables and superconducting quantum communications.

Advances in power electronics are changing the manner in which communication will be used. Power electronic devices are discrete, with many more dimensions of control than their previous, analog counterparts. In other words, devices are becoming more flexible and capable of accomplishing more tasks. There is also increasing resolution along the dimensions in which power can be controlled and transformed. This implies that communication bandwidth will need to be greater and reaction times will have to be faster.

One of the first, primary applications of power electronics were power converters, devices that transform alternating current to direct current or direct current to alternating current. These devices are one of the important components of DG and microgrids because they typically interface renewable energy to the power grid. Converting alternating current to direct current is a means of transferring large amounts of power over long distance safely and without interrupting grid stability. Increasingly "high-bandwidth" bidirectional power converters will be required for the so-called prosumer, who both produces and consumes power from the grid, as well as for energy storage devices. Converters are also a key building block in many other power grid devices. The solid-state transformer will change the operating principal of one of the most ubiquitous grid components, the transformer, enabling greater efficiency and reduced weight, volume, and cost. There will be a significant impact on the passage of power line carrier as well, as one example of the relationship between power electronics and communication. Power electronics will enable high-resolution control of power so that power quality can be tightly controlled in power conditioning systems (PCSs). Voltage sags and flicker, as well as harmonics, can be removed, allowing clean, uninterrupted power for power-quality-sensitive consumers, such as data centers and semiconductor chip manufacturers, as well as in extending the life of heavy motors and other electronic equipment. FACTS are power electronic components capable of controlling power flow in transmission lines. FACTS devices can thus direct flow as required to meet regulatory and business needs and they can reduce the amount of loop or circulating current flows. FACTS devices can increase the transmission capacity of existing power lines, thus saving money. Finally, FACTS devices can aid in grid stability and dampen inter-area oscillations. Power electronics can also provide more efficient control of reactive power support, particularly when generators are stressed. Power electronics also play a role in many other aspects of grid operation; for example, low-voltage ride-through, the ability of renewable energy devices to continue operation when they must be taken off the grid due to low voltage or an apparent fault. Detecting islanded operation is another capability of power electronics. It is interesting to note that some of the problems being addressed today via communication and control may be solved by advances in power electronics; so again, it will be important for communication experts to be aware of advances in power electronics. Power stabilizers can be used to dampen the impact of oscillations, due to transmission line faults or other perturbations (Power Engineering Society, 2003; Yang *et al.*, 2010). Energy storage devices typically utilize direct-current input and direct-current output, and thus conversion from and to alternating current is required by power electronics. Power system stabilizers have been suggested to smoothen renewable DG. Advances in power systems are also leading to new forms of fault isolation devices for distribution systems (Vodyakho *et al.*, 2011); namely, solid-state circuit breakers that can open and close with extreme speed and efficiency. Advances in power electronics have engendered the notion of a new communication technology: power electronic signaling. The concept addresses how to pass communication and control signals through advanced power electronics devices utilizing new modes of operation that power electronics switching provides (Xu and Wang, 2010). Finally, power electronics are used directly within industrial and consumer products to increase efficiency and extend the lifetime of equipment. It is clear that power electronics play a large role in today's power grid and their role will continue to increase as new advances are made.

14.2 Power System Electronics

As power systems electronics technology advances, more of the electric power grid will be operated by power electronic devices. These are the devices that smart grid communication will be interfacing with, supporting, and controlling. Thus, it is important to have at least a cursory understanding of their evolution, operation, and future evolution. As has been noted throughout this book, communication and power electronics are more similar in many respects than is often assumed. Both are transporting a product, in one case information and in the other power. This has driven communication circuitry to become focused on greater complexity and reducing power consumption in order to efficiently modulate, encode, and transmit information. Communication signals thus tend to be lower power and have higher entropy. Power electronics, on the other hand, has focused on transmission of larger amounts of power more efficiently while keeping the signal clean and consistent.

The focus on efficiency for power electronics has driven it in a similar direction as digital communications; namely, a switched, fully-on or fully-off mode of operation. A power device that is completely on or completely off consumes little power, while one that is partially on consumes power that typically results in heat buildup. Thus, rather than creating analog power devices that vary through a continuum of values, power-efficient devices switch rapidly between an on and off state such that the rate of on and off states has the desired average value. Whereas communication engineers desire a signal that is easily detectable and switches rapidly, a power engineer wants the final result to average out to the appropriate analog voltage and current waveforms.

Power electronics technology begin in 1901 when Peter Cooper Hewitt invented the glass-bulb, mercury-arc rectifier (Bose, 2009). The period from approximately 1930 to 1947 saw the development of gas-filled tubes that contained a control grid, or wire mesh separating cathode from anode, in order to control electric power. This included the gas-filled thyratron.

Meanwhile, the semiconductor field was born. The field-effect principle, first disclosed in a patent in 1930, is the ability to modulate the conductivity of a semiconducting resistor by a voltage or electric field applied perpendicular to the length of the resistor. In 1947 the transistor was demonstrated by Walter H. Brattain and John Bardeen with William Shockley as an intensely interested observer. Semiconductor electronics took off in 1948 with the invention of the bipolar junction transistor.

It was not until 1950 that semiconductor power electronics began with simple diodes. In 1957 the silicon controlled rectifier or thyristor was first developed. This is when modern power electronics began; thyristors are still in use today. In 1964, power semiconductor electronics first began to move beyond the diode and to apply a solid-state transistor approach, suggesting a field-effect transistor (FET) for use in power systems in papers by Zuleeg and Teszner. The obvious approach was to increase the physical size of the FET in order to handle greater current and obtain higher gain. But this also increased the parasitic capacitance and increased channel resistance, which limits high-frequency performance: high parasitic capacitance reduces frequency response, while increasing channel resistance limits current flow. Zuleeg called the device a multichannel FET. By the 1970s, power FET development was taking place throughout the world, with new advances taking place rapidly.

High-power semiconductor devices are typically used as switches. As a switch, it is possible to control large amounts of power with relatively low power dissipation through the switching fabric. In fact, in an ideal switch, there should be zero voltage drop when the device is in the "on" state, zero leakage when the device is in the "off" state, and the amount of time to switch from one state to another should be instantaneous. Unfortunately, all practical electronic devices have some power loss and a finite switching time.

There are three fundamental types of power loss in switched device: conduction loss, turn-off loss, and switching loss. Conduction loss appears during the "on" state. This power loss is

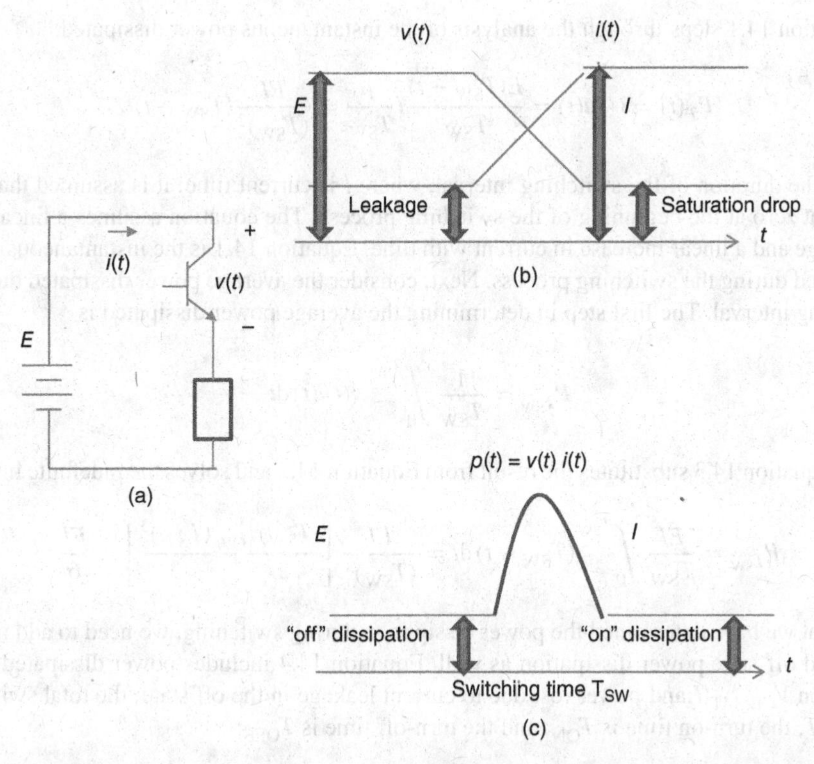

Figure 14.1 Reduction of switching loss has been one of the challenges in high-power semiconductor electronics. A switch is illustrated in (a). Voltage and current are shown as functions of time during a transition of the switch from "off" to "on" in (b). The resulting power consumed during the switch transition is shown in (c). Source: Tzou, 2006. *Power Electronics: An Introduction*. [Online: http://pemclab.cn.nctu.edu.tw/peclub/w3cnotes]

simply determined by the product of the on-state voltage drop across the device and the mean output current. This loss is independent of the switching frequency. Turn-off loss occurs during the "off" period. This a relatively small power loss equal to the off-state voltage drop multiplied by the turn-off leakage current. Switching loss is the power loss when transitioning from an on to off state and from off to on state. These losses are clearly proportional to the switching frequency. The total power loss is the sum of the conduction loss, turn-off loss, and switching loss. Power loss implies heat buildup leading to a vicious cycle in which heat increases power loss, which in turn raises the temperature even further. The result is an inefficient device and a reduced lifetime. Thus, it is important to reduce conduction and switching power losses.

Figure 14.1 shows a representative schematic for a power switching device. When the switch is turned off the circuit is open and no current should flow. However, there is some leakage current. Let us call the voltage drop across the open circuit E and consider the leakage voltage to be negligible for now. When the switch is turned on and the circuit is closed, let us call the current flow I. For simplification, we can assume that voltage and current transitions between the off and on states are linear. This is not precisely accurate in reality, but it is a reasonable approximation for this introductory-level discussion.

Equation 14.1 steps through the analysis of the instantaneous power dissipated:

$$P_T(t) = v(t)i(t) = \frac{E(T_{SW} - t)}{T_{SW}} I \frac{t}{T_{SW}} = \frac{EI}{(T_{SW})^2}(T_{SW} - t)t. \quad (14.1)$$

T_{SW} is the duration of the switching interval, where t is current time; it is assumed that time begins at zero at the beginning of the switching process. The equation assumes a linear drop in voltage and a linear increase in current with time. Equation 14.1 is the instantaneous power dissipated during the switching process. Next, consider the average power dissipated during a switching interval. The first step in determining the average power dissipated is

$$P_{T_{SW}} = \frac{1}{T_{SW}} \int_0^{T_{SW}} v(t)i(t)\,dt. \quad (14.2)$$

Next, Equation 14.3 substitutes the result from Equation 14.2 and solves the indefinite integral:

$$P_{T_{SW}} = \frac{EI}{T_{SW}} \int_0^{T_{SW}} (T_{SW} - t)\,dt = \frac{EI}{(T_{SW})^3}\left[\frac{(T_{SW})^3}{2} - \frac{(T_{SW})^3}{3}\right] = \frac{EI}{6}. \quad (14.3)$$

Now that we have determined the power dissipation during switching, we need to add the on-state and off-state power dissipation as well. Equation 14.4 includes power dissipated while turned on $V_{CE(SAT)}I$ and power lost due to current leakage in the off state; the total switching time is T, the turn-on time is T_{ON}, and the turn-off time is T_{OFF}:

$$P_T = \frac{2(EI/6)T_{SW} + \left(V_{CE(SAT)}I\right)T_{ON} + \left(EI_{leakage}\right)T_{OFF}}{T}. \quad (14.4)$$

Next, we need to consider commutation, the process by which current flow is stopped or forced to take an alternate path. For a thyristor, additional circuitry is required to turn off, or stop, the flow of current. Once the thyristor is turned on, it remains on until a reverse bias is applied long enough to stop the current flow through the thyristor. First, the forward current flow must be stopped. It must remain off long enough to regain its forward blocking capability. Meanwhile current flow must take an alternate path through the device until the thyristor is turned on.

Thermal characteristics are crucial to the efficiency and operation of a power electronics device. The thermal resistance is defined for the volume between the junction and the base or mount of the device and is represented in Equation 14.5 by $\delta T_{j\text{-}b}$. The power dissipation is W.

$$T_j - T_b = \delta T_{j\text{-}b} = WR_{j\text{-}b}. \quad (14.5)$$

Next, let $R_{b\text{-}a}$ be the thermal resistance from the base to the ambient heat sink. Then Equation 14.6 quantifies the total increase in heat:

$$T_j - T_a = \delta T_{j\text{-}a} = W(R_{j\text{-}b} + R_{b\text{-}a}). \quad (14.6)$$

Now let us consider how these devices can help improve power quality.

14.2.1 Power Electronics to Improve Power Quality

Power system devices play a role in addressing power quality. A precise and widely accepted definition of power quality is still lacking; however, it is possible to identify characteristics of power quality. We will define nine types of power quality issues. Voltage sags or dips are decreases in the nominal RMS voltage level by anywhere from 10 to 90% and can last for 0.5 cycles to 1 min. This can be caused by electrical faults or heavy loads such as machinery starting up. The results of a voltage sag can range from causing IT equipment to malfunction, to the unnecessary tripping of protection relays. It can also cause disconnection or loss of efficiency in large electric motors.

Another quality problem is short-duration interruptions. Such interruptions can last from milliseconds to a few seconds. This power quality problem is typically caused by automatic relay operation for protection from a real or assumed electrical fault. This can lead to tripping of additional protection devices, loss of information in data processing, and the interruption of sensitive information technology and plant processing equipment such as adjustable-speed drives, personal computers, and programmable logic controllers. Long interruptions are a total interruption for a duration of greater than 1 or 2 s. This is long enough to ensure stoppage of all equipment and could be due to permanent electrical faults.

A voltage spike is a fast variation of voltage ranging from a few microseconds to a few milliseconds. Such variations may reach up to thousands of volts, even in a low-voltage system. This can be due to lightning, switching of power lines, or power factor correction capacitors, or disconnection of heavy loads. This can lead to destruction of equipment, and of insulation material in particular, as well as data processing errors and data loss, and electromagnetic interference.

A voltage swell is a momentary increase of the voltage at the same power frequency outside the nominal tolerance with a duration of more than one cycle and typically less than a few seconds. It can be caused by starting or stopping heavy loads, poorly dimensioned power sources, or poorly regulated transformers. The result can be data loss in information technology equipment, flickering in lighting and computer screens, and stoppage or damage of sensitive equipment.

Harmonic distortion is characterized by voltage or current waveforms that take on a nonsinusoidal shape. The waveform corresponds to the sum of different sine waves having different magnitudes and phases but with frequencies that are multiples of the power system frequency. This can be caused by many sources, from electric machines to nonlinear loads, such as power electronics equipment. The results are many, including an increase in the probability of resonance, neutral overload in three-phase systems, overheating of cables and equipment, loss of efficiency in electric machines, electromagnetic interference with communication systems, error in measures when using average-reading meters, and nuisance tripping of thermal protection systems.

Voltage fluctuation is the oscillation of a voltage value with an amplitude that is modulated by a signal with a frequency of 0–30 Hz. It can be caused by arc furnaces, frequent starting and stopping of electric motors (for instance, elevators), and oscillating loads. Most consequences are common to undervoltages; for example, the flickering of lighting and screens.

Noise is the superposition of high-frequency signals on the power system frequency waveform. It can be caused by electromagnetic interference provoked by Hertzian waves (such as microwaves and television waves), and radiation due to heavy machinery and electronic equipment, as well as improper grounding. The results are disturbances to sensitive electronic equipment that may cause data loss and data processing error.

Voltage imbalance is variation in a three-phase system in which the three voltage magnitudes or the phase-angle differences between them are not equal. This can be caused by large, single-phase loads as well as by the incorrect distribution of single-phase loads by the three phases of the system. The result can be unbalanced systems, implying the existence of a negative sequence that is harmful to three-phase loads. Three-phase induction machines would likely suffer most from this imbalance.

One question to be addressed is where to implement PCS to remove these power quality problems. For example, one might consider generators that create perfect-quality output and then assume that power quality can be maintained until it reaches the consumer. Another consideration is to place power conditioning equipment within the transmission or distribution systems. Alternatively, power conditioning can be placed on the customer premises. Finally, devices themselves could be designed to include power conditioning as an inherent part of their operation. Of course, DG, a key aspect of the smart grid, could exacerbate power quality problems or help alleviate them, depending on how they are implemented.

Energy storage systems can help mitigate power quality problems by implementing a ride-through capability; they supply quality power during periods of poor power quality from main generators. Devices that implement storage include batteries, flywheels, supercapacitors, and SMES. A battery can be kept fully charged during normal power grid operation and supply power during periods of poor power quality. A flywheel uses power from the grid to store kinetic energy in a rotor. When power quality degrades, energy from the rotor is used to produce direct current which flows through an inverter to generate alternating current. The rotor spins in a vacuum to eliminate drag, and communication is used to remotely monitor and control the flywheel. Flywheels are now being constructed to operatc at 60 000 RPM; the stored energy is equal to the moment of inertia and the square of the rotational speed. Flywheels can typically provide from 1 to 100 s of ride-through time.

A supercapacitor, or ultra-capacitor, has very large capacitance owing to a very small distance between plates, on the order of several angstroms, and a very large plate surface area, 1500-2000 m^2/g. This device can provide power during voltage sags.

In SMES, direct current in a superconducting coil creates a magnetic field. This field can be modulated between an on and off state at the proper rate for input into an inverter, which creates the required alternating current. SMES is typically cooled by liquid helium; however, higher temperature superconductors can be cooled by liquid nitrogen. SMES systems are still rather large and used for short-duration utility switching events.

Certainly, reliable microgrid generators, such as diesel generators, microturbines, and fuel cells, can also provide the short-term power required to mitigate power quality issues. However, these generators take time to come online and thus require one of the previous sources of energy storage to provide power until they power up. The most common solution has been battery storage with a diesel generator; however, a flywheel and diesel generator have also become popular.

Other power electronic devices can be used to mitigate power quality problems. A dynamic voltage restorer is essentially a voltage source in series with the load; it keeps the voltage constant by injecting power from storage through a power converter when needed. A transient voltage surge suppressor will perform the opposite function of injecting power to maintain constant voltage; it will prevent excess voltage from harming a load by clamping the voltage and sending any excess energy to ground. Interestingly, a constant-voltage transformer has been used in the past to maintain constant voltage via resonance and saturation of the transformer core – two events normally avoided. Resonance causes the transformer core to saturate and in this state the transformer maintains a constant voltage. However, it is relatively inefficient and, if not used properly, can cause more power quality issues than it mitigates.

Of course, filters can be created from capacitive–inductive components to filter out noise. They typically act as low-pass filters to remove harmonics. Another means of filtering noise is an isolation transformer. This type of transformer has a grounded, nonmagnetic foil between the primary and secondary coils that effectively filters out harmonics from passing through to the secondary coil. All of these power conditioning and filtering devices could impact or eliminate the benefit of explicitly using power line carrier communication. Finally, SVCs are combinations of capacitors and reactors (inductors) designed to adjust the power factor to unity. They need to respond in a rapid and automated manner to changing load conditions. This was discussed earlier in Section 3.4.1 on integrated volt-VAr control. However, the final result is a regulated voltage level so that voltage sags do not occur or are minimized. Finally, there are passive and active filters that can be used to eliminate harmonics. Passive harmonic filters provide a low-impedance path to ground for the harmonics to be eliminated, while active filters inject current to cancel the harmonics created by a load. The problem with passive filters is that they cannot adapt to changing harmonic conditions and can cause resonance. Active harmonic filters are relatively expensive. Next we consider an advance to the workhorse of the power grid, the transformer.

14.3 Power Electronic Transformer

The power grid exists to transport power and benefits from transforming the power into optimal forms along the way. The typical transformation involves changing the ratio of voltage to current, as we have seen in prior chapters. Since the transformer was invented by Faraday, its concept of operation has changed very little; as every child knows, it is comprised of a primary and secondary coil of wire. It was only a few decades ago that the idea of significantly changing the design of the transformer to a solid-state device was conceived. This section looks at the challenges and advantages of the solid-state transformer and how it will impact communication.

The transformer has a long history. In the summer of 1831, Michael Faraday first demonstrated the principle of electromagnetic induction by constructing the first electric transformer. He constructed a simple transformer to step up the voltage from a battery. He then demonstrated that when the circuit was opened and closed, it caused a compass needle to move. By the mid 1850s, it was recognized that the transformer could be used to create very high, instantaneous voltages that could be used to create sparks. During this time, it was discovered how to improve the elements of the transformer to improve its operation; for example, increasing the area of the loop winding and using iron that had high magnetic permeability. By 1888, the war of the currents, between alternating and direct current, was beginning to take shape. Thomas Edison was backing what looked like the simplicity of direct current since lighting and motors were using direct current. Also, batteries could be directly connected to the system in order to provide backup power. Of course, as we have seen, the inability for the voltage of direct current to be economically increased or decreased made transmission over significant distances impractical and expensive. The advantage of alternating current was precisely the solution to this problem, it could be easily and cheaply stepped up and down as needed for optimal transmission. The main problem was that the alternating-current motor, invented by Tesla, was not in production at the time. The widespread use of an alternating-current lamp further encouraged the use of alternating current. Surprisingly, by the late 1800s, the transformer was already near-100% efficient. Many later improvements were focused on ancillary aspects of the transformer, such as improving power protection or reducing the cost of the transformer.

While the transformer became ubiquitous throughout the power grid, there was room for improvement to the transformer. The transformer, still basically using Faraday's initial design, must be very large in order to handle the amount of power transmitted over the grid. This also makes it tremendously heavy. All of this makes the transformer expensive. The mechanical forces exerted upon the winding and other components of the transformer can be significant and cause loud, annoying vibrations. The transformer can saturate, meaning that the applied magnetic field can no longer increase core magnetization. This can create poor power-quality output and heating. Vibration and heating can wear out the insulation of the windings in the transformer. Mechanical forces, vibration, and heat cause the need for regular and costly maintenance and repair. Many transformers require an efficient cooling system, which adds additional complexity and cost to the system. Also, the electrical properties of the transformer are not ideal. The transformer does not provide perfect isolation: noise on one side of the transformer can pass through to the other side, disrupting power quality. Conversely, the transformer filters power line carrier signals from passing through efficiently. Also, tap changing transformers utilize a mechanical means of adjusting the turns ratio in order to enable the amount of voltage to be adjusted in the distribution system. Frequent adjustment of the tap can cause it to wear out.

The goal of the power electronic transformer is to redesign the transformer using power electronic devices. As this redesign of the transformer has proceeded since the 1980s, the transformer has taken on many different names, including power electronic transformer, solid-state transformer, intelligent universal transformer, flexible power-electronic transformer, and active power-electronic transformer, among many others. The power electronic transformer is becoming metaphorically like an Internet router; as the transformer has been redesigned with power electronics, it has gained more functionality, including independent power ports through which power can be "routed."

At the simplest level, there have been two general approaches to reducing the size and cost and increasing the performance of the transformer. One approach involves alternately switching power flow on and off, thus creating a power duty cycle that can be adjusted so that only a precise amount of power is conveyed through the transformer. Another general approach involves increasing the frequency within the transformer in order to allow the size of the transformer coils to be reduced. A higher frequency allows for a smaller coil to transform a larger amount of power. Afterward, the frequency is converted back to its nominal value.

Consider the relationship between frequency and size of a transformer. Faraday's law of induction helps us to understand how size and frequency are related. Equation 14.7 shows V_s, the instantaneous voltage, N_s, the number of turns in the secondary coil of the transformer, and Φ, the magnetic flux through one turn of the coil:

$$V_s = N_s \frac{d\Phi}{dt}. \tag{14.7}$$

Typically, the frequency is assumed to be constant and the value that is varied is the number of turns in the coil. As is well known, both the primary and secondary windings of the transformer follow the same law. The magnetic flux is the product of the normal component of the magnetic flux density B and the field area A through which the flux "cuts" or flows. The area is typically determined at the time of manufacture by the size of the transformer core and remains constant, while the magnetic field changes with time with the flow of current through the primary winding. Thus, while the subscript "s" in Equation 14.7 refers to the secondary winding, the same flux, changing at the same frequency, passes through the primary winding as well. Thus, the ratio of the equations yields

$$\frac{V_p}{V_s} = \frac{N_p}{N_s}, \tag{14.8}$$

where N_s and N_p are the number of secondary and primary loops, respectively.

In reality, a transformer is a nonideal device, and this is exhibited by the fact that the core of the transformer can saturate. This occurs when the magnetic flux from the coils exceeds the ability of the core of the transformer to become magnetized. The size of the core must be increased in order to avoid saturation. Magnetic saturation causes a large increase in magnetic current and transformer overheating.

Now consider what happens to a transformer as the frequency of the current is increased. First, for a given core size, the power throughput density can be increased before reaching saturation. Second, as can be seen in Equation 14.7, fewer turns in the winding are required. This is one reason that aircraft and many vehicles operate at 400 Hz: to reduce transformer size. Equation 14.9 quantifies the voltage, or electromotive force, in more detail:

$$E_{\text{rms}} = \frac{2\pi fNaB_{\text{peak}}}{\sqrt{2}} = 4.44fNaB_{\text{peak}}. \tag{14.9}$$

The units are RMS volts, the supply frequency is f, the number of turns is N, the core cross-sectional area is a (m^2), and peak magnetic flux density is B_{peak} (Wb/m^2 or T).

Because power semiconductor switches are not perfectly efficient – for example, they have a small leakage current – a typical research goal in the design of power electronic transformers is to reduce the number of switches by examining different switch interconnection topologies.

The concept of resonance, discussed with regard to wireless power transfer in Section 8.4, can be applied quite simply to the transformer. The resonant frequency of an inductor–capacitor circuit is used to create a fast-changing magnetic field that cuts through an adjacent coil, thus inducing current in the adjacent coil. The difference in coil windings determines the change in voltage. A relatively high resonant frequency allows more power to be transferred for a given coil diameter, just as in the case of the previous description of the power electronic transformer.

The power electronic transformer is being proposed as the "Swiss army knife" of the power grid; in other words, capable of doing many different things beyond its original, primary objective of stepping voltage up or down. For communication, this means that the power electronic transformer is becoming an ever-more complicated device to control and monitor.

Power electronic transformers have been designed many different ways with different types of inverters and converters; we will only review a few of the simplest, the buck and boost converter. The buck converter is a direct current to direct current power converter and steps down the voltage. The boost converter is similar and steps up the voltage.

Let us first review the operation of the buck converter shown in Figure 14.2. Notice that there is an inductor in series with the load and a switch that alternately disconnects and reconnects the load. A buck converter transforms direct current as input and yields direct current as

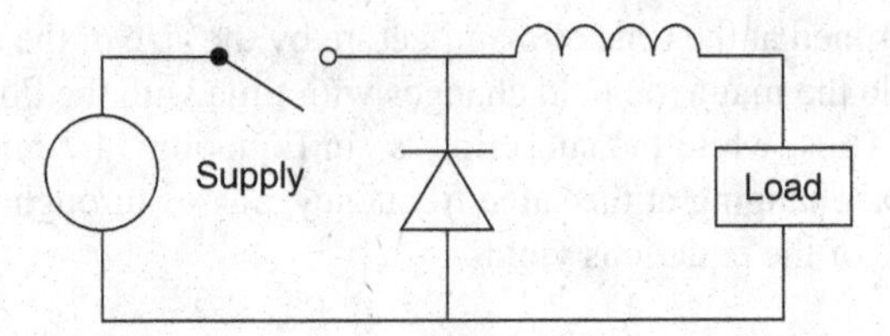

Figure 14.2 A buck converter is comprised of an inductor L in series with the load and in parallel with a diode.

output. Thus, when used in a transformer, power must first be converted to direct current. As an aside, there are some advantages to working with direct current with power electronics that occasionally reignite the old, alternating current versus direct current controversy, or war of the currents, within the power grid. However, that is a wider discussion that is beyond the scope of this section. The inductive element resists a change in current as well as stores energy. When the switch is initially open, there is no current flow. When the switch closes, the inductor charges while resisting the increasing flow of current due to its reluctance. It resists the increase in current flow by creating an opposing voltage across its terminals. This opposing voltage reduces the voltage across the load. Eventually, the current will increase to its maximum value and the opposing voltage will drop to zero. The inductor will then become fully charged by storing energy in its magnetic field. A key point is that if the switch opens before the inductor has fully charged its magnetic field, then the opposing voltage across the inductor will not go to zero and the load will continuously see a lower voltage than that generated by the source. When the switch opens again, the inductor will again resist the change in current. This time the current is decreasing. The inductor will release its stored energy in the form of current by creating a voltage across its terminals that is similar to the voltage supplied by the original power source. If the switch closes again before the inductor fully discharges, the load will not experience a zero voltage.

Now let us consider the operation in a bit more detail. Let V_L be the voltage across the inductor, V_i be the input voltage, and V_o be the output voltage; that is, the voltage across the load. When the switch is closed, $V_L = V_i - V_o$. When the switch is open, $V_L = -V_o$. The current when the switch is closed is

$$\Delta I_{L_{on}} = \int_0^{t_{on}} \frac{V_L}{L}\,dt, \tag{14.10}$$

where $dt = \frac{(V_i - V_o)}{L} t_{on}$ and $t_{on} = DT$. D is the duty cycle, which has a value between zero and one indicating the proportion of time that the switch remains closed, and T is the time duration of one complete on–off cycle. The current when the switch is open is shown in Equation 14.11, and $t_{off} = (1 - D)T$; the value $1 - D$ is the proportion of time that the switch is open and the current is off:

$$\Delta I_{L_{off}} = \int_{t_{on}}^{T=t_{on}+t_{off}} \frac{V_L}{L}\,dt = -\frac{V_o}{L} t_{off}. \tag{14.11}$$

Now we can make a simplifying assumption regarding the energy stored in each component over each full cycle of duration T. This assumption is that the amount of energy stored over each cycle is the same and thus the current in the inductor at the beginning and end of each cycle is also the same. This is captured in

$$\frac{(V_i - V_o)}{L}t_{on} - \frac{V_o}{L}t_{off} = 0. \tag{14.12}$$

DT is the duration of time the switch is in the closed state (that is, the "on" time) and $(1 - D)T$ is the duration of the switch in the open state, or the "off" time. Equation 14.13 shows how the duty cycle impacts the voltage conversion through the inverter:

$$\begin{aligned} (V_i - V_o)DT - V_o(1 - D)T &= 0 \\ V_o - DV_i &= 0 \\ D &= \frac{V_o}{V_i}. \end{aligned} \tag{14.13}$$

The value of D ranges from zero to one; at most, the output voltage can equal the input voltage. However, D is typically less than one, where the output voltage is less than the input voltage and the inverter steps down the voltage as a linear function of D. If the energy required by the load is relatively small, it is possible that the inductor current goes to zero during a portion of the cycle, T. The analysis is similar to that previously described, except that the average value of the inductor voltage is the same at the beginning and end of the cycle; namely, zero. This is represented in Equation 14.14, where the value of δ is shown in Equation 14.15:

$$(V_i - V_o)DT - V_o\delta T = 0, \tag{14.14}$$

$$\delta = \frac{V_i - V_o}{V_o}D. \tag{14.15}$$

It is insightful to realize that the buck converter can also perform impedance matching. Recall from Section 8.2.4 that maximum power point tracking is used in inverters with PV cells that generate electric power. Consider an arbitrary electrical system where V_o is the output voltage, I_o is the output current, η is the power efficiency (ranging from zero to one), V_i is the input voltage, and I_i is the input current. Equation 14.16 shows the output power given the input power:

$$V_o I_o = \eta V_i I_i. \tag{14.16}$$

Now consider that we have Z_o, which is the output impedance, and Z_i, the input impedance. Then, using Ohm's law, it is a simple matter to represent the input and output currents as shown in Equations 14.17 and 14.18:

$$I_o = \frac{V_o}{Z_o}, \tag{14.17}$$

$$I_i = \frac{V_i}{Z_i}. \tag{14.18}$$

Now Equation 14.16 can be changed to Equation 14.19 by replacing the currents from Equations 14.17 and 14.18:

$$\frac{V_o^2}{Z_o} = \eta \frac{V_i^2}{Z_i}. \tag{14.19}$$

Now recall that D is the duty cycle; that is, the proportion of time that the circuit is closed. Then from Equation 14.13 we have

$$V_o = DV_i. \tag{14.20}$$

Equation 14.20 can be placed into Equation 14.19 to derive

$$\frac{(DV_i)^2}{Z_o} = \eta \frac{V_i^2}{Z_i}. \tag{14.21}$$

The input voltage cancels from Equation 14.21 to yield

$$\frac{D^2}{Z_o} = \frac{\eta}{Z_i}. \tag{14.22}$$

Solving for the duty cycle yields

$$D = \sqrt{\eta Z_o / Z_i}. \tag{14.23}$$

The nice thing about Equation 14.23 is that it becomes clear that changing the duty cycle also changes the impedance ratio. In situations such as maximum power point tracking, explained in Section 8.2.4, the maximum power point may be dynamically changing, in which case the duty cycle can be dynamically adjusted to control the impedance and, thus, the maximum power point.

Now we can build upon the previous discussion of the buck converter to discuss the boost converter shown in Figure 14.3. Compare this figure with the buck converter in Figure 14.2. The temporary inductive storage element is located near the power supply and the switch is now parallel to the load, and the rectifier is now in series with the load. This is another direct current to direct current conversion system. However, the difference from the buck converter is

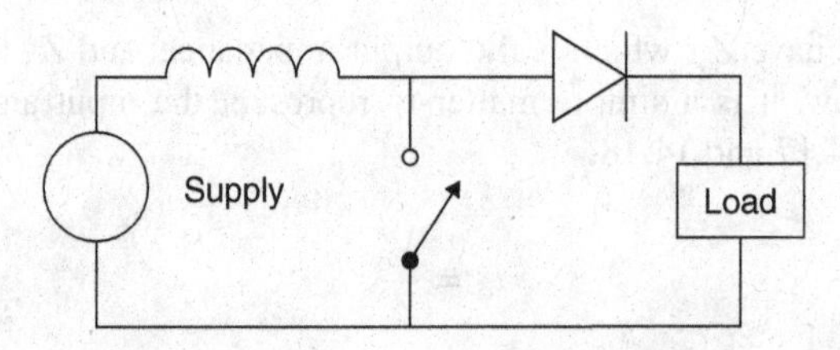

Figure 14.3 A boost converter has an inductor L in series with the power source. A diode is in series with the load and in parallel with the switch.

that it is capable of increasing the output voltage: it is a step-up transformer, whereas the buck converter is a step-down transformer. An interesting use of the boost converter is to extract more energy out of a battery than would otherwise be possible. Normally, once a battery's voltage decreases below a threshold level, current can no longer be drawn from the battery. By boosting the battery storage system's voltage, the remaining energy can be extracted from the depleted battery.

The operation of the boost converter is similar to the buck converter: the inductive element will store energy while tending to resist change in current flow. Now that the inductor is in series with the power source, the inductor will, in effect, aid the power source. While the switch is closed the inductor will store energy. When the switch is open, the load will be the main path for the current, and the impedance seen by the power source and inductor will increase. The inductor will become a power source that is in series with the main power source. If the switch is changed quickly enough, the inductor will not have time to fully discharge and the voltage will always be larger than the power source alone. Assuming the circuit has been initially powered up, then when the switch is closed a change in current flow is experienced by the inductor, as shown by

$$\frac{\Delta I_{\mathrm{L}}}{\Delta t} = \frac{V_{\mathrm{i}}}{L}, \tag{14.24}$$

where V_{i} is the source voltage, I_{L} is the inductor current, and L is the inductance. Δt is the time period during which the switch is on and ΔI_{L} is the change in inductor current.

Solving Equation 14.24 for ΔI_{L}, the total increase in the inductor current can be calculated while the switch is on. This is shown in

$$\begin{aligned} \Delta I_{\mathrm{L_{on}}} &= \frac{1}{L}\int_{0}^{DT} V_{\mathrm{i}}\,\mathrm{d}t \\ &= \frac{DT}{L}V_{\mathrm{i}}. \end{aligned} \tag{14.25}$$

Here, the accumulated current is computed by integrating the voltage over the switch on-time. The switch on-time is determined by the duty cycle D, which represents the fraction of one cycle time T that the switch is on. Thus, the total duration of time during which the switch is on is DT.

Now consider what happens when the switch is opened. The current now must flow through the higher impedance load and V_{o} is the output voltage; that is, the voltage seen by the load. Equation 14.26 shows the relationship between the voltage drop and the current through the inductor:

$$V_{\mathrm{i}} - V_{\mathrm{o}} = L\frac{\mathrm{d}I_{\mathrm{L}}}{\mathrm{d}t}. \tag{14.26}$$

Recalling the definition of the duty cycle explained previously, the switch is open for a duration of time lasting from DT to T for each cycle. Since D represents a fraction of a cycle of duration

T, the switch is open for a duration of time $(1 - D)T$. Solving Equation 14.26 for the inductor current, yields

$$\Delta I_{L_{\text{off}}} = \int_{DT}^{T} \frac{(V_i - V_o)\,dt}{L} = \frac{(V_i - V_o)(1 - D)T}{L}. \tag{14.27}$$

Using a technique similar to that in the buck converter analysis, it is assumed that the system operates normally in a steady-state condition, which implies that no energy is gained or lost during operation and the energy in each component is the same over each cycle of operation. The energy is the same at the beginning and end of each cycle T. The energy in the inductor is shown as a function of current in

$$E = \frac{1}{2} L I_L^2. \tag{14.28}$$

If we assume that the current must then be the same at the beginning and end of each cycle, then Equation 14.29 must hold:

$$\Delta I_{L_{\text{on}}} + \Delta I_{L_{\text{off}}} = 0. \tag{14.29}$$

Equation 14.30 shows the previously derived values substituted into Equation 14.29:

$$\Delta I_{L_{\text{on}}} + \Delta I_{L_{\text{off}}} = \frac{V_i DT}{L} + \frac{(V_i - V_o)(1 - D)T}{L} = 0. \tag{14.30}$$

A little manipulation and simplification yields

$$\frac{V_o}{V_i} = \frac{1}{1 - D}. \tag{14.31}$$

Solving for the duty cycle D in

$$D = 1 - \frac{V_i}{V_o} \tag{14.32}$$

shows that the output voltage will be greater than the input voltage for any nonzero value of D. The output voltage increases with the value of D, and as D approaches one it would appear that the output voltage increases toward infinity. However, it is important to remember that D is the duration of the switch on-time, which means that little current flows through the load while the switch is on. Power must be conserved; in other words, the product of the current and the voltage will remain constant. While the output voltage increases, the output current decreases. It is also important to keep in mind that this analysis assumed all components are ideal; in reality, components are not perfectly efficient and loss will occur.

Finally, there is a combination of the previous two converters known as the buck–boost converter. As might be expected, this converter can both step-up and step-down voltage. The

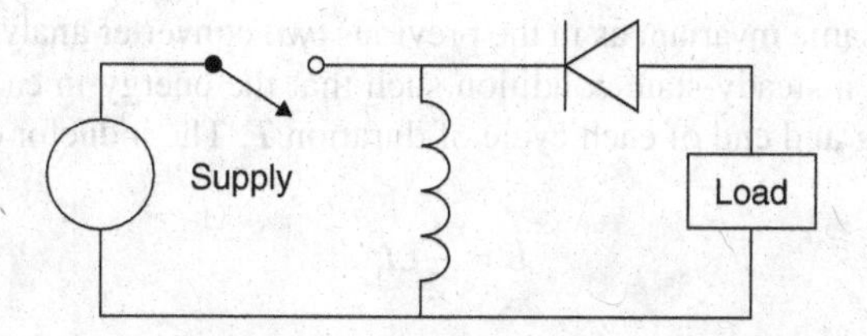

Figure 14.4 A buck–boost converter is comprised of an inductor L in parallel with the power supply and the load.

buck–boost converter is shown in Figure 14.4. The operation of the buck–boost converter is a relatively simple extension of the previous buck and boost converters. When the buck–boost switch is closed, current flows to the inductor and the load in parallel. This allows the inductor to charge. When the switch opens, the inductor supplies energy to the load. Now consider how it can be tuned to allow the voltage to be either stepped up or stepped down. The analysis is similar to the previous two converters, in that the inductor current is the focus of the analysis. Recall that the duty cycle D is a fraction of the total cycle time T during which the switch is closed. Equation 14.33 shows the change in inductor current during the time the switch is closed:

$$\frac{\mathrm{d}I_{\mathrm{L}}}{\mathrm{d}t} = \frac{V_{\mathrm{i}}}{L}. \tag{14.33}$$

The current at the end of the switch on-time increases to a value shown by

$$\Delta I_{\mathrm{L_{on}}} = \int_0^{DT} \mathrm{d}I_{\mathrm{L}} = \int_0^{DT} \frac{V_{\mathrm{i}}}{L}\,\mathrm{d}t = \frac{V_{\mathrm{i}}DT}{L}, \tag{14.34}$$

where $I_{\mathrm{L_{on}}}$ is the value of the inductor current at the end of the switch on-time; that is, just before the switch opens. Recall that D ranges from zero to one, indicating the fraction of cycle time that the switch is closed. Now consider what happens when the switch opens. The current from the input power source is removed and current flows from the inductor through the load. The inductor current at this stage is shown by

$$\frac{\mathrm{d}I_{\mathrm{L}}}{\mathrm{d}t} = \frac{V_{\mathrm{o}}}{L}. \tag{14.35}$$

The current change in the inductor while the switch is open is

$$\begin{aligned}\Delta I_{\mathrm{L_{off}}} &= \int_0^{(1-D)T} \mathrm{d}I_{\mathrm{L}} \\ &= \int_0^{(1-D)T} \frac{V_{\mathrm{o}}\,\mathrm{d}t}{L} \\ &= \frac{V_{\mathrm{o}}(1-D)T}{L}.\end{aligned} \tag{14.36}$$

Now we can use the same invariant as in the previous two converter analyses; namely, that the converter operates in a steady-state condition such that the energy in each component is the same at the beginning and end of each cycle of duration T. The inductor energy is

$$E = \frac{1}{2} L I_{\mathrm{L}}^2. \tag{14.37}$$

Given that the energy is the same at the beginning and end of each cycle, it is apparent that the current must also be the same. This is represented in Equation 14.38.

$$\Delta I_{\mathrm{L_{on}}} + \Delta I_{\mathrm{L_{off}}} = 0 \tag{14.38}$$

Next, the earlier equations can be substituted for the on and off currents as shown in Equation 14.39.

$$\begin{aligned} \Delta I_{\mathrm{L_{on}}} + \Delta I_{\mathrm{L_{off}}} &= \frac{V_{\mathrm{i}} D T}{L} + \frac{V_{\mathrm{o}}(1-D)T}{L} \\ &= 0. \end{aligned} \tag{14.39}$$

Equation 14.39 can then be rearranged to find the ratio of the output to input voltage:

$$\frac{V_{\mathrm{o}}}{V_{\mathrm{i}}} = \left(\frac{-D}{1-D}\right). \tag{14.40}$$

Finally, solving for the duty cycle D, the relationship between input and output voltage can be determined as

$$D = \frac{V_{\mathrm{o}}}{V_{\mathrm{o}} - V_{\mathrm{i}}}. \tag{14.41}$$

Because the duty cycle lies within the range from zero to one, the output voltage is always negative. Also, based upon D, the output voltage can range from being much smaller to much larger than the input voltage, thus stepping the voltage either down or up.

An interesting feature to note with these power electronic transformers is that they are controlled by their switching rate rather than by the number of turns of a winding, the turns ratio, as in a classical transformer. A classical transformer would have taps into different locations within a winding that could be adjusted. However, there are discrete changes in the winding ratio; the switched-mode transformers just discussed allow a continuous voltage adjustment based upon the switching rate. Now let us consider how power electronics is advancing in terms of protection devices.

14.4 Protection Devices and Current Limiters

Power transmission and distribution systems in particular expose large amounts of power to the potential of electrical faults. The system must be designed to mitigate the damage caused by such faults. The technique for mitigating the impact of electrical faults, described in previous chapters, uses relays to detect overcurrent and open, resulting in cutting off power flow to the faulty segment. The problem with this approach is that it also cuts off power to all consumers that happen to be downstream of the fault as well. An ideal solution should detect and mitigate the impact of overcurrent caused by a fault while not interfering with proper current flow to all downstream consumers. A class of devices known as fault current limiters (FCLs) or fault current controllers exists to implement this. These devices "absorb" excess current flow while allowing the proper amount of current to continue flowing. It is also important that the FCL does not reduce power efficiency while in their normal, nonfault, operating mode. A simple approach is to use an inductor, where inductance is $v = L(\mathrm{d}i/\mathrm{d}t)$ and v is voltage, L is inductance in henries, and $\mathrm{d}i/\mathrm{d}t$ is the rate of change in current. This can be rearranged to yield $\mathrm{d}t = (L/v)\mathrm{d}i$. The voltage induced across the inductor is proportional to the rate of change of current and will oppose the change of current, causing it to increase slowly. The inductor can be thought of as a resistor that increases its value the faster the fault current tries to rush through it. It has a high value for $\mathrm{d}i/\mathrm{d}t$, which yields a high resistance. The problem, of course, is that an inductor reduces the power factor and efficiency while in normal mode of operation. Superconducting FCLs have no resistance and do not adversely impact power flow during normal operation. However, they are still relatively expensive and require cooling. But if and when they become economical, they will have a promising future. More will be said about superconducting devices later in this chapter.

There are also solid-state FCLs, currently of two main types: resonance limiters and impedance switch-in limiters. The resonance-based, solid-state devices operate by implementing a "tuned" inductor and capacitor circuit to reduce impedance during normal operation. When a fault occurs, the solid-state device will remove an inductor or capacitor from the circuit, which increases the impedance needed to impede the overcurrent. The solid-state impedance switch-in limiter operates by using thyristors that operate in normal, nonfault mode by continuously switching on during alternate half-cycles to bypass a relatively large impedance. When a fault occurs, the thyristors are turned off, causing the current to flow through the large impedance, reducing the current. The drawback behind this approach is that the normal-mode switching operation causes power loss, as discussed earlier in this chapter.

One of the problems with DG is that it will change the way the operation of protection is currently implemented. Traditional protection mechanisms in the distribution system assume power flows in only one direction. DG and microgrids will enable power to be generated throughout the transmission and distribution system. Relays and reclosers that have been configured to protect various portions of the grid will see different current and impedance values due to the distributed generators. One solution to this problem is to place FCLs on the input from all distributed generators. This would allow the contribution from distributed generators to continue during a fault – that is, allow fault ride-through – and would not require significant changes to existing projection mechanisms.

14.5 Superconducting Technologies

Superconductivity is a large-scale, quantum mechanical phenomenon that can open the door for us to utilize nonclassical physics in many different and fascinating ways that will benefit the power grid, from improved power generation and transport to new forms of communication and computation. First, let us review what is known about the nature of superconductivity and then discuss its potential benefits for the power grid. In Chapter 15 we consider integrating the notion of superconductivity in the grid with communication and computation applications.

The simple definition of superconductivity is a phenomenon exhibited by certain materials such that when the temperature is reduced below a critical value the material has no electrical resistance and within which all magnetic fields are eliminated. Elimination of magnetic fields within the superconducting material is know as the Meissner effect. The temperature below which the material becomes superconducting is known as the critical temperature; this temperature is typically close to absolute zero. However, the critical temperature is specific to different materials; a widely pursued goal of research is to find materials with critical temperatures that are as high as possible in order to reduce the impact on the cooling system for superconducting applications. Note that superconductivity is not just an effect of temperature; many normally conducting materials do not become superconducting even at absolute zero, while some poorly conducting materials at room temperate can become superconducting at temperatures well above absolute zero. This is also indicated by the Meissner effect; that is, the lack of a magnetic field within the superconducting material in its superconducting state.

Along with the critical temperature, there is also a critical magnetic field strength to which the superconductor may be exposed and still remain superconducting. If the critical field strength is exceeded, the superconductor material will leave its superconducting state. Superconductors are classified as Type I and Type II. Type I superconductors lose all superconductivity when exposed to a magnetic field strength above the critical value. Type II superconductors are more resilient; they have two critical magnetic field strength values. They are superconducting below the first value, in a mixed state between the first and the second values, and they lose their superconducting state completely above the second value.

Although superconductivity was discovered in 1911, the question as to how superconductivity works was first understood in 1957 by John Bardeen, Leon Cooper, and Rover Schrieffer and is known as the Bardeen–Cooper–Schrieffer (BCS) theory after the initials of their last names. The concept is that a negative electron flowing through a regularly shaped, positively-charged structure creates a slight tug upon the positive nuclei of the superconducting material. This tug creates a short-duration, positively charged gradient that attracts another electron to follow the original electron that caused the slight deformation. These two electrons continue to follow one another closely, using this technique until their quantum wave functions become aligned with each other. This causes the electron pair to become joined in a collective state known as a condensate; they are also known as a Cooper pair. This is a low-energy formation, and when many of these pairs form the material tends to remain in this low-energy state with the electron pairs flowing efficiently and without interruption. However, the Cooper pair is relatively weak, and any noise, such as thermal vibration, can rip them apart. Thus, temperature and magnetic fields can destroy the electron pairing. While the BCS theory has been successful in explaining and predicting aspects of early, lower temperature superconductors, the newer, higher temperature superconductors, some of which are insulators at room-temperature, are still lacking a widely accepted theory.

The benefits of superconductivity in the power grid may seem obvious; the ability to create superconducting power lines to transport power without resistance would eliminate loss, the power grid would become a perfectly efficient electrical transport system. However, there are at least two interesting reasons this is not so simple. First, there are many challenges to building a superconducting power grid. Second, potential applications for use of superconductivity in the power grid, while numerous, may not be immediately obvious.

The challenges for superconductivity in the power grid include minimizing cost; this includes minimizing the cost of cryogenics, which includes being able to operate in a superconducting state at higher temperatures. This also includes reducing the manufacturing costs of creating strong and reliable superconducting wire, both for power lines and for windings in power equipment. The superconducting material should be able to carry high current density, be flexible and chemically stable, have low alternating-current loss, and be able to continue to operate in a superconducting state when exposed to strong magnetic fields. The benefits are related to the high efficiency and current density of superconducting wires and also in exploiting their ability to switch from a superconducting state to normal state, as we will see.

Superconducting components have long been considered for use in motors and generators. To see why, first consider a simple type of machine known as a homopolar machine. This is the simplest possible type of motor–generator that can be constructed as will be explained shortly. Without loss of generality, the superconducting technique can be applied to other types of motors.

Before we discuss the homopolar motor, recall electric power generation introduced in Section 2.2 and distributed electric power generation discussed in Section 8. While the mechanical force turning the rotor varied greatly in these sections, the concepts always assumed some variation of a synchronous machine that generated the electric power. There is another type of electric machine known as a homopolar motor that converts electric power to mechanical force and a homopolar induction alternator that generates electric power. Homopolar machines are direct-current machines in which a direct-current conductor is positioned within a static magnetic field. The Lorentz force, the right-hand rule from Section 1.3.5, of the magnetic field moves the conductor. This is one of the simplest possible electrical machines; no commutator is required because there is no reversal of current flow required in order for the machine to operate. Just as in a synchronous machine, the homopolar machine can operate as either a motor or a generator depending upon whether electrical or mechanical power is applied. In fact, the term "homopolar" is used to emphasize that the electrical polarity of the conductor and the magnetic field do not change. The homopolar machine is rarely used in practice because it is limited to essentially one turn of a coil; that is, the conductor is a single wire. The homopolar generator and motor are introduced here because they have received consideration by researchers as potentially ideal embodiments for superconducting power.

Consider the parameters impacting the power that an electric machine can either produce or generate (Huang *et al.*, 1998). Equation 14.42 shows the steady-state, continuous-duty, power rating S for an electrical machine:

$$S = \pi^2 n\sigma\beta D^2 L. \tag{14.42}$$

Here, n is the rotor speed, σ is the armature current loading per-unit peripheral dimension, β is the air gap flux density, D the air gap diameter, and L the active length of the machine. Using the laws covered in previous chapters, the equation should be fairly obvious. It should also be intuitively obvious that increasing any value in the equation will increase power output. The power density of the machine can be found by dividing the power rating by the total volume of the machine. This is

$$P_{\mathrm{d}} \propto n\sigma\beta. \tag{14.43}$$

Increasing rotor speed n in the limit leads to a mechanical problem: centrifugal stress can destroy the rotor. Centrifugal stress is proportional to d^2n^2, so Equations 14.44 and 14.45 represent the machine's total power output and power density in terms of rotor stress:

$$S = \frac{\pi^2 \tau_{\text{cent}} \sigma \beta L}{n}, \tag{14.44}$$

$$P_{\text{d}} \propto \frac{\sqrt{\tau_{\text{cent}}}}{D\sigma\beta}. \tag{14.45}$$

Instead of increasing the rotor speed, it is possible to increase the air gap flux density σ and the armature current loading per-unit peripheral dimension σ. Both of these can be increased by employing superconductivity.

Another interesting application of superconducting motors with a direct relationship to the power grid is superconducting electric motors for plug-in electric vehicles (Sekiguchi *et al.*, 2012). In one experiment, a superconducting motor was designed to fit the dimensions of a 2004 Prius engine. Both the primary and secondary windings of a synchronous induction motor, known as a high-temperature superconducting induction-synchronous machine (HTS-ISM), were constructed with superconducting wire. One of the added benefits of a strong superconducting motor would be potential elimination of the requirement for transmission gears. The transmission system in an automobile is one of the prime reasons for low efficiency. Note that the primary winding of the HTS-ISM, although superconducting, carries alternating current; special consideration must be given to selecting a superconducting material that can handle alternating current.

Clearly, high-temperature superconducting technologies are still in an emerging stage. They are finding use in applications where high power density is required and in specialized applications where development cost justifies the unique properties of superconductivity (Hassenzahl *et al.*, 2004). Accordingly, there are two general classes of superconducting applications: (1) those that replace classical power grid components with low-resistance, low-loss superconducting counterparts and (2) those that utilize superconductivity to obtain unique new power grid properties and features. Superconducting components, although initially more expensive, are expected to provide benefits that, over the long term, exceed those of classical components.

HTS power cables enable two immediate benefits: (1) lower resistive loss and (2) greater power density; that is, the capability to transfer more power over a smaller diameter cable. Typically, cables are installed underground or in special ducts. They maintain a constant voltage and the current is adjusted to change the power load as needed. There are two general types of superconducting power cable: (1) warm dielectric design and (2) cold dielectric design. In the warm dielectric design, the HTS wires are wrapped within a channel containing liquid nitrogen. The outer wall of the cable is wrapped with dielectric insulation at room temperature. It uses the least HTS wire but has a higher inductance and requires more cooling stations at closer intervals than the cold dielectric design. In the cold dielectric design, there are two layers of HTS wire separated by a cold dielectric insulating layer. The advantage of the cold dielectric design is that it has lower inductance, a higher current-carrying capacity, and less alternating-current loss, and has less electromagnetic leakage than the warm dielectric design. Of course, the cold dielectric design uses more HTS wire and has a higher initial cost to construct.

The impedance of the HTS cables have 1/6th to 1/20th the impedance of equivalent non-superconducting cables. The lower impedance and higher current-carrying capacity mean that they not only carry more power through a smaller space (for example, in urban areas), but they also relieve bottlenecks in the power grid by reducing impedance, providing an alternate path for power to flow instead of flowing through power lines that have reached their full capacity.

A growing concern with superconducting power lines and other superconducting components in the power grid is that higher fault currents will be generated. In other words, the power grid will be capable of carrying much more current than it otherwise would, and this can result in larger and more prolonged fault events than would otherwise be possible. One solution is to incorporate a stabilizer; that is, a nonsuperconducting pathway through the cable. When a fault occurs, the current would temporarily flow through the stabilizer and the superconducting power cable would transition to a normal, conducting state, thus raising the impedance and limiting the fault current. It should also be noted that, in the worst case, the liquid nitrogen, which is used as the HTS coolant, is nontoxic if accidently released into the air. Another complicating factor is that HTS systems require careful consideration when interfacing to room-temperature components. The severe change in temperature can cause materials that interface with the HTS system to undergo severe temperature gradients that can cause the materials to distort their shape and the properties that they would normally have at ambient, room temperatures.

Recall that Section 3.4.4 discussed HVDC power transmission. HTS power lines can play a role in both alternating-current extra-high voltage (EHV) and HVDC transmission. EHV alternating-current power transmission attempts to minimize the I^2R power loss by reducing current and increasing voltage. However, there are trade-offs due to the higher voltage. More insulation is required, the transmission line must be constructed higher above the ground, transformers must be larger and more expensive to perform the high step up and step down in voltage, and more transformers are required than would otherwise be needed. In addition, alternating current requires reactive power compensation, typically in the form of capacitors along the line, in order to keep the voltage and current in phase with one another. At such high voltages, such capacitors are large and costly. HVDC, by using direct current instead of alternating current, eliminates some of the problems. There are only two lines required, instead of three (assuming three-phase power), and there is no reactive power. Because there is no need to supply reactive power, current can be reduced in the HVDC system while still delivering the same amount of power. The disadvantage is that conversion between alternating and direct current is required at both ends of the HVDC transmission line, which adds significant cost. HTS power cables improve both EHV and HVDC transmission, but will have a different impact than classical transmission. While HTS power lines have lower impedance, they have similar parasitic capacitance to non-HTS lines. Thus, addressing capacitance, rather than inductance, could be an issue for HTS power lines.

As previously mentioned, high fault currents are a growing concern as the power grid continues to expand both in terms of number of consumers and generators and the power capacity that must be transported. Fuses, circuit breakers, and reclosers are a common form of protection against faults. They provide sudden, infinite impedance to the current flow. The problem with this approach is that they cause a sudden disruption of power to consumers and, for large faults, the sudden change can cause the power grid to become unstable. Another approach is to, in a sense, more gently react to the fault by simply limiting the current flow to sustainable amounts. Thus, power is not immediately lost to consumers and the power grid is not subject to a large, sudden change in impedance that could cause it to become unstable. Instead, the fault current limiting approach involves gradually raising the impedance of the faulted power line in order to reduce current flow. Superconductors provide an ideal means to accomplish this change of impedance. These are known as superconducting fault current limiters (SCFCLs).

There are several different types of SCFCLs. A purely resistive SCFCL R_{SC} operates in series with the main current flow and also has a conventional resistance R_p in parallel with it. Thus, the current changes with the change in SCFCL resistance as shown in

$$i_{SC}(t) = i_{sc}(t)\frac{R_p}{R_p + R_{SC}(t)}. \tag{14.46}$$

While the superconductor operates normally, it has near-zero resistance and the full input current flows to the output. When a fault occurs, the superconductor temperature will rise above its critical temperature and begin to create significant resistance. As can be seen Equation 14.46, as the resistance rises, current decreases. The nonsuperconducting, parallel resistance is known as a stabilizer and is typically implemented as a thin layer of conducting material residing over the superconductor. Remember that the power lines naturally have inductance; this means that there will be a temporary voltage buildup across the resistors as the superconductor resistance rises. If it rises too quickly, the build up can exceed safe values.

The inductive SCFCL is a little more interesting. It is embodied in a device that is similar to a transformer. However, the concept of operation is different from a transformer. In this device, both the inner and outer windings are supplied with current. The outer winding is the main current and the inner winding is comprised of a superconducting circuit. As long as the superconductor is in a superconducting state, current flow creates a magnetic shield for the yoke of the transformer. This brings the inductance, as seen by the main current flow, to near zero. The result is a low impedance for the main current. However, as a fault current will cause the superconducting resistance to rise, this reduces the shielding current flow within the transformer, allowing the main current to experiencc the inductance.

There is also a bridge-type SCFCL. In general, a bridge circuit is comprised of two branches that emanate from a common point and a circuit element connects, or "bridges," the two branches. In this case, the branches are comprised of rectifiers and the bridge connecting the branches is a superconducting coil. The branches then join together through additional rectifiers. During normal operation, the main current flows through the parallel branches, with a bias current passing through the superconducting inductor in the bridge. A fault current will cause a pair of rectifiers to become reverse-biased and remain closed, causing the main current to flow directly through the superconducting inductor in the bridge, thus increasing the impedance seen by the fault current.

Another use of superconductivity in the power grid is for SMES systems. Because a superconducting material has no electrical resistance, current will flow indefinitely, creating an associated magnetic field along with it. This field is a form of energy storage. One of the unique features of the SMES is that it can store and release energy nearly instantaneously. Compare this with a battery, which takes a considerable amount of time to recharge and releases energy at a limited rate. This makes SMES ideal when large amounts of power are required quickly. This can include injecting power when necessary to keep the power grid stable, as well as maintaining power quality and providing power when the main power system is interrupted.

An SMES is comprised of a superconducting coil and power conversion system. The power conversion system is required because energy is stored as direct current and many applications use alternating current. As mentioned, power is stored within the magnetic field created by the infinitely circulating current of the superconducting coil. As we know, the energy stored in an inductor is

$$E = \frac{1}{2}LI^2. \tag{14.47}$$

The inductance L is determined by the size and geometry of the coil and the diameter of the wire:

$$L = \mu_0 N^2 \frac{A}{l} \tag{14.48}$$

where μ_0 is the magnetic constant, N is the number of turns in the coil, A is the cross-section area, and l is the length of the coil.

The energy stored as a function of the magnetic field B is

$$E = \oint \frac{B^2}{2\mu_0}\,\mathrm{d}x\,\mathrm{d}y\,\mathrm{d}z. \tag{14.49}$$

Note that $B^2/2\mu_0$ is the local energy density and must be integrated over the three-dimensional space of the magnetic field. Notice that energy storage is based upon the square of the magnetic field; thus, increasing field strength has a significant impact on energy storage. Clearly, it is important for the superconductor to be able to remain in a superconducting state in the presence of a strong magnetic field. But recall that superconductors have a magnetic field threshold beyond which they lose their superconducting state.

In most storage systems, such as batteries or flywheels for example, the amount of energy stored increases linearly with size. However, SMES is different. Consider the magnetic field in a solenoid as shown in

$$\vec{B} = \mu_0 N \frac{I}{l}. \tag{14.50}$$

Here, I is the current flowing through the conductor, N is the number of turns of the conductor in the coil, and l is the length of the coil. It is found that the energy is related to the volume, V_{SC} as shown by

$$V_{SC} \propto E^{\frac{2}{3}}. \tag{14.51}$$

Another consideration in the construction of an SMES as in a transformer is the interaction force between the current and the magnetic field. This creates an outward mechanical force that must be compensated for in the construction of the superconducting coil.

Another general issue related to superconductors is the problem of quenching. If the magnetic field becomes too strong, the rate of change of the field is too rapid, or there are any other of a number of abnormal conditions, eddy currents can form and resistive heating, also known as Joule heating, can occur in certain spots. This localized heating raises the resistance and in turn causes more resistive heating to occur. This creates a vicious cycle in which heating raises resistance that in turn causes more heat. The result is that the entire superconductor quickly leaves the superconducting state and reverts to its room-temperature resistive state. This can be a violent event, resulting in an explosive noise as the energy in the magnetic field is converted to heat and the cryogenic fluid boils away. The sudden decrease in current can cause a large voltage drop across the superconducting device, resulting in sparks and arcing. The evaporated cryogenic fluid can cause asphyxiation. A stabilizer is a room-temperature, conductive pathway for current to flow in order to bypass the superconducting material if a quench occurs.

The SMES also contains an interface that controls when and how power flow between the power grid and the SMES takes place. The SMES communication interface can receive control signals from the power grid as well as generating information about the current state of the coil. The interface can allow the SMES to change power levels within a fraction of an alternating-current cycle; SMES is unique in that it can rapidly release power, changing from full charge to full discharge in milliseconds. This means that control of an SMES should have very low communication latency.

SMES is very efficient at 90%; however, losses do occur. Energy is lost during the charge and discharge process. But, once the device is charged, there is no loss of energy while current flows through the coil. There is also energy loss associated with ancillary equipment; namely, the cryogenic system and the power conversion system.

All of the previously discussed advantages of HTSs can be utilized for transformers. The operation of an HTS transformer is similar to the classical version. However, superconductivity provides no resistive loss and a denser magnetic field so that there is the potential for improved efficiency, reduced size, and less environmental impact because oil is not required.

Superconductivity has been leveraged by power systems researchers primarily for low resistance and dense magnetic fields. However, it is important not to forget that it is also a quantum mechanical phenomenon that has been exploited, at a smaller scale, for computation as well. As is well known, quantum computing holds tremendous potential for high-performance computation. While quantum computation will be used for power grid optimization, where classical power grid computation was discussed in Chapter 11, an interesting concept to consider is the integration of superconducting power and computing applications within the power grid. In this section, we begin by considering superconducting approaches to quantum computing. The discussion of quantum mechanics for computing in the power grid continues in Section 15.5.

14.6 Summary

Power electronic technologies are transitioning power from analog to digital form. They have enabled new dimensions in power control, although it is not without problems, such as leakage current and harmonic noise. At the same time, they are having a significant impact on the operation of the power grid and its communication requirements. Pulse-width modulation for power systems and pulse-code modulation for communications are not too different from one another. It will not be surprising if communication is encoded within variants of pulse-width modulation. A selected set of power electronic devices was presented in this chapter. The chapter ended with an introduction to superconducting technologies. These technologies have the potential to revolutionize power systems, communication, and computation.

Chapter 15 is perhaps the most interesting, because it ponders the longer term future of the power grid. It considers emerging technologies that could radically transform the power grid, communication, and computation. We know this because technologies exhibit trends that allow us to anticipate innovation far beyond the limited smart grid horizon. A few core themes are smaller-scale power and energy generation and management including nanogrids, development of power system information theory, greater ease and flexibility in power transmission, including wireless power transmission, the ability to harness power from geomagnetic storms, and the integration of quantum phenomena with the power grid, including quantum communication, computation, and energy teleportation. Power system information theory enables Maxwell's demon within the power grid opening new possibilities for power and energy. Nanoscale communication networks are discussed for future nanogrids. On the opposite extreme, space-based power generation is also explored. Thus, Chapter 15 will take us into the future of smart grid.

14.7 Exercises

Exercise 14.1 Thyristor

1. What is a thyristor?
2. What is the latching current?
3. What is the holding current?
4. Why is there a finite time delay before the thyristor can operate properly after the current has been removed?
5. What are some typical applications of thyristors in the power grid?

Exercise 14.2 Insulated-Gate Bipolar Transistor

1. What is IGBT?
2. How does an IGBT differ from a thyristor?

Exercise 14.3 Power MOSFET

1. What is a power metal-oxide-semiconductor field-effect transistor (MOSFET)?
2. How does it differ from a thyristor and an IGBT?
3. How similar is a power MOSFET to a computer integrated circuit transistor?

Exercise 14.4 Gate Turn-off Thyristor

1. How does a gate turn-off thyristor differ from a thyristor?

Exercise 14.5 Power Loss

1. What are the components of power loss in a power electronics device?

Exercise 14.6 Superconducting Theory

1. What is a Cooper pair and what role does it play in the BCS theory of superconductivity?
2. What is the de Broglie wavelength of a Cooper pair?
3. What is the coherence length of a Cooper pair and how does it relate to the spacing between atoms in a solid?
4. What are the DC and AC Josephson effects?
5. How narrow is the transition temperature between the normal and superconductive state?
6. What is the relationship between the superconducting transition temperature and the atomic mass of a substance?

Exercise 14.7 Superconducting Communications

1. Explain a few applications of superconducting technology in communications?

Exercise 14.8 Superconducting Generators

1. What is the general equation for the power density of a generator?
2. Why is superconducting technology seen as an ideal solution to wind energy generation problems?

Exercise 14.9 Superconducting Fault Current Limiters

1. What is the general concept of operation for an FCL?
2. How is a superconducting material utilized in an FCL?
3. What is the difference between pure resistive, inductive, and bridge-type SCFCL approaches?

Exercise 14.10 Superconducting Substations

1. What is at least one advantage of locating all superconducting components within a single substation?

Exercise 14.11 Superconducting Synchronous Condensers

1. What purpose do capacitor banks and synchronous condensers serve in the power grid?
2. What is an advantage of a synchronous condenser over a capacitor bank?
3. What is the difference between a synchronous condenser and a motor?
4. What are the advantages of a superconducting synchronous condenser and how is each advantage obtained?

Exercise 14.12 Superconducting Magnetic Energy Storage

1. What is the concept of operation of an SMES?
2. How much energy is stored in the coil of an SMES?
3. Explain what happens during a "magnetic quench."

Exercise 14.13 Superconducting Motors

1. How does one determine the power density of a motor?

Exercise 14.14 Superconducting Power Lines

1. What are the benefits of a superconducting power line?
2. Given that an HTS material is more expensive than a low-temperature superconductor material for cables, what is its advantage?
3. What are the two basic types of HTS cables and what are their general properties?
4. What are the pros and cons of the lower impedance of HTS cables over a traditional cable?
5. How should faults be handled in a superconducting power line?
6. Is there any negative environmental impact caused by liquid nitrogen cooling?
7. What are some of the problems involved in interfacing superconducting cables with ambient-temperature connections?
8. How do inductance and capacitance of superconducting lines compare with ambient-temperature power lines?

Exercise 14.15 Superconducting Quantum Computing

Just as power MOSFETs utilize integrated circuits that may be used for both power systems and computation, so superconducting materials in the grid can leverage quantum computing for both power systems and computational purposes.

1. What are the basic elements of quantum computation using superconductivity?
2. How could quantum computation be used to aid communication and computation in the future power grid?

Exercise 14.16 DG Devices

1. Why are power electronics considered a key enabling component of DG? Give specific examples.

15

Future of the Smart Grid

The energy of the mind is the essence of life.

—Aristotle

15.1 Introduction

This chapter explores the evolution of the power grid beyond what was seen on the horizon for smart grid in mid-decade of 2010. This chapter is a chance to envision what the future might hold. We can extrapolate some of the trends seen in power systems and communication, and learn from the technical evolution that has occurred in other fields as well. Thus, this is both a risky and a fun chapter to think about. There are as many views regarding how the power grid will evolve as there are experts in different fields related to power and communications – which are numerous. In the future view proposed by this chapter, we consider a few core themes. First, both power generation and management will become more ubiquitous and refined to the point that it requires a change in scale, from microgrids to nanogrids. Increasingly smaller scale power management will require small-scale communication, which we also consider here. Aligned with the notion of ubiquitous power, we also consider an increase in ease and flexibility in transporting power; namely, wireless power transmission. The idea that one will have to push a plug into a socket and remain tethered to a power cord will seem strange, quite inconvenient, and primitive. The idea that long cables had to be draped across the planet will also seem equally strange and perhaps ridiculous.

Before considering the future of the power grid in detail, let us first consider the nature of seminal ideas and what we can learn from the evolution of other technologies. There is much to be learned that can be applied here. A disruptive innovation creates a new market by displacing an existing technology. It is disruptive because business and markets do not anticipate the new innovation. On the other hand, the evolution of many technologies has been studied; both technology advances and propagation of the technology demonstrate some level of predictability to those who study such trends. Studies of such trends fall under the heading of "theory of innovation" or "diffusion of innovations." We ask ourselves what those trends will look like for the power grid in this chapter. An example trend is virtual power plants. A virtual

Smart Grid: Communication-Enabled Intelligence for the Electric Power Grid, First Edition. Stephen F. Bush.

power plant is simply a set of distributed power generation sources, often heterogeneous, that are operated as if they were a single, larger, centralized power generation plant. Advantages of a virtual power plant include flexibility and efficiency, because smaller generators can go online and offline more easily to respond to changing demand and the generators may use renewable power. A trade-off is the increase in complexity of managing multiple generators as a single plant. How far can this trend towards managing an increasingly larger number of smaller generators be taken? This leads us to move from virtual power plants, comprised of many small generators, to microgrids, comprised of small generators along with power transport and distribution over a local area. How far can this trend be taken? This leads to the notion of nanogrids namely, power generation and transport managed at the microscopic scale. Next, we review suitable communication for nanogeneration capability. This takes the form of nanoscale communication networks. Then we review a few emerging sciences and technologies related to power systems and the success of the smart grid. These include complexity theory, network science, and advances in machine learning. Next, emerging devices are discussed, including superconducting power cables and current limiters, as well as quantum communication and quantum computation. Quantum communication and computation can have a significant role in the future power grid. We end this last chapter on a topic that hearkens back to the birth of the power grid; namely, wireless power transmission. The chapter ends with a brief summary and a final set of exercises for the reader.

15.1.1 Theory of Innovation

The evolution of the power grid has been slow. However, it should follow the same developmental trends as many technologies. Technology evolution can be analyzed in a manner that is not unlike biological evolution, in which the technology is a biological organism and the market is the environment. The market places pressure on the technology such that all technologies tend to adapt in general ways that can be predicted.

For the smart grid, we first need to clearly identify the producer and customer of the technology. Upon initial consideration, one might say that the utility is the producer, and the end-user (for example, the residential home or business owner) is the customer. However, in this case, the customer, the ultimate consumer, has little direct input into the operation of the system. The utility makes the decisions as to how the grid will evolve on behalf of all end-users. It can make better sense, then, to consider the technical evolution from the perspective of the utility as the customer and the vendors supplying power system components to the power grid as the producers. The value of the power grid and each of its components is defined as the ratio of its "features" over its cost; increasing features and reduced cost lead to improved value.

This notion of value is captured in the "S"-curve plotted in Figure 15.1, which describes the life-cycle of a technology in terms of a relationship between features and cost. Initially, cost is low and features are high for a new technology with little competition. Over time, more competition enters the market and the technology is forced to compete for survival by rapidly increasing features, while cost may increase as well. This is viable as long as the value of the features exceed cost. Eventually, the top of the S-curve is reached. This occurs when no further features can be competitively extracted from the technology. At this point, only cost can be reduced. Finally, as the technology dies, both features and cost are reduced in order to hang on to the market with what remains of the technology. Finally, a new technology, with a new S-curve, will arrive to replace the old, dying technology. As the technology life-cycle follows

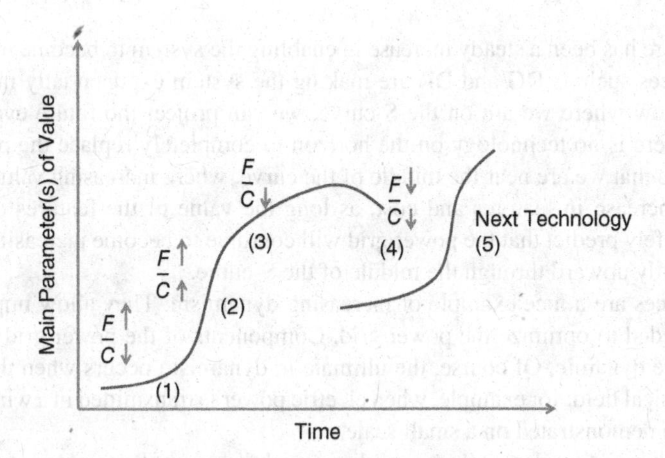

Figure 15.1 The life-cycle of a technology is shown from the perspective of value as a function of time. Different smart grid communication technologies will follow this life-cycle and are at different points along this curve. Value comes from either increasing functionality *F* or decreasing cost *C*. During research and development, initial costs in the lab are high; functionality shows improvement; however, prospects of failure are high (1). Then, in (2), out-of-pocket costs may be recovered while functionality improves along with increasing cost. In (3), the technology reaches maturity; functionality is high and stable. In (4), the technology declines relative to alternative technologies and the only remaining way to maintain value is to reduce cost. Finally, (5) illustrates the end of the original technology and the birth of a new one. Source: Adapted from Cameron, 2005.

its trajectory through the S-curve, there are common trends regarding innovation that can be used to predict how the technology will evolve. Communication technologies follow similar technology trends. Thus, the trick is to understand how communication and power system technologies will evolve in the future, both independently and together within the power grid.

When the power grid originated in 1882, it was a new technology with limited competition; for example, there was competition between Edison and Westinghouse. We might say that cost was moderate, but features, being novel, were high. However, as electric power became more ubiquitous, with the exception of developing a standard socket for electric plugs, the only features recognized from a consumer standpoint were safety and availability. Reducing cost was the only way to increase value. However, from a utility standpoint, many features were continuously added in order to achieve safety, availability, and cost reduction. In this respect, the power grid has been following one of the many predictable product trends. In examining the smart grid, the view of who is the producer and who is the consumer can shift. For example, the utility is the consumer of technology from vendors that produce products for the power grid. Ultimately, individuals who in the past have always been consumers of power may become producers of power for themselves, the utility, or directly for other consumers thus, the producer-consumer perspective is not always obvious or fixed.

A common trend in technology evolution is the move toward increased dynamism; that is, moving from a static, monolithic structure towards a dynamic, flexible one. This is evident in many ways, starting from the very beginning, in moving from direct current to alternating current, incorporating transformers, capacitors, and switches. Now there is an ever-increasing trend toward making the power grid more active with power electronics; for example, power converters. There has been a steady increase in enabling the system to become more dynamic. Clearly, advances such as DG and DR are making the system exponentially more dynamic. Thus, if we know where we are on the S-curve, we can project the future evolution of the power grid. There is no technology on the horizon to completely replace the power grid, so we may assume that we are near the middle of the curve, where increasing value comes from a continuous increase in features and cost, as long the value of the features outweighs the cost. We can safely predict that the power grid will continue to become increasingly dynamic, increasing rapidly upward through the middle of the S-curve.

FACTS devices are a nice example of increasing dynamism. They allow impedance to be changed as needed to optimize the power grid. Components of the power grid will continue to become more dynamic. Of course, the ultimate in dynamism occurs when the technology becomes a physical field; for example, when electric power is transmitted in a wireless manner, which has been demonstrated on a small scale.

When evaluating technology trends, we distinguish between the target system, the power grid and communication in our case, and the environment in which it resides, known as the supersystem. General technology trends indicate that individual power system components will tend to integrate with one another, as they view each other as supersystems as well as merge with the grid supersystem. For example, ubiquitous components such as transformers and power converters will take on more functionality and merge with other capabilities in the grid.

Another general trend in technology evolution is known as integration with the supersystem. The components of a technology will evolve to merge with their immediate environment, known as the supersystem. Ultimately, nanoscale generation will be the complete integration with the environment. Individual PV panels include built-in power converters and storage. They may even merge with each other. The power grid will integrate with electric vehicles such that vehicles take on functionality to help the grid; for example, temporary storage or other ancillary support services. The integration of the power grid with communication is the essence of the smart grid. There are numerous other infrastructures with which the power grid may integrate as it tends to merge with its "supersystem."

Another trend clearly indicated in the power grid is increasing controllability. Systems tend to develop more ways in which they can be controlled. The level of control increases and the number of controllable states increases. DR is the poster child for controllability, but there are more subtle forms of controllability. As mentioned, FACTS as well as many other advances in power electronics are allowing more dimensions to be controlled.

Components become more coordinated with each other and with their supersystem. The balance between supply and demand is one salient form of coordination. Designing the power grid to be stable is another. Knowing precisely when and where to incorporate communication is one of the key forms of coordination for the power grid.

15.2 Geomagnetic Storms as Generators

In early September of 1859, strange events began to occur to one of humanity's earliest, large-scale, technical infrastructures; a communication infrastructure that most resembled the power grid at the time, the telegraph system. Telegraph systems began to fail and some telegraph operators received electric shocks. Then telegraph pylons began to emit sparks and telegraph paper began to ignite into flames. However, one of the most interesting phenomena observed was that some telegraph systems continued to operate after their power supplies had been completely disconnected.

These phenomena and others like them became part of what came to be known as the Carrington Event because the British astronomer Richard Carrington was the first to realize that it was related to a coronal mass ejection that he had witnessed throughout August of 1859 and a particularly large ejection that occurred on September 1. This event is particularly significant because it was the first known use of GIC harnessed for useful purposes; namely, the powering of telegraph systems. Recall that geomagnetic storms were first introduced in Section 3.4.2.

The details for the cause of GIC are complex and not fully understood. The sun emits vast amounts of plasma through coronal mass ejections that form what is called the solar wind passing through the Solar System. Plasma, by definition, is an ionized state of matter; the plasma carries the magnetic field of the Sun with it as it travels. A physical process known as "magnetic reconnection" occurs when the magnetic field line topology becomes rearranged; this can cause the release of large amounts of kinetic and thermal energy, as well as cause particle acceleration (Pulkkinen, 2007). It is possible for the Earth's magnetic field, residing within the Earth's magnetosphere, to undergo magnetic reconnection with the surrounding interplanetary magnetic field (IMF) from the solar wind. This can create electric current within the Earth's ionosphere, an outer layer of the atmosphere that has been ionized by the solar wind. The solar wind can actually push the Earth's magnetosphere toward the night-side of the Earth, distorting its normal topology and causing magnetic reconnection to occur. Strong currents within the ionosphere can then induce current within conductive terrestrial objects. However, the amount of current induced depends upon many factors, including the distance between the ionospheric current and the conductive terrestrial object, the orientation of the ionospheric current and the terrestrial object, and the ground resistivity around the conductive terrestrial object. If the terrestrial object is part of a larger conductive network, then the conductive network topology can have a significant impact on the amount of induced current. Given that most of these parameters are unknown, the ability to predict the amount of induced current has been an extremely challenging task.

The ionosphere has played a significant role in both power and communication. Radio communication has long relied upon the ionosphere for the generation of skywaves, the ability of the ionosphere to reflect radio waves generated from the ground back toward the ground. This enables relatively low-power, high-frequency, line-of-site communication to travel very long distances around the Earth. Electromagnetic waves from a radio transmission cause electrons in the ionosphere to oscillate at the same frequency as the radio wave. Energy is absorbed and thus lost in the process. However, this resonance can cause the RF waves to be re-radiated by the ionosphere back towards the ground.

Electric power utilities have long experienced the effects of GIC regardless of whether it was fully understood (Kappenman, 1996). Transformers and other equipment tend to fail more frequently in regions more highly impacted by solar storms, and these failure trends follow the 11-year solar cycle of sunspot activity, although with an approximate 3-year lag. A simplified explanation views the GIC as primarily a direct-current offset, although GIC is variable and is not, strictly speaking, direct current. However, the offset-current from GIC causes power transformers to saturate every half-cycle. During saturation, alternating current flowing through the transformer loses its ideal waveform. This corruption of the alternating-current waveform can then cause harmonics and other undesirable affects to propagate through the power grid. Once utilities became aware of GIC, it has been recognized as the cause of numerous significant power outages throughout the world. One example is the March 13, 1989, geomagnetic storm that caused the collapse of the Hydro-Quebec power grid as the protection system began to respond by opening relays, causing a cascading power outage.

Simple steps to protect the power grid from GIC have included adding series capacitors to block the direct-current component from GIC and adding strong capacitors to ground to prevent current flow from the Earth entering the neutral wire of the power line. However, the ground capacitor also blocks current flow needed if there is an actual power fault. Thus, a fast and reliable device is required to distinguish fault current from GIC and switch the capacitor into the circuit when needed. Unfortunately, all of these approaches are expensive and often rarely used. Thus, the cost justification can be difficult. Another approach is to attempt to better understand the cause of geomagnetic storms and predict when they will occur and how strong they will be. With enough warning, preventive action can be taken that may be less expensive than building permanent safeguards. It has been estimated that observation of coronal mass ejections can give a 2- to 3-day advance notice of a GIC event. However, this time period is highly variable. Satellites also sit at a stationary point approximately one million miles from the Earth, known as a Lagrangian point, capable of monitoring the Sun and and Earth's magnetic field. A Lagrangian point is a position in orbit where the satellite appears stationary with respect to the Sun and the Earth. This allows the satellite to provide a warning up to an hour beforehand of an impending geomagnetic storm.

Research into the correlation between measurements of geomagnetic storm activity, the interplanetary magnetic field, the Earth's geomagnetic field, and GIC as measured at ground level has been done to help predict and understand GIC (Trichtchenko and Boteler, 2006). Clearly, the impact of the magnetic field is strongest when perpendicular to the direction of a power line and the induced electric field is proportional to the rate of change of the geomagnetic field. Thus, the rate of change of the geomagnetic field is often used as a measure of potential GIC strength. However, ground conductivity also plays a significant role. A closed contour may be formed through the power line and deep underground. The electrical characteristics of the underground conductive path play a significant role in the strength of the induced current. Specifically, as previously mentioned, GIC is not strictly direct current, but has frequency components that may be impacted by the electrical characteristics of geological formations deep underground. It turns out that the strength of the GIC may be proportional either to the derivative of the geomagnetic field or directly to the strength of the field itself.

An example of GIC can be analyzed via a simplified application of the Biot–Savart law. To find the current induced in a loop, consider voltage as determined by the area of the loop multiplied by the magnetic flux through the loop, assume the resistance of the loop can be estimated, and then compute the current using Ohm's law (Pirjola, 2000). The voltage is $g\pi a^2$, where g is the time derivative of the vertical magnetic component across the loop in nT/s and a is the radius of the loop in km, and r (m Ω/km) is the resistance of the loop. Let the electric field parallel to the power line be E, the power line be of length L, and the resistance of the Earth connection be S. Then the GIC is

$$\text{GIC} = \frac{EL}{S + rL}. \tag{15.1}$$

As L becomes larger – that is, for longer power lines – GIC increases. Also, for very large L, GIC tends to become independent of the length.

Geomagnetic storms clearly involve tremendous amounts of power. Why are significant sums of money and effort being spent to mitigate, avoid, and thus waste this power rather than utilize it? In fact, as we have seen in the geomagnetic storm of 1859, the power from a magnetic storm was actually harnessed, although accidentally, to do useful work in operating telegraphs. In fact, it does indeed appear possible to harness such power (Pulkkinen *et al.*, 2009).

As we have seen, the maximum power attainable from geomagnetic storms is determined by the solar wind impacting the magnetosphere. It has been estimated that the total amount of power from both electromagnetic and kinetic energy is on the order of 0.1 to 1 TW with periods of up to 10 TW of energy. However, up to 100 GW of power is lost by Joule heating and being pulled into the ionosphere, known as electron precipitation. It is interesting to compare this source of power with other forms of power, such as wind estimated at 100 TW and solar radiation, estimated at 100 000 TW. It should be noted that electricity generation at the time this was written provides an average power of 2 TW. Thus, while geomagnetic storm energy is interesting, it does not appear to have the raw power of more traditional sources of renewable energy generation. Although we also have to consider that traditional forms of renewable power generation can only extract a fraction of the total possible energy stated. This is because of inefficiency in renewable generation and the fact that it is not possible for renewable generators to be placed everywhere in order to completely extract all energy.

Prior work has considered the energy that could be extracted from a geomagnetic storm using a conventional, but extremely large, coil. Note that no superconducting elements, which could improve efficiency, were considered in this analysis. The coil is assumed to be comprised of the same material as used in ultra-high voltage transmission lines. The hypothetical coil in this analysis is quite large, assuming a circumference of 4200 km, giving it an area of 700 000 km^2. It is assumed to be comprised of 50 turns of the cable. The cable has a resistance of 3×10^{-3} Ω/km. In order to derive the maximum power from the coil, it is assumed that the resistance of the load matches that of the coil for maximum power transfer. The load has resistance R_L and, based upon the length and resistance of the cable, the coil has a resistance of R_I which is 630 Ω. This assumes $R = \rho L$, where L is the length of the conductor and ρ is the unit resistance. The analysis is simplified by assuming that the geomagnetic storm-induced-current is direct current even though in reality it varies over the duration of the storm. This analysis is made more realistic by using actual storm data. The analysis used data from a real geomagnetic storm that occurred on April 6–7, 2000.

First, the vertical component of the magnetic field from the geomagnetic storm B_Z is integrated over the surface area of the giant coil:

$$\Phi_S = \int B_Z \, dS \tag{15.2}$$

This results in the total magnetic flux, which is required to compute the electromotive force in the coil. Note that the magnetic field from the geomagnetic storm varies rapidly. Next, knowing the number of turns in the coil N and the magnetic flux Φ_S that is normal to the surface area of the giant coil from Equation 15.2, the electromotive force ε is

$$\varepsilon = -N\frac{d\Phi_S}{dt}. \tag{15.3}$$

The maximum power transfer theorem helps to determine the optimal load source and load resistance in order to maximize the total power transferred from the source to the load. It turns out that the maximum power is transferred when the source resistance and load resistance are equal. Note that this does not mean the efficiency is higher. If the load resistance is greater than the source resistance, then the efficiency may be higher, but the total power transmitted to the load will be lower since the total resistance is greater than if both source and load resistance were equal. On the other hand, if the source resistance were greater than the load resistance, more power is dissipated within the source and less reaches the load. The optimum amount of power transfer occurs when both are equal. If V_S is the source voltage, R_S is the source resistance, and R_L is the load resistance, then

$$P_L = \frac{1}{2}\frac{V_S^2 R_L}{(R_I + R_L)^2} \tag{15.4}$$

shows the maximum power transferred as a function of voltage and resistance. Simplifying Equation15.4 and replacing the voltage with the electromotive force ε yields

$$P_{max} = \frac{1}{4}\frac{\varepsilon^2}{R_I}, \tag{15.5}$$

where R_I is the internal resistance of the giant coil defined earlier. Peak voltage would vary over the duration of the geomagnetic storm. In this particular storm data, it rises up to 100 kV at times. The power output would rise to 1 MW during the strongest portion of this storm. The use of superconducting technology could significantly increase power output. It should be noted that this is an ideal, simplified analysis; more detail could be incorporated that would result in a more realistic analysis.

We should note that many large coils already exist within the topology of the power grid. It would be a matter of quickly reconfiguring the network to take advantage of the geomagnetic storm energy and communication would be crucial in accomplishing this. Conceptually, we would like a "Maxwell's demon" to exist within the power grid capable of capturing the geomagnetic storm energy. This could someday be a new feature of the "smart grid."

15.3 Future Microgrids

As we know from Section 5.5.1, microgrids are a local grouping of generators, storage, and loads that operate as a single entity. Microgrids can operate in a mode that is either connected or autonomously disconnected from the power grid. The concept of the microgrid has been proposed as a way of constructing a simpler, more-resilient power grid architecture. In an ideal case, the entire power grid would be comprised of interconnected microgrids, each running as autonomous units and each capable of going up or down with minimal impact on neighboring microgrids. They were originally designed to require little, if any, explicit communication (Lasseter, 2002). Instead, localized sensing and control could be implemented to manage the microgrid. Voltage regulation, reactive power control, and frequency could all be handled by means of droop control, as was discussed in Section 5.5.1.

Returning to the topic of microgrids, they were designed to partition the power grid into small, reliable, manageable chunks. Instead of managing entire interconnects that can reach the size of a subcontinent and supplying hundreds of gigawatts of power, the power grid can be partitioned into smaller, more-manageable, and potentially autonomous operating units. An obvious and reoccurring question regards how far this concept can be taken in terms of scaling down to smaller power grid units and the fundamental impact of scale.

15.3.1 From Power Grids to Microgrids to Nanogrids: Continuing the Reduction in Scale

It is obvious that most of our electronic devices have their own power supply. We take for granted that there is a power socket in the wall and that is the definitive interface between the power grid and our electric devices. Typically, the assumption has been that everything impacting the power grid and smart grid takes place on the outside (that is, the power grid side) of the wall socket. The question that this section raises is what happens if that interface or barrier is broken. In other words, what happens if individual devices were to become more active components in the operation of the power grid? This is the realm of the so-called "nanogrid": the idea that current, often strong psychological barriers, such as the power socket, are broken and the power grid is divided into ever-smaller autonomous units. Imagine individual power electronic devices such as laptops, televisions, and toasters all take a bit more active role in power control than they do as of the time this book was written. Each device could monitor typical power grid operating parameters such as frequency, current, voltage, and phase. The devices could regulate their individual duty cycles to reduce stress on the power grid. They might each control a small portion of reactive power using their local, and ever-more sophisticated, power supplies and power converters. Each device would decide when to use power based upon price controls, as in the existing DR paradigm. Ultimately, each device would seek for, and possibly even generate or scavenge for, its own power. As communication becomes more ubiquitous among devices, the integration between communication and power will become tighter, with more devices sharing the same physical channels for power and communication. Eventually, each device will become its own small "microgrid" or nanogrid.

It is possible that, as this happens, the need for ubiquitous alternating current will need to be reconsidered. Direct current may make more sense in many such applications. The number of power converters and transformers could be reduced by having devices share direct current just as easily as many devices today share information with one another. Each device becomes a "prosumer." The micro- to nano-grid transition can be a seen as a transition to a swarm-like power architecture, with many, small devices operating under simple, local rules that lead to global optimal power generation, transport, and management.

But why stop at this scale? Why not think from the bottom up, the very bottom, instead of the top down? In other words, consider power generation and distribution at the smallest possible scale and consider what such a real nanogrid might look like and what type of communication it might require (Bush 2013b). Section 15.3.2 considers nanogeneration and power harvesting.

15.3.2 Nanogeneration

> Electric power is everywhere present in unlimited quantities and can drive the world's machinery without the need of coal, oil, gas, or any other of the common fuels.
>
> —Nikola Tesla

A key element of a real nanogrid is the nanogenerator. Any generator is really a power converter, converting power from one type to another; for example, thermal or mechanical to electrical. Nanogenerators are no different: they convert thermal, chemical, electromagnetic, and mechanical power to electricity, but on a much smaller scale. This is similar to energy harvesting used today for small electronic devices. Energy sources surround us, emitting continuous sources of power that we never use. These ambient power sources range from small thermal gradients to the electromagnetic energy from television and radio stations. There are a myriad nanogeneration and energy-harvesting devices that are operational, in development, and proposed. While providing an exhaustive list of these devices would be tedious and invariably skip some important future devices, it is instructive to visit a few in more detail.

It is possible to use nanostructures that exhibit the piezoelectric effect to generate electricity in the form of a piezoelectric nanogenerator. The piezoelectric effect is the electric charge that accumulates when mechanical stress is applied to some materials. The effect is reversible, which means that applying a charge to the piezoelectric material will induce stress in the material and cause its shape to expand. As an aside, this technique can be used for many applications, including the creation of ultrasonic sound waves. In simple terms, the piezoelectric effect comes from changes in dipole alignment in piezoelectric materials induced by stress in the material. As a specific example, a piezoelectric nanogenerator harnesses the charge buildup in a piezoelectric nanowire. Forces exerted perpendicular to the nanowire can be extracted from the charge distribution along the top tip of the nanowire. Also, a group of vertically grown nanowires can generate current when force is applied along the top and/or bottom of the wires. For example, pressing down on the nanowire produces current in one direction and pulling up on the nanowires creates current in the opposite direction. Thus, a nanoscale alternating-current generator can be formed.

Another type of nanogenerator converts mechanical energy to electricity via creating and capturing static electricity, known as the triboelectric effect. Whereas "piezo" in piezoelectric means "pressure," the word "tribo" in "triboelectric" means to rub. It is the generation of electric charge caused by rubbing certain different materials together. The ancient example of rubbing glass with fur is a classic macroscopic example. The strength of the charge produced depends upon the material, roughness of the contact area, temperature, and strain, among others. The triboelectric generator uses this concept with materials at the nanoscale in particular, bringing two sheets of material together with electrodes on their backs to gather the charge. When the sheets are pressed, causing them to come into contact with one another, current is created in one direction; when the pressure is released and the sheets come apart, current flows in the opposite direction. Thus, this nanogenerator is also an alternating-current generator.

Switching from mechanical to thermal energy, the pyroelectric nanogenerator can utilize the spontaneous polarization of certain anisotropic materials due to temperature fluctuations. This differs from the Seebeck effect, which relies upon a temperature difference between the two sides of a material to drive diffusion of charge carriers. Instead, the mechanism for the pyroelectric effect is related to the behavior of dipoles in a material. When there is a constant temperature, the dipoles take their natural orientation with only a small amount of wobble in their position and no current is produced. When the temperature rises, the dipoles have a greater amount of wobble in their alignment, creating a charge. When cooled, the dipoles have less wobble around their normal alignment, become more aligned, and this also creates current in the opposite direction from the previous case. Thus, again, we have a potential alternating-current generator given regular temperature fluctuations.

There are, of course, many other forms of energy harvesting that can be used to create electricity at the nanoscale. As previously mentioned, this includes extracting ambient electromagnetic energy from radio waves, much of which is wasted. We have covered the typical renewable energy sources such as wind and PV. For these, the concept will be to determine whether nanogenerator versions of these renewable DG sources will have advantages; for example, greater efficiency, improved power transfer, or greater power density while leading to lower overall cost.

The nantenna is an interesting approach for capturing ambient wireless power. Recall from Section 8.4 that the rectenna is a rectifying antenna that has been used to receive wireless power at microwave wavelengths. The nantenna is a nanoscale rectifying antenna. Theoretically, light is electromagnetic radiation at a higher frequency than radio waves, and so an antenna that is small enough should be capable of receiving power in a similar manner to the rectenna. The concept is to use large numbers of nantennas to turn light into usable amounts of electric current. The theory of operation of the nantenna is the same as that for the rectifying antenna. The electromagnetic light wave will induce current in the nantenna that oscillates with the electromagnetic field from the light, yielding an alternating current. A rectifier in the nantenna converts the alternating current to direct current. Since light from the sun has a spectrum that ranges from 0.3 to 2.0 μm, the nantenna should be hundreds of nanometers long. This will allow the nantenna to resonate with lowest impedance with the light.

There are complicating factors in using the nantenna. For example, the "skin effect" occurs when, at high frequencies, current tends to flow through the nantenna only near the surface. This creates greater resistance than expected. Another problem is creating diodes that operate efficiently at the required size and frequency. Large parasitic capacitance occurs using conventional diodes at the necessary size and frequency.

Taking the above challenges into account, claims have been made that the nantenna has a much higher efficiency than conventional single-junction solar cells. Another advantage of the nantenna is that by adjusting its length it can become more efficient at receiving energy from different frequencies of light. In other words, it could be possible to dynamically adjust the sensitivity of the nantenna to different frequencies of light. This degree of freedom is currently difficult to achieve in current single-junction solar cell technology. Conventional solar-cell technology uses a fixed semiconductor band gap to convert photons of light into electricity. Changing the band gap could require changing the semiconductor material and would be difficult to accomplish dynamically.

Another potential form of nanogeneration is electrostatic or capacitive. The main concept is to utilize a structure similar to a charged capacitor; that is, oppositely charged plates. The charge exerts a force between the plates. Mechanical force exerted on the plates changes their separation distance and is converted to electrical energy. The disadvantage of this approach is that some form of initial electromotive force must be present to create the initial charge on the plates.

Magnetostatic nanogenerators are somewhat similar in concept to conventional electric power generators. Magnetostatic nanogenerators are very small rotating magnets that, as they wobble due to vibration, can induce current in nearby conductors. There are numerous other types of potential nanogenerators, including biological ones that draw energy from chemical process within living organisms to create electric power. However, these would take us beyond the scope of this book.

15.3.3 Real Nanogrids

One might consider power supplies and management within integrated-semiconductor computer chips as the closest thing in existence today to a real nanogrid. However, semiconductor chips are carefully designed, controlled platforms in which loads are small, interconnects are fixed, and components are reliable, carefully controlled, and well understood. A power line rarely goes down or a widespread blackout rarely cascades inside a semiconductor chip (except perhaps when used improperly or exposed to static electricity or other abnormal conditions).

The type of nanogrid we are considering is one that involves millions of small nanogenerators accumulating enough power to run a heavy household device such as a washing machine. The nanogenerators and their supporting electrical transport system, or nanogrid, are not assumed to be either fixed or robust. In fact, the nanogenerators may be mobile, able to adjust their position in order to extract the maximum power from their environment. For example, they should be able to turn toward the sun or move toward sound and vibration, or simply move in a better position to "feed" from their sources of power. These nanogenerators would have the capability of swarm-like nodes, following simple rules and communicating only with their nearest neighbors. As we have seen, some nanogenerators produce alternating current and some direct current; therefore, nanoscale power conversion may be needed. The generators would have to decide how to optimally route their power. Let us consider what communication would look like in such a real nanogrid.

15.4 Nanoscale Communication Networks

A nanoscale communication network provides communication channels among nanomachines typically assumed to be on the order of a few hundred nanometers in size. The vision, definition, and framework for a nanoscale communication network are currently being developed in the IEEE P1906.1 standards working group. Such a network will be a very small "Internet of Things" or an M2M communication network. In a real nanogrid, the "things" or machines will be nanogenerators, discussed earlier, and potentially power interconnection channels and power accumulation and storage devices and nanoloads. One of many challenges will be addressing how the nanogenerated power is aggregated and interfaced with the human-scale world. The nanoscale communication standard explicitly excludes fixed, integrated-circuit interconnects; it focuses upon a framework enabling ad hoc nanoscale communication. As illustrated in Figure 15.2, the physical layer includes several types of nanoscale networks, including: (1) molecular motors (walking), (2) cell signaling (flow and diffusion), (3) carbon nanotube (electromagnetic), and (4) quantum nanoscale networks.

As often happens with seminal concepts, the inspiration and early results come from nature. Many early developments in nanoscale networking have come from observing and understanding nanoscale communication used in biological systems. This is clearly the case for molecular motors and the large, general category of cell signaling mechanisms. Molecular motors are complex molecular structures found in living cells that are capable of "walking"

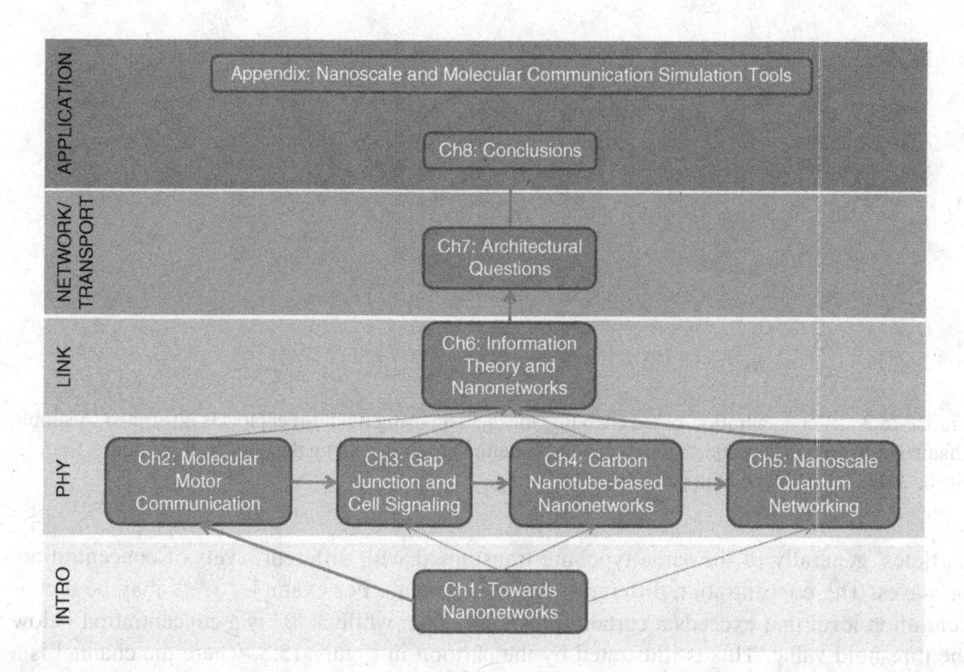

Figure 15.2 A nanoscale communication network allows high-resolution management of energy resources, including nanogenerators and the extraction of small amounts of energy from the environment. There are many potential physical layers; a framework for nanoscale communication networks is being developed by the IEEE P1906.1 standards working group. Source: Bush, 2010b.

while carrying a cargo. They "walk" along small, rail-like structures that form the cytoskeleton of a cell, called microtubules. Microtubules are long cylindrical structures, slightly larger than carbon nanotubes. The cargo carried by a molecular motor is another molecule, often viewed as a packet of information in nanoscale communication. Variations of biological molecular motors have been artificially created to perform different functions. Any molecular structure that is capable of generating its own motion by walking or actively ratcheting itself along a pathway can be considered a nanoscale communication channel.

Other, general, cell-signaling techniques include flow and diffusion-based techniques for nanoscale communication. These are passive mechanisms in which molecules representing information have no active carrier, such as a molecular motor, but passively rely upon the environment for motion. This includes, for example, molecules in a micro- or nanofluidic medium in which fluid flow and turbulence are present in order to "carry" molecules representing packets of information. Another passive means for channel operation is the reliance upon diffusion to provide information transport. Calcium signaling in biological organisms is a classic example.

Several different mechanisms for encoding information have been proposed within the above categories of nanoscale communication channel mechanisms. One approach is molecular encoding, which is simply assigning a symbol to unique types of molecules or variants of a molecular structure. The receiver interprets the symbol based upon the type of molecule detected. Another approach is concentration-based encoding. In this approach, many small

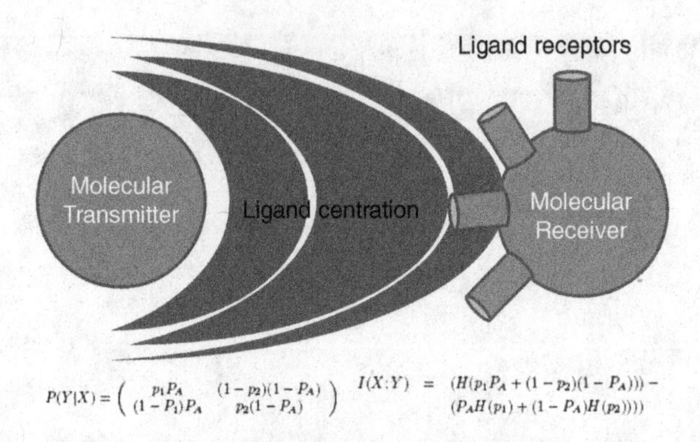

$$P(Y|X) = \begin{pmatrix} p_1 P_A & (1-p_2)(1-P_A) \\ (1-P_1)P_A & p_2(1-P_A) \end{pmatrix} \qquad I(X:Y) = (H(p_1 P_A + (1-p_2)(1-P_A))) - (P_A H(p_1) + (1-P_A)H(p_2))))$$

Figure 15.3 A concentration-based encoding mechanism using ligand-receptors is illustrated. A simple channel transition matrix and channel capacity equation are shown along the bottom of the figure. Source: Bush, 2011. Reproduced with permission of Springer.

particles, generally of the same type, are transmitted with different levels of concentration, or waves. The concentration difference encodes a value. For example, a "1" may be a concentration level that exceeds a certain threshold value, while a "0" is a concentration below the threshold value. This is illustrated by the cartoon in Figure 15.3, where the channel is a ligand-receptor nanoscale communication system. The channel transition matrix and mutual information are shown beneath the figure, where P_A is the probability of emission of the ligand, p_1 is the probability of successful delivery and reception of the ligand, and P_2 is the probability that the receiver detects no ligand when none was transmitted. The channel transition matrix is simply a square matrix whose elements are the probability of symbol transmitted along the rows and symbol received across the columns. Thus, the diagonals of the channel transition matrix are the probability of symbols transmitted and received properly. The off-diagonal elements are symbols that have been confused. From this matrix, the mutual information $I(X; Y)$ and the total theoretical channel capacity of such a communication channel can be determined. Determining precise values for the above probabilities for a specific, realistic nanoscale communication network is the challenging part.

There are also electromagnetic-based approaches to nanoscale communication. These tend to be inspired by existing macroscale communication technologies, but scaled down to the order of hundreds of nanometers. Examples of these techniques include the single-carbon-nanotube radio, which is basically a mechanically vibrating carbon nanotube whose vibrations are cause by a transmitted radio signal. A single carbon nanotube is able to simultaneously serve several radio receiver functions, including antenna and demodulator. Carbon nanotube antennas have also been explored for terahertz wave transmission and reception, as well as ad hoc random communication interconnects. The single carbon nanotube radio is an excellent example of the technology trends of integrated functionality and merging with the supersystem.

Finally, quantum mechanical phenomena are being explored for nanoscale communication networks. As we reach down to the very small scale, the wavelike nature of matter begins to exhibit more distinctly, and this may be leveraged to design entirely new forms of communication. General concepts being explored in this category of nanoscale communication include leveraging quantum entanglement and quantum superposition along with quantum operators for transporting energy states over a distance.

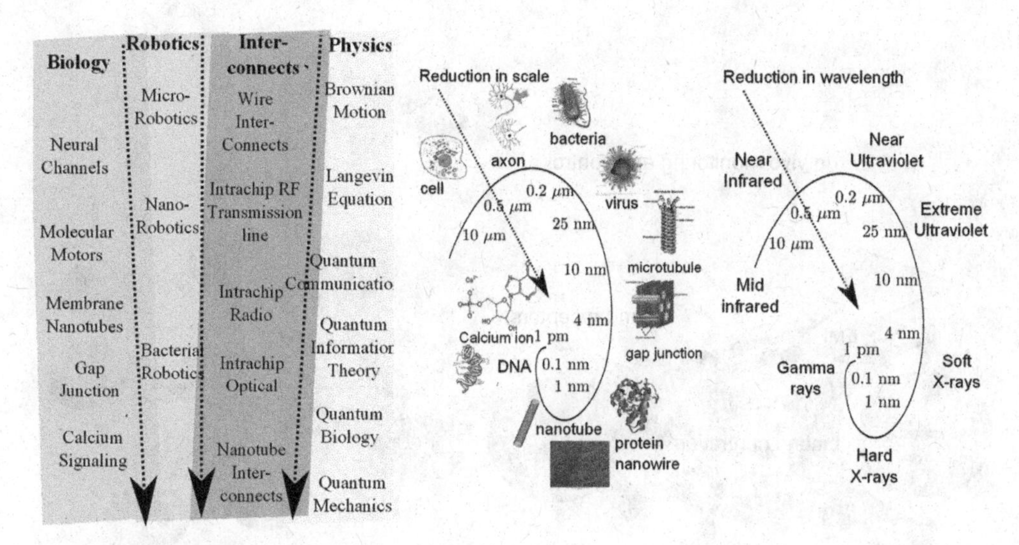

Figure 15.4 The origin of communication at decreasing scale has been driven by biology, nanorobotics, semiconductor interconnections, and basic physics, as illustrated on the left. On the right, components of nanoscale communication networks are compared with electromagnetic wavelengths. Source: Bush, 2011. Reproduced with permission of Springer.

Some of the drivers and inspiration for nanoscale communication, as well as the change in scale, are illustrated along the left side of Figure 15.4. Biology, nanorobotics, leveraging understanding of existing chip interconnects, and fundamental notions from physics have all pointed toward a smaller-scale form of communication. The right side of the figure provides an overview of the scale involved in nanoscale communication. The scales are in the form of a spiral, indicating that the scale can extend outward toward the infinite or inward toward the infinitesimal. The left spiral shows various entities that have been explored for implementing nanoscale communication channels. The spiral on the right shows the wavelengths that exist at precisely the same scale as that on the left. Entire nanoscale networks can fit within the wavelength of some of today's typical communication systems. A few applications of nanoscale communication networks are sketched in Figure 15.5. Examples include potentially leveraging in vivo biological-signaling mechanisms for an in vivo Internet that could be used to monitor and control biological functions on an individual molecular level. The ligand-receptor system represents the general fact that nanoscale communication often utilizes physical principles not typically used in communication, and thus not susceptible to problems such as electromagnetic interference. Also, smart materials have long been a dream; in this case, nanoscale communication provides the communication mechanism by which composite elements of such materials could communicate with one another (Bush and Goel, 2006).

The IEEE P1906.1 *Recommended Practice for Nanoscale and Molecular Communication Framework* (http://standards.ieee.org/develop/project/1906.1.html) is a standards working group sponsored by the IEEE Communications Society Standards Development Board whose goal is to develop a common framework for nanoscale and molecular communications. The framework is designed to balance room for innovation in this emerging technology with the clear definition required for progress to occur by determining a common definition,

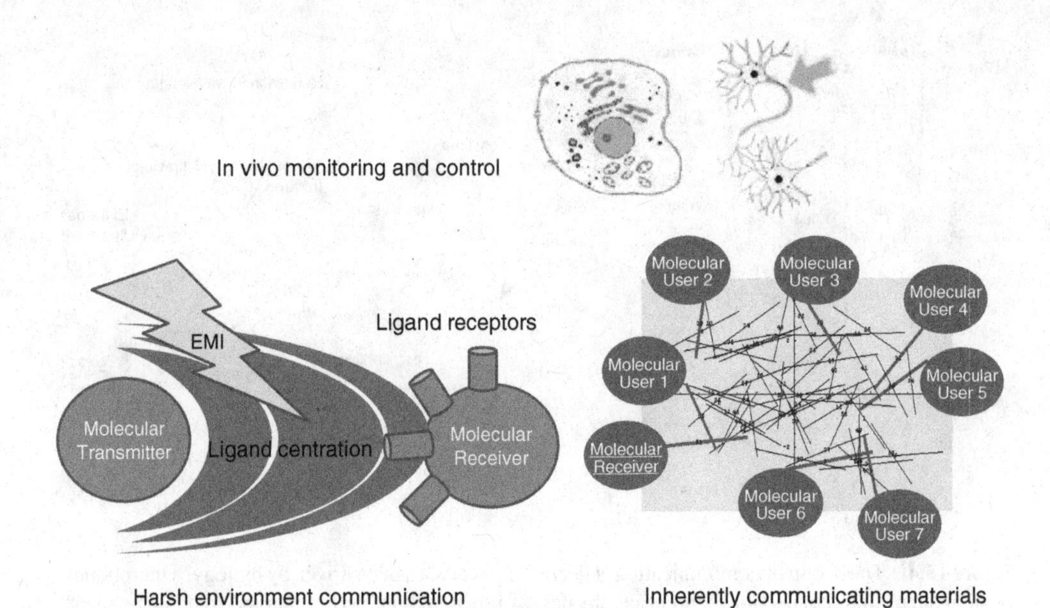

Figure 15.5 Three typical nanoscale communication network applications are illustrated: (1) communication in unique and harsh environments, (2) in vivo nanoscale communication, and (3) smart-material communication.

goals, use-cases, and framework for nanoscale communication networking. Simply providing something as simple as a common definition was required because researchers interpreted the meaning in many different and incompatible ways. Because researchers in the working group are from industry and academia with a wide variety of disciplines in mathematical modeling, engineering, physics, economics, and biological sciences, common terminology and definitions were needed. Finally, the standards effort has organized and classified the components discussed in this section required for a nanoscale communication network.

15.5 Emerging Technologies

As we have seen, the macroscopic quantum effects of superconductivity are typically leveraged for the purpose of maximizing or minimizing fairly obvious power grid phenomena, such as creating dense magnetic fields in generators and motors, decreasing resistance in order to minimize power loss, and enabling controlled resistance in the case of FCLs. However, the macroscopic quantum phenomena of superconductivity could also be used to provide auxiliary support within the power grid, such as for communication and computation. This could change the nature of the smart grid, which generally assumes classical techniques will be used for communication and computation.

Quantum computing has been able to move from the atomic world of individual nuclei and electrons to the macroscopic world of of integrated-circuit technology using the same superconducting mechanism we have just discussed for power systems (Devoret *et al.*, 2004). Removing the potential for dissipation of energy is a requirement for quantum computing because it enables quantum coherence over time intervals long enough to accomplish computation. Quantum coherence means that particle wave functions remain intact and do not mix with the environment; preconfigured superposition and entanglement of states remain in place in order for computation to take place. Quantum superposition is one of the key elements of quantum computation. While a complete discussion of quantum computation would take us beyond the scope of this book, the basic ideas can be explained simply. While a classical binary computer can only be in one of 2^n states at any one time, a quantum computer with n qubits can be in an arbitrary superposition of up to 2^n different states *simultaneously*. Quantum operators have been developed that allow probabilistic computation on the superposition of these states. Ideally, this means that every individual quantum operation can impact 2^n states simultaneously, resulting in extremely efficient and fast results. A single quantum bit that is in a superposition between a 0 and 1 is known as a qubit.

Returning to our quantum integrated-circuit, which is a macroscale element compared with typical quantum phenomena, the quantum circuit must be cooled to a temperature such that coherence is maintained, as mentioned earlier, and so that the energy of thermal fluctuations is less than the energy quantum of the transition between quantum states of the qubit.

The superconducting tunnel junction, or Josephson junction, has been the target of much research as a key element of a quantum computer. It is a nonlinear and nondissipative element that is suitable for a qubit. It is comprised of two superconducting thin films separated by an insulating layer that is thin enough to allow the tunneling of discrete charge through the insulating barrier. Tunneling involves the wavefunction of a particle being capable of passing through the tunnel barrier.

A key concept of interest here is that of quantum phenomena in electrical circuits. For example, the charge in a capacitor may be represented by a wavefunction representing all possible charge configurations. A specific example is that the charge may be in a superposition of being positive and negative, or current in a loop may be flowing in a superposition of both directions simultaneously. Research is leading toward demonstrating more quantum phenomena in larger scale systems.

15.5.1 Quantum Energy Teleportation

Quantum phenomena have long been associated with *information* transport and computation. However, just as we have seen throughout this book, particularly in Section 6.3, that information and power transmission are one-and-the-same, the same is true in the quantum realm. Just as teleportation of state – for example, the information within a qubit – has been theorized and demonstrated (Lloyd *et al.*, 2004), the same should be true for energy. Delving into a full explanation would take us too far beyond the scope of this book; however, it is important to note this concept for future benefit.

The concept, following the notion of quantum teleportation of a state, is to use entanglement of a ground state along with commonly used, local operations and classical communication to transport energy from one system to another (Hotta, 2008a, 2008b, 2009). It does not utilize classical energy diffusion and can be much faster. Just as in information teleportation, it does not violate laws of physics; it does not allow for faster-than-light transport and energy is neither created nor destroyed.

The concept involves the utilization of zero-point energy fluctuations, which are related to the Heisenberg uncertainty principle. Since a particle's speed and momentum cannot both be known with certainty, it can never come to complete rest; even at its lowest, ground state there is always a small, but finite, energy associated with a particle that is included with its total energy but is not normally accessible. However, if there is ground-state entanglement between two systems, then information about some particular local zero-point energy can be teleported to a remote system. The measurement operation in the teleportation process enables energy transmission. Clearly, individual zero-point energy transmissions are quite small, but as in many of the ideas presented in this chapter, large-scale application of the concept could be significant.

15.6 Near-Space Power Generation

It is always interesting to consider large changes in scale when attempting to think creatively about how a technology may evolve. In this case, we transition from the quantum and nanoscale to the astronomical and consider the idea of beaming power from space to Earth. Wireless power transmission is required for this; wireless power transport was covered in Section 8.4. This section extends that concept to space-based power generation. Also note that satellite communication was discussed in Section 8.5.

A vision for space-based solar power generation is comprised of potentially thousands of gigawatt satellite generators that, aggregated together, are capable of generating terawatts of power. The challenge with this approach lies in reducing the cost of implementation so that it approaches the cost of a ground-based generator of equivalent power capacity. This can be broken down into several optimizations: increasing the power conversion efficiency, managing unconverted power, and minimizing antenna size. A low-power conversion efficiency requires a larger collector area, which increases construction cost and transport cost of the satellite. The unconverted power must be removed or reflected back into space in order to avoid overheating, and the antenna size increases with both height of the orbit and wavelength of the energy collected.

A formula for the feasibility of such a system has been proposed:

$$k = \frac{25\,000 P \eta S}{C}, \tag{15.6}$$

where P (dollars/kWh) is the selling price for the power (Komerath *et al.*, 2012). The efficiency of transmission to the ground is η. S (kW/kg) is the specific power. Specific power is the power generated in orbit per-unit mass required in orbit to generate the power. C (dollars/kg) is the launch cost into LEO. The factor of 25 000 comes from architectural analysis and is chosen to incorporate the scale and assumed time for the system to pay for itself. This equation provides the ratio of benefit to cost; in other words, the value of the technology. Clearly, the goal is to maximize the value by reducing the cost to launch the satellite or increasing power generation efficiency or its beamed transmission efficiency to the ground. This assumes that the cost of power P is driven by the cost of traditional, terrestrial power generation.

The specific power S and the efficiency of power transmission to the ground η are of key interest. As mentioned, the cost of the cooling system detracts from the specific power and adds to the launch cost. The need for cooling comes from the nature of current PV technology. Current single-junction semiconductor technology requires that photons of light must penetrate through the surface of the semiconductor material before driving electrons into the conduction band to create electric power. This penetration generates heat that is absorbed within the semiconductor material; if not removed, the heat will accumulate, reaching destructive levels. Instead of expending energy to remove this heat, it has been suggested that the heat could be used in a form of co-generation on board the satellite to produce additional power via a gas turbine that is also located on board the satellite. Another solution is to filter the light into its spectral components before it hits the semiconductor surface. Then, only the most efficiently converted spectra will contact the semiconductor material; more of the energy will be converted to electric power and less absorbed as heat.

As mentioned early in this section, there would be many satellites in LEO. The low orbit would be ideal in order to improve wireless power transmission efficiency between the satellites and Earth. In order to ensure power to all locations on Earth, and given the fact that there would be many satellites, there would be a widely spread constellation of such satellites and each would be capable of relaying power to each other. This is similar to the nanogrid concept, where many small generators would be aggregating power among themselves. In fact, there are several similarities between the ad hoc nanogrid and space-based power generation. This also includes the ad hoc nature of the power and communication networks; the nodes would be mobile and the channels formed dynamically.

One concept to improve efficiency and reduce cost is to place two types of satellites into orbit. Lightweight, mirror satellites would be placed into a high Earth orbit. These satellite would open a flexible, mirror-like material that would reflect the light onto lower orbit, but heavier, power conversion and beamforming satellites. These satellites would carry the power electronics required to convert light to electricity and beam it where needed.

As previously mentioned, the satellites would be capable of transmitting power to each other as required to reach a target destination on Earth that would receive the accumulated power. However, it is also proposed that there be a reverse power channel in which power can be beamed from locations on Earth to the satellite constellation. This would take the place of terrestrial transmission lines and allow intercontinental power transmission. For example, excess power generated on the side of the Earth in daylight could be shared in real time with consumers on the night-side of the Earth. In this case, the space-based system is an ancillary power grid transmission service.

Regardless of whether the space-based system is used for power generation or transport, it opens up new challenges for communication. Communication and control will have to deal with long distances and relatively long propagation times. Also, control of a wireless power transmission system will be an interesting challenge, given the high latencies and the fact that there should be no interference between the wireless power and communication systems.

The beam capture formula

$$\frac{D_r D_t}{\lambda S} = 2.44 \tag{15.7}$$

illustrates some of the challenges of beaming power between Earth and space, where the relationship is between a transmitting phased-array of diameter D_t, a receiver main beam-lobe spot diameter D_r, the wavelength λ and transmitter-to-receiver distance separation S. As the distance or wavelength increases, the transmitter and/or receiver size must increase correspondingly. Given that there must be a large separation between transmitter and receiver, there is pressure to reduce the wavelength in order to keep the array as small as possible.

However, smaller wavelengths and corresponding higher frequencies are less able to penetrate the atmosphere efficiently. Some work has considered lower orbit satellites and using millimeter wavelengths as a good compromise between array size and efficiency.

15.7 Summary

This book has hopefully provided a unique perspective on the evolution of the power grid with respect to communication. In this final summary, a vision for the future of the power grid, and specifically power grid communication, is revealed. One of the fundamental principles throughout, and one that should be reasonably obvious, is that if smart grid communication is to aid in the operation of power transport, then the communication media should meet the requirements imposed by the characteristics of monitoring and control of electric power. Communication is used to either reduce or react to power entropy within the power grid. This can take many forms, including reducing uncertainty of change in demand, enforcing stability by reducing oscillation in the power angle, and responding to sudden change due to faults. Put more succinctly, communication is an equal and opposite reaction to power entropy. From a communications perspective, there are four standard components of the logical partition of the grid. First, there is power entropy. Power entropy is the "complexity" of change in electric power. It drives the communication necessary to control power, where power entropy is defined analogously to information entropy. Second, there is power area or density. This is exhibited, on the one hand, by bulk power control in the transmission system versus small amounts distributed over a wide area. Widely dispersed power monitoring and control requires more data routing and aggregation. Third, there is, of course, power efficiency. Specifically, this is the target efficiency to be achieved in which we theorize that, all else being equal, higher power efficiency requires more communication to achieve for any given telecommunication or power grid architecture. Finally, the fourth component is that the amount of power drives communication latency. Large, misdirected power flows – for example, typical power faults – require lower latency communication. Much of power grid communication technology derives from these components, which comprise the principle that communication is an equal and opposite reaction to power entropy. Figure 15.6 illustrates communication and power events on a space-time diagram. This allows one to see whether communication meets the requirements posed by a power anomaly.

We have shown how information theory may be considered a mature theory within its original core application of communications source and channel coding. The application of information theory becomes less mature as it moves further away from its original core use in source and channel coding, and is applied to other areas such as graph theory and machine learning. Information theory will be applied not only to its core application of channel optimization for communication networks, but also to new areas within the smart grid. As we have seen, fundamentals of information theory will allow tighter integration between power systems and communications, as discussed throughout this book.

From a more applied perspective, we can see four general parts to the future of the power grid with relation to information theory and network analysis. First, there will be a tendency toward decentralization. Control of the grid will become more decentralized, which will require more concepts relevant to network science, such as graph spectra, random matrix theory and information theory. In these techniques, large numbers of relatively simple components converge toward desired global behavior as the power grid architecture divides into

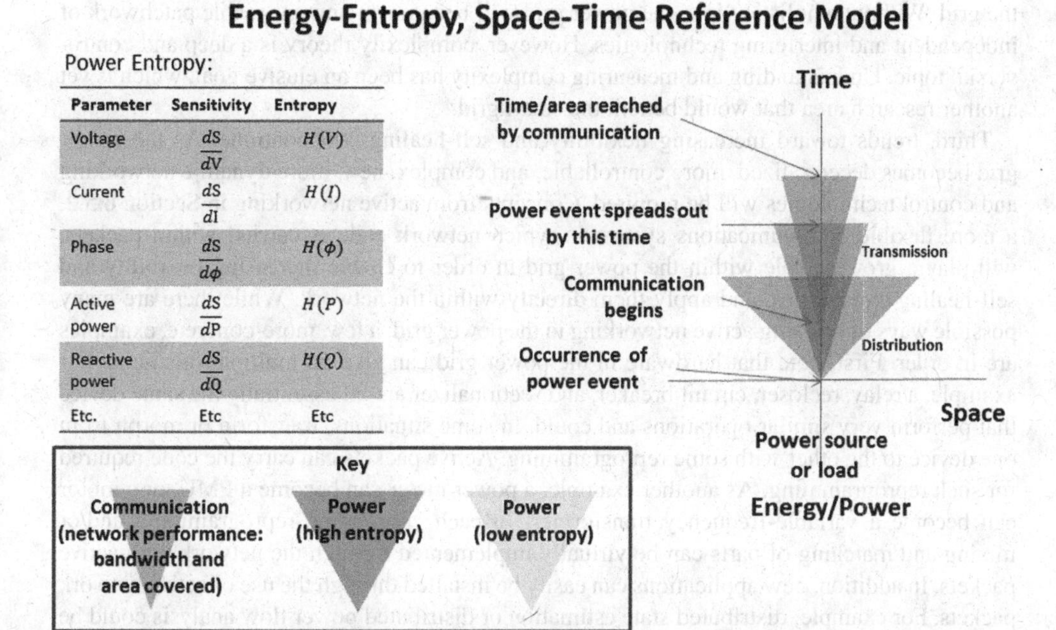

Greater power entropy requires more bandwidth to transmit:

- Slows down the space/time propagation of information

Figure 15.6 A space-time diagram illustrates the spread of an anomaly through space and time in the power grid. Communication must be able to "overlap" the anomaly in both space and time in order to enable the ability to manage it.

micro- and nanogrids. Machine learning and agent-based approaches will be implemented, and communication will be required to implement inter-agent communication. The core of the power grid will remain a form of M2M communication. Aspects of M2M communication, through information theory, may be utilized to improve communication efficiency. As the grid becomes more decentralized, its graph structure will become more "complex" and there will be a larger impact from network effects.

Second, the trend towards increased controllability and complexity will continue. As decentralization becomes more prevalent, the grid will become both more dynamic and controllable by adding more types of control systems. Analysis from an information-theoretic perspective, including its integration with market forces, will be required. Ideas that more closely combine machine communications and machine learning (for example, Bush and Hughes, 2003) would be beneficial. Concepts from the related and emerging field of network science will come to the fore at this point. Centralized generation will yield to more decentralized generation, and the transmission portion of the grid will become less utilized as more power is generated and used locally. Large, prime movers will no longer set the frequency of operation. Instead, a large-scale interconnection of microgrids will become more prevalent. Power protection mechanisms will adapt to more complex, two-way power flows. Volt-VAr control will become even more challenging. Synchrophasors and inverters will be nearly ubiquitous throughout the grid. Without a holistic view, the power grid will become an unmanageable patchwork of independent and interfering technologies. However, complexity theory is a deep and controversial topic. Understanding and measuring complexity has been an elusive goal, which is yet another research area that would benefit the smart grid.

Third, trends toward increasing flexibility and self-healing will continue. As the power grid becomes decentralized, more controllable, and complex, new, more dynamic networking and control technologies will be required. Concepts from active networking in Section 6.5.2, a more flexible communications system in which network code is carried within packets, will play a growing role within the power grid in order to enable increasing flexibility and self-healing capabilities, and apply them directly within the network. While there are many possible ways of utilizing active networking in the power grid, a few, more-concrete, examples are in order. First, note that hardware in the power grid can take on multiple functions. For example, a relay, recloser, circuit breaker, and sectionalizer are all essentially the same device that perform very similar operations and could, in some situations, transform or morph from one device to the other with some reprogramming. Active packets can carry the code required for such reprogramming. As another example, a power meter can become a PMU, or a motor can become a variable-frequency transformer. In each case, some reprogramming and/or mixing and matching of parts can be virtually implemented through the network with active packets. In addition, new applications can easily be installed through the use of active network packets. For example, distributed state estimation or distributed power flow analysis could be implemented within the network by active packets. Code within active packets would process power information *while* the information is being routed through the network. This opens up a new dimension in coupling communication with the processing required for power systems. An extreme form of flexibility in the power grid will come from wireless power transmission, which was discussed in detail in Section 8.4 as well as by Chowdhary and Komerath (2010); Garnica *et al.* (2011). The potential for store-and-forward forms of power transmission may also become possible. There may eventually be a delay-tolerant network form of power routing through the power grid, enabled by advances such as vehicle-to-grid communication.

Fourth, new forms of efficiency from advances at the micro- and nanoscale will arise. A common, long-term trend for technologies in general is to become so small and flexible that they eventually merge with their environment. Thus, in the long term, it is anticipated that DG will continue to evolve to high power densities and very small scales – for example, nanogrids approaching the molecular-level – for which new forms of small-scale networking, such as nanoscale networks, will be required (Bush, 2011a). Examples of such small-scale, power-generating approaches are the ability to capture power from radio transmissions to recharge batteries, and graphene-based power generation, among many mentioned in this book (Dhiman *et al.*, 2011; Vyas *et al.*, 2011). These are only examples; there are others being researched today, and many more will appear in the future. Wireless power transmission, as previously mentioned, is another example of the power grid merging with the environment. One is left to imagine where the advances in a new classical and quantum "power system information theory" will lead. A final section, the appendix, addresses those researchers who are often curious about how to obtain public domain simulators for smart grid communication. Simulation tools are undergoing rapid change and development, thus, it should be noted that this information, while highly-valuable to some researchers, may quickly become out-of-date, thus it has been placed in this appendix. There is an introduction to the concept of co-simulation and a description of freely-available power system and communication network simulators.

15.8 Exercises

Exercise 15.1 Power Grid Evolution

1. What are some ways in which we might expect the power grid to evolve based upon common technology trends?
2. What is the power grid's supersystem and how might the power grid evolve to integrate with its supersystem?

Exercise 15.2 Power from Geomagnetic Storms

1. How could power be harnessed from geomagnetic storms?
2. How does the length of the power line relate to the amount of current induced by a geomagnetic storm?
3. How could the smart grid help to increase the power extracted from a geomagnetic storm?

Exercise 15.3 Future Microgrids

1. What was the architectural concept behind the conceptual notion of a microgrid?
2. How are microgrids likely to evolve in the future?

Exercise 15.4 Nanogeneration

1. Describe three types of nanogenerators.
2. What would be some of the challenges in using massive numbers of nanogenerators to create power for the power grid?
3. Describe the nantenna and how it relates communication and power generation.

Exercise 15.5 Nanogrids

1. How would real nanogrids compare and contrast with today's power grid?

Exercise 15.6 Nanoscale Communication Networks

1. What are three different information encoding mechanisms for nanoscale communication networks?

Exercise 15.7 Quantum Energy Teleportation

1. What is the concept behind quantum energy teleportation?
2. What are zero-point energy fluctuations?

Exercise 15.8 Near-Space Power Generation

1. What elements comprise the benefit-to-cost ratio for near-space power generation?
2. What are some of the key challenges involved in near-space power generation?
3. How does wavelength impact the beamed power?

Exercise 15.9 Future Power Grid

1. How small can autonomous, self-sufficient, interacting, grid components, such as micro-grids, become?
2. How large can a single, global, power grid, able to optimize world power, become?
3. What new changes are on the horizon that could disrupt current trends in a smart grid?

Exercise 15.10 Boundaries of the Future Power Grid

New advances in nanotechnology are understanding and controlling energy processes on a small scale deep within materials. Yet, we tend to have an artificial bias towards when and where power is managed by the power grid and when and where it is managed locally.

1. At what scale does the management of power end for the power grid and why?
2. As we have discussed throughout the book and particularly in Section 6.5.2, the power grid is becoming more "active" in the way that it transports and manages power. Should the communication network also become more active in its ability to rapidly change and flexibly deploy new protocols? What are the benefits and challenges of active smart grid communications?

Appendix: Smart Grid Simulation Tools

The conventional view serves to protect us from the painful job of thinking.
—John Kenneth Galbraith

It does not take long for a person new to the smart grid, particularly those with a communication networking background, to begin asking, "Where can I find smart grid simulators?" The goal of this appendix to help answer this question. The danger in attempting to answer this question is that any answer written now will undoubtedly be out of date by the time you read this, given that simulation tools, particularly public-domain tools, are rapidly evolving. Both power systems and communication researchers have developed simulation tools, including real-time and hardware-in-the-loop simulation tools, for many decades. There are a myriad of tools available in each field. The problem has been that, until the smart grid concept came along, very few simulation tools combined both power grid and communication simulation capability in one package; just as these two fields have progressed independently from one another, so too have their tools been developed in isolation from one another.

However, progress is being made in the development of joint power system and communication simulation tools. A caution is in order: it is generally a good idea to read this book first and understand the fundamentals as well as perform the appropriate analysis before attempting any serious simulation activity. Relying solely on a simulation to provide solutions is never a good idea; simulation should be only one form of experimental validation of an idea or architecture.

With the above caution out of the way, joint power grid and communication networking simulation is necessary and useful to validate the operation of both highly complex systems. The physics of the operation of the electric power grid needs to be studied when coupled with the communication performance of its control systems. The interaction of the physics of operation with the complexity of communication and control can be difficult to predict. The power grid requires simulators capable of complex, continuous, differential equation solutions. Communication network simulators tend to involve discrete, packet-oriented computation. Thus, the fundamental nature of the simulators in each field is considerably different, making

Smart Grid: Communication-Enabled Intelligence for the Electric Power Grid, First Edition. Stephen F. Bush.

direct integration of the two an awkward and potentially inefficient process. This has led to the notion of co-simulation, or simply leveraging the work done in both fields to develop advanced simulators and connecting them together.

Simulators

OpenDSS

OpenDSS stands for open Distribution Simulation Software that was released by EPRI to provide a free, open-source, distribution system simulator. It is a power system simulator for electric distribution systems and was developed to run on Microsoft Windows. One of the goals of the open-source simulator is to encourage users to contribute models that they develop. OpenDSS is a comprehensive tool from a power systems perspective; it includes support for distribution planning and analysis, general alternating-current circuit analysis, the analysis of DG interconnections, annual load and generation simulations, wind plant simulations, and annual power flow simulations. The simulator can handle power flow analysis, harmonics, dynamics, and fault studies. The basic simulator does not explicitly handle communication; however, the simulator could be extended to incorporate communication with varying levels of fidelity or the simulator could be used in a co-simulation with a communication simulator.

PowerWorld Simulator Version 16

The PowerWorld Simulator Version 16 is a free software demo of the PowerWorld simulation product. It can be obtained from http://www.powerworld.com/download-purchase/demo-software. The free version limits the size of a power grid that can be analyzed, but includes optimal power flow, *PV*/*QV* curves, transient stability analysis, and analysis of GICs. Explicit communication and networking are not part of the simulator.

MATPOWER

Similar to PowerWorld, MATPOWER is also a power flow analysis package. MATPOWER is a freely available MATLAB M-file package that solves power flow problems. Again, communication and networking are not explicitly within the scope of this package.

Network Simulator: ns-2 and ns-3

The "ns" in ns-2 and ns-3 stands for "network simulator." Note that there was also an ns-1, now obsolete. These are a series of discrete-event, network simulators used in research and education, which are freely available under the GNU GPLv2 license. The network simulator series has been designed as an open simulator for the communication network research community. There are a large number of ns-2 and a growing number of ns-3 simulation modules that have been contributed to the public domain. It should be noted that many of these contributed modules, having been developed in the academic community primarily by graduate students, are often not fully tested and are sparsely documented. The network simulator series has been useful for those knowledgeable in the communication networking community for testing a wide variety of communication protocols that could possibly be used in the smart grid. However, it is unlikely that ns-2 or ns-3 developers will incorporate significant power system functionality into the simulator. This means that the network simulator series can at best be used in co-simulation with a power systems simulator.

What follows in this appendix are a few power grid simulators that meet the criteria of (1) being public domain, (2) incorporating smart grid elements, and (3) that either include or are likely to include communication as part of the simulation tool.

GridSim

GridSim has an unfortunate name for this simulator because there are many other tools and simulators in unrelated fields that have the same name, causing confusion. While GridSim is not available for download at the time this is being written, the tool has enough unique features to be interesting (Anderson *et al.*, 2012). GridSim appears to have been motivated by the need to experimentally validate the impact of communication upon the use of wide-area, synchrophasor measurements. Thus, this tool directly addresses the issue of both the power grid and communication.

GridLab-D

GridLab-D is another power system simulator focused on the distribution system. It features power flow analysis using the Gauss–Seidel method and has detailed models of power and transmission lines, transformers, voltage regulators, fuses, switches, shunt capacitor banks, and so on. It also models customer metering and can incorporate reclosers, islanding, and DG models. It also supports thermal parameter models. The tool does not yet claim to support support communication, although it is anticipated to incorporate integration with ns-3 in the near future.

References

Abur, A. (2009), Impact of phasor measurements on state estimation, in *International Conference on Electrical and Electronics Engineering, 2009. ELECO 2009*, IEEE, pp. I3–I7.

Ackermann, T., Andersson, G., and Söder, L. (2001), Distributed generation: a definition, *Electric Power Systems Research*, **57** (3), 195–204, doi:10.1016/S0378-7796(01)00101-8.

Adams, J. (2006), An introduction to IEEE STD 802.15.4, in *2006 IEEE Aerospace Conference*, IEEE, pp. 1–8, doi:10.1109/AERO.2006.1655947.

Ahmed, S. S., Yeong, T. W., and Ahmad, H. (2003), Wireless power transmission and its annexure to the grid system, *IEE Proceedings – Generation, Transmission and Distribution*, **150** (2), 195–199.

Albadi, M. H. and El-Saadany, E. F. (2007), *Demand Response in Electricity Markets: An Overview*, IEEE, doi:10.1109/PES.2007.385728.

Alizadeh, M., Scaglione, A., and Wang, Z. (2010), On the impact of smartgrid metering infrastructure on load forecasting, in *2010 48th Annual Allerton Conference on Communication, Control, and Computing (Allerton)*, IEEE, pp. 1628–1636.

Amin, M. (2004), Balancing market priorities with security issues, *IEEE Power and Energy Magazine*, **2** (4), 30–38.

Anderson, D., Zhao, C., Hauser, C. H. *et al.* (2012), "Intelligent design" real-time simulation for smart grid control and communications design, *IEEE Power and Energy Magazine*, **10** (1), 49–57, doi:10.1109/MPE.2011.943205.

Anderson, R. N., Boulanger, A., and Powell, W. B. (2011), Adaptive stochastic control for the smart grid, *Proceedings of the IEEE*, **99** (6), 1098–1115.

Anwar, A., Roy, N., and Pota, H. (2011), Voltage stability analysis with optimum size and location based synchronous machine DG, in *2011 21st Australasian Universities Power Engineering Conference (AUPEC)*, IEEE, pp. 1–5.

Aravinthan, V., Karimi, B., Namboodiri, V., and Jewell, W. (2011), Wireless communication for smart grid applications at distribution level – feasibility and requirements, in *2011 IEEE Power and Energy Society General Meeting*, IEEE, pp. 1–8.

Arya, V., Hazra, J., Kodeswaran, P. *et al.* (2011), CPS-Net: In-network aggregation for synchrophasor applications, *2011 Third International Conference on Communication Systems and Networks (COMSNETS 2011)*, 1–8, doi:10.1109/COMSNETS.2011.5716510.

Azmy, A. and Erlich, I. (2005), Impact of distributed generation on the stability of electrical power system, in *2005. IEEE Power Engineering Society General Meeting*, IEEE, volume 2, pp. 1056–1063.

Bak, P., Tang, C., and Wiesenfeld, K. (1987), Self-organized criticality: an explanation of the $1/f$ noise, *Physical Review Letters*, **59** (4), 381–384, doi:10.1103/PhysRevLett.59.381.

Barbato, A., Capone, A., Carello, G. *et al.* (2011a), House energy demand optimization in single and multi-user scenarios, in *2011 IEEE International Conference on Smart Grid Communications (SmartGridComm)*, IEEE, pp. 345–350.

Barbato, A., Capone, A., Rodolfi, M., and Tagliaferri, D. (2011b), Forecasting the usage of household appliances through power meter sensors for demand management in the smart grid, in *2011 IEEE International Conference on Smart Grid Communications (SmartGridComm)*, IEEE, pp. 404–409.

Barber, P., Kitroser, I., and Koo, C. (2004), Air interface for fixed and mobile broadband wireless access systems – management PLANe procedures and services, http://www.ieee802.org/16/netman/}Doc16g.

Smart Grid: Communication-Enabled Intelligence for the Electric Power Grid, First Edition. Stephen F. Bush.

Barron, A., Rissanen, J., and Yu, B. (1998), The minimum description length principle in coding and modeling, *IEEE Transactions on Information Theory*, **44** (6), 2743–2760.

Bastani, S., Yousefi, S., Mazoochi, M., and Ghiamatyoun, A. (2009), Delay and throughput trade-off in WiMAX mesh networks, in *International Conference on Communication Software and Networks, 2009. ICCSN '09*, IEEE, pp. 283–286, doi:10.1109/ICCSN.2009.106.

Beardow, P., Barber, J., Owen, R., and Bell, J. (1993), The application of satellite communications technology to the protection of the rural distribution networks, in *Fifth International Conference on Developments in Power System Protection, 1993*, IEEE, pp. 17–20.

Beer, J. (2007), High efficiency electric power generation: the environmental role, *Progress in Energy and Combustion Science*, **33** (2), 107–134, doi:10.1016/j.pecs.2006.08.002.

Belhomme, R., De Asua, R., Valtorta, G. *et al.* (2008), ADDRESS – active demand for the smart grids of the future, in *SmartGrids for Distribution, 2008. IET-CIRED. CIRED Seminar*, IET, pp. 1–4.

Bellman, R. (1952), On the theory of dynamic programming, *Proceedings of the National Academy of Sciences of*, **38** (8), 716–719.

Benford, J., Swegle, J., and Schamiloglu, E. (2007), *High power microwave applications*, Taylor & Francis Group, LLC, chapter 3, 2nd edition, pp. 43–108, doi:10.1201/9781420012064.ch3.

Benmouyal, G., Meisinger, M Burnworth, J., Elmore, W. *et al.* (1999), IEEE standard inverse-time characteristic equations for overcurrent relays, *IEEE Transactions on Power Delivery*, **14** (3), 868–872.

Bennett, C. H. (2002), Notes on Landauer's principle, reversible computation and Maxwell's demon, http://arxiv.org/abs/physics/0210005.

Bhattacharjee, S., Calvert, K., and Zegura, E. (1997), Active networking and the end-to-end argument, in *1997 International Conference on Network Protocols, 1997. Proceedings.*, IEEE Computer Society, pp. 220–228, doi:10.1109/ICNP.1997.643717.

Blaabjerg, F., Teodorescu, R., Liserre, M., and Timbus, A. (2006), Overview of control and grid synchronization for distributed power generation systems, *IEEE Transactions on Industrial Electronics*, **53** (5), 1398–1409, doi:10.1109/TIE.2006.881997.

Borozan, V., Baran, M., and Novosel, D. (2001), Integrated volt/VAr control in distribution systems, in *IEEE Power Engineering Society Winter Meeting, 2001*, IEEE, volume 3, pp. 1485–1490, doi:10.1109/PESW.2001.917328.

Bose, B. K. (2009), Power electronics and motor drives – recent progress and perspective, *IEEE Transactions on Industrial Electronics*, **56** (2), 581–588.

Brown, L. (2011), ITU-T and smart grid, http://www.smartgrid.com/wp-content/uploads/2011/09/8____Les.pdf.

Brown, W. and Eves, E. (1992), Beamed microwave power transmission and its application to space, *IEEE Transactions on Microwave Theory and Techniques*, **40** (6), 1239–1250.

Brown, W. C. (1984), The history of power transmission by radio waves, *IEEE Transactions on Microwave Theory and Techniques*, **32** (9), 1230–1242, doi:10.1109/TMTT.1984.1132833.

Bu, S., Yu, F., and Liu, P. (2011), A game-theoretical decision-making scheme for electricity retailers in the smart grid with demand-side management, in *2011 IEEE International Conference on Smart Grid Communications (SmartGridComm)*, IEEE, pp. 387–391.

Budimir, D. and Marincic, A. (2006), Research activities and future trends of microwave wireless power transmission, in *Sixth International Symposium Nikola Tesla*, http://ebooks.z0ro.com/ebooks/Nikola_Tesla-2006-Serbian_Symposium-Lecture_Papers/papers/Tesla-Symp06_Budimir.pdf.

Bush, S. and Goel, S. (2006), Graph spectra of carbon nanotube networks, in *1st International Conference on Nano-Networks and Workshops, 2006. NanoNet '06*, IEEE, pp. 1–10.

Bush, S., Mahony, M., and Devarajan, A. (2011a), An analysis of smart grid fault mitigation aided by wireless communication, Technical Report 2011GRC182, February 2011, GE Global Research.

Bush, S., Mahony, M., and Devarajan, A. (2011b), Network theory and smart grid fault detection, isolation, and reconfiguration, Technical Report 2011GRC288, GE Global Research.

Bush, S. F. (2000), Islands of near-perfect self-prediction, in *Proceedings of VWsim'00: Virtual Worlds and Simulation Conference, WMC'00: 2000 SCS Western Multi-Conference, San Diego, SCS*, http://www.research.ge.com/~bushsf/an/vwsim00.pdf.

Bush, S. F. (2002), Active virtual network management prediction: complexity as a framework for prediction, optimization, and assurance, in *Proceedings DARPA Active Networks Conference and Exposition*, Santa Fe Institute, Santa Fe, NM, pp. 534–553, doi:10.1109/DANCE.2002.1003518.

Bush, S. F. (2005), A simple metric for ad hoc network adaptation, *IEEE Journal on Selected Areas in Communications*, **23** (12), 2272–2287, doi:10.1109/JSAC.2005.857204.

Bush, S. F. (2010a), IEEE SmartGridComm 2010 Panel Session, October 5, National Institute of Standards and Technology, Gaithersburg, MD.

Bush, S. F. (2010b), *Nanoscale Communication Networks*, Artech House, ISBN: 978-1608070039.

Bush, S. F. (2011a), Communications for the smart grid. IEEE Distinguished Lecture Tour.

Bush, S. F. (2011b), Toward in vivo nanoscale communication networks: utilizing an active network architecture, *Frontiers of Computer Science in China*, **5** (1), 1–9, doi:10.1007/s11704-011-0116-9.

Bush, S. F. (2013a), Information theory and network science for power systems, in *IEEE Smart Grid Vision for Computing: 2030 and Beyond*, edited by S. Goel, S. F. Bush, and D. Bakken, IEEE, New York, NY, chapter 5, pp. 128–161.

Bush, S. F., Goel, S., and Simard, G. (2013b), IEEE Vision for Smart Grid Communications: 2030 and Beyond Roadmap, ISBN(s):9780738186467, 9780738186474.

Bush, S. F., Hershey, J., and Vosburgh, K. (1999), Brittle system analysis, http://arxiv.org/pdf/cs/9904016.pdf.

Bush, S. F. and Hughes, T. (2003), On the effectiveness of Kolmogorov complexity estimation to discriminate semantic types, http://arxiv.org/ftp/cs/papers/0512/0512089.pdf.

Bush, S. F. and Kulkarni, A. B. (2001), *Active Networks and Active Network Management: A Proactive Management Framework*, Springer.

Bush, S. F. and Kulkarni, A. B. (2007), *Active Networks and Active Network Management: A Proactive Management Framework. Solution Manual*, Kluwer Academic/Plenum, http://www.research.ge.com/~bushsf.

Bush, S. F. and Li, Y. (2006), Network characteristics of carbon nanotubes: a graph eigenspectrum approach and tool using Mathematica, Technical report 2006GRC023, GE Global Research, http://www.research.ge.com/~bushsf/pdfpapers/2006GRC023_Final_ver.pdf.

Bush, S. F. and Smith, N. (2005), The limits of motion prediction support for ad hoc wireless network performance, http://arxiv.org/abs/cs/0512092.

Cameron, G. (2005), Trends of engineering system evolution trends of increasing dynamization, *TRIZ Journal*, (December), 1–10, http://www.triz-journal.com/archives/2005/12/01.pdf.

Candès, E. and Wakin, M. (2008), An introduction to compressive sampling, *IEEE Signal Processing Magazine*, **25** (2), 21–30, doi:10.1109/MSP.2007.914731.

Cao, M., Ma, W., Zhang, Q., and Wang, X. (2007), Analysis of IEEE 802.16 mesh mode scheduler performance, *IEEE Transactions on Wireless Communications*, **6** (4), 1455–1464, doi:10.1109/TWC.2007.348342.

Caron, S. and Kesidis, G. (2010), Incentive-based energy consumption scheduling algorithms for the smart grid, in *2010 First IEEE International Conference on Smart Grid Communications*, IEEE, pp. 391–396, doi:10.1109/SMARTGRID.2010.5622073.

Carpenter, K. H. (2004), The Poynting vector: power and energy in electromagnetic fields, Technical report, Kansas State University.

Carroll, R., Trachian, P., Affare, S. *et al.* (2008), Development of TVA SuperPDC: phasor applications, tools, and event replay, in *2008 IEEE Power and Energy Society General Meeting – Conversion and Delivery of Electrical Energy in the 21st Century*, IEEE, pp. 1–8, doi:10.1109/PES.2008.4596276.

Chang, B.-J., Liang, Y.-H., and Su, S.-S. (2009), Adaptive competitive on-line routing algorithm for IEEE 802.16j WiMAX multi-hop relay networks, in *2009 IEEE 20th International Symposium on Personal, Indoor and Mobile Radio Communications*, IEEE, pp. 2197–2201, doi:10.1109/PIMRC.2009.5450380.

Cheney, R. M., Thorne, J. T., and Hataway, G. (2009), Distribution single-phase tripping and reclosing: overcoming obstacles with programmable recloser controls, in *2009 62nd Annual Conference for Protective Relay Engineers*, IEEE, pp. 214–223, doi:10.1109/CPRE.2009.4982514.

Choi, S., Park, S., Kang, D. *et al.* (2011), A microgrid energy management system for inducing optimal demand response, in *2011 IEEE International Conference on Smart Grid Communications (SmartGridComm)*, pp. 19–24.

Chowdhary, G. and Komerath, N. (2010), Innovations required for retail beamed power transmission over short range, http://www.iiis.org/CDs2010/CD2010SCI/IMETI_2010/PapersPdf/FA664AB.pdf.

Ci, S., Qian, J., Wu, D., and Keyhani, A. (2012), Impact of wireless communication delay on load sharing among distributed generation systems through smart microgrids, *IEEE Wireless Communications*, **19** (3), 24–29.

Cicconetti, C., Akyildiz, I., and Lenzini, L. (2009), WiMsh: a simple and efficient tool for simulating IEEE 802.16 wireless mesh networks in ns-2, in *Proceedings of the Second International ICST Conference on Simulation Tools and Techniques*, ICST, doi:10.4108/ICST.SIMUTOOLS2009.5679.

Cicconetti, C., Mingozzi, E., and Stea, G. (2006), An integrated framework for enabling effective data collection and statistical analysis with ns-2, in *Proceeding from the 2006 Workshop on Ns-2: The IP Network Simulator*, ACM, New York, NY. Article No. 11.

Clancy, C., Hecker, J., Stuntebeck, E., and O'Shea, T. (2007), Applications of machine learning to cognitive radio networks, *IEEE Wireless Communications*, **14** (4), 47–52.

Cleveland, F. (2012), IEC 62351 Security Standards for the Power System Information Infrastructure.

Cohen, R., Erez, K., ben-Avraham, D., and Havlin, S. (2000), Resilience of the internet to random breakdowns, *Physical Review Letters*, **85** (21), 4626–4628, doi:10.1103/PhysRevLett.85.4626.

Committee on Network Science for Future Army Applications (2006), *Network Science*, National Research Council.

Conti, J. (2010), Annual energy outlook 2010: with projections to 2035, http://www.eia.gov/oiaf/aeo/pdf/0383(2010).pdf.

Dagle, J. (2006), Postmortem analysis of power grid blackouts – the role of measurement systems, *IEEE Power and Energy Magazine*, **4** (5), 30–35, doi:10.1109/MPAE.2006.1687815.

Dahal, N., King, R., and Madani, V. (2012), Online dimension reduction of synchrophasor data, in *2012 IEEE PES Transmission and Distribution Conference and Exposition (T&D)*, pp. 1–7.

Damsky, B., Gelman, V., and Frederick, E. (2003), A solid state current limiter, http://www.epa.gov/electricpower-sf6/documents/conf02_damsky_paper.pdf.

Deese, A. (2008), Analog methods for power system analysis and load modeling, PhD, Drexel University, http://144.118.25.24/handle/1860/2822.

Del Valle, Y., Venayagamoorthy, G. K., Mohagheghi, S. *et al.* (2008), Particle swarm optimization: basic concepts, variants and applications in power systems, *IEEE Transactions on Evolutionary Computation*, **12** (2), 171–195, doi:10.1109/TEVC.2007.896686.

Delille, G., Francois, B., and Malarange, G. (2010), Dynamic frequency control support: a virtual inertia provided by distributed energy storage to isolated power systems, in *2010 IEEE PES Innovative Smart Grid Technologies Conference Europe (ISGT Europe)*, IEEE, pp. 1–8, doi:10.1109/ISGTEUROPE.2010.5638887.

Dessanti, B., Komerath, N., and Flournoy, D. (2012), Wireless transfer of power: proposal for a five-nation demonstration by 2020, *Online Journal of Space Communication*, (17), http://spacejournal.ohio.edu/issue17/fivenation.html.

Devoret, M. H., Wallraff, A., and Martinis, J. M. (2004), Superconducting qubits: a short review, http://arxiv.org/abs/cond-mat/0411174.

Dhiman, P., Yavari, F., Mi, X. *et al.* (2011), Harvesting energy from water flow over graphene, *Nano Letters*, **11** (8), 3123–3127, doi:10.1021/nl2011559.

Dierks, T. (2008), The transport layer security (TLS) protocol version 1.2, http://wiki.tools.ietf.org/html/rfc5246.

Dionigi, M. and Mongiardo, M. (2011), CAD of efficient wireless power transmission systems, in *2011 IEEE MTT-S International Microwave Symposium Digest (MTT)*, pp. 1–4, http://ieeexplore.ieee.org/xpls/abs_all.jsp?arnumber=5972606.

Domingos, P. (2012), A few useful things to know about machine learning, *Communications of the ACM*, **55** (10), 78–87.

Driesen, J. and Visscher, K. (2008), Virtual synchronous generators, in *2008 IEEE Power and Energy Society General Meeting – Conversion and Delivery of Electrical Energy in the 21st Century*, IEEE, pp. 1–3, doi:10.1109/PES.2008.4596800.

Drozdovski, N. and Caverly, R. (2002), GaN-based high electron-mobility transistors for microwave and RF control applications, *IEEE Transactions on Microwave Theory and Techniques*, **50** (1), 4–8.

D'Souza, M., Bialkowski, K., Postula, A., and Ros, M. (2007), A wireless sensor node architecture using remote power charging, for interaction applications, in *10th Euromicro Conference on Digital System Design Architectures, Methods and Tools (DSD 2007)*, IEEE, pp. 485–494, doi:10.1109/DSD.2007.4341513.

Dugan, R. and McDermott, T. (2002), Distributed generation, *IEEE Industry Applications Magazine*, **8** (2), 19–25, doi:10.1109/2943.985677.

Dupont, B., Meeus, L., and Belmans, R. (2010), Measuring the "smartness" of the electricity grid, in *2010 7th International Conference on the European Energy Market (EEM)*, IEEE, pp. 1–6, doi:10.1109/EEM.2010.5558673.

Dy-Liacco, T. (2002), Control centers are here to stay, *IEEE Computer Applications in Power*, **15** (4), 18–23.

Edris, A., Adapa, R., Baker, M. *et al.* (1997), Proposed terms and definitions for flexible AC transmission system (FACTS), *IEEE Transactions on Power Delivery*, **12** (4), 1848–1853, doi:10.1109/61.634216.

Elizondo, D., Gentile, T., Candia, H., and Bell, G. (2010), Overview of robotic applications for energized transmission line work – technologies, field projects and future developments, in *2010 1st International Conference on Applied Robotics for the Power Industry (CARPI 2010)*, IEEE, pp. 1–7, doi:10.1109/CARPI.2010.5624478.

Engelberg, S. (2008), Tutorial 15: control theory, part I, *IEEE Instrumentation & Measurement Magazine*, **11** (3), 34–40.

EPRI (2010), Smart grid executive summary.

Faria, P., Vale, Z., Soares, J., and Ferreira, J. (2011), Demand response management in power systems using a particle swarm optimization approach, *IEEE Intelligent Systems*, (99), 1–9, doi:10.1109/MIS.2011.35.

Figueredo, L. F. C., Santana, P. H. R. Q. A., Alves, E. S. *et al.* (2009), Robust stability of networked control systems, in *IEEE International Conference on Control and Automation, 2009. ICCA 2009*, IEEE, pp. 1535–1540.

FitzPatrick, G. J. and Wollman, D. A. (2010), NIST interoperability framework and action plans, in *2010 IEEE Power and Energy Society General Meeting*, IEEE, pp. 1–4, doi:10.1109/PES.2010.5589699.

Freeth, T., Bitsakis, Y., Moussas, X. *et al.* (2006), Decoding the ancient Greek astronomical calculator known as the Antikythera mechanism, *Nature*, **444** (7119), 587–951, doi:10.1038/nature05357.

Garnica, J., Casanova, J., and Lin, J. (2011), High efficiency midrange wireless power transfer system, in *2011 IEEE MTT-S International Microwave Workshop Series on Innovative Wireless Power Transmission: Technologies, Systems, and Applications (IMWS)*, pp. 73–76.

Ge, Y., Tham, C.-k., Kong, P.-y., and Ang, Y.-H. (2008), Capacity estimation for IEEE 802.16 wireless multi-hop mesh networks, in *IEEE Wireless Communications and Networking Conference, 2008. WCNC 2008*, IEEE, pp. 2651–2656, doi:10.1109/WCNC.2008.465.

GE Digital Energy (2007), Protection basics: introduction to symmetrical components, Technical report, http://www.gedigitalenergy.com/smartgrid/Dec07/7-symmetrical.pdf.

General Electric (1977), Distribution system feeder overcurrent protection, Technical report, GE Power Management, Ontario, Canada, http://www.geindustrial.com/publibrary/checkout/GET-6450?TNR=White Papers|GET-6450|generic.

Giacomelli, G. and Patrizii, L. (2003), Magnetic monopole searches, http://arxiv.org/abs/hep-ex/0302011.

Goudarzi, H., Hatami, S., and Pedram, M. (2011), Demand-side load scheduling incentivized by dynamic energy prices, in *2011 IEEE International Conference on Smart Grid Communications (SmartGridComm)*, pp. 351–356.

Grover, P. and Sahai, A. (2010), Shannon meets Tesla: wireless information and power transfer, in *2010 IEEE International Symposium on Information Theory Proceedings (ISIT)*, IEEE, pp. 2363–2367, doi:10.1109/ISIT.2010.5513714.

Gruber, F. K. and Marengo, E. A. (2008), New aspects of electromagnetic information theory for wireless and antenna systems, *IEEE Transactions on Antennas and Propagation*, **56** (11), 3470–3484.

Gundogdu, A. E. and Afacan, E. (2011), Some experiments related to wireless power transmission, in *2011 Cross Strait Quad-Regional Radio Science and Wireless Technology Conference (CSQRWC)*, IEEE, volume 1, pp. 507–509.

Guo, X., Rouil, R., Soin, C. *et al.* (2009), WiMAX system design and evaluation methodology using the ns-2 simulator, in *First International Communication Systems and Networks and Workshops, 2009. COMSNETS 2009*, IEEE, pp. 1–10.

Hassenzahl, W., Hazelton, D., Johnson, B. *et al.* (2004), Electric power applications of superconductivity, *Proceedings of the IEEE*, **92** (10), 1655–1674.

Hataway, G., Warren, T., and Stephens, C. (2006), Implementation of a high-speed distribution network reconfiguration scheme, in *59th Annual Conference for Protective Relay Engineers, 2006*, IEEE, pp. 134–140, doi:10.1109/CPRE.2006.1638697.

Hayajneh, M. and Gadallah, Y. (2008), MAC 25-2 – throughput analysis of WiMAX based wireless networks, in *IEEE Wireless Communications and Networking Conference, 2008. WCNC 2008*, IEEE, pp. 1997–2002, doi:10.1109/WCNC.2008.355.

Hekland, F. (2004), A review of joint source-channel coding, Technical report, Norwegian University of Science and Technology.

Hernández-Orallo, J. and Dowe, D. L. (2010), Measuring universal intelligence: towards an anytime intelligence test, *Artificial Intelligence*, **174** (18), 1508–1539, doi:10.1016/j.artint.2010.09.006.

Hobby, J. (2010), Constructing demand response models for electric power consumption, in *2010 First IEEE International Conference on Smart Grid Communications (SmartGridComm)*, pp. 403–408.

Holbert, K., Heydt, G., and Ni, H. (2005), Use of satellite technologies for power system measurements, command, and control, *Proceedings of the IEEE*, **93** (5), 947–955.

Hotta, M. (2008a), A protocol for quantum energy distribution, *Physics Letters A*, **372** (35), 5671–5676.

Hotta, M. (2008b), Quantum measurement information as a key to energy extraction from local vacuums, *Physical Review D*, **78** (4), 045006.

Hotta, M. (2009), Quantum energy teleportation in spin chain systems, *Journal of the Physical Society of Japan*, **78** (3), 034001.

Huang, B. (2006), Stability of distribution systems with a large penetration of distributed generation, Doktor der Ingenieurwissenschaften, Universität Dortmund.

Huang, S., Luo, J., Leonardi, F., and Lipo, T. (1998), A general approach to sizing and power density equations for comparison of electrical machines, *IEEE Transactions on Industry Applications*, **34** (1), 92–97.

Hui Wan, Li, K., and Wong, K. (2005), An multi-agent approach to protection relay coordination with distributed generators in industrial power distribution system, in *Fourtieth IAS Annual Meeting. Conference Record of the 2005 Industry Applications Conference, 2005*, IEEE, volume 2, pp. 830–836, doi:10.1109/IAS.2005.1518431.

IEEE (1995), IEEE Standard for Withstand Capability of Relay Systems to Radiated Electromagnetic Interference from Transceivers.

IEEE (2010), IEEE Standard for Electric Power Systems Communications – Distributed Network Protocol (DNP3), doi:10.1109/IEEESTD.2010.5518537.

IEEE (2011), IEEE Standard for Synchrophasor Data Transfer for Power Systems, doi:10.1109/IEEESTD.2011.6111222.

Ishiba, M., Ishida, J., Komurasaki, K., and Arakawa, Y. (2011), Wireless power transmission using modulated microwave, in *2011 IEEE MTT-S International Microwave Workshop Series on Innovative Wireless Power Transmission: Technologies, Systems, and Applications (IMWS)*, pp. 51–54.

Jain, R. and Al Tamimi, A.-k. (2008), System-level modeling of IEEE 802.16E mobile WiMAX networks: key issues, *IEEE Wireless Communications*, **15** (5), 73–79, doi:10.1109/MWC.2008.4653135.

James, G. C. (2008), Analytical methods for scientific demand response, in *Australasian UniversitiesPower Engineering Conference, 2008. AUPEC '08*, pp. 1–4.

Jiang, Y. (2010), *A Practical Guide to Error-Control Coding Using MATLAB*, Artech House.

Jøsang, A. (2012), Subjective logic, Technical report, University of Oslo, http://persons.unik.no/josang/papers/subjective_logic.pdf.

Joseph, III, M. (2000), Cognitive radio: an integrated agent architecture for software defined radio, Doctor of technology, Royal Institute of Technology (KTH), http://web.it.kth.se/~maguire/jmitola/Mitola_Dissertation8_Integrated.pdf.

Kallitsis, M. G., Michailidis, G., and Devetsikiotis, M. (2011), A decentralized algorithm for optimal resource allocation in smartgrids with communication network externalities, in *2011 IEEE International Conference on Smart Grid Communications (SmartGridComm)*, IEEE, pp. 434–439, doi:10.1109/SmartGridComm.2011.6102361.

Kappenman, J. (1996), Geomagnetic storms and their impact on power systems, *IEEE Power Engineering Review*, **16** (5), 5–8.

Kas, M., Yargicoglu, B., Korpeoglu, I., and Karasan, E. (2010), A survey on scheduling in IEEE 802.16 mesh mode, *IEEE Communications Surveys & Tutorials*, **12** (2), 205–221, doi:10.1109/SURV.2010.021110.00053.

Katti, S., Rahul, H., Katabi, D. *et al.* (2008), XORs in the air: practical wireless network coding, *IEEE/ACM Transactions on Networking*, **16** (3), 497–510, doi:10.1109/TNET.2008.923722.

Kefayati, M. and Baldick, R. (2011), Energy delivery transaction pricing for flexible electrical loads, in *2011 IEEE International Conference on Smart Grid Communications (SmartGridComm)*, IEEE, pp. 363–368, doi:10.1109/SmartGridComm.2011.6102348.

Klump, R., Agarwal, P., Tate, J. E., and Khurana, H. (2010), Lossless compression of synchronized phasor measurements, in *2010 IEEE Power and Energy Society General Meeting*, IEEE, pp. 1–7, doi:10.1109/PES.2010.5590156.

Ko, Y.-S. (2009), A self-isolation method for the HIF zone under the network-based distribution system, *IEEE Transactions on Power Delivery*, **24** (2), 884–891, doi:10.1109/TPWRD.2009.2014482.

Kolmogorov, A. (1965), Three approaches to the definition of the quantity of information, *Problems of Information Transmission*, **1**, 3–11.

Komerath, N. and Chowdhary, G. (2010), Retail beamed power for a micro renewable energy architecture: survey, in *2010 International Symposium on Electronic System Design (ISED)*, pp. 44–49, doi:10.1145/0000000.0000000.

Komerath, N., Dessanti, B., and Shah, S. (2012), A gigawatt-level solar power satellite using intensified efficient conversion architecture, in *2012 IEEE Aerospace Conference*, IEEE, pp. 1–14, doi:10.1109/AERO.2012.6187079.

Komerath, N. and Komerath, P. (2011), Implications of inter-satellite power beaming using a space power grid, in *2011 IEEE Aerospace Conference*, IEEE, pp. 1–11.

Komerath, N., Komerath, P., and Creek, J. (2011), The case for millimeter wave power beaming, in *The 4th International Multi-Conference on Engineering and Technological Innovation: IMETI 2011*.

Komerath, N., Venkat, V., Fernandez, J., and Robertson, G. A. (2009), Near-millimeter wave issues for a space power grid, in *AIP Conference Proceedings*, AIP, volume 1103, pp. 149–156, doi:10.1063/1.3115489.

Koutitas, G. (2012), Control of flexible smart devices in the smart grid, *IEEE Transactions on Smart Grid*, **3** (3), 1333–1343, doi:10.1109/TSG.2012.2204410.

Krauter, S. and Depping, T. (2003), Monitoring of remote PV-systems via satellite, in *Proceedings of 3rd World Conference on Photovoltaic Energy Conversion, 2003*, volume 3, pp. 2202–2205.

Kristoffersen, T. (2007), Stochastic programming with applications to power systems, PhD, University of Aarhus, http://data.imf.au.dk/publications/phd/2007/imf-phd-2007-tkk.pdf.

Kron, G. (1945), Electric circuit models of the Schrödinger equation, *Physical Review*, **67** (1–2), 39–43.

Kroposki, B., Lasseter, R., Ise, T. *et al.* (2008), Making microgrids work, *IEEE Power and Energy Magazine*, **6** (3), 40–53, doi:10.1109/MPE.2008.918718.

Kubo, Y., Shinohara, N., and Mitani, T. (2012), Development of a kW class microwave wireless power supply system to a vehicle roof, in *2012 IEEE MTT-S International Microwave Workshop Series on Innovative Wireless Power Transmission: Technologies, Systems, and Applications (IMWS)*, pp. 205–208.

Kulkarni, R. V. and Venayagamoorthy, G. K. (2011), Particle swarm optimization in wireless-sensor networks: a brief survey, *IEEE Transactions on Systems, Man, and Cybernetics, Part C (Applications and Reviews)*, **41** (2), 262–267, doi:10.1109/TSMCC.2010.2054080.

Kumar, S., Patel, P., Mittal, A., and De, A. (2012), Design, analysis and fabrication of rectenna for wireless power transmission – virtual battery, in *2012 National Conference on Communications (NCC)*, pp. 1–4.

Kundur, P., Paserba, J., Ajjarapu, V. *et al.* (2004), Definition and classification of power system stability – IEEE/CIGRE Joint Task Force on Stability Terms and Definitions, *IEEE Transactions on Power Systems*, **19** (3), 1387–1401, doi:10.1109/TPWRS.2004.825981.

Landauer, R. (1961), Irreversibility and heat generation in the computing process, *IBM Journal of Research and Development*, **5** (3), 183–191.

Langbein, P. (2009), Lessons learned from real-life implementation of demand response management, in *IEEE/PES Power Systems Conference and Exposition, 2009. PSCE '09*, IEEE, p. 1, doi:10.1109/PSCE.2009.4840240.

Lasseter, R. (2002), MicroGrids, in *IEEE Power Engineering Society Winter Meeting, 2002*, IEEE, volume 1, pp. 305–308, doi:10.1109/PESW.2002.985003.

Lee, S.-H. and Lorenz, R. D. (2011), A design methodology for multi-kW, large airgap, MHz frequency, wireless power transfer systems, in *2011 IEEE Energy Conversion Congress and Exposition (ECCE)*, pp. 3503–3510.

Li, B., Qin, Y., Low, C., and Gwee, C. (2007), A survey on mobile WiMAX [wireless broadband access], *IEEE Communications Magazine*, **45** (12), 70–75, doi:10.1109/MCOM.2007.4395368.

Li, H. and Han, Z. (2011), Synchronization of power networks without and with communication infrastructures, in *2011 IEEE International Conference on Smart Grid Communications (SmartGridComm)*, IEEE, pp. 463–468, doi:10.1109/SmartGridComm.2011.6102367.

Li, H. and Qiu, R. (2010), Need-based communication for smart grid: when to inquire power price?, in *2010 IEEE Global Telecommunications Conference (GLOBECOM 2010)*, pp. 1–5.

Li, H., Xu, C., and Banerjee, K. (2010), Carbon nanomaterials: the ideal interconnect technology for next-generation ICs, *IEEE Design & Test of Computers*, **27** (4), 20–31, doi:10.1109/MDT.2010.55.

Li, J.-W. (2011), Wireless power transmission: state-of-the-arts in technologies and potential applications, in *2011 Asia-Pacific Conference Proceedings (APMC)*, pp. 86–89.

Liang, Y. and Campbell, R. H. (2008), Understanding and simulating the IEC 61850 standard, Technical report, University of Illinois at Urbana-Champaign.

Lin, J. (2002), Space solar-power stations, wireless power transmissions, and biological implications, *IEEE Microwave Magazine*, **3** (1), 36–42.

Lisovich, M., Mulligan, D., and Wicker, S. (2010), Inferring personal information from demand-response systems, *IEEE Security & Privacy*, **8** (1), 11–20.

Lloyd, S., Shahriar, M., and Hemmer, P. (2004), Teleportation and the quantum internet, http://lapt.ece.northwestern.edu/files/2000-002.pdf.

Loyka, S. (2005), Information theory and electromagnetism: are they related?, *Microwave Review*, (November), 38–46.

Lu, S., Samaan, N., Diao, R. *et al.* (2011), Centralized and decentralized control for demand response, in *2011 IEEE PES Innovative Smart Grid Technologies (ISGT)*, Ieee, pp. 1–8, doi:10.1109/ISGT.2011.5759191.

Lui, A. (2000), Tutorial on geomagnetic storms and substorms, *IEEE Transactions on Plasma Science*, **28** (6), 1854–1866, doi:10.1109/27.902214.

Maccari, L., Paoli, M., and Fantacci, R. (2007), Security Analysis of IEEE 802.16, in *IEEE International Conference on Communications, 2007. ICC '07*, IEEE, pp. 1160–1165, doi:10.1109/ICC.2007.197.

Mackiewicz, R. (2006), Overview of IEC 61850 and benefits, in *2005/2006 IEEE PES Transmission and Distribution Conference and Exhibition*, IEEE, pp. 376–383, doi:10.1109/TDC.2006.1668522.

Madani, V., Vaccaro, A., Villacci, D., and King, R. L. (2007), Satellite based communication network for large scale power system applications, in *2007 iREP Symposium – Bulk Power System Dynamics and Control – VII. Revitalizing Operational Reliability*, Ieee, pp. 1–7, doi:10.1109/IREP.2007.4410572.

Makarov, Y. V., Reshetov, V. I., Stroev, V. A., and Voropai, N. I. (2005), Blackouts in North America and Europe: analysis and generalization, in *2005 IEEE Russia Power Tech*, IEEE Press, Piscataway, NJ, pp. 1–7.

Marculescu, D., Marculescu, R., and Pedram, M. (1996), Information theoretic measures for power analysis [logic design], *IEEE Transactions on Computer-Aided Design of Integrated Circuits and Systems*, **15** (6), 599–610, doi:10.1109/43.503930.

Marihart, D. (2001), Communications technology guidelines for EMS/SCADA systems, *IEEE Transactions onPower Delivery*, **16** (2), 181–188.

Markushevich, N. and Chan, E. (2009), Integrated voltage, Var control and demand response in distribution systems, in *IEEE/PES Power Systems Conference and Exposition, 2009. PSCE '09*, IEEE, pp. 1–4, doi:10.1109/PSCE.2009.4840256.

Martigne, P. (2011), ITU-T Smart Grid Focus Group Activity, http://docbox.etsi.org/workshop/2011/201104_SMARTGRIDS/02_STANDARDS/ITUTSGFocusGroup_Martigne.pdf.

Martin, K. (2006), IEEE Standard for Synchrophasors for Power Systems, doi:10.1109/IEEESTD.2006.99376.

Martin, K. and Carroll, J. (2008), Phasing in the technology, *IEEE Power and Energy Magazine*, **6** (5), 24–33, doi:10.1109/MPE.2008.927474.

Martin, K. E. (2010), Synchrophasors in the IEEE C37.118 and IEC 61850, in *2010 5th International Conference on Critical Infrastructure (CRIS)*, Ieee, pp. 1–8, doi:10.1109/CRIS.2010.5617573.

Marvel, K. and Agvaanluvsan, U. (2010), Random matrix theory models of electric grid topology, *Physica A*, **389** (24), 5838–5851, doi:10.1016/j.physa.2010.08.009.

Marwali, M., Jung, J., and Keyhani, A. (2007), Stability analysis of load sharing control for distributed generation systems, *IEEE Transactions on Energy Conversion*, **22** (3), 737–745.

Marwali, M. N., Jung, J.-w., and Keyhani, A. (2004), Control of distributed generation systems – part II: load sharing control, *IEEE Transactions on Power Electronics*, **19** (6), 1551–1561.

Massa, N. (2000), Fiber optic telecommunication, http://spie.org/Documents/Publications/00%20STEP%20Module%2008.pdf.

McCann, R., Le, A. T., and Traore, D. (2008), Stochastic sliding mode arbitration for energy management in smart building systems, in *IEEE Industry Applications Society Annual Meeting, 2008. IAS '08*, IEEE, pp. 1–4, doi:10.1109/08IAS.2008.8.

McDonald, J. (2008), Adaptive intelligent power systems: active distribution networks, *Energy Policy*, **36**, 4346–4351, doi:10.1016/j.enpol.2008.09.038.

Medina, J., Muller, N., and Roytelman, I. (2010), Demand response and distribution grid operations: opportunities and challenges, *IEEE Transactions on Smart Grid*, **1** (2), 193–198, doi:10.1109/TSG.2010.2050156.

MEF (2011), Microwave technologies for carrier Ethernet services, Technical Report February, Metro Ethernet Forum, http://metroethernetforum.org/index.php.

Mendel, J. (1995), Fuzzy logic systems for engineering: a tutorial, *Proceedings of the IEEE*, **83** (3), 345–377.

Modbus (2006), MODBUS Application Protocol Specification V1.1b.

Mohagheghi, S., Parkhideh, B., and Bhattacharya, S. (2012), Inductive power transfer for electric vehicles: potential benefits for the distribution grid, in *2012 IEEE International Electric Vehicle Conference (IEVC)*, pp. 1–8.

Mohagheghi, S., Stoupis, J., and Wang, Z. (2009), Communication protocols and networks for power systems-current status and future trends, in *IEEE/PES Power Systems Conference and Exposition, 2009. PSCE '09*, IEEE, pp. 1–9, doi:10.1109/PSCE.2009.4840174.

Mohammed, S. S., Ramasamy, K., and Shanmuganantham, T. (2010), Wireless power transmission – a next generation power transmission system, *International Journal of Computer Applications*, **1** (13), 100–103.

Mohd, A., Ortjohann, E., Schmelter, A. *et al.* (2008), Challenges in integrating distributed energy storage systems into future smart grid, in *IEEE International Symposium on Industrial Electronics, 2008. ISIE 2008*, IEEE, pp. 1627–1632, doi:10.1109/ISIE.2008.4676896.

Mohsenian-Rad, A.-H. and Leon-Garcia, A. (2010), Optimal residential load control with price prediction in real-time electricity pricing environments, *IEEE Transactions on Smart Grid*, **1** (2), 120–133, doi:10.1109/TSG.2010.2055903.

Montambault, S. and Pouliot, N. (2010), About the future of power line robotics, in *2010 1st International Conference on Applied Robotics for the Power Industry (CARPI 2010)*, IEEE, pp. 1–6, doi:10.1109/CARPI.2010.5624466.

Myrda, P. T., Taft, J., and Donner, P. (2012), Recommended approach to a NASPInet architecture, *2012 45th Hawaii International Conference on System Science (HICSS)*, 2072–2081, doi:10.1109/HICSS.2012.496.

Nelson, T. L. and FitzPatrick, G. J. (2010), NIST role in the interoperable smart grid, in *2010 IEEE Power and Energy Society General Meeting*, IEEE, pp. 1–3, doi:10.1109/PES.2010.5589733.

NERC Steering Group (2004), Technical analysis of the August 14, 2003, blackout: what happened, why and what did we learn?, Technical report, North American Electric Reliability Council.

Neves, A., Sousa, D. M., Roque, A., and Terras, J. M. (2011), Analysis of an inductive charging system for a commercial electric vehicle, in *Proceedings of the 2011-14th European Conference on Power Electronics and Applications (EPE 2011)*, IEEE, pp. 1–10.

Nguyen, P. H., Kling, W. L., and Myrzik, J. M. A. (2008), The interconnection in active distribution networks, in *International Conference on Energy Security and Climate Change: Issues, Strategies and Options*.

Nguyen, P. H., Kling, W. L., and Myrzik, J. M. A. (2009), Power flow management in active networks, in *2009 IEEE Bucharest PowerTech*, IEEE, pp. 1–6, doi:10.1109/PTC.2009.5282094.

NIST (2009), NIST framework and roadmap for smart grid interoperability standards, Technical report, NIST, http://www.nist.gov/public_affairs/releases/upload/smartgrid_interoperability_final.pdf.

NIST FIPS (2002), 198: The keyed-hash message authentication code (HMAC).

Niyato, D. and Wang, P. (2012), Cooperative transmission for meter data collection in smart grid, *IEEE Communications Magazine*, **50** (4), 90–97, doi:10.1109/MCOM.2012.6178839.

Nuqui, R. F., Zarghami, M., and Mendik, M. (2010), The impact of optical current and voltage sensors on phasor measurements and applications, in *2010 IEEE PES Transmission and Distribution Conference and Exposition*, IEEE, pp. 1–7, doi:10.1109/TDC.2010.5484271.

Nwankpa, C., Deese, A., Liu, Q. *et al.* (2006), Power system on a chip (PSoC): analog emulation for power system applications, in *IEEE Power Engineering Society General Meeting, 2006*, pp. 1–6.

Nyeng, P. and Ostergaard, J. (2011), Information and communications systems for control-by-price of distributed energy resources and flexible demand, *IEEE Transactions on Smart Grid*, **2** (2), 334–341.

Ondrej, S., Zdenek, B., Petr, F., and Ondrej, H. (2006), ZigBee technology and device design, in *International Conference on Networking, International Conference on Systems and International Conference on Mobile Communications and Learning Technologies, 2006. ICN/ICONS/MCL 2006*, IEEE, p. 129, doi:10.1109/ICNICONSMCL.2006.233.

O'Neill, D., Levorato, M., Goldsmith, A., and Mitra, U. (2010), Residential demand response using reinforcement learning, in *2010 First IEEE International Conference on Smart Grid Communications (SmartGridComm)*, pp. 409–414.

Osepchuk, J. (2010), The magnetron and the microwave oven: a unique and lasting relationship, in *2010 International Conference on the Origins and Evolution of the Cavity Magnetron (CAVMAG)*, pp. 46–51.

Ostendorp, M., Gela, G., and Hirany, A. (1997), Fiber optic cables in overhead transmission corridors: a state-of-the-art review, Technical Report TR-108959, EPRI, http://nocapx2020.info/wp-content/uploads/2010/10/fiber-optic-tr-108959.pdf.

Ott, A. (2010), Evolution of computing requirements in the PJM market: past and future, in *2010 IEEE Power and Energy Society General Meeting*, pp. 1–4.

Palensky, P. and Dietrich, D. (2011), Demand side management: demand response, intelligent energy systems, and smart loads, *IEEE Transactions on Industrial Informatics*, **7** (3), 381–388.

Parvania, M. and Fotuhi-Firuzabad, M. (2010), Demand response scheduling by stochastic SCUC, *IEEE Transactions on Smart Grid*, **1** (1), 89–98, doi:10.1109/TSG.2010.2046430.

Peeters, E., Six, D., Hommelberg, M. *et al.* (2009), The ADDRESS project: an architecture and markets to enable active demand, in *6th International Conference on the European Energy Market, 2009. EEM 2009*, IEEE, pp. 1–5, doi:10.1109/EEM.2009.5207145.

Peirson, G., Pollard, A., and Care, N. (1955), Automatic circuit reclosers, *Proceedings of the IEE Part A: Power Engineering*, **102** (6), 749, doi:10.1049/pi-a.1955.0156.

Phadke, A. and Thorp, J. (2006), History and applications of phasor measurements, in *2006 IEEE PES Power Systems Conference and Exposition, 2006. PSCE '06*, IEEE, pp. 331–335, doi:10.1109/PSCE.2006.296328.

Philipps, H. (1999), Modelling of powerline communication channels, in *Proceedings of the 3rd International Symposium on Power-Line Communications and its Applications*, pp. 14–21.

Pierce, Jr, R. (2012), Primer on demand response and a critique of FERC Order 745, *Journal of Energy and Environmental Law*. GWU Legal Studies Research Paper No. 577.

Pignolet, G., Hawkins, J., Kaya, N. *et al.* (1996), Results of the Grand-Bassin case study in Reunion Island: operational design for a 10 kW microwave beam energy transportation, in *47th International Astronautical Congress, Beijing*, pp. 7–11.

Pipattanasomporn, M., Feroze, H., and Rahman, S. (2009), Multi-agent systems in a distributed smart grid: design and implementation, in *IEEE/PES Power Systems Conference and Exposition, 2009. PSCE '09*, IEEE, pp. 1–8, doi:10.1109/PSCE.2009.4840087.

Pirjola, R. (2000), Geomagnetically induced currents during magnetic storms, *IEEE Transactions on Plasma Science*, **28** (6), 1867–1873.

Plenio, M. B. and Vitelli, V. (2001), The physics of forgetting: Landauer's erasure principle and information theory, *Contemporary Physics*, **42** (1), 25–60, doi:10.1080/00107510010018916.

Popović, Z. (2006), Wireless powering for low-power distributed sensors, *Serbian Journal of Electrical Engineering*, **3** (2), 149–162.

Pourmousavi, S. and Nehrir, M. (2011), Demand response for smart microgrid: initial results, in *2011 IEEE PES Innovative Smart Grid Technologies (ISGT)*, pp. 11–16.

Power Engineering Society (2003), Overview of power system stability concepts, in *IEEE Power Engineering Society General Meeting, 2003*, volume 3, doi:10.1109/PES.2003.1267424.

Prokopenko, M., Boschetti, F., and Ryan, A. J. (2009), An information-theoretic primer on complexity, self-organization, and emergence, *Complexity*, **15** (1), 11–28, doi:10.1002/cplx.20249.

Pulkkinen, A., Viljanen, A., and Pirjola, R. (2009), Harnessing celestial batteries, *American Journal of Physics*, **77** (7), 610, doi:10.1119/1.3119172.

Pulkkinen, T. (2007), Space weather: terrestrial perspective, *Living Reviews in Solar Physics*, **4**, 1–60.

Rahimi, F. and Ipakchi, A. (2010), Overview of demand response under the smart grid and market paradigms, in *2010 Innovative Smart Grid Technologies (ISGT)*, IEEE, pp. 1–7, doi:10.1109/ISGT.2010.5434754.

Reid, E., Gerber, S., and Adib, P. (2009), Integration of demand response into wholesale electricity markets, in *IEEE/PES Power Systems Conference and Exposition, 2009*, IEEE, p. 1, doi:10.1109/PSCE.2009.4840219.

Rengaraju, P., Lung, C.-H., Qu, Y., and Srinivasan, A. (2009), Analysis on mobile WiMAX security, in *2009 IEEE Toronto International Conference on Science and Technology for Humanity (TIC-STH)*, IEEE, pp. 439–444, doi:10.1109/TIC-STH.2009.5444459.

Reza, M., Schavemaker, P. H., Slootweg, J. G. *et al.* (2004), Impacts of distributed generation penetration levels on power systems transient stability, in *IEEE Power Engineering Society General Meeting, 2004*, volume 2, pp. 2150–2155.

Roncolatto, R. A., Romanelli, N. W., Hirakawa, A. *et al.* (2010), Robotics applied to work conditions improvement in power distribution lines maintenance, in *2010 1st International Conference on Applied Robotics for the Power Industry (CARPI 2010)*, IEEE, pp. 1–6, doi:10.1109/CARPI.2010.5624436.

Roossien, B., Hommelberg, M., Warmer, C. *et al.* (2008), Virtual power plant field experiment using 10 micro-CHP units at consumer premises, in *CIRED Seminar on SmartGrids for Distribution, 2008. IET-CIRED*, pp. 23–24.

Rosas-Casals, M. (2010), Power grids as complex networks: topology and fragility, in *2010 Complexity in Engineering. COMPENG '10*, IEEE, pp. 21–26, doi:10.1109/COMPENG.2010.23.

Ruff, L. (2002), Economic principles of demand response in electricity, http://content.knowledgeplex.org/ksg-test/cache/documents/484.pdf.

Russell, C. (1956), *The Art and Science of Protective Relaying*, John Wiley & Sons.

Rye, E. (2007), Tales of the unexpected [ferroresonance], in *IEE Colloquium on Warning! Ferroresonance Can Damage Your Plant*, pp. 1/1–1/7.

Saad, W., Han, Z., Poor, H., and Basar, T. (2011), A noncooperative game for double auction-based energy trading between PHEVs and distribution grids, in *2011 IEEE International Conference on Smart Grid Communications (SmartGridComm)*, pp. 267–272.

Saleem, A., Honeth, N., and Nordstrom, L. (2010), A case study of multi-agent interoperability in IEC 61850 environments, in *2010 IEEE PES Innovative Smart Grid Technologies Conference Europe (ISGT Europe)*, pp. 1–8.

Saltzer, J., Reed, D., and Clark, D. (1984), End-to-end arguments in system design, *ACM Transactions on Computer Systems (TOCS)*, **2** (4), 277–288.

Samadi, P., Schober, R., and Wong, V. (2011), Optimal energy consumption scheduling using mechanism design for the future smart grid, in *2011 IEEE International Conference on Smart Grid Communications (SmartGridComm)*, pp. 369–374.

Samardzija, D. (2007), Some analogies between thermodynamics and Shannon theory, in *41st Annual Conference on Information Sciences and Systems, 2007. CISS '07*, IEEE, pp. 166–171.

Sample, A. P., Meyer, D. T., and Smith, J. R. (2010), Analysis, experimental results, and range adaptation of magnetically coupled resonators for wireless power transfer, *IEEE Transactions on Industrial Electronics*, **58** (2), 544–554.

Samuelsson, O., Repo, S., Jessler, R. *et al.* (2010), Active distribution network – demonstration project ADINE, in *2010 IEEE PES Innovative Smart Grid Technologies Conference Europe (ISGT Europe)*, IEEE, pp. 1–8, doi:10.1109/ISGTEUROPE.2010.5638988.

Sanders, M. P. and Ray, R. E. (1996), Power line carrier channel & application considerations for transmission line relaying, Technical Report Pulsar Document Number C045P0597, Pulsar Technologies, Inc., Coral Springs, FL.

Sankar, L., Kar, S., Tandon, R., and Poor, H. V. (2011), Competitive privacy in the smart grid: an information-theoretic approach, http://arxiv.org/abs/1108.2237.

Santillan, V., Baircenas, G., and Capel, G. (2006), Modular design in reclosers, in *IEEE/PES Transmission & Distribution Conference and Exposition: Latin America, 2006. TDC '06*, IEEE, volume 1, pp. 1–6, doi:10.1109/TDCLA.2006.311367.

Sauter, T. (2010), The three generations of field-level networks – evolution and compatibility issues, *IEEE Transactions on Industrial Electronics*, **57** (11), 3585–3595, doi:10.1109/TIE.2010.2062473.

Saxena, D., Singh, S., and Verma, K. (2010), Application of computational intelligence in emerging power systems, *International Journal of Engineering, Science and Technology*, **2** (3), 1–7.

Schwartz, M. (2009), Carrier-wave telephony over power lines: early history [history of communications], *IEEE Communications Magazine*, **47** (1), 14–18, doi:10.1109/MCOM.2009.4752669.

Schweitzer, E. O., Whitehead, D. E., Guzmán, A. *et al.* (2010), Applied synchrophasor solutions and advanced possibilities, in *2010 IEEE PES Transmission and Distribution Conference and Exposition*, pp. 1–8.

Schweppe, F. and Wildes, J. (1970), Power system static-state estimation, part I: exact model, *IEEE Transactions on Power Apparatus and Systems*, **PAS-89** (1), 120–125, doi:10.1109/TPAS.1970.292678.

Sekiguchi, D., Nakamura, T., Misawa, S. *et al.* (2012), Trial test of fully HTS induction/synchronous machine for next generation electric vehicle, *IEEE Transactions on Applied Superconductivity*, **22** (3), 5200904, doi:10.1109/TASC.2011.2176094.

Shahidehpour, M. and Wang, Y. (2003), Special topics in power system information system, in *Communication and Control in Electric Power Systems: Applications of Parallel and Distributed Processing*, Wiley-IEEE Press, chapter 12, pp. 439–475, doi:10.1002/0471462926.ch12.

Shannon, C. (1956), The zero error capacity of a noisy channel, *IRE Transactions on Information Theory*, **2** (3), 8–19.

Shannon, C. (2001), A mathematical theory of communication, *ACM SIGMOBILE Mobile Computing and Communications Review*, **5** (1), 55.

Shannon, C. E. (1948), A mathematical theory of communication, *The Bell System Technical Journal*, **27**, 379–423, 623–656.

Shinohara, N. (2011), Power without wires, *IEEE Microwave Magazine*, **12** (7), S64–S73.

Shinohara, N. (2012), Wireless power transmission for solar power satellite (SPS), http://www.sspi.gatech.edu/wptshinohara.pdf.

Shinohara, N. and Ishikawa, T. (2011), High efficient beam forming with high efficient phased array for microwave power transmission, in *2011 International Conference on Electromagnetics in Advanced Applications (ICEAA)*, volume 1, pp. 729–732.

Siddique, A., Firdaus, A., Lamont, L., and Chaar, L. (2012), Rural electrification using wireless power transmission system, in *2012 2nd International Conference on the Developments in Renewable Energy Technology (ICDRET)*, pp. 7–12.

Siddiqui, O. (2008), The green grid: energy savings and carbon emissions reductions enabled by a smart grid, http://www.smartgridnews.com/artman/uploads/1/SGNR_2009_EPRI_Green_Grid_June_2008.pdf.

Silva, P., Karnouskos, S., and Ilic, D. (2012), A survey towards understanding residential prosumers in smart grid neighbourhoods, in *2012 3rd IEEE PES International Conference and Exhibition on Innovative Smart Grid Technologies (ISGT Europe)*, pp. 1–8.

Simovski, C. and Morits, D. (2013), Enhanced efficiency of light-trapping nanoantenna arrays for thin film solar cells, http://arxiv.org/abs/1301.3290.

Skowyra, R., Lapets, A., Bestavros, A., and Kfoury, A. (2013), Verifiably-safe software-defined networks for CPS, in *HiCoNS '13 2nd ACM International Conference on High Confidence Networked Systems*, pp. 101–110.

Smakhtin, A. and Rybakov, V. (2002), Comparative analysis of wireless systems as alternative to high-voltage power lines for global terrestrial power transmission, in *Proceedings of the 31st Intersociety Energy Conversion Engineering Conference, 1996. IECEC 96*, IEEE, volume 1, pp. 485–488.

So-In, C., Jain, R., and Tamimi, A.-K. (2010), Capacity evaluation for IEEE 802.16e mobile WiMAX, *Journal of Computer Systems, Networks, and Communications*, **2010**, 1–13, doi:10.1155/2010/279807.

Solé, R., Valverde, S., and Sol, R. V. (2004), Information theory of complex networks: on evolution and architectural constraints, *Complex Networks*, **207**, 189–207.

Solé, R. V., Rosas-Casals, M., Corominas-Murtra, B., and Valverde, S. (2007), Robustness of the European power grids under intentional attack, *Physical Review E*, **77** (2), 7, doi:10.1103/PhysRevE.77.026102.

Solomonoff, R. (1964), A formal theory of inductive inference. Part I, *Information and Control*, **7** (1), 1–22.

Spencer, Q. H. (2008), An information-theoretic analysis of electricity consumption data for an AMR system, *2008 IEEE International Symposium on Power Line Communications and Its Applications*, 199–203, doi:10.1109/ISPLC.2008.4510423.

Starr, C. (1986), The Electric Power Research Institute, *Science*, **219**, 1190–1194.

Stragier, J., Hauttekeete, L., and De Marez, L. (2010), Introducing smart grids in residential contexts: consumers' perception of smart household appliances, in *2010 IEEE Conference on Innovative Technologies for an Efficient and Reliable Electricity Supply*, IEEE, pp. 135–142, doi:10.1109/CITRES.2010.5619864.

Su, C. and Kirschen, D. (2009), Quantifying the effect of demand response on electricity markets, *IEEE Transactions on Power Systems*, **24** (3), 1199–1207.

Subedi, L. and Trajković, L. (2010), Spectral analysis of internet topology graphs, in *Proceedings of 2010 IEEE International Symposium on Circuits and Systems (ISCAS)*, pp. 1803–1806.

Sui, H., Wang, H., Lu, M.-s., and Lee, W.-j. (2009), An AMI system for the deregulated electricity markets, *IEEE Transactions on Industry Applications*, **45** (6), 2104–2108, doi:10.1109/TIA.2009.2031848.

Taylor, T., Marshall, M., and Neumann, E. (2001), Developing a reliability improvement strategy for utility distribution systems, in *2001 IEEE/PES Transmission and Distribution Conference and Exposition. Developing New Perspectives*, IEEE, pp. 444–449, doi:10.1109/TDC.2001.971275.

Timbus, A., Liserre, M., Teodorescu, R., and Blaabjerg, F. (2005), Synchronization methods for three phase distributed power generation systems – an overview and evaluation, in *IEEE 36th Power Electronics Specialists Conference, 2005. PESC '05*, IEEE, pp. 2474–2481, doi:10.1109/PESC.2005.1581980.

Tomsovic, K., Bakken, D., Venkatasubramanian, V., and Bose, A. (2005), Designing the next generation of real-time control, communication, and computations for large power systems, *Proceedings of the IEEE*, **93** (5), 965–979.

Trichtchenko, L. and Boteler, D. (2006), Response of power systems to the temporal characteristics of geomagnetic storms, in *Canadian Conference on Electrical and Computer Engineering, 2006. CCECE '06*, pp. 390–393.

Tympas, A. and Dalouka, D. (2008), Metaphorical uses of an electric power network: Early computations of atomic particles and nuclear reactors, http://www.metaphorik.de/sites/www.metaphorik.de/files/journal-pdf/12_2007_tympasdalouka.pdf.

Uluski, R. (2008), Interactions between AMI and distribution management system for efficiency/reliability improvement at a typical utility, in *2008 IEEE Power and Energy Society General Meeting – Conversion and Delivery of Electrical Energy in the 21st Century*, pp. 1–4.

Vaessen, P. (2009), Wireless power transmission, Technical Report September, Leonardo Energy.

Vaughan-Nichols, S. J. (2011), OpenFlow: the next generation of the network?, *Computer*, **44** (8), 13–15.

Verboomen, J., Van Hertem, D., Schavemaker, P. *et al.* (2005), Phase shifting transformers: principles and applications, in *2005 International Conference on Future Power Systems*.

Virk, G. S., Loveday, D. L., and Cheung, J. Y. M. (1990), Model-based controllers for BEMS, in *International Conference on Control, 1994. Control '94*, volume 2, pp. 901–905.

Vodyakho, O., Widener, C., Steurer, M. *et al.* (2011), Development of solid-state fault isolation devices for future power electronics-based distribution systems, in *2011 Twenty-Sixth Annual IEEE Applied Power Electronics Conference and Exposition (APEC)*, pp. 113–118.

Voloh, I. and Johnson, R. (2005), Applying digital line current differential relays over pilot wires, in *2005 58th Annual Conference forProtective Relay Engineers*, IEEE, pp. 287–290.

Von Meier, A. (2006), *Electric Power Systems: A Conceptual Introduction*, Wiley-IEEE Press.

Vos, A. (2009), Effective business models for demand response under the smart grid paradigm, in *IEEE/PES Power Systems Conference and Exposition, 2009. PSCE '09*, IEEE, p. 1, doi:10.1109/PSCE.2009.4840261.

Vu, H. L., Chan, S., and Andrew, L. L. H. (2010), Performance analysis of best-effort service in saturated IEEE 802.16 networks, *IEEE Transactions on Vehicular Technology*, **59** (1), 460–472, doi:10.1109/TVT.2009.2033191.

Vyas, R., Lakafosis, V., Tentzeris, M. *et al.* (2011), A battery-less, wireless mote for scavenging wireless power at UHF (470–570 MHz) frequencies, in *IEEE Antennas and Propagation Society International Symposium*, pp. 1069–1072.

Waffenschmidt, E. and Staring, T. (2009), Limitation of inductive power transfer for consumer applications, in *13th European Conference on Power Electronics and Applications, 2009. EPE '09*, pp. 1–10.

Wall, D. W. (1982), Messages as active agents, in *Proceedings of the 9th ACM SIGPLAN-SIGACT Symposium on Principles of Programming Languages – POPL '82*, ACM Press, New York, NY, pp. 34–39, doi:10.1145/582153.582157.

Walling, R., Saint, R., Dugan, R. *et al.* (2008), Summary of distributed resources impact on power delivery systems, *IEEE Transactions on Power Delivery*, **23** (3), 1636–1644, doi:10.1109/TPWRD.2007.909115.

Wang, B., Han, L., Zhang, H. *et al.* (2009), A flying robotic system for power line corridor inspection, in *2009 IEEE International Conference on Robotics and Biomimetics (ROBIO)*, IEEE, pp. 2468–2473, doi:10.1109/ROBIO.2009.5420421.

Wang, B., Teo, K. H., Nishino, T. *et al.* (2011a), Experiments on wireless power transfer with metamaterials, *Applied Physics Letters*, **98** (25), 254101, doi:10.1063/1.3601927.

Wang, H. (2010), The advantages and disadvantages of using HVDC to interconnect AC networks, in *2010 45th InternationalUniversities Power Engineering Conference (UPEC)*, pp. 4–8.

Wang, J., Biviji, M., and Wang, W. (2011b), Lessons learned from smart grid enabled pricing programs, in *2011 IEEE Power and Energy Conference at Illinois (PECI)*, pp. 1–7.

Wang, S., Meng, X., and Chen, T. (2012), Wide-area control of power systems through delayed network communication, *IEEE Transactions on Control Systems Technology*, **20** (2), 495–503, doi:10.1109/TCST.2011.2116022.

Weaver, W. and Wirth, R. (2004), Science and complexity, *Science*, **6** (3), 65–74.

Webber, G., Warrington, J., Mariethoz, S., and Morari, M. (2011), Communication limitations in iterative real time pricing for power systems, in *2011 IEEE International Conference on Smart Grid Communications (SmartGridComm)*, pp. 463–468.

Wells, J. (2009), Faster than fiber: the future of multi-G/s wireless, *IEEE Microwave Magazine*, **10** (3), 104–112.

Wijetunge, U., Perreau, S., and Pollok, A. (2011), Distributed stochastic routing optimization using expander graph theory, in *2011 Australian Communications Theory Workshop (AusCTW)*, pp. 124–129.

Witte, J., Mendis, S., Bishop, M., and Kischefsky, J. (1992), Computer-aided recloser applications for distribution systems, *IEEE Computer Applications in Power*, **5** (3), 27–32, doi:10.1109/67.143271.

Wollenberg, B. and Sakaguchi, T. (1987), Artificial intelligence in power system operations, *Proceedings of the IEEE*, **75** (12), 1678–1685, doi:10.1109/PROC.1987.13935.

Wollman, D. A., FitzPatrick, G. J., Boynton, P. A., and Nelson, T. L. (2010), NIST coordination of smart grid interoperability standards, in *2010 Conference on Precision Electromagnetic Measurements (CPEM)*, IEEE, pp. 531–532, doi:10.1109/CPEM.2010.5544388.

Wu, F., Moslehi, K., and Bose, A. (2005), Power system control centers: Past, present, and future, *Proceedings of the IEEE*, **93** (11), 1890–1908, doi:10.1109/JPROC.2005.857499.

Wu, F. Y. (2004), Theory of resistor networks: the two-point resistance, *Journal of Physics A: Mathematical and General*, **37**, 6653–6673, doi:10.1088/0305-4470/37/26/004.

Xhafa, A., Kangude, S., and Lu, X. (2005), MAC performance of IEEE 802.16e, in *2005 IEEE 62nd Vehicular Technology Conference, 2005. VTC-2005-Fall*, IEEE, volume 1, pp. 685–689, doi:10.1109/VETECF.2005.1558000.

Xiao, Y., Song, Y., and Sun, Y.-Z. (2000), Application of stochastic programming for available transfer capability enhancement using FACTS devices, in *IEEE Power Engineering Society Summer Meeting, 2000*, volume 1, pp. 508–515.

Xu, W. and Wang, W. (2010), Power electronic signaling technology – a new class of power electronics applications, *IEEE Transactions on Smart Grid*, **1** (3), 332–339.

Yadav, R., Das, S., and Yadava, R. (2011), Rectennas design, development and applications, *International Journal of Engineering Science and Technology*, **3** (10), 7823–7841.

Yamanaka, Y. and Sugiura, A. (2011), Possible EMC regulations for wireless power transmission equipment, in *2011 IEEE MTT-S International Microwave Workshop Series on Innovative Wireless Power Transmission: Technologies, Systems, and Applications (IMWS)*, pp. 97–100.

Yang, H., Raza, A., Park, M. *et al.* (2009), Gigabit Ethernet based substation, *Journal of Power Electronics*, **9** (1), 100–108, http://www.jpe.or.kr/On_line/admin/paper/files/9-12.pdf.

Yang, J.-l., Han, R.-c., Yu, S.-j., and Zhang, Y.-j. (2010), Design of a nonlinear power system stabilizer, in *2010 International Conference on Computational Aspects of Social Networks (CASoN)*, IEEE, pp. 683–686.

Yao, A. C.-C. (1979), Some complexity questions related to distributive computing(preliminary report), in *Proceedings of the Eleventh Annual ACM Symposium on Theory of Computing – STOC '79*, ACM Press, New York, New York, USA, pp. 209–213, doi:10.1145/800135.804414.

Yap, K., Huang, T., Dodson, B. *et al.* (2010), Towards software-friendly networks, in *APSys '10 Proceedings of the First ACM Asia-Pacific Workshop on Systems*.

Yeganeh, S. (2013), On scalability of software-defined networking, *IEEE Communications Magazine*, **51** (2), 136–141.

Yu, Y. and Cioffi, J. (2001), On constant power water-filling, in *IEEE International Conference on Communications, 2001. ICC 2001*, IEEE, volume 6, pp. 1665–1669, doi:10.1109/ICC.2001.937077.

Yue, S., Chen, J., Gu, Y. *et al.* (2011), Dual-pricing policy for controller-side strategies in demand side management, in *2011 IEEE International Conference on Smart Grid Communications (SmartGridComm)*, pp. 375–380.

Zander, J. and Forchheimer, R. (1980), Preliminary specification of a distributed packet radio system using the amateur band, Technical Report LiTH-ISY-I-408, University of Linköping.

Zander, J. and Forchheimer, R. (1983), SOFTNET – an approach to higher level packet radio, in *Proceedings, AMRAD Conference*.

Zhang, J., Huang, Y., and Cao, P. (2012), Harvesting RF energy with rectenna arrays, in *2012 6th European Conference on Antennas and Propagation (EUCAP)*, pp. 365–367.

Zhang, L., Zhao, J., Han, X., and Niu, L. (2005), Day-ahead generation scheduling with demand response, in *2005 IEEE/PES Transmission and Distribution Conference and Exhibition: Asia and Pacific*, pp. 1–4.

Zhang, P., Li, F., and Bhatt, N. (2010), Next-generation monitoring, analysis, and control for the future smart control center, *IEEE Transactions on Smart Grid*, **1** (2), 186–192, doi:10.1109/TSG.2010.2053855.

Zhao, L., Park, K., and Lai, Y.-C. (2004), Attack vulnerability of scale-free networks due to cascading breakdown, *Physical Review E*, **70** (3), 2–5, doi:10.1103/PhysRevE.70.035101.

Zhong, W. X., Lee, C. K., and Hui, S. Y. R. (2011), General analysis on the use of Tesla's resonators in domino forms for wireless power transfer, *IEEE Transactions on Industrial Electronics*, **60** (1), 261–270.

Zhong, Z., Xu, C., Billian, B. *et al.* (2005), Power system frequency monitoring network (FNET) implementation, *IEEE Transactions on Power Systems*, **20** (4), 1914–1921, doi:10.1109/TPWRS.2005.857386, http://ieeexplore.ieee.org/lpdocs/epic03/wrapper.htm?arnumber=1525121.

Zhou, Y. and Fang, Y. (2006), Security of IEEE 802.16 in mesh mode, in *IEEE Military Communications Conference, 2006. MILCOM 2006*, IEEE, pp. 1–6, doi:10.1109/MILCOM.2006.302083.

Zhu, J. and Abur, A. (2007), Effect of phasor measurements on the choice of reference bus for state estimation, in *IEEE Power Engineering Society General Meeting, 2007*, Ieee, pp. 1–5, doi:10.1109/PES.2007.386175.

Zou, Y., Huang, X., Tan, L. *et al.* (2010), Current research situation and developing tendency about wireless power transmission, in *2010 International Conference on Electrical and Control Engineering (ICECE)*, pp. 3507–3511, doi:10.1109/iCECE.2010.853.

Index

Smart Grid: Communication-Enabled Intelligence for the Electric Power Grid, First Edition. Stephen F. Bush.